Environmental History

Volume 15

Series Editor

Mauro Agnoletti, Florence, Italy

The series intends to act as a link for ongoing researches concerning the historical interrelationships between man and the natural world, with special regard to the modern and contemporary ages. The main commitment should be to bring together different areas of expertise in both the natural and the social sciences to help them find a common language and a common perspective. Interdisciplinarity and trans-disciplinarity are needed for more and better understanding of the environment and its history, with new epistemological frameworks and methodological practices. The links between human activities and flora, fauna, water, soil, are examples of the most debated topics in EH, while established discipline like forest history, agricultural history and urban history are also dealing with it. The human impacts on ecosystems and landscapes over time, the preservation of cultural heritage, studies of historical trajectories in pattern and processes, as well as applied research on historical use and management of landscapes and ecosystems, are also taken into account. Other important topics relate to the history of environmental ideas and movements, policies, laws, regulations, conservation, the history of immaterial heritage, such as traditional knowledge related to the environment.

Mauro Agnoletti · Saša Dobričič · Tessa Matteini ·
Juan Manuel Palerm
Editors

Cultivating Continuity of the European Landscape

New Challenges, Innovative Perspectives

Springer

Editors
Mauro Agnoletti
DAGRI
University of Florence
Florence, Italy

Saša Dobričič
Graduate School
University of Nova Gorica
Nova Gorica, Slovenia

Tessa Matteini
Department of Architecture
University of Florence
Florence, Italy

Juan Manuel Palerm
Proyectos Arquitectónicos
University of Las Palmas de Gran Canaria
Las Palmas de Gran Canaria, Spain

ISSN 2211-9019 ISSN 2211-9027 (electronic)
Environmental History
ISBN 978-3-031-25715-5 ISBN 978-3-031-25713-1 (eBook)
https://doi.org/10.1007/978-3-031-25713-1

This Springer imprint is published by the registered company Springer Nature Switzerland AG
The registered company address is: Gewerbestrasse 11, 6330 Cham, Switzerland

Paper in this product is recyclable.

Preface

This book collects the contributions of more than 90 experts from all Europe who attended the meeting to celebrate the 20th anniversary of the European Landscape Convention, occurred in Florence (Italy) in October 2020. The meeting was held in the beautiful Salone de Cinquecento, inside Palazzo Vecchio, the town hall of the city of Florence, where the convention was signed in the year 2000. The organization of a meeting, partly in presence and partly on line, required an extraordinary effort considering that in those times Europe was still heavily affected by COVID-19. We have to thank the town of Florence, the Landscape Observatory of the Region of Tuscany and the Project GIAHS Capacity Building of the Italian Agency for Development Cooperation, for the support offered in order to make the meeting possible, as well as UNISCAPE for the scientific content.

Considering the times we are living, characterized by the outbreak of the COVID-19 pandemic in early 2020, climatic and political shocks, together with many rapid and intense socio-economic changes, it may seem trivial to discuss about landscape. We think it's the right time instead for considering the value of a landscape approach. The essential aspect of landscape, as a democratic right, suggests that we have to consider it as a project capable of including in one vision all the cultural, social, economic and environmental diversity characterizing the European territory. Whatever is the judgment on the level of the application of the convention in each signatory states, the recognition of all the cultural diversities existing in the European territory is the only way to reduce the power of the forces favoring the distance between cultures instead of the process of integration. It is also worth remembering that, according to genetic research, the European continent shows a very high correlation between genetic diversity and geographic diversity, which brings about the great landscape diversity associated to different cultures, that any political process should take into utmost consideration. This is also the reason why taking the landscape into account requires to continue the work of identification and assessment of all the European landscapes, as well as the monitoring of the changes occurring in urban, natural and rural areas.

The COVID-19 promoted a greater attention to landscape quality. Many rural and urban people were unable to travel for work reasons, immobilized by lockdowns; this has contributed to a renewed attention to rural areas and green areas, especially in the surroundings of urban settlements. The COVID-19 pandemic is exceptional in that it has shown how a shock of global proportions can occur suddenly, spread rapidly and compromise the status and livelihoods of billions of people to an unprecedented degree and over a long period generating a new attention to environmental resources and rural areas that were less influenced by the infections.

Another effect of the COVID-19 was the development of smart work, an act of production performed independent of time and place. In its ultimate form, the "office" no longer exists and traditional work conventions such as work hours are no longer so relevant, removing thousands of people from the daily routine of going to office in the morning and back home in the evening. This is one of the effects of COVID-19 that will stay, independently from the reduction of the infections. The greater amount of time spent at home has resulted in a growing attention to many services associated to landscape, as also to the current booming of tourism due to the reduction of the limitations to mobility. This new situation would require a better attention to landscape policies. It is worth remembering that there is still a long way to go before we can speak of a real integration among cultural, urban, environmental and agricultural policies that are often acting independently from each other. Even in the most advanced examples of landscape plans applying to entire territories, the integration between development plan and landscape plan shows the difficulties of promoting rules that combine a sectorial approach with an integrated approach responding to the needs of the entire society.

According to the definition of landscape proposed by the convention, the perception of the population is a fundamental aspect to consider, and this brings about the need to analyze the different experiences of the people living in different territories. While most of the European citizens are concentrated in urban and periurban areas, with a relatively limited extension, a much smaller amount of the population lives in the countryside and an even smaller percentage is directly connected with agriculture and forest activities. This means that when planning and management policies are undertaken, it should also be considered that people in rural areas may be limited in number, but have a closer contact with the environment, while they create and manage landscape. Therefore, in times of extreme climatic events, affecting both environmental and agricultural activities, a closer attention to rural territories is necessary. Agricultural activities not only produce food and create landscapes, but also prevent the effects of extreme climatic events, through a careful ongoing management of the territory that serves the entire society, an activity of utmost importance for our fragile environment.

When thinking of landscape policies, it should also consider the differences existing between the concepts of territory, landscape, nature, environment and related conservation strategies. These terms are often overlapped one to another, even in the scientific arena, but, conceptually, they mean different things. The case of environment is very useful to explain these diversities, especially in times when environment is the main concern at world level. Environmental quality responds mainly to indicators that can be measured and transferred into political action and monitored. Ideally, if we can achieve a certain quality of the air, soil, water and biodiversity, we can have an acceptable environmental quality in any country. Therefore, in the case of many countries accepting common standards in terms of indicators, we might achieve a good environmental quality in many parts of the world. When speaking of landscape quality instead, there is no way to indicate one landscape type that fits for all, because landscape is always a local project. Although we can measure many features of the landscapes, the overall values associated to landscapes are not always depending from measurable indicators. The cultural identity represented by the structure of one landscape may exist even in an area with no forests, although forests are always seen as valuable environmental element. Well-established scientific approaches are useful tools for landscape assessment, but the beneficiaries' perception, as well as the cognitive and emotive impact of direct perception, expressed by the local culture, should be taken into consideration in order to capture the entirety of a system. This is also a way to study landscape which is a very challenging activity, especially in a world dominated by sectorial knowledge. Many different skills are needed in order to perform the systemic approach required to apply a transdisciplinary approach to such a fascinating theme.

Florence, Italy

Mauro Agnoletti
Editor of the Series

Acknowledgments

We are deeply indebted to all authors for their contributions, infinite patience and support throughout the book's publishing process.

Our deepest gratitude goes to Sebastião Salgado and his studio for their generous and precious contribution and to Antonino Abrami and the International Academy of Environmental Sciences for the facilitation process.

We would like to thank all university members of the UNISCAPE network for the collaborative and thought-provoking environment in which we are able to cultivate attention on landscape research, education and practice, having landscape as a privileged place where disciplines meet.

Tessa Matteini and Saša Dobričič would like to thank Tommaso Zannaica for his constant effort and support in imagining and implementing the key activities of UNISCAPE, including the finalization of this book. We thank Ann des Jardins for editing and proofreading of this book and Catherine Goodrich for her help in editing.

We would like to thank Markus Spaeth for the kind guidance and support throughout the challenging moments of setting up the material and Rajan Muthu for collaborative assistance in the finalization of this book.

Refugees from the former Yugoslavia, Camp Ivankovo, Croatia, 1994.

Introduction: Preserving Continuity in Uncertain Times

Ineluctable modality of the visible: at least that if no more, thought through my eyes.
James Joyce

Side Note

The uncommon title of this introduction aims at revealing the uncertain, even desolating, scenario in which the UNISCAPE's 20th anniversary celebration of the European Landscape Convention took place and which has substantially affected the reflections expressed in the following contributions. We could never have imagined that less than one year after setting the scene for the celebration, all our rationale and most concrete certainties would tremble and ultimately crash because a tiny, invisible being had entered our lives, influencing it irremediably. Indeed, nothing suggested that *cultivation*, the driving concept of the conference, intended as the art of inventing "the common garden of life", combined with the most minute species, would actually become not only a concrete reality, but also a key agent of discovery revealing many other uncomfortable truths embedded in our society. To make things worse, a mere instant after what was imagined, at last, as a global ecological awakening from the long capitalist dreamland and a step forward from the pandemic emergency with its proliferation of fuzzy metaphors of war, killer, enemy, fight, Europeans would again collapse disastrously, in this very moment and with no accompanying metaphors, into another tragic scenario of limitless death and destruction.

Hence, it is with great discomfort but also with a glimmer of hope, that we invite readers to ignore, for an instant, the omnipresent martial lexicon and try to imagine how the meaning of *conflict*, *problem*, *challenge*, *obligation*, *will*, etc. could be disentangled with the use of different words like *interaction*, *observation*, *persuasion*, *attention*, *care*, *cultivation* and *emotion*; in other words with a landscape focused lexicon.

Because landscape has nothing to do with the land of fairy tales, but it is the first vision of belonging that we recall at our awakening once our disenchantment of the old world is over.

An Uncommon Introduction

Initially, the 20th anniversary of the European Landscape Convention created an occasion for UNISCAPE member universities to discuss the main achievements and lessons learned in the two decades following its adoption in Florence. However, what was intended as a single celebratory event turned out to be a much longer journey of reflection undertaken jointly with many members of the UNISCAPE community. This over one-year long adventure, bouncing between different COVID-19 emergencies, emerged as an extraordinary continuity of different actions and initiatives. The unexpected rise of public dialogue of an unprecedented intensity encouraged open debate, the formation of different working groups to process key public statements and the proactive opening up of UNISCAPE to a wider public: ultimately remaking UNISCAPE's political *garden*.

Indeed, between the opening and closing conferences, in October of 2020[1] and 2021[2] respectively entitled *Cultivating the Continuity of the European Landscape*, many singular forms of reflection carried out in different formats of discussion have nourished UNISCAPE's open forum: in addition to conference contributions, followed by a series of online lectures *Landscape, a place where disciplines meet*[3] and the international workshop W.A.VE on *Islands, Architecture and Landscapes of water*,[4] the works concluded fortunately "in person" with a range of diverse and inspiring landscape ateliers (*Woods go urban*[5]; *Isolario*[6]; *Perception conception*

---

[1] *Cultivating continuity of the European Landscape*, first edition, October, 16–17, 2020, Palazzo Vecchio, Florence.

[2] *Cultivating continuity of the European Landscape. One year after*, October, 22–24, Bardini Garden and Villa, Florence.

[3] This online lecture series is part of the program of UNISCAPE's Year of Landscape 2020-21 launched in October to mark the 20th anniversary of the European Landscape Convention. In 2023, the series is at its third edition.

[4] Promoted with the Università IUAV in Venice and dedicated to Enrico Fontanari, June 28, July 16, 2021.

[5] Coordinated by Lisa Diedrich (SLU Swedish University of Agricultural Sciences) and Catherine Szanto (ENSA La Villette).

[6] *Landscape of Islands and International Carlo Scarpa Prize for Garden* (draft title) coordinated by Luigi Latini (Fondazione Benetton Studi e Ricerche, Treviso, IUAV).

and visions for terraced landscapes[7]; *UAV landscapes*[8]; *Reinventing public spaces,*[9] *Cultivating energies: in pursuit of a new Ektopy*[10]), four round tables (*Cultivating the landscape*[11]; *Landscape policy and governance*[12]; *Landscape design and time*[13]; *Observing landscape*[14]) and a *Challenges/Living Lab*[15].

To manifest our will to openly discuss, share future challenges and find common solutions for the European Landscape together with wider society, the drafting of the second UNISCAPE manifesto on the *Future of the European Landscape* and the formation of *Uniscape's Green strategy* have been initiated. These distinctly opened, shared and evolving initiatives were essentially set up as a vivid public reminder that European Landscape (the most authentic expression of the relationship between people and places) and the European Landscape Convention (the driver of key democratic principles) can both truly frame the new European green shift by contributing to a more inclusive, sustainable and beautiful Europe, a "place open for everyone".

Hence, this publication, originated during a year of intensive and numerous "in person" and "online" meetings, should not be intended simply or solely as conference proceedings, but rather as a blend of contributions, that express through different visions and approaches the genuine desire for landscape to reactivate and animate the impassioned reasoning that will enable us to face, through a multiplicity of local actions, some of the most urgent, common and global world challenges. Because landscape in its essence equals *diversity*, it imposes an engagement with singularities of both types: with individual perceptions that permanently change as well as with unpredictable character of nature in its perpetual movement and transformation.

---

[7] Curators: Timmi Tillmann, Maruja Salas (ITLA, International Terraced Landscapes Alliance), Lucka Azman (University of Lubljiana).

[8] Coordinated by Margherita Vanore (Università Iuav di Venezia, Department of Architecture and Arts) with PRIDE Laboratory (Pro Research in Integral Design Environment), Research Infrastructure IR.IDE.

[9] *Landscape design exercises on Largo Annigoni, Florence*, coordinated by Anna Lambertini and Antonella Valentini (DIDA University of Florence) supported by AIAPP (Italian Association of Landscape Architects) and Municipality of Florence.

[10] *Curators* Paolo Picchi, Dirk Oudes, Sven Stremke, Academy of Architecture, Amsterdam University of the Arts, High Density Energy Landscapes Research Group; Roberta Pistoni, LATTS école d'urbanisme de Paris, Université Gustave Eiffel; Auréline Doreau, Chaire Paysage et Énergie auprès de l'ENSP Versailles, Marseille; Alessandra Scognamiglio, ENEA, Italian National Agency for New Technologies, Energy and Sustainable Economic Development; Angela Grassi, ETA Florence, Envision, Transform, Accelerate; supported by AIAPP (Italian Association of Landscape Architects).

[11] Chaired and coordinated by Mauro Agnoletti (University of Florence).

[12] Chaired and coordinated by Gian Franco Cartei (University of Florence), C. Cassatella (Politecnico of Turin).

[13] Chaired and coordinated by Rita Occhiuto (University of Liège), Margherita Vanore (Università IUAV, Venice).

[14] Chaired and coordinated by Juanjo Galan (Universidad Politécnica de Valencia), Juan Manuel Palerm (ULPGC).

[15] Chaired and coordinated by Saša Dobričič (University of Nova Gorica).

Already at the draft stage of this momentum of gathering, we have decided that the celebratory events should reflect upon the achievements and challenges of ELC with a judicious dose of "philosophy of suspicion" entangled within different debate formats as well as within diverse forms of expression, to ensure critical dialogue between multiple approaches but most of all to avoid any entrapment into commemorating achievements. Discussion, lectures, seminars, workshops, statements, design and art, all contribute to setting up a new stage for different expressions that knowledge, experience and imagination, when embracing landscape, can produce.

It was clear that we needed not just to organize events and exchange ideas but to enact a process that would move us a step forward from the necessary proliferation and cultivation of micro-areas of study disciplines to uncover some "concrete universality" that binds the multitude of diverse and specific problems: in the words of our keynote guest Emanuele Coccia "to move from the 'point of view' to the 'point of life'". In other words: to completely rethink, through the eyes of landscape, the predominant narratives, models and categories that affect our perception and understanding of the world, and that somehow we continue to cultivate when perhaps, we ought instead to challenge or, at least, reconsider. Specifically, in recalling the assumptions of Franco Farinelli's contribution: facing a great return of *Places* and evident fading of the logic of *space*, can landscape dominate as a strategic model for understanding the world and rethinking globalization? In a world that has in many ways become evanescent and immaterial, can we trust ourselves to such a fleeting and intangible model as landscape?

Indeed, the two main challenging dimensions of our initiatives, *Landscape as a* (restless) *Project* and *Landscape as a Common*, framed the session of keynote contributions, by recalling the above-mentioned dilemmas: *Landscape as a project* seriously reconsiders our capacity to give body to something that does not exist yet, like the future. It seems indeed that in an unpredictable world imbued with uncertainties, mainstream "futurology" delivers nothing but weak and easily digestible certainties. We are indeed an anxious collective that, under the umbrella of fear or pressure of multiple predictions and prefabricated solutions, is incapable of envisaging any common scenarios, to give a concrete body to a future. So, Landscape Project, considered as a powerful thinking tool able to restlessly *project*, is to literally eject into the future a comprehensive image of a collective and "interspecific" body of thought. Recalling the suggestion of Frédérique Aït-Touati, we could consider Landscape as a perfomative agent, a model of human–nature performance on common ground.

Conversely but intertwined with the Landscape Project: *Landscape as common* considers landscape as a privileged playground to practice and test our capacity to inhabit, to dwell among the extended community of living beings, in other words to be a craftsman of the same fabric of life. Are we able to nourish those forms of reciprocal use that create "wealth" as a distinctly shared dimension, wealth that does not end as an appropriation? As Amy Strecker points out, we have to re-align the concept of property with the landscape and re-imagine the way in which property is conceived. So, beyond private or public property, a belonging that is experienced as relationship, as reciprocal use, as a co-designing process of the common dimension by treating all forms of existence as genuine and authentic: as to say making the history also with those who apparently do not have the power to make it; and this is indeed what further challenges both the theory of action and our projects, that result literally *restless*.

That said, any celebration of an anniversary, even more so if it is that of the European Landscape Convention, must acknowledge the progress that has been made in the implementation of ELC and in the overall empowering of the European Landscape dimension. National and international relations have gradually taken it into account, contributing to the introduction of landscape issues in different sectors of society. This in turn has led to increased involvement in the *protection, management* and *planning* of both *outstanding* and *everyday* landscapes.

The European Landscape Convention has indeed been a catalyst in the generation of an all-encompassing and systematic effort to build a broadly defined and shared dimension of multiple landscape preservation and design strategies with the aim of progressing toward true inclusiveness at national and international levels. But most of all, the *landscape model* has stimulated traditional scientific, educational and professional categories to systematically acknowledge the need to redefine their boundaries, to step out and to engage with paradigms that fully embrace the complexity of life in a globalized world so clearly characterized by persistent environmental and social issues.

On the other hand, for many of us, teachers, researchers and, most of all, students, landscape has become a powerful thinking model that can ignite cross-sectorial debate and empower truly collaborative imagination. Landscape has pervaded traditional academic disciplines with a new "ecology of thought" that actually reframes the way we see our relationship with other facets of Earth and widens the focus from single elements to bind them into a coherent ensemble of many different voices that make up our mosaic of "art of living".

We have all acknowledged that new efforts have involved the educational and research sectors in developing those forms of continuity that are cultivated entirely within and in-between diversity.

In essence, perhaps the most fortunate and explicit expression of the implementation efforts of the ELC today, 20 years after its adoption, lies in its rich and diverse constellations of protected areas, natural reservoirs, heritage sites, etc.

But can these cultural landscapes, considered as authentic and diverse articulations of fortunate coexistence, migrate out of the protective belt and meld with our world to re-design it and enact new relationships with animated and non-animated actors that make up our extended community? Is there a risk that this *migration* only reproduces the format thereby creating an outlet of standardized visions of how landscape should look, after all landscapes are often explained before they are really seen?

Hence, if landscape is a relational model driven by a seamless bond between the observed and the observer, a "background" to the actors, in a reciprocal cycle of influence, then how should new patterns of connection between the different patches composing the European Landscape be established and integrated? And how should these forms of *continuity* be adopted while keeping all the components of the landscape model united and interrelated albeit in friction with the world and history?

Maybe some hints for the future debate can be found in the reading of the volume.

An Uncommon Structure

Lastly, we would like to expend a few brief words as to the conception and the laborious creation of this book. At the outset, this volume was intended as a traditional conference proceedings book, in other words a faithful report of the first UNISCAPE Florence Conference (held on October, 16–17, 2020).

In this sense, originally the core of the volume was composed of three complementary sections (*Landscape Policy and Governance*, edited by Claudia Cassatella and Gian Franco Cartei; *Landscape design and Time*, edited by Rita Occhiuto and Margherita Vanore; and *Observing Landscape* by Juan Manuel Palerm and Juanjo Galan). The crucial goal of this first part was the broader, innovative and shared exploration, beyond national and disciplinary boundaries, of the *landscape dimension* (and related landscape-oriented issues) as proposed by the ELC.

Then, as previously mentioned, during these last two years, the global context and the specific framework have been continuously transforming and evolving.

And so too this volume has been progressively enriched in an attempt to preserve the continuity of a landscape-centered discourse, *restlessly* developed within our UNISCAPE community, where the strong exigence of building new cultural places of sharing and awareness emerged.

In this sense, the first and last chapters have been conceived to complete the structure of the book, reporting the continuity of the reflections brought on over time.

So, Part I, dedicated to *New Challenges for European Landscapes*, is edited by Saša Dobričič and hosts a selection of the papers proposed by the keynote speakers who gave their lectures in both Florence conferences, whereas Part V, edited by Tessa Matteini and Conor Newman includes a set of reflections for the future challenges and landscape-oriented perspectives, drafted by some of the UNISCAPE members.

Saša Dobričič
Tessa Matteini

Contents

Part I New Challenges for European Landscapes

1 **Performing Landscape** 3
Frédérique Aït-Touati

2 **Policies, Landscape and Politics** 9
Franco Farinelli

3 **Global Landscape Governance on the 20th Anniversary of the European Landscape Convention** 17
Amy Strecker

Part II Landscape Policy and Governance

4 **"Landscape Policy and Governance: An Introduction"** 29
Gian Franco Cartei

5 **Landscape Planning Policy as a Measure of Implementation of the European Landscape Convention** 33
Stefano Civitarese Matteucci

6 **Policy and Governance Challenges Facing European Rural Landscapes** 43
Jørgen Primdahl

7 **Landscape Policies of Europe** 51
Clara Queiroz da Costa and Maria José Curado

8 **Landscape Policy in Portugal. Where Are We?** 63
Isabel Joaquina Ramos and Maria Freire

9 **A Country Through Crisis and Beyond: The Case of the Greek Landscape** 71
Theano S. Terkenli and Nikolaos Skoutelis

10 The ELC, a Game Changer in Scottish Policy Discourse 81
Sarah Govan

11 One Territory, One Plan, Many Authorities: The Future of Really Protecting the Landscape 91
Luca Di Giovanni

12 The Daily Governance of Landscape: The Contribution of Local Landscape Commissions 99
Raffaella Laviscio

13 Enclosing the Horizon: Land Privileges and the Right to Landscape 109
Marius Fiskevold

14 Landscape Observatory Västra Götaland & Balaton Ecomuseum—A Swedish–Hungarian Interaction for Sustainable Landscape Management 115
Bosse Lagerqvist, Sándor Némethy, and Anders Nilsson

15 Landscape and Public Participation in Norwegian Local Planning—An Example of Implementing the European Landscape Convention 125
Marte Lange Vik

16 Implementing Landscape Policies by Promoting Sustainability, Democracy, and Participation. Awareness Raising and Landscape Education in Trentino, Italy 133
Gianluca Cepollaro and Bruno Zanon

17 Common Property in Europe: Landscape Design Evolution Over Time 141
Monica Pantaloni, Giovanni Marinelli, and Davide Neri

18 Collaborative Governance for Reviving Tiber Landscapes Downstream of Rome 155
Anna Laura Palazzo and Romina D'Ascanio

19 Planning the Adaptation of European Landscapes to Climate Change: The Experience of the EIT Climate-KIC AELCLIC Project 165
Daniele Torreggiani, Juanjo Galan, Francisco Galiana, Bas Pedroli, and Emilio Servera-Martínez

20 The European Baykal—Regional Development Inspired by the European Landscape Convention 171
Nataliya Pozhidaeva, Alexandra Kruse, and Victor Pozhidaev

21 Twenty years of Landscape Policy and Governance in Europe, and the Way Ahead 201
Claudia Cassatella

Part III Landscape Design and Time

22 Landscape Design and Time 209
Rita Occhiuto and Margherita Vanore

23 Geologic Architecture 215
Marco Navarra

24 Time-Tested Design and Implementation: Differing Temporalities in Landscape Design 227
Michel Hössler

25 The Art of Failing Urban Planting Precariousness as Landscape Project Ecology 233
Denis Delbaere

26 Seascape Design. A Limit for the European Landscape Convention 243
Daniela Colafranceschi

27 Water-Related Projects for the Urban Environment. The Case Studies of Lisbon Metropolitan Area 251
Caterina Anastasia

28 Soil Displacement. Landscape Project as an Infrastructure Across Building Geography and Grounding Metabolism. The Case Study of Pialassa Piomboni Constructed Wetland in Ravenna 263
Vittoria Mencarini, Gianni Lobosco, Luca Emanueli, and Massimo Tondello

29 Mapping the Rhythm in Wetlands 275
Stefano Tornieri

30 The Creative Time of Gardens: From Microcosm to Landscape Infrastructure 285
Antonia Di Lauro

31 Underground Heritage Valorization of Camerano's Caves in Center Italy: A Case of Transition Toward Projects Integrating the Local Community and Landscape 293
Alici Antonello, Andreoni Luca, Chiapparino Francesco, De Seriis Marco, Fioretti Ilaria, Galli Andrea, Malinverni Eva Savina, Marcheggiani Ernesto, Paci Francesco, and Pierdicca Roberto

32 **Compromised Landscapes** ... 305
Barbara Angi and Alberto Soci

33 **An Open Project for the Trembling Landscape of the Inner Areas: Fragile and Wounded** ... 311
Angela D'Agostino and Giovangiuseppe Vannelli

34 **Abandoned Army Barracks in Friuli Venezia Giulia (Italy) as New Time-Shaped Community Landscape Potential** ... 321
Luca Maria Francesco Fabris and Federico Camerin

35 **Rhythm and Continuity in Interventions to Reactivate Vacant Urban Spaces. Recognizing the Stratification of Places to Foster an Autonomous Process of Integration Between the Parts** ... 331
Chiara Caravello and Elisa Baldin

36 **What Future for Italian Historical Urban Landscape—A Veneto Case Study** ... 337
Daniela Ruggeri

37 **Green Infrastructure for the Landscape Reconfiguration of the VII Municipio of Roma: An Open Project, Beyond Urban Regeneration** ... 341
Giorgio Caprari, Matej Gulic, and Elio Trusiani

38 **The De-structuring of Infrastructure Landscapes** ... 353
Josep Lluís Miralles-Garcia

39 **Venice Marathon: From Landscape Brand to Healthy Infrastructure** ... 361
Massimo Triches

40 **KM0. Merging Public Space and Nature** ... 373
Enric Batlle i Durany and Javier Rivera Linares

41 **Project Perspectives for Making Landscape** ... 383
Rita Occhiuto and Margherita Vanore

Part IV Observing Landscape

42 **Observing Landscape: General Reflections and Linkages with the UNISCAPE2020 Conference** ... 391
Juanjo Galan and Juan Manuel Palerm

43 **The Landscape Observatory of Catalonia. A Journey of Fifteen Years** ... 397
Pere Sala i Martí

44 **Aesthetically Performed Landscape Criticism** ... 403
Nina Marie Andersen

45 Cultivating Drystone Walls—Regional Observatories for European Landscapes 411
Torben Dam

46 Landscape Observatory in Alqueva, Alentejo Region (Portugal). Why It Matters? 423
Maria Freire, Isabel Joaquina Ramos, and Maria da Conceição Rego

47 Landscape Observatories: An Analytical Framework and a Case Study at the Landscape Observatory of Finland 431
Juanjo Galan

48 Regional Coordination of Large-Scale Infrastructure Projects—A New Role for Landscape Observatories? 441
Anders Larsson

49 The Observation of Informal Landscape Public Participation 447
Elena Lorenzetto

50 Governance Challenges of a Local Landscape Observatory Initiative in a Remote Landscape: Ways Ahead 455
Maria Graça Saraiva, Isabel Loupa-Ramos, and Veerle Van Eetvelde

51 The Atlas of Terraced Landscapes of Trentino, Italy 465
Giorgio Tecilla and Alberto Cosner

52 The Museum as an Active Observatory: Protecting and Shaping the Landscape 477
Maciej Żołnierczuk

Part V Cultivating Innovative Perspectives

53 Landscape Continuity: Cultivating Different Allegories of the Inseparable 489
Saša Dobričič

54 Cultivating Landscape Continuity. The Projects " Design" in the Architecture of the Landscape 499
Juan Manuel Palerm

55 Cultivating the Landscape Dimension 515
T. Matteini

56 Living in an Anthropocene: Reconciling Culture and Nature for a Sustainable Future 525
Conor Newman

57 The European Landscape as a Task: Understanding the Past, Engaging in the Present, Living the Future? 533
Bas Pedroli

58 Where the Disciplines Meet: Uniscape as a Laboratory 539
Conor Newman

Contributors

Frédérique Aït-Touati CNRS-EHESS, Paris, France

Caterina Anastasia CIAUD, Research Centre for Architecture, Urbanism and Design, Lisbon School of Architecture, Universidade de Lisboa, Lisbon, Portugal

Nina Marie Andersen Associate Professor, Norwegian University of Life Sciences, Ås, Norway

Galli Andrea D3A Department of Agricultural, Food and Environmental Sciences, Marche Polytechnic University, Italy LRC/CIRP, Marche, Italy

Barbara Angi DICATAM Department, Università degli Studi di Brescia, Brescia, Italy

Alici Antonello DICEA Department, Università Politecnica delle Marche (Italy), LRC/CIRP, Ancona, Italy

Elisa Baldin Université de Liège – Lab VTP, Liège, Belgium

Enric Batlle i Durany Batlle I Roig Associates, Barcelona, Spain

Federico Camerin Departamento de Urbanística y Ordenación del Territorio - Grupo de Investigación en Arquitectura, Urbanismo y Sostenibilidad (GIAU+S), Universidad de Valladolid - Universidad Politécnica de Madrid, Madrid, Spain

Giorgio Caprari School of Architecture and Design, University of Camerino, Camerino, Italy

Chiara Caravello Université de Liège – Lab VTP, Liège, Belgium;
Politecnico di Milano, DAStU, Milano, Italy

Gian Franco Cartei University of Firenze, Firenze, Italy

Claudia Cassatella Interuniversity Department of Regional and Urban Studies and Planning, Politecnico di Torino, Turin, Italy

Gianluca Cepollaro Scuola per il Governo del Territorio e del Paesaggio (step), Trento, Italy

Daniela Colafranceschi University of Naples Federico II, Napoli, Italy

Maria da Conceição Rego Department of Economics, Centre for Advanced Studies in Management and Economics–CEFAGE–UÉ, School of Social Sciences, University of Évora (UÉ), Evora, Portugal

Alberto Cosner President Research Group TeSto, Trento, Italy

Maria José Curado Faculty of Sciences, University of Porto, Porto, Portugal

Angela D'Agostino Department of Architecture - University of Naples Federico II, Naples, Italy

Romina D'Ascanio Department of Architecture, Roma Tre University, Rome, Italy

Clara Queiroz da Costa Landscape Architecture Master, University of Porto, Porto, Portugal

Torben Dam Landscape Architect MDL, Landscape Architecture and Planning, KU-Science, University of Copenhagen, Copenhagen, Denmark

Denis Delbaere Laboratoire Conception Territoire Histoire (LaCTH), Ecole Nationale Supérieured'Architecture et de Paysage de Lille (ENSAPL), Villeneuve-d'Ascq, France

Luca Di Giovanni University of Urbino, Urbino, Italy

Antonia Di Lauro Politecnico Di Milano—Scuola Di Architettura, Urbanistica Ingegneria Delle Costruzioni, Milan, Italy

Saša Dobričič University of Nova Gorica, Nova Gorica, Slovenia

Luca Emanueli Sealine Research Center, University of Ferrara Department of Architecture, Ferrara, Italy

Marcheggiani Ernesto D3A Department of Agricultural, Food and Environmental Sciences, Marche Polytechnic University, Italy LRC/CIRP, Marche, Italy

Luca Maria Francesco Fabris DASTU, Politecnico di Milano, Milano, Italy; Beijing University of Civil Engineering and Architecture, Beijing, China

Franco Farinelli Professor Emeritus, University of Bologna, Bologna, Italy

Marius Fiskevold The Norwegian University of Life Sciences, Ås, Norway

Chiapparino Francesco DISES Department, Università Politecnica delle Marche (Italy), LRC/CIRP, Marche, Italy

Paci Francesco D3A Department of Agricultural, Food and Environmental Sciences, Marche Polytechnic University, Italy LRC/CIRP, Marche, Italy

Maria Freire Department of Landscape, Environment and Planning, Centre for Art History and Artistic Research–CHAIA-UÉ, School of Sciences and Technology, University of Évora (UÉ), Évora, Portugal

Juanjo Galan Department of Architecture, Aalto University, Espoo, Finland; Department of Urban Planning, Universidad Politécnica de Valencia, Valencia, Spain

Francisco Galiana Universitat Politècnica de València, Valencia, Spain

Sarah Govan University of Edinburgh, Edinburgh, Scotland

Matej Gulic Environmental Protection Department - Green Areas Design Office, Roma Capitale, Italy

Michel Hössler Agence Ter, Paris, France

Fioretti Ilaria Municipality of Camerano, Ancona Province, Italy

Alexandra Kruse insitu World Heritage Consulting, Buc, France

Bosse Lagerqvist Department of Conservation, University of Gothenburg, Gothenburg, Sweden

Anders Larsson Department of Landscape Architecture, Planning and Management, SLU, Alnarp, Sweden

Raffaella Laviscio Department of Architecture Built Environment and Construction Engineering, Politecnico di Milano, Milano, Italy

Gianni Lobosco Sealine Research Center, University of Ferrara Department of Architecture, Ferrara, Italy

Elena Lorenzetto Fondazione Benetton Studi Ricerche, Treviso, Italy

Isabel Loupa-Ramos IST, CiTUA, University of Lisbon, Lisbon, Portugal

Andreoni Luca DISES Department, Università Politecnica delle Marche (Italy), LRC/CIRP, Marche, Italy

De Seriis Marco Municipality of Camerano, Ancona Province, Italy

Giovanni Marinelli Department of Materials, Environmental Sciences and Urban Planning, Polytechnic University of Marche, Ancona, Italy

Pere Sala i Martí Landscape Observatory of Catalonia, Catalonia, Spain

T. Matteini Department of Architecture, University of Florence, Florence, Italy

Stefano Civitarese Matteucci Department of Legal and Social Studies, The University of Chieti-Pescara (IT), Chieti, Italy

Vittoria Mencarini IDAUP International Doctorate in Architecture and Urban Planning, Department of Architecture, University of Ferrara, Ferrara, Italy

Josep Lluís Miralles-Garcia Departament d'Urbanisme, Universitat Politècnica de València, Valencia, Spain

Marco Navarra Department of Civil Engineering and Architecture, University of Catania, Catania, Italy

Sándor Némethy Department of Conservation, University of Gothenburg, Gothenburg, Sweden

Davide Neri Department of Agricultural, Food, and Environmental Sciences, Polytechnic University of Marche, Ancona, Italy

Conor Newman National University of Ireland, Galway, Ireland;
University of Galway, Galway, Ireland

Anders Nilsson Region Västra Götaland, Gothenburg, Sweden

Rita Occhiuto Department of Architecture, University of Liège, Liège, Belgium

Anna Laura Palazzo Department of Architecture, Roma Tre University, Rome, Italy

Juan Manuel Palerm EA_ULPGC, Las Palmas, Spain;
UNISCAPE, Florence, Italy;
University of Las Palmas de Gran Canaria, Las Palmas, Spain

Monica Pantaloni Department of Agricultural, Food, and Environmental Sciences, Polytechnic University of Marche, Ancona, Italy

Bas Pedroli Land Use Planning Group, Wageningen University, Wageningen, The Netherlands;
JustLandscape, Amsterdam, The Netherlands

Victor Pozhidaev The Interregional Public National Landscape Salvation Fund—NLSF, Moscow, Russia

Nataliya Pozhidaeva The Interregional Public National Landscape Salvation Fund—NLSF, Moscow, Russia

Jørgen Primdahl Department of Natural Resources and Natural Resource Management, University of Copenhagen, Copenhagen, Denmark

Isabel Joaquina Ramos Department of Landscape, Environment and Planning, Interdiscliplinary Centre of Social Sciences–CICS.NOVA.UÉvora, School of Sciences and Technology, University of Évora (UÉ), Évora, Portugal

Javier Rivera Linares Universidad Cardenal Herrera-CEU, CEU Universities, València, Spain

Pierdicca Roberto DICEA Department, Università Politecnica delle Marche (Italy), LRC/CIRP, Ancona, Italy

Daniela Ruggeri Università Iuav di Venezia, Venice, Italy

Maria Graça Saraiva CIAUD, Research Center for Architecture, Urbanism and Design, Lisbon School of Architecture, Universidade de Lisboa, Lisbon, Portugal

Malinverni Eva Savina DICEA Department, Università Politecnica delle Marche (Italy), LRC/CIRP, Ancona, Italy

Emilio Servera-Martínez Universitat Politècnica de València, Valencia, Spain

Nikolaos Skoutelis Technical University of Crete, Kounoupidiana, Greece

Alberto Soci DICATAM Department, Università degli Studi di Brescia, Brescia, Italy

Amy Strecker UCD Sutherland School of Law, Dublin, Ireland

Giorgio Tecilla Landscape Observatory, Autonomous Province of Trento, Trento, Italy

Theano S. Terkenli University of the Aegean, Lesvos, Greece

Massimo Tondello Sealine Research Center, University of Ferrara Department of Architecture, Ferrara, Italy

Stefano Tornieri IUAV University, IR.IDE (Infrastructure Research Integral Design Environment), Venice, Italy

Daniele Torreggiani University of Bologna, Bologna, Italy

Massimo Triches Dipartimento di Culture del Progetto, Infrastruttura di Ricerca IR.IDE, Laboratorio di Progettazione PRIDE, Università Iuav di Venezia, Venice, Italy

Elio Trusiani School of Architecture and Design, University of Camerino, Camerino, Italy

Veerle Van Eetvelde Department of Geography, Ghent University, Ghent, Belgium

Giovangiuseppe Vannelli Department of Architecture - University of Naples Federico II, Naples, Italy

Margherita Vanore Department of Architecture and Arts, Università Iuav di Venezia, Venice, Italy

Marte Lange Vik Department of Environmental Sciences, Western Norway University of Applied Sciences, Sogndal, Norway

Bruno Zanon Università degli Studi di Trento and Scuola per il Governo del Territorio e del Paesaggio (step), Trento, Italy

Maciej Żołnierczuk Development Department, Museum of King Jan III's Palace at Wilanów, Warsaw, Poland

Part I
New Challenges for European Landscapes

Chapter 1
Performing Landscape

Frédérique Aït-Touati

Abstract This article describes the redefinition of landscape in relation to current transformation of cartography and speculative design.

Keywords Landscape · Cartography · Speculative design · Ecology · Planet · Earth · Description · Theatre

Reading about the different histories of the notion of landscape, particularly in art history, I can see that all of them associate the emergence of landscape and the moment in Western culture when sensitivity and knowledge, or the subjective and objective approach to nature, began to divorce. According to the French geographer Augustin Berque, for instance, the relationship between the history of the notion of landscape and the world is inseparable from the objectifying distancing of Western scientific rationality. Landscape is the world viewed from a distance by a subject who feels and rationalizes from a distant point of view, outside the object he is focusing on. But while this process of distancing allowed nature to exist as landscape, according to Berque the development of science would soon lead to the inevitable dissociation of a factual world, studied rationally, from a phenomenal world, perceived subjectively and taken over by art. It is this phenomenal world that landscape was to have the task of embodying.

It is no surprise then if the rise of the concept of landscape is linked to the history of painting and that of the theatre. Landscape as a scenic space has largely been identified with the development of linear perspective and the application of its techniques to pictorial art. Within the realm of theatre, the same constellation of Ptolemaic geography, perspective and the arts is particularly well exemplified by the early seventeenth-century court masques (a predecessor of opera) produced by the author Ben Jonson (the leading author of the day after Shakespeare) together with the pioneering scenographer and architect Inigo Jones.

F. Aït-Touati (✉)
CNRS-EHESS, Paris, France
e-mail: f.aittouati@gmail.com

T. Matteini et al. (eds.), *Cultivating Continuity of the European Landscape*, Environmental History 15, https://doi.org/10.1007/978-3-031-25713-1_1

Jonson's preface to the first of these masques, the 1605 Masque of Blackness, contains an early use of the word *landscape* meaning scenery and a very early identification of landscape with nature. From this link to the history of theatre, the meaning of the term landscape has maintained the idea of backdrop.

I explored this issue in a book entitled *Terra Forma, Manual of Potential Cartographies,* which I coauthored with two landscape architects, Alexandra Arènes and Axelle Grégoire. The book develops seven models in seven chapters. Each chapter presents a map resulting from the invention of a new model to visualize a specific issue. The seven chapters are points of view on reality, possible visions of the world sketched through different prisms. Written in the literary genre of an exploration narrative, this book is also a drawing handbook; it attempts to sketch new types of landscapes with the aim of gradually and collectively constituting a new kind of atlas. In this article, I will attempt to share some of the tools that helped us shift from vision to other means of perception—not only transforming gazing but also the position of the body by broadening the perceptual tools and thus moving from optics to haptics. What emerges is a new definition of the concept of landscape: as an inhabited territory, as a model of human–nature performance on common ground which is built together. What I would like to suggest in this article is not that the notion of landscape has disappeared or should be discarded because of its link to the division of nature from culture but rather that the very definition of landscape is changing in various ways, making it a powerful concept helping us reflect on our current situation. Here, landscape is seen not as content, a backdrop, but as a performative agent.

1.1 Living Landscapes

One of the starting points of *Terra Forma* was the question: Where in our classical maps are the living beings? Where are the entities we talk about every day in our reflections on ecology? How can we make room for them? One answer is to question the parameters that structure the map, or hack the classic tools of cartography and architecture. We started by removing the background of the maps so that the topographical elements no longer exist. The question then arises: *What will build the landmarks?* This empty map background crystallizes the relationship between living things and their territory. The moving bodies of "animated beings", as we call them, become the driving force behind the map. By removing the map's background, we obtain a liquid space. This leads to other circulations, new spaces of encounters. The map becomes the expression of the multiplicity of interactions inhabiting it. It invites us to reflect on the links between the points of life co-inhabiting it. This map can help in making a diagnosis by integrating the data of a territory, but it can also promote negotiation by questioning the community it brings together.

The map called "Living Landscapes" explores the way in which living beings trace their territories, leave footprints on the surface, cohabit, develop thick borders, i.e., environments, or adjust their rhythms of life. The cartographer must question the movement of these entities and the forms of territory they generate: this is what we are attempting by going from a point of view to a point of life. A point of view, as we know, is a distanced gaze, one that embraces a landscape to construct a perspective image of reality. The idea of point of life, a term we borrowed from Emanuele Coccia, starts with the living and considers that each living person integrates the world in his own way: a second inversion. Living things create their own survival space so landscape can be understood as an extra-anatomical body. The resulting mix of body and territory can be found in the "Living Landscape" map. We have made the earth's crust and our skin coincide as two surfaces undergoing the same types of transformation as they face climate change.

Our reflection on the point of view or viewpoint led us to re-examine the methods of noting and drawing. How can the invisible be drawn, be it time, movement, or different scales? The idea was therefore to proceed by shifting and transforming the gaze, graphic conventions, notation principles and drawing techniques. In addition to eliminating the map background, we played with multiplying the points of view, the articulation of the different representation systems such as the cut, the plane and the axonometry, or simplifying the colour code to the monochrome. We wanted to focus on the cartographic experience rather than on the result by focusing on the link between narrative and cartographic writing, whose new aesthetic has both a cognitive and a political aim befitting the long history of representing the Earth. By questioning cartographic conventions, putting them once again into play to collectively invent new ones, we free ourselves from their normative power to redeploy resources, reconsider ways of living and represent our territories of life. This "geoaesthetics" arises from the need to produce artefacts in order to understand reality that which surrounds us and what we do with it. Like in a laboratory, as the Science and Technology Studies (STS) would say, reality is constructed with artifice through experimentation.

The "Living Landscapes" map is based on the principle of portolan charts, those very important navigational tools produced up to the Renaissance. Originally, a portolan was a map mainly defining coasts which allowed ships to find their way at sea thanks to the notation of winds and currents. We started from the observation that space is increasingly fluid, liquid, moving, made up of various felicitous or infelicitous associations (think viruses and humans). In any case, space is absolutely not this stable, neutral, fixed and quiet thing that we thought we had mastered so the idea is to chart what we have called the "living territories" of the different beings: people, but also micro-organisms, cats... Then we try to find places of intersection, geometric signatures specific to each living thing. It's a question we can ask everyone: What does your living space look like? Does it look more like an owl's or a rabbit's living ground? We were interested in the fact that by putting very different scales on the same map, we see links between beings, ways of moving and relationships to space, with the project of making a never-exhaustive typology of the different movements, these different ways of relating to the earth. *Terra Forma* invites the reader to re-anchor himself by drawing on the ground, on the Earth, to "land" on and in this famous critical area where living beings and their resources are concentrated. For us, the challenge of using cartography was to understand the superimposition of humans and non-humans, to comprehend the continuum from body to space and to observe the links between beings. A map can tell a story and bodies compose the map; it's a choreography.

1.2 Landscape as a Critical Zone

We can neither deny nor remain indifferent to the fact that humans have become a major geological force that is not only rapidly transforming the chemical and physical composition of the Earth but also disrupting ecosystems at such a rate that they do not have time to adapt. Humans exploit the depths, disturb the strata of eras, create layers of plastic and concrete[1] and shape the Earth. They are terraforming the Earth, they are creating new kinds of landscapes. The "ground" map attempts therefore to picture not only the surface, but also the depths and the atmosphere—a broadened landscape. The map works like a glove turned inside out: the atmosphere has been placed in the centre, the deep earth on the periphery, and what occupies most of the map is the thin layer of the earth's crust, which extends like a skin—what scientists call the "critical zone" because it is in danger and all life is concentrated there. By means of a kind of anamorphosis, the aim is to show what we usually don't see due to a problem of scale: the critical zone is infinitesimal in relation to the globe. But this thing we don't see is the place where life, and its resources, are concentrated.

On the inner side of the glove, the ground suddenly reveals an alternative cosmology: a subarchitecture with "inhabiting organisms." Seen from below, the soil that was thought to be homogeneous and solid is moving, porous, liquid, composed of particles and agents that constantly imprint and modify its structure and shape. The skin of the Earth and the multiple agents that produce it are visible through this lens, indicating the almost equal contribution of human and non-human actors. This is why we suggest exploring the Earth as earth—soil, humus—rather than as a globe.

This is our *terra incognita*: the ground right under our feet. The entities made visible by this map of the soil "hold" us in a common history, entangled in the compost of the Earth. However, each entity, each element is discernible. Each has its own ways of becoming and responding to other entities which may affect its living space. Therefore, if we can grasp its activities and living space, it is possible to draw its map. This map is a prerequisite for any political negotiation between these entities, enabling them to define a territory—their living territory—in order to confront it with those of others and to provide a space to discuss how they overlap. The soil map therefore enables us to shift our view, and by exploring the soil, it reveals an invisible part: a living part of the world.

---

[1] Zalasiewicz (2016).

1.3 Ruined Landscapes

Traditionally, ruins herald the end of a story, an accumulation of traces that no longer make sense. But one might equally consider ruins to be part of a process that stretches from construction, to demolition, to the engendering of new soil. It is this process that interested us: considering ruins as a new topography, their assimilation into soil to restore its habitability and shape. Map of Memories is a tool for the regeneration of ruined landscapes. The compass is a spatio-temporal matrix that transforms former ruins into a potential future depending on the action undertaken, the nature of the ruins (their degree of potential habitability after regeneration), and the time required for their regeneration. This process draws new territories onto formerly dislocated lands. The purpose of this map is to move from the conquering *terra formation* that damages the world that it transforms to processes of rereading, rewriting, and transforming ruined landscape into new signs to be interpreted. The Map of Memories is more prospective and fosters political and architectural reflection on the use and habitability of ruins.

This procedure makes remains visible that have become useless—clumsy architectural and industrial fossils. However, these ruins may also become tools for action.

Just like an organism, a territory has its own DNA, a memory in which a mass of identity information, traces and footprints are encoded. This mass may be left passive, buried under ever-increasing ruins, or it may be activated. The degree of habitability of territories is now very often measured by the ability of its inhabitants to revitalize its memory. Memory, like ours, which is largely produced, crafted and hybridized. Some traces are valued, while others are obscured. The remains, fossils and relics of the past are sometimes staged, as in a memorial. More often, however, we try to forget them. The earth underground is paradoxically a material and our collective unconscious to bury and hide what we would like to ignore. We gladly forget the past of some sites without acknowledging the scope of the problem left by their traces: an exhausted soil, for example, is a dead soil, an invisible but burdensome ruin. Nuclear waste and other infrastructure networks are buried in the subsurface so as not to hinder the movements of living beings but also, in a way, to be forgotten. It is necessary to revitalize them by drawing the ruins and fossils of our geohistory, this very recent human geology made of the slow sedimentation of our artefacts.

Still following Anna Tsing, we hope that these traces are the foundations of a possible future. In a more prospective way than for the other maps, we imagine a graphic tool which recycles the waste of the past for use in possible futures, and which provokes mutations. The model is conceived as a device to connect the past and the future and to activate, transfer, encode or reverse the processes of making ruins to help create the conditions for inhabitants to live in them again. The aim is to encourage the making of new foundations to share a vivid memory instead of an archaeology of disaster.

Originally, "terra forming" was a term used in science-fiction: it's the idea of transforming another planet (for instance Mars) in order to inhabit it as a new earth.

While our project is precisely the contrary—showing how the Earth itself has been terraformed by living beings—from this science-fictional reference, we keep the notion of "potential": the freedom to anticipate, to test scenarios. We have attempted to pass from speculative fiction to speculative visualization. In the book's succession of images, it is clear that we have gone on to draw on cosmic imagination. We have tried to preserve the power of fiction, the excitement of discovery, and to reinject them into the discovery of a new land, a new landscape—our own. It is a counter-conquest of space. And finally, more than science-fiction, what brought the three of us together was firstly the capacity of maps to unfold worlds and, even more so, the will to open up paths of reflection different from those of tales of collapse or dystopias.

As an ending, allow me to quote a play I have written with Bruno Latour—because theatre is of course another way to cherish, re-invent and re-animate our terrestrial landscapes, to change landscape from backdrop to actor:

> Imagine being told that the world in which you thought you were situated is not at all like the one you imagined. There you are, quietly surrounded by objects with clearly defined shapes, a table, a chair, each separate from the others and the whole collection inserted as though in a painting, inside a space independent of them. And you, you are there, in front of this spectacle of objects that move only if they are pushed or pulled, Galilean objects. In your head you obviously have a whole set of deep or superficial thoughts and you feel perfectly distinct from this assemblage of things, at the right distance, again as if in front of a canvas.
>
> You may see the leaves of a tree through a window, the clouds that the wind blows across the sky, the contours of a mountain… but they too are at a conventional distance, an object among objects, all inserted into infinite space, all distinctly delineated.
>
> Now, little by little, things get more complicated. The rain clouds that cover the horizon are seeded by bacteria. The skies where the clouds course are held together because microbes have been piling up their droppings in the sky for billions of years. The average temperature you enjoy, the very air you breathe, you owe to the trees you see outside.
>
> Do you realize what's going on? The living beings are no longer side by side but are starting to overlap, drooling on each other, mingling, intertwining. These beautiful and deep thoughts that you thought you were having only happen thanks to the billions of bacteria in your intestines. The wood used to make the chair comes from a distant forest, the movement of all these superimposed beings carries you away. The landscape unravels, the painting fades away. You no longer see anything as a spectacle.[2]

References

Aït-Touati F, Latour B (2022) Moving earths. in terrestrial trilogy, B42, Paris

Zalasiewicz J (2016) The extraordinary strata of the anthropocene. In: Serpil Oppermann et Serenella Iovino (dir.), Environmental humanities: voices from the anthropocene, Lanham (Maryland), Rowman & Littlefield

[2] Aït-Touati and Latour (2022).

Chapter 2
Policies, Landscape and Politics

Franco Farinelli

Abstract Elias Canetti warned us that behind any model there is always a metamodel whose nature is malign. The nature of the landscape model is thus political, a specific form of malignance. Just think of Alexander von Humboldt's project, without which landscape would never have moved from the purely artistic to the scientific sphere: a decisive transition because it was preparatory and even essential to the current equating of landscape and territory, a term that at the end of the Roman empire in the Code of Justinian is explained as falling within the scope of political power because it derives not from "*terra*" (earth) but rather from "*terrore*" (terror). How this extraordinary backflip leads from aesthetics through the reinvention of nature to the political, and its effects, has yet to be fully elucidated. In any case, Humboldt's problem was to ensure the success of the *Weltbürgerplan,* the bourgeoisie's plan for world domination against the interests of the originally feudal aristocracy, and he chose the concept of landscape as his preferred vehicle for the indispensable structural mutation of the culture of that era's civil society who were called to abandon their "vacuous poetic games", as Humbolt himself would say, in favour of knowledge that was not only contemplative but also functional to action, control and the management of the world, for knowledge no longer of only arts and the humanities but finally scientific.

Keywords Landscape · Globalisation · Politic · Earth's models

1. The origin of the strategy: landscape and the politicization of aesthetic elements.

Elias Canetti warned us that behind any model there is always a metamodel whose nature is malign. The nature of the landscape model is thus political, a specific form of malignance. Just think of Alexander von Humboldt's project, without which landscape would never have moved from the purely artistic to the scientific sphere: a decisive transition because it was preparatory and even essential to the current

F. Farinelli (✉)
Professor Emeritus, University of Bologna, Bologna, Italy
e-mail: francofarinelli84@icloud.com

T. Matteini et al. (eds.), *Cultivating Continuity of the European Landscape*, Environmental History 15, https://doi.org/10.1007/978-3-031-25713-1_2

equating of landscape and territory, a term that at the end of the Roman empire in the Code of Justinian is explained as falling within the scope of political power because it derives not from "*terra*" (earth) but rather from "*terrore*" (terror). How this extraordinary backflip leads from aesthetics through the reinvention of nature to the political, and its effects, has yet to be fully elucidated. In any case, Humboldt's problem was to ensure the success of the *Weltbürgerplan,* the bourgeoisie's plan for world domination against the interests of the originally feudal aristocracy, and he chose the concept of landscape as his preferred vehicle for the indispensable structural mutation of the culture of that era's civil society who were called to abandon their "vacuous poetic games", as Humbolt himself would say, in favour of knowledge that was not only contemplative but also functional to action, control and the management of the world, for knowledge no longer of only arts and the humanities but finally scientific.

In cultural terms, Humbolt's project won because he was clever enough to operate along internal lines, that is, to speak the same language as his interlocutors who were sometimes extraordinary *connoisseurs* of painting, music or literature but wholly unfamiliar with the great "book of nature" and talk of science that, on the contrary, they were able to read and so were functional to a given political power, that power that saw its historic adversary in civil society and opinion or the nascent public sphere. Without the victory of the latter, we would not still speak of landscape, even though the meaning and function of that model appear quite different today, even inverted compared to the original terms of Humbolt's project.

2. The cognitive nature of the strategy: from emotion to science.

In the second volume of his main work, the *Cosmos*, published in Berlin in 1847, the year before the first of the uprisings that would bring the bourgeoisie to power in Prussia, Humboldt outlines the history of the models governing humanity's view of the world from their origins. And his whole reconstruction revolves around the strategic value of the landscape model. Humboldt distinguishes three stages of awareness, three stages in the cognitive relationship between man and his environment, valid not only for phylogeny, so humankind as a species, but also regarding ontogeny, the history of a single individual. The first stage is that of impression (*Eindruck*) that arises in the human soul in its original manifestation as a primogenial sentiment towards the grandeur and beauty of nature. Its form of awareness is, in fact, that of landscape, which corresponds to the world understood as an aesthetic-sentimental harmonic whole (still) extraneous to any rational analysis and which thus regards only the psychic faculty of the subject. *Eindruck* is a compound word, only apparently simple. *Druck* means impression and is also used to indicate that of typographical characters on a white sheet of paper. For Humboldt, this relates to the sensitivity of the subject who is looking: the white sheet is her soul, the characters printed on it are landscape features. But of equal importance is the other half of the term, the prefix *Ein*. It means "one" but actually functions dually. On the one hand, it refers to singularity, to the individuality of the subject who is looking, and whose looking triggers a cognitive process. On the other, it signals the attitude of the subject to reduce the slew of impressions to a unit so that from the very beginning, although only from an aesthetic and impression point of view, the sphere of awareness is configured as a whole, as a single entity ready to reveal "the order hidden under the skin of phenomena".

In the following stage, that of *Einsicht* or examination, the sentimental whole is disarticulated so that it can be translated into scientific terms. In the term *Einsicht*, the prefix, which would seem to be identical, actually means the opposite of its meaning in *Ein-druck*. Here *Sicht* means "view", sight closely connected to reflection, to rational thought. And the singularity expressed by the prefix regards not only the subject but also the object; it refers to the intimate initial impression, to the concentration of thought on a single element or on a single relationship between elements from among those present, as an entirety. In this intermediate stage, that of scientific analysis, there is no longer landscape (sentiment, aesthetic impression) nor, consequently, totality, but only cold dissection and the rational detection of relationships between individual components.

The eclipse of the whole is only temporary however; it regards only the second, intermediate, level of awareness. Totality is completely re-established in the third and last stage, what Humboldt identifies with the concept of *Zusammenhang*, which indicates the totality constituted by being together *(Zusammen)* where all the elements are mutually interdependent, what today we would call complexity. Thus, for Humbolt the development of any awareness is nothing more than the translation of the auroral impression expressed by landscape into finally scientific terms, landscape that is absolutely not scientific in itself but without which science would not be possible.

3. The success of the strategy: from painting to the "face of the Earth".

Humboldt's strategy was aimed at transforming "the man of taste" into an observer of nature by taking advantage of the "distinguished humanistic education" of the "highest working classes" and "all the cultured classes" to give them learning that could ensure world control through scientific knowledge. And *Views of Nature* (*Ansichten der Natur*) is in fact what he called the book, first published in Tubingen in 1808, that would convince the entire European bourgeoisie to study the physical world. It is in this work that, for the first time in Europe, the concept of landscape conceptually morphs from the aesthetic to the scientific; it is literally revolutionary from a history of knowledge point of view. Today the term *Ansicht* (literally: view) still means two things: what can be seen and what one thinks, which makes sense in light of their shared sense, their indissociable origin, their absolutely coinciding functionality. Humboldt's project was directed at the entire bourgeois, or civil if you prefer, public sphere. Here Humboldt worked on a revolution that was not directly political but rather cultural with the intention of leading the protagonist of the publicly humanistic dimension, the connoisseur of works of art, towards a vision of the world that could develop into scientific comprehension of the world itself and that no longer stopped at the stage of mere contemplation. The idea was to transform the bourgeois culture of the time starting from its aesthetic matrix to morph the limited culture of poetic and pictorial knowledge into the science of nature, ready for concrete domination of the "face of the Earth" instead of its mere representation. According to Humboldt, the pictorial view of landscape was the vehicle of this transformation.

In short, for Humboldt, the strategist of European civil thought at the turn of the nineteenth century, entering into the sphere of scientific knowledge meant completely traversing the "kingdom of aesthetic appearance", as Franz Mehring called it. And it is precisely the concept of landscape, known to all the bourgeoisie because of their knowledge of paintings and other artistic descriptions, that was conceived of as the most suitable means for ensuring this transition.

4. The inversion of the strategy: the European landscape convention and the Aesthetization of political elements.

To a certain extent, two centuries on from the launch of the cultural engineering action conceived by Humboldt, of his cultural project that permitted the victory of civil society over what the bourgeoisie of the time called "the court of the old truth", we are now seeing a programmatic reversal of this strategy with the European Landscape Convention. On the one hand, the Convention moves in the direction of the spirit and intention of the original programme, for example by refusing any substantialist conception of landscape, any idea that landscape can be reduced to a simple set of elements, that it is only something material. Even for the Convention, and precisely in Humboldt's sense, landscape is a model whose nature, like all models, is not a set of things but a way of looking at and conceiving of them, a type of relationship among them. In logical terms: landscape does not mean the set of components relating to the "face of the Earth" but only their meaning; it corresponds, that is, (in the words of Gottlob Frege) to the specific way in which the components in question "are proffered", i.e. as they present to our eyes. As Leibniz explained in the 1700s, looking at something means judging it. The term that, from the start, the Convention uses to express this is "perception": precisely due to its fruits, i.e., the outcome of the perception process, a territory and environment are transformed into landscape and thus deprived of their own ontological or cognitive legitimacy. This is a decisive metamorphosis, that so far has not been adequately focused or reflected upon, and encapsulates the entire issue of the relationship between landscape and policy, that is to say, of the possible forms of existence of a landscape policy and its function. On the one hand, this mutation is completely in line with Humboldt's original operation because it too is based on the distinction between a thing and the image of that thing. On the other, though, it reverses and contradicts the whole spirit and entire intention (in short, the sign) of the original cultural and political valence of landscape because it moves in exactly the opposite direction: with the Convention, the political element (a territory) assumes an aesthetic nature (it becomes landscape). But the difference between the political and aesthetic sphere is crucial and threatens the very existence of landscape policies: the political sphere in fact functions according to rules and regulations—it is the realm of law; on the contrary, the aesthetic sphere cannot be reduced, by nature or customarily by tradition, to any rule or regulation, at least not in legal systems informed by liberalism.

5. Landscape and globalization.

The European Landscape Convention contains another important innovation that is completely coherent (probably unwittingly) with the original spirit of the model:

the subject perceiving is also a mobile subject that corresponds to not only "inhabitant" but also to "tourist", that by definition moves. Just observe any painting by Caspar David Friedrich, the painter who fixed the canons of romantic landscape, to understand how in his large paintings all the human figures he paints onto the background of nature which dominates the representation are travellers, people caught in an extemporaneous attitude that does not imply any duration or temporal stability. The mysterious person who turns away from us to look at the sea of clouds beneath him is visibly leaning on an alpenstock and certainly does not live on the peak of the mountain, and the same can be said for all his other subjects, the monk next the sea at dusk, the group of travellers on the edge of the white cliff rising from the Baltic and so on. "Tourists", in fact, in the language of the Convention, transhumant beings destined to shortly move their ephemeral presence somewhere else, to swiftly and programmatically remove their static condition according to a political project that at the time of Friedrich, as for Humboldt, involved moving from an aristocratic-feudal to a republican form, that modern centralized.

Territorial state in which we still live. Hence, based on this analogy, a question arises: according to what political project, to what territorial regime change does the perception modality of the perceiving subject—and with it, the perception modality of landscape or of land as Lucio Gambi would have said—return to the spotlight today? It is difficult to imagine, at least as an initial hypothesis, that this change is not extraneous to the momentous changes underway on a planetary scale whose various processes are summarily referred to as globalization and whose most obvious manifestation is precisely the difficulty states have in producing a policy to contain international and intercontinental human mobility.

6. Mapping, Space, Place.

In effect, the material constitution of the modern state, the territory of modernity, is founded on a series of premises that we could actually call spatial, i.e., deriving from the model of space elaborated by Euclid, the space of classical geometry which, starting with Ptolemy, later became the syntax of the terrestrial surface. In this view, the entire construction of modernity conforms to a single, successful project: the transformation of the face of the earth into a single gigantic space, that is to say, into a single gigantic geographical map. As if to say that to truly understand modernity, we need to overturn what we have implicitly been led to believe since infancy and understand that for the last five-six centuries, a map is not a copy of the face of the earth but rather it is the face of the earth that has become a copy of a map. Without this inversion of what we normally take for granted, the advent of the modern, spatial functionality of the world would never have been possible. And the first of the premises allowing the modern triumph of space is precisely the immobility of the subject who is looking, as is still evident for those who visit the portico of the Spedale degli Innocenti in Florence where the modern era was birthed in the early 1500 s with Brunelleschi's invention of linear prospective (what Panofskij called "artificial" to distinguish it from the "natural" perspective of the ancients). In order for the perspective trick to work, the observer must stay still. If not, there would be no rule to calculate the distances and consequently the size of what is being observed. Otherwise, a place could never transform into space.

The difference between place and space is fundamental. Conceived as a single space (according to modern criteria), the face of the earth has become "the realm of general equivalence" in the sense that all its component parts are interchangeable with any other, commutable, as mathematicians would say, susceptible, in other words, to being substituted one for the other without any change in the working of the world because it is subject to a single quantitative logic and a single linear metric standard. In functional terms, space implies a reduction in the time needed to cross the Earth's surface, and if the problem is only the speed of crossing it, no one part is distinguishable from any other. A place, on the contrary, is a part of our planet's surface that cannot be interchanged with any other because conceptually it is a function of its specific singular qualities. A place is the opposite of space and landscape is the form of a place.

7. Why landscape today.

The modern crisis of the state model depends on the crisis of the spatial model since the advent of the Net has energetically questioned, if not outright annihilated, the importance of linear distance between one point and another as a decisive factor in the working of the world. But what is the Net? It is a set of elements whose nature is quite heterogeneous. Firstly, it is made up of machines (*hardware*) containing a thought (*software*). But as Manuel Castells explains, this is not enough. You need to add women and men who unite with it and are functionally (not ontologically) indistinguishable from it because materially inseparable from it. The revolutionary character of this condition is comprehensible only if you think of what is one of the cardinal assumptions, after the immobility of the subject, of the modern existence of space: the clear distinction between subject and object, between mind and matter (Descartes), between what is animate and what, instead, is inanimate. A separation that, as has just been recalled, no longer exists on the Net. And that has never been valid, from Humboldt onwards, for our cognitive model of landscape based, in fact, on noting the impossibility of deporting or extracting the subject from the totality of the natural elements surrounding him, of putting distance between the former and the latter. The German romantics called this sentimental condition *Stimmung*, a term that meant the unified vibrating of the observer with everything surrounding him, the acknowledgement of a whole within which it was impossible to separate what was human from what was not. Which is exactly what happens today to operators on the Net for whom, from the point of view of the working of the world, there is no interval, that is, no possibility of distancing themselves from the surrounding machines. It is thus in virtue of its selfsameness with the Net's logic founded on the disappearance of the distance between subject and object that, differently but just as decisively as it did in nineteenth century, landscape dominates once again as a strategic model for understanding the world in these times of globalization: a world that works based on places rather than the logic of space.

Globalization started at the end of the 1960s with the advent of the net, with the event that by dismantling the structure of modern space has forced us to revise all our cognitive models. In the fifteenth century amidst the general uncertainty, the humanist from Forlì, Biondo Flavio, author of *L'Italia illustrata* (Illustrated Italy), invited his readers to trust space, or more precisely, to hang onto geographical maps as if they were rafts able to navigate the tempestuous waves of history. Instead, today we must entrust ourselves to such an evanescent and immaterial model as landscape because it is the working of the world that has in many ways become evanescent and immaterial. The certainty remains, shared by Flavio, that only by firstly adhering to a generalized and sentient model of the world can a political choice present itself that is at least plausible, that is to say, potentially shared by the populace, even if not always feasible. As if to say that it is exactly the outcome of issues relating to landscape and its conformation that will determine not only future policies but even more so, future politics.

Chapter 3
Global Landscape Governance on the 20th Anniversary of the European Landscape Convention

Amy Strecker

Abstract This chapter argues that reflecting on the continuity of the European landscape requires us to look beyond the European Landscape Convention as a legal instrument, and the borders of Europe, as a region. This is because the challenges facing European landscapes are increasingly global in nature, and the continuity of the European landscape is inextricably tied up with the wider world we inhabit. The chapter highlights the contradictory role played by international law in landscape governance and explores some of the legal obstacles to, and potential avenues for, achieving spatial justice.

Keywords ELC · International law · Spatial justice · Landscape governance · Human rights

3.1 Introduction

The European Landscape Convention was a paradigm-changing instrument for the way in which it conceived of landscape. This chapter argues that reflecting on the continuity of the European landscape, from a legal perspective, requires us to look beyond the European Landscape Convention as an instrument and the borders of Europe, as a region. This is because the challenges facing the European landscape are increasingly global in nature, and the continuity of the European landscape is inextricably tied up with the wider world we inhabit. While the European Landscape Convention and other instruments that make up the body of "landscape law" have aspirational norms as their aims, there also exists a wider legal framework that causes damage to the progress made in landscape governance. For example, the law governing international trade and investment linked to agriculture, resource extraction, infrastructure, and development plays a major role in shaping landscapes and yet, these are often overlooked in wider discussions and policy fora on landscape.

A. Strecker (✉)
UCD Sutherland School of Law, Dublin, Ireland
e-mail: amy.strecker@ucd.ie

T. Matteini et al. (eds.), *Cultivating Continuity of the European Landscape*, Environmental History 15, https://doi.org/10.1007/978-3-031-25713-1_3

This paper discusses the dual role of law in landscape governance and argues that landscape, in the sense of it implying legal rights and obligations, still faces substantial obstacles in the pursuit of spatial justice. I will discuss both the limits of the law as regards use and access rights to landscape, and explore the possibility of advocating a broader conceptualisation of property to include spatial justice, advocating a re-alignment of property with landscape. I propose that if the law is to play a more meaningful role in landscape governance, it is incumbent upon us to re-imagine the way in which property is conceived for communities in Europe and beyond.

3.2 Landscape Law and the European Landscape Convention

"Landscape law" refers to a body of law encompassing cultural heritage law, environmental law, and certain human rights relevant to landscape (see Strecker 2018a, b). It is a hybrid legal field which, taken as a whole, more adequately reflects the holistic nature of landscape itself. Landscape law includes UNESCO cultural heritage treaties, particularly cultural landscapes within the World Heritage Convention, Council of Europe treaties, most notably the European Landscape Convention, human rights provisions, environmental law pertaining to impact assessment, as well as a number of other relevant soft law instruments.

The European Landscape Convention (hereafter ELC) was the first international instrument to give explicit expression to landscape in its title. Although not a human rights treaty, it was paradigm-changing for the way in which it emphasised the human dimension to landscape and the diverse ways in which landscape is perceived by people (Article 1). The ELC goes further in humanising landscape than the World Heritage Convention, which attempted to shift the focus of cultural landscapes towards to people and communities (Rossler 2006), but still relies on top-down policies in which the lives of the inhabitants living in World Heritage landscapes are often curtailed. The Convention has greatly influenced both the discourse around landscape over the past two decades, as well as the activity at academic, civil society and governmental levels continent-wide. This is undoubtedly due to the dynamic conceptualisation of landscape espoused by the Convention as well as its scope: the ELC focuses not only on outstanding landscapes but also on every day or degraded landscapes—the places where most people live and work—hence emphasis not only on protection but also on management and planning (Article 2). The preamble states that the protection, management, and planning of landscape entail "rights and responsibilities for everyone," and it is this aspect of the Convention that has been taken up and linked to ideas of spatial justice, human rights, and democracy (see, for example, Egoz et al. 2011; Strecker 2011; Wall and Waterman, 2018; Egoz et al. 2018).

However, despite igniting conversations about the links between landscape and human rights, from a legal perspective, the Convention is limited in terms of its capacity to facilitate access to justice in regions where public participation remains an aspirational goal. This is due to several factors. The first is that the ELC is very much tailored to the European model of governance. It was borne out of a frustration on the part of local and regional authorities in Europe and was seen in part as a reaction to the Common Agricultural Policy (CAP) and other pan-European policies that impacted on the character of landscapes. The ELC requires a certain level of resources, technology (for Landscape Character Assessment), and participatory governance that are often beyond the capacity of, and diverge from, the political structure of many of the world's regions. It is no coincidence that the best practice examples of landscape democracy come from regions with both resources and a strong regional identity, coupled with political will (Catalonia is a case in point). In relation to case law, to my knowledge there has only been one example where the ELC has been successfully invoked before a court of law to protect public access to a locally significant landscape. The Blitse Duinen case (Netherlands) involved a group of local residents who successfully challenged the blocking of access to a forest path by private landowners, relying on the ELC.[1] This case is promising for (civil law) jurisdictions such as the Netherlands, where public interest proceedings are more accepted. For dualist states (common law jurisdictions), such as Ireland, however, international law does not have direct effect in national law and may not therefore be invoked before national courts.

The second is that while the ELC recognises that people have participatory rights when it comes to landscape, this presupposes a functioning deliberative democracy, where public participation will be engaged with and lead to a just solution to future planning. For many regions (including in Europe), especially those on the periphery, this remains an aspirational goal, miles away from the reality of the planning process. Disputes often arise later in the process and communities can find themselves in a legal lacuna in terms of challenging a land-use decision. Although the ELC refers to the Aarhus Convention (on access to information, public participation and access to justice) in its preamble, it makes no actual reference to access to justice in its text. It can be seen that the emphasis of the Convention is on the proactive planning and development of landscape at local and regional level—as a territorial project, but does not offer a solution with regard to solving disputes that arise in later stages of the planning process when the participation process has lacked in transparency or equity (Strecker 2018a, b). Several studies have shown that there exists a prevailing gap between participation rhetoric on paper and participation at the operational level (Jones and Stenseke 2011; Conrad et al. 2011; Jorgensen et al. 2016). This means that access to justice remains a problem, a fact even more discernible in peripheral regions and amongst marginalised communities, as the former case of the Shell pipeline in Erris, in the West of Ireland, and the current case of a gold mine application in the Sperrins Mountains, Northern Ireland, illustrate.

[1] For further information, see <http://vriendenbiltseduinen.simpsite.nl/natuurmonument> accessed 29 September 2021.

3.3 Global Landscape Governance Beyond the ELC

Due to the popularity of the ELC beyond Europe, a number of initiatives have taken place since its adoption with a view to furthering the Convention's aims beyond Europe. Ten years after the ELC's adoption there was a proposal initiated by the International Federation of Landscape Architects (IFLA) to generate support for a global landscape convention, based on the principles of the ELC (I wrote a reasoned opinion at the time outlining the pros and cons of such a proposal). In 2010, an expert meeting was convened by the Director General of UNESCO following a request by IFLA to review its feasibility.[2] The meeting was attended by 23 experts from a variety of regions and fields (anthropologists, ecologists, geographers, landscape architects, planners, and lawyers) as well as representatives from a number of interested organisations (including IFLA, IUCN, ICOMOS, UNESCO and the Council of Europe). It appears from the report of the meeting that the European Landscape Convention provided the well-spring for discussion; its implementation over the first ten years was examined and its new conceptual approach praised. It was noted how no such integrated instrument existed for landscape at the global level, and how an international landscape convention would "capture an extraordinary but fragile renaissance that is taking place, as government authorities, civil society but also investors are beginning to appreciate the true value and complexity of the landscape."[3] The meeting of experts discussed the challenges of such a convention, and concluded that a global convention was urgently needed. It was also proposed that UNESCO be the most appropriate body for its adoption and supervision. The proposal did not gain the support of UNESCO's Executive Council on that occasion, although there have been other developments in the area of cultural landscapes since then, namely a Recommendation on the Historic Urban Landscape.

Even if the idea of a global landscape convention had progressed, however, I am not convinced that UNESCO would have been the most appropriate body for its undertaking. While IFLA's proposals undoubtedly stem from the desire to enhance awareness of the importance of landscape for well-being at the global level, it nevertheless poses certain problems if viewed in the context of international law. I have argued elsewhere that in order to be effective and truly add something to the existing framework, an international landscape convention would have to engage with and address some of the root causes of landscape deterioration caused by global policies, rather than rehash some of the principles of existing instruments (Strecker 2018a, b, 111). Furthermore, it is strongly suggested that a list not be included in any such convention, since, aside from the likely duplication with already existing World Heritage List (and other instruments), this would detract from the real challenges such as food and energy production, land use management, climate change and desertification, infrastructure, and development. "Landscape is a reflection of a process, so it is a process that needs to be addressed in order to combat negative

[2] Expert meeting on "The International Protection of Landscapes," Paris, UNESCO Headquarters, Room IV, 18 April 2011, Agenda.

[3] Ibid.

changes in the landscape" (Strecker 2018a, b, 111). Interestingly, the report on the feasibility of an International Landscape Convention mentions a "right to landscape" twice, not envisaged to be enforceable, but rather as a recognition of people's right to decent landscape and quality of life. In 2012, a meeting organised by UNESCO on the international protection of landscapes adopted the Florence Declaration on Landscape, which called upon intergovernmental agencies and secretaries responsible for United Nations programmes, together with NGOs, to work towards an international mechanism. Amongst other things the Declaration reaffirms landscape as a common good and states that the "right to landscape" is a human necessity.[4] There is thus a growing awareness of the links between landscape and human rights. More recently, in 2016, a protocol was adopted amending the European Landscape Convention to the Council of Europe Landscape Convention. Article 2 of the Protocol adds a new paragraph to the text of the ELC expanding its geographic scope to those non-members of the Council of Europe. "wishing to enable the application of the values and principles formulated in the Convention to non-European states who so desire."[5] The idea of extending the ELC's principles beyond Europe likely comes from well-intentioned aspirations based on the interest in the ELC from outside of Europe. However, the ELC is very much a product of its time and place, and its application beyond Europe might be more complicated than initially considered. Although the ELC "democratises" landscape, the Convention nevertheless pays a great deal of attention to administrative arrangements, training for landscape specialists, and committees of experts (Jorgensen 2016, 43), which could be viewed as problematic from a decolonial perspective.

3.4 The Janus Face of International Law

International law has an increasing impact on the ways in which communities in many parts of the world interact with and access landscapes—from changing farming patterns due to international trade agreements, to collective tenure and customary rights in landscapes that are simultaneously the location of large-scale resource extraction projects. The last decades have marked a dramatic increase of foreign investment in agricultural land, for example. According to one report, the size of land affected by land acquisition agreements signed between 2008 and 2009 alone was more than ten times what it had been in the previous decade (Perez et al. 2011), and the trend is moving in the same direction (Sassen 2014; Von Bernstorff 2016; Romanin et al. 2016). International investment is based on the assumption that protection of investments will stimulate foreign direct investment (FDI) and that FDI will lead to economic growth and development in the host state. Yet this is not always the case. Research has shown that many large-scale land investments increasingly

[4] Florence Declaration on Landscape.

[5] Council of Europe, Protocol Amending the European Landscape Convention, European Treaty Series No. 219.

result in the displacement of local communities, violations of economic and social rights, and destruction of the natural and cultural landscape (Cotula 2012; Borras and Franco 2012; Von Braun and Meinzen-Dick 2009; Von Bernstorff 2016). This is corroborated by geospatial analysis from applied geography demonstrating that contrary to popular belief, land deals do not target idle or marginalised land, but instead often target populated croplands, remote forests, and grasslands (Messerli and Eckert 2014). It emerges therefore that there are multiple spheres of international law, private and public, operating in opposition to one another.

I have argued previously (Strecker 2018a, b) that the emerging landscape law represented in the European Landscape Convention and other international texts recovers some of the substantive nature of landscape defined by geographer Kenneth Olwig in his seminal paper *Recovering the Substantive Nature of Landscape*, where he shows how "a substantive understanding of landscape is more concerned with social law and justice than natural law and aesthetics" (Olwig 1996). The incongruity of international norms advocating more localised versions of landscape governance is one of the paradoxes of our modern globalised world, where global assemblages of territory, authority, and rights cut across the binary of national versus global (Sassen 2008). An example of such assemblages include the "Right to Landscape," "Defining Landscape Democracy," and "Landscape Citizenships" initiatives, which gathered together scholars from a variety of disciplines and regions concerned with the social justice dimension of landscape, transcending inter-state boundaries (see, for example, Egoz, Makhzoumi and Pungetti 2011; Egoz 2015; Egoz et al. 2018; Wall and Waterman, 2018). These movements contributed to my thinking on landscape as a term that was re-imagined in part as a result of the social demise of property.

However, it became clear through my research that, while landscape is recognised as having a strong human rights dimension, legal claims for rights to landscape, while based on real issues of concern—and sometimes on genuine violations of national law—are not a viable means of accessing justice for land rights violations or for preventing destructive development in cultural landscapes (Strecker 2018a, b). This is because landscape is still within the realm of public policy (sovereignty over natural resources); in practice, it is often considered in preservationist terms rather than substantive ones dealing with peoples' relationships to the places where they live; and while human rights are for the most part considered as individual rights, landscape is a collective good, and difficult to articulate within the current framework.

As a landscape law scholar, I am reticent to admit the legal limits of landscape as a tool for accessing justice. For all the normative developments made in environmental justice and cultural rights, property still dominates the way in which cases concerning land are being interpreted at national and international levels. An applicant's standing or sufficient interest is often equated with land ownership rather than any other form of relation to land, and access to justice for communities facing destruction of their local landscapes (and consequent dispossession and/or environmental degradation) can be difficult to prove either due to the lack of substantive environmental rights or to the way in which access to cultural heritage is narrowly construed by the judiciary.

By contrast, the right to property has been interpreted in the Inter-American context to include communal customary tenure and the collective rights of indigenous peoples, as well as spiritual and cultural links with land, even in the absence of title. In the case of *Maya Indigenous Communities of Toledo District v Belize*, for example, the Inter-American Commission found that Belize had violated the Mayan Communities' right to use and enjoy their property by granting concessions to third parties to exploit natural resources without informed consent (Case 12.053, IACtHR Report 40/04). The Commission noted that indigenous peoples' right to property is based on international law, does not depend on domestic recognition of property interests, and is grounded in custom and tradition (paras. 153, 194). Likewise, in *Xákmok Kásek Indigenous Community v. Paraguay*, the Inter-American Court repeated that for indigenous communities their relationship with the land is not merely a matter of possession and production, but rather a material and spiritual element that they must fully enjoy (IACtHR Series C No. 214, para. 86). The Court also noted that while this concept of property does not necessarily correspond to the classic concept of property, it nevertheless deserves equal protection under the American Convention of Human Rights, and that failing to recognise the specific versions of the right to use and enjoyment of property would be equivalent to maintaining that there is only one way of using and enjoying property and this, in turn, would make Article 21 meaningless for millions of individuals (para. 87).

This case law is significant because it recognises the importance of values in land beyond that of private ownership. Yet two problems emerge here. One is that indigenous peoples still face enormous challenges in the implementation of these decisions due to, amongst other things, the way in which property is enshrined and conceptualised in national constitutional law. Two, issues of land access not only affect indigenous peoples, but also millions of marginalised communities worldwide depending on land for common survival. Limiting a cultural and social interpretation of property to an exception is sidestepping the fundamental injustice, as well as the emancipatory potential, at the heart of the concept as it applies to disenfranchised communities more generally.

In addition, focusing on aspirational norms only captures half of the picture. Behind most major landscape disputes lurks another field of international law that often plays a much more influential role in terms of its capacity to affect local environments and facilitate destructive development. Take the Keystone XL (Dakota) Pipeline for example: what was happening behind the scenes was a major investment to the tune of billions of dollars facilitated by international economic law. When the Obama administration refused to approve the project based on environmental concerns, TransCanada sued the US government for US$15 billion under NAFTA's dispute settlement mechanism (ICSID Case No. ARB/16/21). Trump subsequently reversed the decision and gave the green light to the pipeline (TransCanada then dropped the case).

International investment dispute settlement mechanisms give foreign investors the right to claim for compensation for the expropriation of their "property rights" in investments, even where decisions were made on the basis of environmental, public interest or other human rights concerns. Given that land-based investment in agriculture, resource extraction or infrastructure covers a very significant amount of global investment, the abstract notion of property rights in international investment law (land as commercial asset) frequently collides with the "lived-in" property rights of people and communities on the ground (see Cotula 2017). Imagine the asymmetry whereby major investors can be compensated for loss of (potential) revenue, but local communities who have been dispossessed or whose landscape has been degraded, receive nothing. Yet that is often the reality. The disparity between who gets to have property rights and who does not is still substantial.

3.5 Conclusion: Law as Problem, Law as Solution

The discursive influence of the ELC over the past two decades illustrates the societal role of law in acting as a reference point and providing a common vocabulary for collective action. Although the ELC lacks the character of a substantive human rights treaty, it has nevertheless ignited a wave of creative thinking amongst scholars on how to strengthen people–place relations. Parallel movements on the rights of nature, rights to a healthy environment, and rights to cultural heritage can be seen as complementary to articulations of a right to landscape. However, as a tool for *accessing justice,* I argue that we need to look beyond the ELC and insert landscape within the vocabulary of human rights. In legal terms, I propose that one way of recovering the substantive nature of landscape in law is to revisit the concept of property itself.[6] I argue that analysing property rights through a spatial justice lens recognises the unequal distribution of resources and rights between different places and landscapes. It therefore adds the much-neglected variable of "place" to international law scholarship, a necessary addition given the role of land in peoples' lives. While much attention has been paid to land grabbing, the privatisation of land holdings and spatial injustice from disciplines such as political and spatial geography, sociology and anthropology (e.g., Sassen 2014; Sikor and Lund 2009), international law scholarship has not engaged to the same extent with the role of property in facilitating and fighting these global phenomena. The rationale is to critically reflect on the role of law by starting with the problem of property itself. Although property was excluded from the two Human Rights Covenants of 1966[7] (having originally been included in the Universal Declaration of Human Rights), it was subsequently included in the regional human rights frameworks and is one of the most litigated provisions at regional level. If property can include abstract economic interest such as shares, why can't it also include substantive interest such as community use? Some national laws recognise land rights outside ownership, but these depend largely on political will and are open to exploitation, which is one of the reasons why supranational human rights remain important.

[6] See for further information, http://landlawandjustice.eu.

[7] International Covenant on Civil and Political Rights and International Covenant on Economic, Social and Cultural Rights.

Interestingly, Nicole Graham has shown that in the early etymology of property, land had significance greater than the sum of its economic production value and was also an important component of identity (Graham 2011). The early notion of property entailed the mutual identification of the owner and the owned, whereas the modern meaning of the word divorces property from identity. Early substantive understandings of property are being partially recovered in the case law dealing with indigenous and tribal peoples, but there is no reason why a socially just interpretation of property cannot also be applied in the case of other communities, especially in the case of marginalised ones. Property is at the root of much landscape injustice, but can it also be part of the solution? I argue that it can, provided that its interpretation is considered (or subverted) to include the elements originally contained within the meaning of the term, when land had significance greater than the sum of its economic value and was also a matter of social, cultural, and ecological life.

Acknowledgements This research has received funding from the European Research Council (ERC), under the EU's Horizon2020 research and innovation programme, grant agreement no. 853514.

References

Borras SM, Franco JC (2012) Global land grabbing and trajectories of agrarian change: a preliminary analysis. J Agrar Chang 12:34–59

Conrad E, Cassar LF, Jones M et al (2011) Rhetoric and reporting of public participation in landscape policy. J Environ Policy Plann 13:23–47

Cotula L (2017) Land, property and sovereignty in international law. Cardozo J Int Comparat Law 25(2):219–286

Egoz S, Makhzoumi J, Pungetti G (2011) The right to landscape: contesting landscape and human rights, Ashgate.

Egoz S (2015) Defining Landscape Democracy, Norwegian University of Life Sciences (NMBU): Centre for Landscape Democracy

Egoz S, Jorgensen K, Ruggeri D (2018) Defining Landscape Democracy. A path to spatial justice. Elgar

Graham N (2011) Lawscape: property. Routledge, Environment and Law

Jones M, Stenseke M (2011) The european landscape convention. Challenges of participation. Springer

Jorgensen K, Clemetsen M et al (2016) Mainstreaming landscape through the European landscape convention. Routledge

Olwig K (1996) Recovering the substantive nature of landscape. Ann Assoc Am Geogr 86(4):630–653

Romanin J, Bonfanti A, Seatzu F (2016) Natural resources grabbing: an international law perspective, Brill

Rossler M (2006) World heritage cultural landscapes: a UNESCO Flagship Programme 1992–2006. Landsc Res 31:333–353

Sassen S (2014) Expulsions: brutality and complexity in the global economy. Harvard University Press

Sikor T, Lund C (2009) The politics of possession. property, authority and access to natural resources. Wiley-Blackwell

Strecker A (2018a) Landscape Protection in International Law. Oxford University Press
Strecker A (2018b) The law is at Fault? Landscape rights and 'Agency' in international law. In: Wall ED, Waterman T (eds) Landscape and agency: critical essays, Routledge
Strecker A (2011) The 'Right to Landscape' in International Law. In: Egoz S et al (eds) The Right to Landscape: Contesting Landscape and Human Rights, Ashgate
Von Bernstorff J (2018) Community interests and the role of international law in the creation of a global market for agricultural land. In: Benvenesti E, Nolte G (eds) Community interests across international law, Oxford University Press, pp 278–296
Von Bernstorff J (2016) Who is entitled to cultivate the land? sovereignty, land resources and foreign investment in agriculture in international Law. In: Romanin J, Francesca D, et al (eds) Natural resources grabbing: an international law perspective. Brill, pp 55–74

Part II
Landscape Policy and Governance

Chapter 4
"Landscape Policy and Governance: An Introduction"

Gian Franco Cartei

Abstract Although the European Landscape Convention is the first international act adopted by a European institution with the aim of promoting the protection and planning of European landscape is not clear how to implement its contents. The Convention doesn't contain prescriptions but only principles. And these principles are very important. The relevance of principles like subsidiarity, participation, sustainable development, and integration.

Keywords Landscape · Participation · Subsidiarity · Sustainable development

1. For those familiar with the European Landscape Convention and its contents, it is no surprise to hear that the Convention was the first international act adopted by a European institution with the aim of promoting the protection, management and planning of European landscape.

What do we mean by the title of this section 'Landscape Policy and Governance'? Neither are easy subjects to study, but both are fundamental to implementing the Convention.

The starting point is the Convention's definition of landscape—'*Landscape means an area, as perceived by people, whose character is the result of the action and interaction of natural and/or human factors*'—and its scope—'*... the Convention applies to the* ***entire territory*** *and covers natural, rural, urban and peri-urban areas.* [...] *It concerns landscapes that might be considered outstanding as well as everyday or degraded landscapes*'—These elements emphasize that in the Convention's perspec-

G. F. Cartei (✉)
University of Firenze, Firenze, Italy
e-mail: gianfranco.cartei@unifi.it

T. Matteini et al. (eds.), *Cultivating Continuity of the European Landscape*, Environmental History 15, https://doi.org/10.1007/978-3-031-25713-1_4

tive, all landscapes deserve attention, regardless of their value. All types of landscape "decisively influence the quality of Europeans' surroundings".[1]

On the other hand, the question of how to implement the Convention's objectives arises. The Convention contains no prescriptions regarding this problem. In fact, its contents consist of principles, not rules; of objectives, not means. This lack of clear procedural schemes contributes to weakening the legal structure of the Convention. Besides, there are some ambiguities in the text which cast doubts on fundamental aspects.

Think, for example, of the aforementioned definition of landscape in article 1: 'Landscape' means an area, as perceived by people, whose character is the result of the action and interaction of natural and/or human factors. The expression 'area' is not the equivalent of territory as a whole. Can the Convention accommodate a model, like the Italian one, which still distinguishes between landscape management and urban planning?

Think also of article 12 according to which the Convention '*shall not prejudice stricter provisions concerning landscape protection, management and planning contained in other existing or future binding national or international instruments*'. The absence of any hint as to the parameters defining 'stricter provisions' permits each State to claim that its own legislation is in accordance with the Convention.

2. Despite some ambiguities in the Convention's text, we have some fundamental principles to consider. The principle of 'subsidiarity' is one of them. According to article 4, the Convention will be implemented 'in conformity with its constitutional principles and administrative arrangements, and respecting the principle of subsidiarity, taking into account the European Charter of Local Self-government'. The subsidiarity principle's content is clear: any decision and action concerning landscape has to be taken by the authorities closest to the population concerned, which means that local authorities have to be involved in the implementation process.

This clarifies the importance of the principle of 'participation'. According to article 5, it is fundamental 'to establish procedures for the participation of the general public, local and regional authorities, and other parties with an interest in the definition and implementation of the landscape policies'. The weight and the scope of participation means that it involves 'taking into account the social perception of landscape and popular aspirations in choices regarding landscape protection, management and planning'.[2]

[1] Explanatory Report (45), according to which "This comprehensive coverage is justified for the following reasons: every landscape forms the setting for the lives of the population concerned; urban and rural landscapes interlock in complex ways; most Europeans live in towns and cities (large or small), the quality of whose landscapes greatly affects their lives; and rural landscapes occupy an important place in the European consciousness".

[2] Recommendation CM/Rec(2008)3 of the Committee of Ministers to member states on the guidelines for the implementation of the European Landscape Convention.

Actually, subsidiarity and participation can be seen as two sides of the same coin. Nonetheless, there are other principles no less important in this context. The principle of '**sustainable development**' stands out. It is mentioned in the Convention's Preamble and in the Committee of Ministers Recommendation CM/Rec(2008)3 to member states on guidelines for the implementation of the European Landscape Convention as a concept "fully integrating the environmental, cultural, social and economic dimensions in an overall and integrated fashion, that is, by applying them to the entire territory". There are grounds to consider this principle the basis of every other principle enshrined in the Convention. Even though the text of the Convention refers to this principle, it doesn't offer any specific definition. Still, we know that it originated in the mid-1980s when the *Brundtland Commission* formulated it in its report '*Our Common Future*'. Since then it has always stood for the integration of environmental concerns with development. Accordingly, the Rio Conference of 1992 affirmed that "*environmental protection shall constitute an integral part of the development process*".

In this perspective, the importance given to the principle of 'integration' is not surprising. According to article 5 d, each Party undertakes "to integrate landscape into its regional and town planning policies and in its cultural, environmental, agricultural, social and economic policies, as well as in any other policies with possible direct or indirect impact on landscape". The meaning seems clear: if the management of landscape focuses on 'the territory as a whole' and it is seen as 'a territorial project', it is necessary to include the landscape dimension in all spatial management policies and integrate it into sectoral policies.[3]

Applied to the Convention, these principles make landscape policy the result of a dynamic process of comparing and balancing interests—whether public, private, individual or collective—, quality objectives and planning processes. This confirms the idea that "In seeking the right balance between protection, management and planning of a landscape, it should be remembered that the aim is not the preservation or "freezing" of the landscape at a particular point in its lengthy evolution. Landscapes have always changed and will continue to change, both through natural processes and through human action".[4]

3. In this perspective, the title of this section can be deceptive if interpreted too narrowly

When we refer to the action of 'Landscape policy', we probably cannot entirely rely on the text of the Convention. The latter defines such a policy as "an expression by the competent public authorities of general principles, strategies and guidelines that permit the taking of specific measures aimed at the protection, management and planning of landscapes". Is there a concept which encompasses all such objectives? In other words, using the same expressions and notions laid down in article 1, what is the common denominator of "Landscape protection", "Landscape management" and

[3] Recommendation CM/Rec(2008)3 of the Committee of Ministers to member states on the guidelines for the implementation of the European Landscape Convention, Part I—I.1.

[4] Explanatory Report to the European Landscape Convention.

"Landscape quality objective"? Probably, the notion which best addresses the concept is "Landscape planning". In most European countries, landscape policy is essentially articulated through plans. The notion of planning, however, appears to be relatively neglected in the text of the Convention where it is limited to forward-looking actions to enhance, restore or create landscapes.

At the same time, some questions can be raised over the term 'governance', which is different from 'government'. The fact that article 4 provides that each Party shall implement the Convention according to its own division of powers in conformity with its constitutional principles and administrative organizations does not exclude the obligation to allocate competences in line with the subsidiarity principle. It means, on the one hand, that any relevant decisions should be taken by the authorities closest to the community concerned and that, on the other hand, any decision and activity must involve the members of the public.

It is not an easy task to accomplish. It implies not only that the people concerned be the population at large, including "all the relevant stakeholders: national, regional and local authorities, the population directly affected, the general public, non-governmental organisations, economic operators and landscape professionals and scientists", but also that participation "should be a feature of all the different phases in the processes of drawing up and implementing landscape policies, in particular those of landscape appraisal, definition of landscape quality objectives, decision-making and implementation of actions over time".[5]

The uncertainty surrounding such concepts as Landscape Policy and Governance explains why UNISCAPE has decided to focus a thematic session on the topic.

[5] Recommendation CM/Rec(2008)3 of the Committee of Ministers to member states on the guidelines for the implementation of the European Landscape Convention, Part II.2.3.

Chapter 5
Landscape Planning Policy as a Measure of Implementation of the European Landscape Convention

Stefano Civitarese Matteucci

Abstract The paper asks whether and to what extent the Italian planning system conforms to the European Landscape Convention's objectives and rationale. To do it, the complex nature of the European Landscape Convention implementation, due to its ambiguous legal nature and structure, is first discussed. Then, the discussion moves to how the manifold meanings of the landscape within the Convention react to planning practices. Given the duplicity of the core purpose of landscape in the Convention, either as subjective perception or as objective scenery, the answer to the question above is affirmative as long as we recognise the tension running through the text of the Convention between these two notions.

5.1 Introduction

In this paper, I would like to touch upon three issues which our topic somehow evokes. The first is the actual implementation of the European Landscape Convention (ELC), which is very complex due to the legal nature and structure of this instrument. Secondly, the place of planning in the Convention's framework and particularly the manifold meanings of the landscape it encompasses. Thirdly, whether and to what extent the Italian planning system conforms to the Convention's objectives and rationale. Given the duplicity of the core meaning of landscape in the Convention, my conclusion is affirmative as long as we recognise the tension running through its text between landscape as subjective perception and landscape as objective scenery.

S. C. Matteucci (✉)
Department of Legal and Social Studies, The University of Chieti-Pescara (IT), Chieti, Italy
e-mail: stefano.civitarese@unich.it

T. Matteini et al. (eds.), *Cultivating Continuity of the European Landscape*, Environmental History 15, https://doi.org/10.1007/978-3-031-25713-1_5

5.2 The Implementation of the Convention

Art. 6E of ELC deals with its implementation by providing that "*to put landscape policies into effect, each Party undertakes to introduce instruments aimed at protecting, managing and/or planning the landscape*". This is not much more than a reiteration of the general and specific measures listed in the Convention (articles 5 and 6).

The guidelines adopted by the Committee of Ministers on 6 February 2008 distinguish between *regulatory implementation* through legislation and policy documents, and *voluntary implementation* through agreements between stakeholders. They add that the choice of method can depend on the local situation, which will vary even within the same country. The implementation may be included in general and sectoral instruments at the different administrative, programming and spatial-planning levels. Moreover, regulatory implementation depends on existing legislation and the type of objective as well as the desired outcome in terms of protection, management or planning. Measures, in turn, may be included either in spatial-planning documents or in bespoke instruments.

In brief, a key issue emerging around Europe is that national governments of member states need to develop procedures to address how the implementation of the ELC may be achieved within and through national policy.[1] Implementing the ELC is subtler and more challenging than many other international law instruments because it hinges on integrating landscape (and people's relationship with it) with landscape quality objectives, national spatial strategies and planning processes. According to the explanatory report, this requires following a kind of progression from education and awareness to landscape character assessment and classification, and an array of other protection, management, and planning measures. All this makes assessing legal implementation beyond a superficial comparative analysis of landscape laws a pretty puzzling task. There are as many different ways of implementing it as any state can think of and even different ways of monitoring and assessing the implementation process as well as evaluating whether and to what extent the ELC has brought about change.

Looking at the policy/regulatory aspect, let me take the two examples of the UK and Spain.

In the UK, "Natural England", an executive non-departmental public body sponsored by the Department for Environment, Food and Rural Affairs is in charge of the ELC's implementation. In 2009, Natural England issued guidelines aiming to make the ELC more meaningful and relevant to a wide range of organisations by describing the possible actions, including Action Plans that each organisation can take. Such measures, which are not based on a specific legislative mandate, are firmly rooted in a vision of policy-planning integration and aim to favour a cooperative and dialogical pattern over a command and control regime.

The implementation of the ELC in the Catalonia Region is a recurrent example of a best practice. My impression, however, is that centralised power (vested in a

[1] Roe (2013).

powerful political institution) plays a robust role in this model. Within the framework of the Spanish quasi-federal government system, Catalonia enjoys autonomy and exclusive regulatory competence regarding land zoning and landscape planning. The 2005 Landscape Act is almost a photocopy of the ELC's definition of landscape: "*any part of the territory, as perceived by the community, whose character results from the action of natural or human factors and their interrelationships*". Based on the ELC's aims and objectives, Catalonia has developed a system of "landscape catalogues" issued by the Department of Land and Sustainability with the collaboration of the Landscape Observatory, defined by law as instruments of descriptive and programmatic character. They determine landscape types, identify their value and state of conservation and propose quality objectives to be pursued (art. 10, 2005 Landscape Act). Based on these catalogues, landscape guidelines are provisions specifying and normatively incorporating proposals into regional and sub-regional master plans for the implementation of landscape quality objectives. The sub-regional master plans determine the cases in which the guidelines are to be "translated" into rules of direct application, directives for the revision of urban plans, or norms requiring a mandatory assessment from the competent landscape authority. Public participation is considered an essential asset in the process of setting up the guidelines. The law, however, only establishes that participation means and procedures be the same as those established by legislation concerning urban planning and administrative procedure in general.

It is interesting to note that more recently, the English approach has also shifted towards landscape character and sensitivity assessment tools without, however, changing its integrated and cooperative approach. These assessments—which somehow remind us of the EU strategic environmental assessment—are defined as tools to be used to inform strategies, plans, and policies at a range of scales (from national/regional to local), where many a scenario can be put forward.

The ELC's ample discretion and variety of implementation methods and tools chimes with the notion that the ELC lacks strength as a legal tool in itself, being characterised "by the absence of specific rules to be applied".[2]

5.3 The Multiple Meaning of Landscape and the Place of Planning

As has been noted by Amy Strecker in her excellent book on landscape in international law, the absence of "legal bite" is partly due to the lack of a system of protection and procedural rights in the development of plans and policies.[3] She fleshes out this point by explaining that the role of the ELC in legal proceedings is negligible. In virtually none of the cases before international or national fora dealing with landscape is the

[2] Cartei (2012).

[3] Strecker (2018) 106.

ELC mentioned as a relevant treaty with binding obligations.[4] This remains valid both for the European Court of Human Rights case-law, even when cultural heritage is involved as a means of legitimising restrictions on fundamental freedoms, and national courts. However, I also agree with Amy Strecker that this is above all a consequence of the exceedingly broad definition of landscape provided in the text: the explanatory report reads that the Convention applies to the entire territory and covers natural, rural, urban and peri-urban areas. [...] It concerns landscapes that may be considered outstanding as well as ordinary landscapes. In other words, just land.

Landscape as land/territory reminds us of the theory of Alberto Predieri, a prominent Italian public law scholar who wrote in the Sixties that landscape ("paesaggio"), which the Italian Constitution includes among its fundamental principles as something to cherish and protect, was to be understood as the "form of the land".[5] As such, Predieri claimed, the form of the land should be, in turn, the fundamental subject of urban planning laws and institutions. This implies that the (local) urban planning system in charge of landscape management must endeavour to see land from the point of view of landscape when making decisions about development, zoning, etc. The Italian legislative model had never embraced such a view until very recently (I'll come back to this further on) but rather stuck to a framework where landscape management and planning were separate matters from spatial/urban planning. The former is in the remit of the central government, the latter a local government policy matter.

This point touches upon one of the defining characteristics of the ELC which is its emphasis on localism. The Landscape Convention is a creation of the Congress of Local and Regional Authorities which represents distinctly local and regional interests and practices that are not necessarily those of the nation-states. Tellingly, the Italian Landscape Code establishes that landscape shall be protected regarding those aspects and features which materially and visibly represent the national identity by expressing cultural values. Opposite to this, Kenneth Olwig has even claimed that the very notion of landscape is "very much a regional affair, with roots going deep into the identities of the historical regions from which the Mediterranean Landscape Charter and the Landscape Convention emerged. This is the landscape of place to which people become attached through history".[6] In his view, this notion is, in turn, linked to democracy (and justice). For, as we read in the explanatory report, if people are given an active role in decision-making on landscape, they are more likely to identify with the areas and towns where they spend their working and leisure time as, supposedly, they have more influence on their surroundings. In this case, they can reinforce local and regional identity and distinctiveness, and this is rewarding in terms of individual, social and cultural fulfilment (Europe 2000b, II, × 24).

Nonetheless, even if a regional/local perception of landscape seems to predominate in the Convention's definition, the perception of landscape as a catalyst for other

---

[4] Strecher (n 3) 107.

[5] Predieri (1969).

[6] Olwig (2007).

values (aesthetic, ecological, historical, etc.) plays an important role as well. Thus, the idea that the definition of "landscape" rests on people's perception of an area means *people* refers to "actual people" (almost in a kind of opinion poll fashion) and not people as embodied in representative institutions, which is, at the very least, disputable. In other words, we can detect a tension running through the ELC expressible by three pairs of contrasting notions of landscape: landscape either as subjective perception or objective scenery; landscape as the expression of either "direct" or representative democracy; and landscape as local or national policy.

This tension—which surfaces in reading the actual text of the Convention and its interpretation in the explanatory report as they lead in different directions—has also been presented as that between technical/expert knowledge and the knowledge of the general populace as expressed by their values, conventions, customs and practices. This generalised notion of landscape would mainly rest on immaterial things, such as particular local food that confers a perceived uniqueness on each part of Europe.[7] In this regard, we should take heed of Olwig's proposition that the ELC boils down to a process of coming together to gain awareness about the many different perspectives on landscape so "it is thus the practice of implementing the Convention that is important, not the letter of the Convention understood as if it were statutory law".[8]

Nonetheless, the objective of safeguarding something having both a physical substratum (land as soil) and some quality that makes it stands out implies that the planning process must also incorporate the protection of landscape in its technical/objective meaning. The mention of the principle of subsidiarity in the Convention as a regulating criterion of implementation is significant in this regard. As has been noted, again by Amy Strecker,[9] it is essential to distinguish between local democracy on the one hand and the entire delegation of power to local administrations on the other. Indeed, subsidiarity means that governance should be exercised as close as possible to citizens but that it be awarded to higher levels of government if that provides a better outcome. The Italian constitutional version of this is that power should be scaled up if the need for differentiation and adequacy requires it, including being better equipped to make prudent planning decisions.

5.4 The Italian Route to the Implementation of the Convention

In the light of the preceding, does the Italian planning system conform to the Convention's objectives and rationale? My hunch is that the Italian landscape planning model reflects the same tension noted above and as such it represents a viable example of regulatory/policy implementation by mixing its long-standing approach to safeguarding landscape with new input from the ELC. In this final part, I'll give evidence

[7] Olwig (n 6) 590-1.

[8] Olwig (n 6) 591.

[9] Strecker (n 3) 102.

for this claim by referring to both the legislative framework and the latest generation of landscape plans.

The Italian legislative planning model,[10] which revolves around regional landscape plans, is divided into two regimes meshed into one instrument. Under the first regime, plans must provide specific prescriptions to safeguard, revamp, manage, etc. those portions of Italian territory which have been already recognised in the law as "landscape patrimony" and the consequent administrative implementation measures. There are two fundamental types of such patrimony: (a) areas, premises, gardens, historical centres, etc. of outstanding beauty identified by ministerial orders over the years since the beginning of the last century; (b) geographical entities having environmental and ecological interest, such as Alpine mountain ranges over 1600 mt. and Apennines over 1200 mt.; forests; lakes; rivers; coastal areas up to 300 m from the sea and others. These are common patrimony of public interest by default in virtue of the direct effect of law and as such, they cannot be altered without a special permit from the landscape authority. For all such landscape patrimony, planning measures must be decided by an agreement between the relevant state department (the Ministry of Cultural Heritage) and each region. Here the objective/technical meaning of landscape seems to take precedence.

The second regime potentially concerns the regional territory as a whole, which must be studied and catalogued in consideration of its diverse and peculiar character and divided into landscape units to assign landscape quality objectives. It is here that one can sense the terminology of the ELC. Significantly, this part of the planning process is not subject to agreement with the Ministry of Cultural Heritage. Somehow, this second regime of landscape planning allows us to glimpse the notion of a merge between landscape planning and general spatial planning at the regional level. In other words, and a bit optimistically, one can envisage a trend towards a model of landscape-driven regional spatial planning. For example, a provision of the Apulia landscape plan reads that "through the PPTR, the Region achieves the integration of landscape policy with urban and spatial planning, cultural heritage, environmental, agricultural, and socio-economic policies, as well as any other policy with territorial effects". Several other plans contain provisions in the same vein. The Tuscany region plan is possibly the most explicit and developed in this respect: an entire chapter of the regulations attached to the plan—under the title "Land Development Strategy"—deals with the articulation of every land management and planning policy, included infrastructure.

To test the extent and effectiveness of this trend, each of the regional plans should be examined in the context of overall land management and planning legislation and praxis. In any case, by looking at regional plans that have already been approved or are in the process of approval, we can at least seek confirmation of the influence of the ELC on the actual planning process as well as the persistence of the aforementioned

[10] See the *Code of Cultural Heritage and Landscape*, http://www.normattiva.it/uri-res/N2Ls?urn:nir:stato:decreto.legislativo:2004-01-22;42!vig=2020-11-29.

Indicator	Short denomination	Concern
(1) Governmental agency	Agency	Denomination of the administrative body responsible for the management of the plan
(2) Complete spatial coverage	Completeness	Does the plan cover the entire extension of the administration?
(3) Landscape units	Units	Does the plan include landscape units?
(4) Atlas of landscape units	Atlas	Does the plan include an atlas of the landscape units?
(5) Photographic representation	Photo	Does the landscape atlas show photographic documentation?
(6) Ecological network	Net	Does the plan explicitly refer to the institution of an ecological network?
(7) Rural and agricultural landscapes	Rurality	Does the plan refer to rural and agricultural landscapes?
(8) Observatory of landscape quality	Observatory	Is an observatory of landscape quality described and (going to be) instituted?
(9) Institutional geographic information system Spatial data infrastructure	GIS/SDI	Are institutional geographic information system and spatial data infrastructure conceived as tools available for supporting landscape planning?
(10) SEA	SEA	Does the planning procedure imply the activation and development of a SEA?

Fig. 5.1 Indicators selected for the analysis of last generation landscape planning tools

Outline of Regional Landscape Planning in Italy

	Region/ province	Landscape Plan (Acronym), Year	Agency	Completeness	Units	Atlas	Photo	Net	Rurality	Observatory	GIS/ SDI	SEA
→	Apulia	PUTTP, 2001 PPTR, 2013	Area Policies for the Environment, the Networks and the Urban Quality	√	√	√	√	√	√	√		√
	Calabria	QTRP, 2013	Department of Urban Planning and Regional Government	√	√	√	√	√	√	√	√	√
	Lazio	PTPR, 2007	Area Landscape and Regional Planning	√	√	√	√		√		√	
	Lombardy	PTR, 2013 PPR, 2010	General Department of Urban and Regional Defence and Planning	√	√	√	√	√	√	√	√	√
→	Piedmont	PPR, 2009	Regional Division for Strategic Programming, Regional Policies and Housing	√	√	√		√	√		√	√
	Sardinia	PPR, 2006 PPS, 2013	General Department of Urban and Regional Planning and Building control		√	√	√		√	√	√	√
	Trento	PUP, 2008	Section Urban Planning and Landscape Protection	√	√			√	√		√	√
→	Tuscany	PIT, 2009 New PIT, 2014	General Department of Territorial and Environmental Policies	√	√	√	√	√	√	√	√	√
	Umbria	PPR, 2012	Department for Programming, Innovation and Competitiveness	√	√	√	√	√	√	√	√	√
	Veneto	NPTCR, 2007	Division Regional Planning and Parks	√	√	√		√	√			√
→	Friuli VG	PPR 2018	Division for Landscape and Biodiversity	√	√			√	√	√	√	√

Adapted from A. De Montis, 2016

→ The arrow indicates that the plan is in force (approved)

Fig. 5.2 Outline of Regional Landscape Plans in Italy

dual regime. As to the former, I refer to a grid of indicators proposed by de Montis (2016) (Fig. 5.1).[11]

In my view, the following three characteristics stand out regarding the alleged influence of a broader concept of landscape in line with the ELC: the territorial scope of the plans, which cover the whole region in almost all cases; the provision of landscape units, and often of atlases, to designate and identify landscape change and dynamics in all circumstances; and the numerous provisions of ecological networks (see Fig. 5.2).

Just to give an example, in the Friuli-VG regional plan, a landscape network is an interconnected system of landscapes covering the whole region which detects any type of terrain (natural, rural, urban, etc.) to preserve, improve, and manage them.

[11] De Montis (2016).

Another relevant feature is the provision of an observatory of landscape quality in several cases, which the literature deems an indicator of a policy for fostering social perceptions of local landscapes and bottom-up practices which facilitate participation. Innovative tools for involving people are generally considered particularly in tune with the ELC. A strong emphasis on participation and subsidiarity as a way of enacting procedures for the social production of landscape can be observed in the Apulia plan but not as much in other regions.

Alongside these innovative landscape planning process characteristics, the other planning regime is alive and well. In all plans, there are chapters—drawn up in agreement with the central state cultural heritage department—devoted explicitly to regulating the use of "landscape patrimony" of national interest. There is however a distinct sense that the two sides (even in the same document) remain somewhat distant from each other, expressing the inevitable tension characterising the notion itself of landscape. Nevertheless, they also contaminate each other, and so the strain somehow produces a new equilibrium enriching landscape protection as a whole. My view is that we should not try to make one approach prevail over the other such as when it is claimed that by not grounding the process for identifying, managing and planning landscape on the complete equivalency of state and local authority competences, Italian landscape law breaches international obligations.[12] Another ground for claiming deviation from the ELC regards the failure to allow people to participate in a deliberative capacity regarding landscape decisions.[13]

To conclude, I think that over these two decades, the ELC has undoubtedly influenced landscape safeguarding in a positive way in so far as it has encouraged states and other institutions to develop policies and tools aimed at making landscape—in its manifold meanings—a cornerstone of land use and management. However, this regards the policy/institutional dimension. As always, it is quite another matter to effect change on the ground, and for that, I guess we should be much more occupied with the educational and coming together side of the ELC.

References

Cartei GF (2012) The Implementation of the European Landscape Convention and Public Participation. Eur Public Law 18:271

Council of Europe (CoE) (2000) European Landscape Convention. Explanatory Report (CETS 2000) no 176

De Montis A (2016) Measuring the performance of planning: the conformance of Italian landscape planning practices with the European Landscape Convention. Eur Plan Stud 24:1727–1745

Di Giovanni L (2019) L'attività conoscitiva della pubblica amministrazione nel procedimento di pianificazione paesaggistica. Rivista Giuridica dell'Edilizia 559

Olwig K (2007) The practice of landscape 'conventions' and the just landscape: the case of the European landscape convention. Landsc Res 32:585

[12] Di Giovanni (2019).

[13] Di Giovanni (n 12).

Predieri A (1969) Significato della norma costituzionale sulla tutela del paesaggio. In: Studi per il XX Anniversario dell'Assemblea Costituente, vol II, Le libertà civili e politiche. Vallecchi, Firenze, pp 381–428

Roe M (2013) Policy change and ELC implementation: establishment of a baseline for understanding the impact on UK national policy of the European landscape convention. Landsc Res 38:768–798

Strecker A (2018) Landscape protection in international law (OUP)

Chapter 6
Policy and Governance Challenges Facing European Rural Landscapes

Jørgen Primdahl

Abstract European rural landscapes are highly diverse and rich in cultural and natural values. They attract a large number of international tourists and, to a large extent, they contribute to making many European regions attractive living places for an increasingly mobile population. Currently European landscapes are facing a number of policy and planning challenges which must be dealt with including the need for policy integration, pro-active place making approaches and collaborative governance. These challenges are illustrated through two examples: urban sprawl and land use conflicts associated with farming and nature conservation. Different institutional solutions are briefly outlined and a framework for collaborative landscape strategy making is presented together with a concrete example of such as strategy.

European rural landscapes hp are highly diverse and rich in cultural and natural values (Stanners and Bourdeau 1995). They provide more food and timber than is consumed in Europe. They contain a rich diversity of habitats and species. They attract a large number of international tourists and, to a large extent, they contribute to making many European regions attractive living places for an increasingly mobile population.

Currently, these landscapes are facing profound changes caused by a combination of global change (including climate change) and local adaptations and pressure from people striving for a better place to live (Pinto-Correia 2018). Moreover, the changes are complex and intersect, which means that individual technical fixes for sustainable development are not an option. Consequently, landscape policies must be able to cope with wicked problems—they must be integrated, more pro-active and built into collaborative governance practices. Why and how landscape policy and planning must have these characteristics is the focus of this paper.

J. Primdahl (✉)
Department of Natural Resources and Natural Resource Management, University of Copenhagen, Copenhagen, Denmark
e-mail: jpr@igh.ku.dk

T. Matteini et al. (eds.), *Cultivating Continuity of the European Landscape*, Environmental History 15, https://doi.org/10.1007/978-3-031-25713-1_6

6.1 Integration and Place Making

Besides the values and intentions of owners and users, landscapes are regulated through various combinations of market and public policy including spatial planning. Urban areas have traditionally been subject to detailed interventions and for many years, urban planning has been dominated by integrated planning with a focus on place making. Most large European cities have gone through profound renewal processes making them much more attractive as places to live and visit than they were in the past, which means that most of them have increased in terms of density and population size in recent decades. These developments in urban planning are closely linked to an overall shift from a highly sectoral planning approach in which major functions such as traffic, housing, recreation and business centres were planned in separate processes to a much more integrated approach. There has also been a shift from planning practices mainly focused on conflict management to a more development oriented, place making approach (Healey 2004). It must be added, however, that these processes have not changed the overall trend towards urbanisation: low density urban development. Urban sprawl has also increased (EEA 2016).

For rural landscapes, developments have been different. Public policy for European rural landscapes is highly sectoral and much more focused on conflict management than place making. In line with this, rural landscape policy has remained centralised with relatively limited stakeholder involvement. It has been concerned with agricultural development, forestry, water resource management, mining, nature conservation, cultural heritage, outdoor recreation and other issues. Designations, whether they be international, national or regional, have often been a main instrument. Consequently, rural landscape planning has contributed to the functional segregation of the countryside, often unintentionally.

However, in most countries, spatial planning is focused on urbanised regions and, with some exceptions, there is a limited tradition of integrated planning for rural areas. To a large extent, rural landscapes are seen mainly as a domain for agriculture and forestry and these sectors have traditionally been left out of spatial planning, which is a problem because changes affecting rural landscapes intersect and cannot be addressed effectively in isolation.

Agricultural change, urbanisation (including counter-urbanisation) and climate change—to mention three main driving forces—affect the same landscapes in different combinations but rarely in isolation from each other. Moreover, pro-active interventions are often required to respond to changes, for example, in terms of promoting efficient and sustainable food production, habitat restoration, flood protection, recreational networks, peri-urban landscape management and village renewal. Old fashioned, reactive planning based on designations and zoning schemes is not enough. What is needed is a planning approach that can integrate sectorial policy goals and support projects aimed at creating well-functioning and attractive rural places.

6.2 Lack of Space and Competition for Land

When climate change alters the conditions for agriculture and when urbanisation processes and demands for more land to be allocated to nature conservation increase, competition for land increases. A recent study of sectoral claims for space in Denmark calculated that 40% more land would be needed if all claims were to be met (Teknologirådet 2016).

In practice, it is mainly a combination of landowners' decisions (although these are often adaptations to external driving forces) and public policy that will be decisive in land-use changes. In a European context, this has been the case for centuries, as there is a long history of public land-use regulation in all European countries (Jepsen et al. 2016). Land-use policy and planning can be applied to allocate the different land-use types to separate areas based on combinations of conditions and the potential of the land, on the one hand, and the political reality behind policy decisions, on the other. Alternatively, policies aimed at establishing multifunctional uses can be applied to reduce the total area needed for the functions in question. Using two examples, I argue that, in practice, what is needed is a combination of a focused land allocation approach and application of the principle of multifunctionality.

The first example is urban sprawl and how to deal with it in landscape planning. Urban sprawl represents a problem because it fragments agricultural land and natural habitats and consumes more resources (land, energy, water) than dense urban areas and it appears that avoiding or reducing urban sprawl is a common aim of spatial planning (EEA 2016). Therefore, the avoidance of urban sprawl is an argument for separating urban and rural land uses and aiming for relatively sharp boundaries between urban and rural landscape. In contrast, the old dream of combining the best of the country with the best of the city, as in Ebenezer Howard's vision of the Garden City (Howard 1965), is very much alive and is one of the drivers behind urban sprawl. Furthermore, the current interest in urban farming is linked to this desire to bring the rural and urban together. Combinations of relatively dense urban neighbourhoods with green networks including recreational open space, food production, water management (supply, retention) and habitats may represent a planning solution to such combinations in urban fringe landscapes. However, it will require a planning approach able to integrate different objectives and support pro-active projects to establish green networks (Rolf et al. 2019). Therefore, this example can be used to argue both for relatively clear boundaries between (built-up) urban structures and rural land use and for multiple use.

The other example is agricultural land-use versus land needed for nature conservation purposes including biodiversity protection. In this discussion, the so-called 'land sparing' principle has been compared to 'land sharing' (Phalang et al. 2011, Fischer et al. 2014). In land sparing, farming and nature conservation are separate in order to optimise food production and biodiversity protection, whereas in land sharing, the two functions overlap in space so agriculture and habitats occur in the same landscape. As with the first example, it can be argued that both principles

Fig. 6.1 A river valley in Central Jutland, Denmark. Calls for future changes include rewetting of the peat lands in the valley floor to reduce CO2-emissions, restoration and extensive grazing of overgrown semi-natural meadows and bogs, better drainage to improve agricultural production, end better conditions for outdoor recreation

are not mutually exclusive—they can be combined. Ecologists have argued convincingly about the need for large habitats in which natural processes can occur relatively undisturbed, which means that farming in such areas cannot be allowed. On the other hand, biodiversity in many parts of Europe is intimately linked with extensive forms of farming such as the Iberian savanna landscape with varying densities of holm and cork oak named ‘dehesa’ in Spain and ‘montado’ in Portugal. In a policy and planning context, the challenge here is to find specific solutions which effectively ensure long term biodiversity through designations and then find ways (including appropriate spatial scales) to integrate other functions (farming, recreation, drinking water supply, heritage conservation, etc.) wherever such functions are demanded and do not violate long term protection. Again, the example shows both the need for segregation and multifunctionality, as well as the need for actions to provide a network of protected habitats. The river valley shown in Fig. 6.1 illustrates a Danish context in which claims are made for very different changes. These changes must been seen together, preferable on a regional scale. Some changes may be integrated others will have to take place on different areas.

6.3 The Challenge of Collaborative Landscape Governance

Summing up, policy and planning challenges facing Europe landscapes include the need for policy integration, pro-active place making and complex forms of land uses. If we add the requirement of democracy and therefore, of ways to support fair combinations of rights and duties, we are searching for ways to develop a collaborative governance approach to landscape protection, management and change.

In terms of organisational solutions to these challenges, a number of different ways to support and coordinate landscape actions have evolved over time. Ian Hodge (2016) distinguishes between three forms: (1) Organisations working on behalf of broader groups through non-profit organisations. These organisation often own large areas of land such as the National Trust in England. In other situations, such organisations work to guide farmers' and foresters' landscape management into more desired and coordinated directions through subsidies and/or advisory services. (2) In another situation, land owners get together to organise collective agreements with pubic authorities on landscape issues. The Dutch so-called agri-environment cooperatives are an example of such an institution, whose members also include public bodies, organisations and individuals who do not own farmland. The Dutch co-operatives evolved from so-called environmental cooperatives (Renting et al. 2001) to a national system of agri-environment cooperatives, which are developing their own schemes in collaboration with Dutch and EU authorities and signing collective agreements under the Rural Development Programme on behalf of their members (Dutch Ministry of Economic Affairs 2016). (3) A third institutional way to cope with integration and pro-active projects is through public–private partnerships working together to pursue policy objectives; a way that is widely used in water resource management climate policy.

In practice, these different institutional constructions overlap considerably and can be found in numerous different versions throughout Europe. It would undoubtedly be fruitful if the networks of the European Landscape Convention systematically investigated such organisations and their effects on the rural landscapes in question.

Turning to collaborative processes, the so-called 'landscape approach' has gained currency in recent years. Referring to the unifying meaning of landscape as a concept, Sayer et al. (2013) argue for applying 10 principles to guide decision making concerning landscapes including involvement of stakeholders, learning, awareness building, user friendly monitoring, multifunctionality and land-use trade-offs. In a literature review of the landscape approach, Arts et al. (2017, p. 456) note a high diversity in methodological approaches to multidisciplinary landscape governance but welcome it and warn against reliance on '... one overarching framework, for example a social-ecological system framework'.

Inspired by Patsy Healey's (1998, 2009) and Albrecht's (2006) work on collaborative planning and spatial strategy making, a framework for landscape strategy making shown in Fig. 6.2 has been developed through a number of experimental projects in which the author of this paper has participated (Kristensen and Primdahl 2020; Primdahl et al. 2020). In the context of central plans and policies, regional and central programmes, municipal planning and land owners' and users' own values and plans, a landscape strategy process typically consists of four partly overlapping dimensions: (1) creation of a common understanding and interest among stakeholders of the landscape in question. (2) capturing the situation between the past and the future landscape, that is, bringing past developments together with goals for the future, (3) mobilising resources including knowledge and ideas and (4) framing the strategy and selecting priority projects. Through interview surveys, landscape analyses, public lectures, excursions and workshops, the strategy is developed in collaboration with

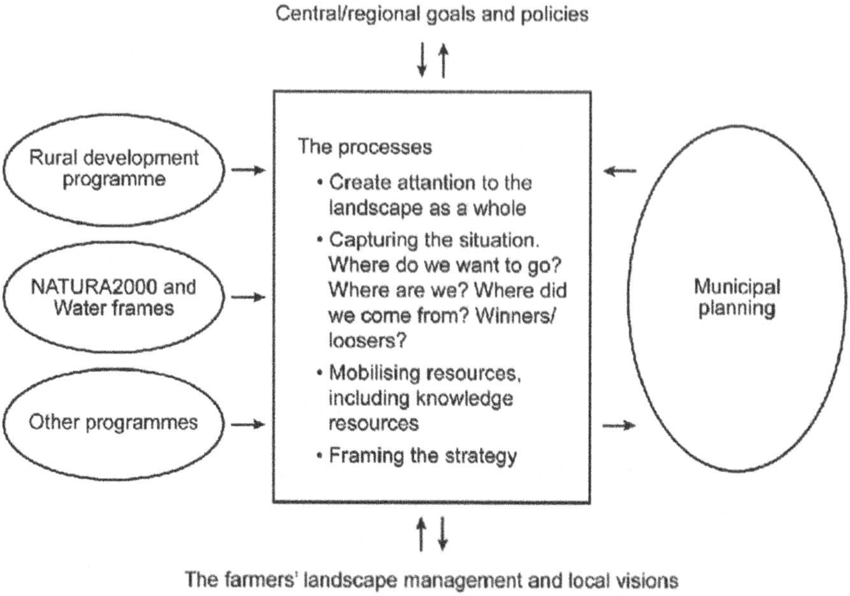

Fig. 6.2 A framework for landscape strategy making. Adapted from Kristensen and Primdahl (2020)

public agencies (typically municipal and central offices for planning, environmental management and nature conservation), farmers and other large land owners, local communities of different kinds and rural residents. Out of such processes comes a strategy document with a broad ownership. Often the strategy document consists of three parts: visions and objectives, a spatial frame, and a number of priority projects, that is, projects which have a good chance of being realised and a fair chance of starting or enhancing a virtuous circle of development.

The spatial frame from a recent experimental strategy making process in a coastal landscape in Danmark is shown in Fig. 6.3.

Fig. 6.3 An example of the spatial frame from a landscape strategy for a Danish coastal landscape affected by population decline. Landscape conservation and enhancements are seen as development factors partly aiming at attracting new residents and tourists. From Primdahl et al. (2020, p. 7)

6.4 Conclusion

New approaches to landscape governance are needed and no simple universal model of institutional organisation or process framework is available. Relative to conventional landscape policy and planning, more integrated, pro-active and collaborative approaches are needed. Many experiences are currently underway throughout Europe and they should be studied and evaluated systematically from different perspectives.

References

Albrechts L (2006) Bridge the gap: from spatial planning to strategic projects. Eur Plan Stud 14:1487–1500

Arts B, Buizer M, Horlings L, Ingram V, van Oosten C, Opdam P (2017) Landscape approaches: a state-of-the-art review. Ann Rev Environ Resour 42:439–463

Dutch Ministry of Economic Affairs (2016) The cooperative approach under the new Dutch agri-environment-climate scheme. Background, procedures and legal and institutional implications. Retrived from http://toekomstglb.nl/wp-content/uploads/LR_95078_Dutch_Agri_Enviroment_V2jgd-2.pdf

EEA (2016) Urban sprawl in Europe. European Environmental Agency Report no. 16, Copenhagen

Fischer J et al (2014) Land sparing versus land sharing: moving forward. Conserv Lett 7(3):149–157
Healey P (1998) Collaborative planning in a stakeholder society. Town Plann Rev 69(1):1–21
Healey P (2004) The treatment of space and place in the new strategic spatial planning in Europe. In: Müller B, Löb S, Zimmermann K (eds) Steuerung und Planung in Wandel. VS Verlag für Sozialwissenschaften, Wiesbaden, pp 297–329
Healey P (2009) In search of the 'strategic' in spatial strategy making. Plan Theory Pract 10(4):439–457
Hodge I (2016) The governance of the countryside. Property, planning and policy. Cambridge University Press, Cambridge
Howard E (1965) Garden cities of tomorrow. Edited, with a preface by F. J. Osborne, 23rd edn. MIT Press, Cambridge, MA
Jepsen et al (2016) Transitions in European land-management regimes between 1800 and 2010. Land Use Policy 49:53–64
Kristensen LS, Primdahl J (2020) Landscape strategy making as a pathway to policy integration and involvement of stakeholders: examples from a Danish action research programme. J Environ Planning Manage 63(6):1114–1131
Phalan B et al (2011) Reconciling food production and biodiversity conservation: land sharing and land sparing compared. Science 333:1289
Pinto-Correia T, Primdahl J, Pedroli B (2018) European landscapes in transition: implications for policy and practice. Cambridge University Press, Cambridge
Primdahl J, Folvig S, Kristensen LS (2020) Landscape strategy-making and collaboration. The hills of Northern Mors, Denmark; a case of changing focus and scale. Land 9:189
Renting H, Ploeg JD (2001) Reconnecting nature, farming and society: environmental cooperatives in the Netherlands as institutional arrangements for creating coherence. J Environ Planning Policy Manage 3:85–101
Rolf W, Pauleit S, Wiggering H (2019) A stakeholder approach, door opener for farmland and multifunctionality in urban green infrastructure. Urban Forestry Urban Greening 40:7383
Sayer J, Sunderland T, Ghazoul J, Pfund JL, Sheil D, Meijaard E et al (2013) Ten principles for a landscape approach to reconciling agriculture, conservation, and other competing land uses. PNAS 110:8349–8356
Stanners D, Bourdeau P (eds) (1995) Europe's environment. The Dobris assessment. European Environment Agency, Copenhagen
Teknologirådet (2016) Anvendelsen af Danmarks Areal I fremtiden. Fonden Teknologirådet, København

Chapter 7
Landscape Policies of Europe

Clara Queiroz da Costa and Maria José Curado

Abstract Signed in October 2000, the European Landscape Convention was the first treaty exclusively dedicated to landscape as a holistic unit of territory. It is an international commitment by European governmental organisations to plan instruments for the progressive development of places while safeguarding their natural and cultural dynamics. The Convention includes provisions that should be understood uniformly by the signatory parties at all administrative levels. This article looks at the different approaches to landscape characterisation and identification developed by these parties according to the definitions and specific measures of the Convention. Its purpose is to comprehend the most significant accomplishments and challenges of its implementation. It also aims to disseminate methodologies in the design of landscape tools concerning political and cultural contexts.

Keywords Landscape policies · Landscape Governance · Landscape planning and regulation

7.1 Introduction

The present article considers the Convention's definitions while assessing signatory parties' policies based on case-studies—Andorra, Catalonia, Hungary, Ireland, Mallorca, Portugal, Switzerland, and The Netherlands. It aims to analyse the Convention's performance through procedures methodologically designed to assess these cases according to the following chapters.

C. Q. da Costa (✉)
Landscape Architecture Master, University of Porto, Porto, Portugal
e-mail: claraqmdacosta@gmail.com

M. J. Curado
Faculty of Sciences, University of Porto, Porto, Portugal
e-mail: mcurado@fc.up.pt

T. Matteini et al. (eds.), *Cultivating Continuity of the European Landscape*, Environmental History 15, https://doi.org/10.1007/978-3-031-25713-1_7

7.2 Methodology

The Convention's specific measures were the starting point for case-studies selected for their geographical distribution, scale, diversity, and distinctiveness of policies. The database consisted of legislative acts, official reports, and scientific articles. The creation of inventory tables referencing descriptive parameters attempted to reassume the collected data by providing further detail on the landscape approach (Table 7.1).

7.2.1 *Comparison of Case-Studies*

The parameters were associated with rhetorical questions for the definition of comparison criteria. These questions sought to organise the parameters into conceptual steps as the main stages of a policy. The procedure's results correlate the criteria and the case-studies. Through *Yes* or *No* answers, the chart shows the differences and similarities between them by focusing on the mere existence of provisions (Table 7.2).

7.2.2 *Evaluation of the Case-Studies*

The evaluation of a policy's performance through indicators is based on the parameters and the European Landscape Convention Information System report questions,

Table 7.1 Proposal of parameters for the inventory tables (da Costa 2019, p. B2–B4)

Parameters from the inventory tables	
Name	Description
Territorial legal framework	Laws in which landscape is subject
Territorial management instruments	Programmes or plans aimed at spatial planning
Landscape management instruments	Tools in which landscape is subject
Authorities and entities	The ones responsible for laying down landscape and sectoral tools
Awareness-rising	Campaigns that promote the value of landscapes
Training and education	Specialist training and school courses on the landscape
Identification and evaluation	The applied methods on the study and analysis of landscapes
Landscape quality objectives	Objectives aimed at protection, management, and planning
Implementation	Legal instruments designed for landscape
Monitoring	Method of the landscape assessment
Financing	The existent funds for landscape policy or strategy

Table 7.2 Proposal of criteria the comparison of case-studies (da Costa 2019, p. B5–B7)

Proposal of criteria for comparison
STEP ONE: WHERE ARE WE NOW? I Describes the current frame work for landscape identification, characterisation, and training. PARAMETER: territorial legal framework, authorities and entities, awareness—raising, training and education, identification and evaluation
C1—Recognition of a landscape in the legal framework C2—Existence of awareness-raising campaigns C3—Existence of multidisciplinary training courses C4—Existence of school and university courses C5—Existence of a landscape characterisation approach (existence of landscape units or types)
STEP TWO: WHERE DO WE WANT TO GO? I Describes the ambitions of the desired landscape. PARAMETER: landscape quality objectives
C6—Existence of spatial management tools with landscape provisions C7—Existence of objectives aimed at landscape protection C8—Existence of objectives aimed at landscape management C9—Existence of objectives aimed at landscape planning C10—Integration of landscape provisions in sectoral policies
STEP THREE: HOW WILL WE GET THERE? I Details the operational approach of the defined objectives. PARAMETER: implementation, financing
C11—Existence of a landscape policy C12—Existence of a landscape strategy C13—Existence of measures or plan of action C14—Existence of a financial fund
STEP FOUR: HOW WE WILL KNOW WE HAVE ARRIVED? I Describes the periodical assessment of the operational approach within measures or actions of the quality objectives. PARAMETER: implementation, monitoring
C15—Existence of a monitoring programme

from the Council of Europe. These indicators are meant to be adjustable, refutable, and valid in all cases without being too specific or vague. Table 7.3 references the indicators to the report questions[1] and each of the parameters:

All indicators are crucial for the evaluation of cases, though the ones that assess the efficiency of some tools were omitted since monitoring actions have not been executed yet. For the actual evaluation, the correlation of indicators and case-studies displays the following rating scale:

0 Not mentioned and affects the policy's quality
1 Poorly executed or poorly understandable
2 Mentioned and has the potential for improvement
3 Effective and strengthens the policy
ND NO DATA (available for rating)

[1] According to the questions from the Information System of the Council of Europe of the Implementation of the European Landscape Convention National Report from Andorra, downloaded from the ELCIS website on 15th March 2019 (https://elcl6.coe.int/WebForms/Public_List.aspx).

Table 7.3 Proposal of indicators for the evaluation of case-studies (da Costa 2019, p. B8–B10)

Proposal of indicators for evaluation	
El. Cooperation at the international scale	Parameter: authorities and entities
	Evaluates on the partnership between governments from the signatory parties, international authorities and NGO's
E2. Cooperation at national, regional and local scales	Parameter: awareness-raising
	Evaluates on the partnership between the administrative governments; authorities; public and private entities and, professionals
E3. Public access to information related to landscape matters	Parameter: awareness-raising
	Evaluates the variety of information on landscape matters and how it is disseminated (conferences, social network, websites, television)
E4. Training and education effectiveness	Parameter: training and education
	Evaluates the type of sponsors that promote training and education on landscape matters (university or technical courses) and the target audience
E5. Public participation in decision-making processes	ELCIS questions: **39, 40, 41, 42**. *Are there any specific procedures for regional/local authorities or the general public or other parties to participate in public decisions about landscape?*
	Evaluates the type of public participation mechanisms (interviews, surveys, workshops) that exist and on when they are applied (characterisation, quality objectives, monitoring stages)
E6. Landscape provisions in sectoral objectives	Parameter: implementation
	Evaluates if the sectoral plan includes provisions on landscape protection, management, and planning
E7. The efficiency of spatial management tools	Parameter: territorial management instruments and landscape management instruments
	Evaluates on the coherency and legibility of the spatial management system
E8. The efficiency of the plan of action	Parameter: implementation
	Evaluates on the operational approach of a plan—schedule, specific tasks and the responsible agents
E9. The efficiency of the monitoring array	Parameter: monitoring
	Evaluates on the reliability and consistency of the monitoring approach
E10. The efficiency of the financial fund	Parameter: financing
	Evaluate the scope of action and management of landscape funds

The result is an average obtained by summing the scores of the evaluated indicators and dividing by the number of indicators assessed. This procedure results in a relative grading[2] useful for collaborative purposes where parties need feedback on how they compare to other groups.

This article is a summary of a six-month research project restricted by deadlines; thus, it was not possible to contact the stakeholders in time. Therefore, it shows preliminary scores given by the authors under the supervision of the Landscape Observatory of Catalonia, an entity with vast experience in the execution of tools. The stakeholders will be allowed to rectify in the future, according to the current status of the policies.

7.3 The Landscape Approach to the Case-Studies

See Fig. 7.1.

7.3.1 Andorra

The Andorran landscapes are identified in an apparently simple way, given that its area of action is smaller than other regions. Its geographical proximity to Catalonia reflects on the approach adopted in its analysis, based on the existence of the Landscape

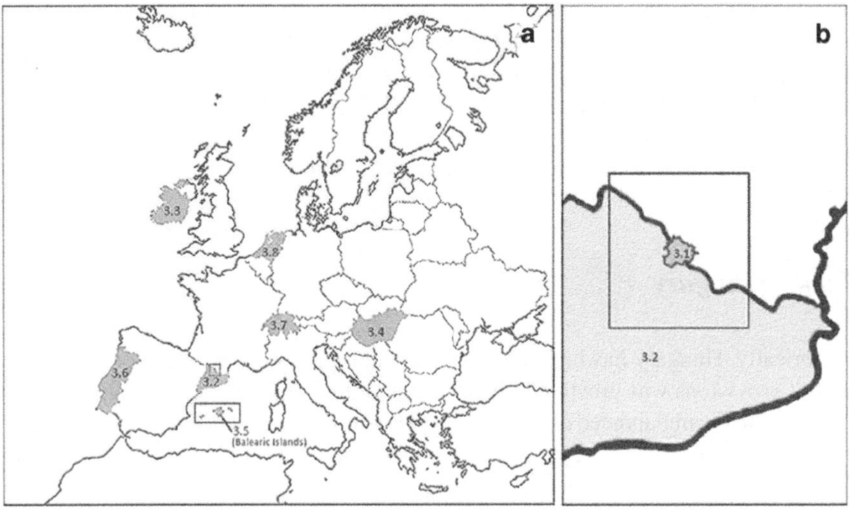

Fig. 7.1 Location of the selected case-studies at European level (**a**), including Andorra (**b**)

[2] University of Washington (2010).

Catalogue of Andorra. The Catalogue has no legal precedent, but its results were a significant contribution to its strategy. The evaluation of landscape status is done annually and gives insights regarding the schedule of actions. Some of them do not conform their predicted timeline. Still, is an expeditious way of monitoring the landscape.

7.3.2 Autonomous Region of Catalonia, Spain

Catalonia's adherence to the Convention in 2000 and its implementation in 2005 was the first step for the definition of a landscape policy at a regional level. Although the State is responsible for defining landscape and nature directives, some of the regions have already created their policies. The Catalan tools–such as the Landscape Catalogues–characterise units while defining objectives towards the achievement of the desired landscape. These tools permit the integration of landscape directives into spatial planning as they guarantee the coherence of Catalonia's territorial development.

7.3.3 Ireland

Ireland has already developed studies on landscape at a local and regional scale. However, at a higher level, provisions were not accompanied by the execution of measures that had impacts on landscape dynamics. In terms of approach, the Central Government is attempting to make amends by strengthening the legal framework towards its recognition. The Convention's implementation at the national level is particularly important since the strategy follows its specific measures as driving forces for its execution. The national strategy results from a bottom-up action strengthening commitment to landscape appreciation.

7.3.4 Hungary

Historically, Hungary has had experience in landscape studies; however, the definition of provisions was practically non-existent until the Convention's ratification in 2007, due to the unbalanced expropriation of lands that contributed to society's underestimation of landscape. The strategy accompanies international tools—notably, from Switzerland and Ireland—that have a research background but also follow the Convention's specific measures as quality objectives for the execution of an initial national strategy. It shows potential for improvement through the inclusion of provisions in sectoral policies currently in action.

7.3.5 Autonomous Region of the Balearic Islands—Mallorca, Spain

Mallorca's strategy recognises landscape value in local development. However, it does not have a specific law for landscape since its spatial planning tools have the capacity to control processes that can modify character. The proposal of directives merely in spatial planning might be an appropriate response within the Balearian context for a non-redundant framework. The strategy follows long-term guidelines that aim to improve landscape assessment, such as the protection of landscapes envisaged in law, the improvement of degraded areas, the acknowledge of landscape in sectoral policies and the promotion of landscape accessibility.

7.3.6 Portugal

Portugal shows efforts in the design of landscape instruments although it still has an ambiguous policy as it is seen more as a sectoral action rather than a holistic element. Before it entered into force in 2005, studies were done at a continental level[3]; quality objectives have been defined but not yet integrated into spatial management. Still, the Constitution has acknowledged landscape protection in spatial planning since 1976.[4] This reference is proof that its landscape policies have the potential for development through landscape directives that accompany the Convention and the State's interests.

7.3.7 Switzerland

By the time of the Convention's signing, Switzerland had a robust policy on landscape planning. Since the nineteenth century, the development of nature observation studies has fostered the promotion of nature and heritage as driving forces for spatial development. The envisagement of landscape in 1966[5] encouraged new training and financial investments aimed at biotope conservation and cooperation between administrations. The appreciation of landscape and nature as interchangeable terms is the

---

[3] Contributos para a Identificação e Caracterização da Paisagem em Portugal Continental (Volumes I, II, III, IV, V + Cartography map + CD-ROM): D'ABREU, Alexandre Cancela (UE); CORREIA, Teresa Pinto (UE); OLIVEIRA, Maria do Rosário (UE). Direção-Geral de Ordenamento do Território e Desenvolvimento Urbano. Lisboa, 2004. ISBN: 972-8569-28-9.

[4] Constitution of the Portuguese Republic of 1976, 2005 amendment (Official Gazette of the Portuguese Republic Assembly—DR no. 86—Series I—10.04.1976): landscape is subject of Article no. 66 ("Environment and Life Quality").

[5] Federal Act of July 1st of 1966 on the Protection of Nature and Landscape (NLP—identification number: CC 451).

basis of the Swiss polices, expressive and significant in its way for its ability to combine into a whole.

7.3.8 *The Netherlands*

Part of the Dutch landscape is "land reclaimed from water" constructed as a preventive measure to mitigate rising sea-level, but soon appreciated for its cultural and agricultural value. Landscape directives later integrated the spatial planning framework. Dutch policies gradually recognised the difficulty of linking landscape with spatial quality, realising that the cooperation between ministries was necessary. The Convention's implementation strengthened social power in the coordination of strategies for the achievement of a cohesive and participative policy.

7.4 Discussion of Results

The comparison and evaluation processes give a general perspective on the Convention's performance by referencing provisions currently in place through the actions of the signatory parties. Tables 7.4 and 7.5 display the results, as explained in the methodology of the article:

Table 7.4 Result from the Comparison of case-studies (da Costa 2019, p. B159)

Case-study	SCALE	C1	C2	C3	C4	C5	C6	C7	C8	C9	C10	C11	C12	C13	C14	C15	TOTAL
Andorra	N	L															13/15
Catalonia	R	C/L															14/15
Hungary	N	C															11/15
Ireland	N	L															12/15
Mallorca	I	C															11/15
Portugal	N	C															11/15
Switzerland	N	L															15/15
The Netherlands	N	L															15/15

Criteria: **C1**—Recognition of landscape in the legal framework; **C2**—Existing of awareness-rising campaigns; **C3**—Existence of multidisciplinary courses; **C4**—Existence of school and university courses; **C5**—Existence of a Landscape Characterisation Approach (landscape units or types); **C6**—Existence of Spatial Management Tools with Landscape Provisions; **C7**—Existence of objectives aimed at Landscape Protection; **C8**—Existence of objectives aimed at Landscape Management; **C9**—Existence of objectives aimed at Landscape Planning; **C10**—Integration of Landscape Provisions in Sectoral Policies; **C11**—Existence of a Landscape Policy; **C12**—Existence of a Landscape Strategy; **C13**—Existence of measures or plan of action; **C14**—Existence of a Financial Fund; **C15**—Existence of a Monitoring Programme

Table 7.5 Results from the Evaluation of case-studies (da Costa 2019, p. B160)

Case-study	SCALE	E1	E2	E3	E4	E5	E6	E7	E8	E9	E10	TOTAL
Andorra	N	2	2	2	2	3	2	2	2	3	2	2.2
Catalonia	R	3	2	3	2	3	2	3	ND	ND	1	2.4
Hungary	N	2	1	2	1	1	3	1	ND	ND	2	1.6
Ireland	N	2	2	2	2	3	2	1	ND	ND	0	1.8
Mallorca	I	2	2	2	2	1	3	2	ND	ND	0	1.8
Portugal	N	2	1	2	2	2	1	1	ND	ND	0	1.4
Switzerland	N	2	3	3	2	3	3	3	2	3	3	2.7
The Netherlands	N	2	3	2	3	3	3	3	2	2	3	2.6

LEGEND OF THE TABLES

[green] Yes [red] No

C Landscape in the Constitution

L Landscape in law

N National

R Regional

I Insular

Indicators: **E1**—Cooperation at the International scale; **E2**—Cooperation at National, Regional, and Local scales; **E3**—Public access to information related to landscape matters; **E4**—Training and Education effectiveness; **E5**—Public Participation in decision-making processes; **E6**—Landscape provisions in Sectoral Objectives; **E7**—Efficiency of Spatial Management Tools; **E8**—Efficiency of the Plan of Action; **E9**—Efficiency of the Monitoring Array; **E10**—Efficiency of the Financial Fund. **Scores**: **0**—Not mentioned and effects the quality of the implementation; **1**—Poorly executed or poorly understandable; **2**—Mentioned and has potential for improvement; **3**—Effective and strengths the policy and/or strategy; **ND**—NO DATA (available for rating)

The results show that all case-studies follow most of the provisions successfully. However, only a few have monitoring and financial arrays, which is not surprising since the Convention does not detail these topics much. It also reflects the difficulty that Central Governments have in designing landscape budgets. The lack of financing is, without a doubt, one of the most significant weaknesses in the execution of tools.

Switzerland and the Netherlands stand out for fulfilling all the criteria items, which can be justified by their long tradition in landscape management. The envisagement of landscape tools in territorial planning earned Catalonia a high score as well. Although the addressee of actions is in distinct scales, Mallorca and Ireland have the same score since they have new strategies. Hungary has a slightly worse score probably due to its lack of landscape provisions until recently. Andorra has the highest score on the efficiency of the monitoring array at the local level. At a national level, Portugal has the lowest score. In this case, the regional and local agents must encourage well-designed strategies that contribute to the strength of a national landscape policy.

The Convention calls on parties to act towards bettering quality living. Administrative cooperation through bottom-up and top-down tactics, the provision of training and public participation tools plus the design of operational approaches to spatial planning can contribute to higher policy performance in which a significant breakthrough is necessary: the cultural acceptance of landscape as a heritage of values that needs investment and protection.

Scheme 7.1 *New deal* in landscape (BSLA Journal 2016)

7.5 Final Considerations

Throughout the recent evolution of landscape dynamics, new obstacles have appeared that are difficult to deal with from a traditional sectoral point of view, developed over time. Thus, a reform on landscape perception is necessary. Such reform is vital in a globalised era that defies us to be innovative towards territorial development and differentiation in terms of content and forms of actions (Nogué and Sala 2018, p. 265) (Scheme 7.1).

Acknowledgements The authors would like to thank the Landscape Observatory of Catalonia team for their support in the writing of the master's thesis *Proposal of a National Landscape Policy for Portugal* (da Costa 2019), the primary reference for this article.

References

European Treaties

Council of Europe (2000) Explanatory report on the European landscape convention. Florence, 20.XX.2000. European Treaty Series—No. 176

Governmental Reports

Consell de Mallorca. Estratègia de Paisatge del Consell de Mallorca. Olmo, Rafael. Palma: Consell de Mallorca

Departament de Territori I Infraestructures. 2019. https://www.conselldemallorca.cat/media/55724/09_EP_Estrategia_Paisaje_CIM_cat.pdf

Govern d' Andorra (2011) Estratègia Nacional del Paisatge de Andorra. Moles, Anna, and Nàtalia Rovira. Andorra la Vella: Govern d' Andorra (Ministeri de Turisme i Medi Ambient, Departament de Medi Ambient). ISBN: 978-99920-0-658-0

Government of Ireland (Department of Arts, Heritage and the Gaeltacht) (2015) National landscape strategy for Ireland 2015–2025. Government Publications, Dublin. https://www.chg.gov.ie/app/uploads/2015/07/N-Landscape-Strategy-english-Web.pdf

Kormany (Government of Hungary) (2017) National landscape strategy (2017–2026). Kormany (Department of National Parks and Landscape Protection, Ministry of Agriculture), Budapest. https://www.kormany.hu/download/f/8f/11000/Hungarian%20National%20Landscape%20Strategy_2017-2026_webre.pdf

Office fédéral de l'environnement OFEV (1997) Conception 'Paysage Suisse' (CPS)

Office fédéral de l'environnement (2003) Paysage 2020. Commentaires et programme. Synthèse réalisée pour les principes directeurs Nature et Paysage de l'OFEFP. Berne: Confédération Suisse (Office fédéral de l'environnement OFEV). Numéro: DIV-8408-F

Rijksoverheid (Government of the Netherlands) (2008) Agenda Landschap, landschappelijk verantwoord ondernemen voor iedereen. Den Haag: Ministerie van Landbouw, Natuurbeheer en Visserij LNV and Ministerie van Volkshuisvesting, Ruimtelijke Ordening en Milieu VROM

Legislative Acts

Diari Oficial de la Generalitat de Catalunya no. 4407. Llei 8/2005, de 8 de juny, de protecció, gestió i ordenació del paisatge, June 16, 2005. https://portaljuridic.gencat.cat/eli/es-ct/l/2005/06/08/8

Diário da República n.^{2}130/2015, Série I. Resolução do Conselho de Ministros n.^{2} 45/2015, July 7, 2015. https://data.dre.pt/eli/resolconsmin/45/2015/07/07/p/dre/pt/html

Scientific Reports

da Costa CQ (2019) Proposal of a national landscape policy for Portugal. Master's internship report, Faculdade de Ciências, Universidade do Porto. https://hdl.handle.net/10216/124680

Nogué J, Sala P (2018) Landscape, local knowledge and democracy. In: Routledge handbook of landscape character assessment, pp 265–278

Websites

BSLA Journal (2016) New Deal—La qualité du paysage comme tâche d'intérêt commun. Congrès annuel du Forum Paysage

University of Washington (2010) Faculty resource on grading—pros and cons of typical grading practices. https://depts.washington.edu/grading/practices/prosandcons.html

Chapter 8
Landscape Policy in Portugal. Where Are We?

Isabel Joaquina Ramos and Maria Freire

Abstract The Landscape European Convention was ratified by Portugal in 2005. Since then, the legal framework of land use planning and the environment has changed considerably and several policies and programs have been published. Major revisions were made in 2014 and 2015. Despite the extensive legal framework concerning land use planning, environment and nature conservation in Portugal and the emergence of a landscape policy, the legal system still does not provide specific instruments for landscape and continues to address landscape in a sectoral way. A comprehensive public landscape policy is essential, based on the concept of sustainable development. The aim of this paper is to analyze and discuss how Landscape and European Landscape Convention principles have been considered in the Portuguese legal spatial planning system, before and after the European Landscape Convention.

Keywords European landscape convention · Landscape · Spatial planning system · Portugal

This work is financed by national funds through FCT—Foundation for Science and Technology, I.P., within the scope of the project «UIDB/04647/2020» of CICS.NOVA—Centro Interdisciplinar de Ciências Sociais da Universidade Nova de Lisboa.

This work is financed by national funds through FCT—Foundation for Science and Technology, I.P., within the scope of the project (UI&D 112)/2020—[Ref.ª UIDB/00112/2020] of CHAIA/UÉ.

I. J. Ramos (✉)
School of Sciences and Technology/Department of Landscape, Environment and Planning; Researcher at Interdisciplinary Centre of Social Sciences, CICS.NOVA.UÉvora, University of Évora (UÉ), Évora, Portugal
e-mail: iar@uevora.pt

M. Freire
School of Sciences and Technology/Department of Landscape, Environment and Planning, Researcher at Centre for Art History and Artistic Research, CHAIA-UÉ, University of Évora (UÉ), Évora, Portugal
e-mail: mcmf@uevora.pt

T. Matteini et al. (eds.), *Cultivating Continuity of the European Landscape*, Environmental History 15, https://doi.org/10.1007/978-3-031-25713-1_8

8.1 Introduction

The European Landscape Convention (ELC) was ratified by Portugal in 2005. The concept of landscape was reinforced, and landscape policy emerged as a central concern. Landscape must become an integral part of land use, urban planning, cultural, environmental, agricultural, social and economic policies, as well as any other policies with potential direct or indirect impacts on landscape. The implementation of these policies requires not only a planning and development system that provides specific instruments for landscape but also that landscape be integrated into the whole sectoral and spatial planning framework. Since 2005, several policies and programs have been published and the legal framework has considerably changed, particularly concerning land use planning and environment in Portugal. And how have they incorporated landscape and European Landscape Convention principles?

The aim of this paper is to analyze and discuss how landscape has been considered in the Portuguese system of law, mainly concerning the spatial planning framework.

8.2 Before ELC 2000 in Portugal

Portugal is one of the European countries where training in landscape architecture started early (in 1942). The first generation of landscape architects initiated their career and practice influencing the construction of our landscapes, the use of specific concepts, terminology and methodologies, public policies and governance as well as research in the field of landscape.

The first national protected area dates from 1971. From then until Rede Natura 2000 in 2005, several others were classified at National, European and International level (Fig. 8.1).

The concept of **landscape** was defined by law in the first Portuguese Environmental Law (Law 11/87, 7th April, art. 5, n. 2c.):

> Landscape is the geographical, ecological and aesthetic unit resulting from the action of Man and the reaction of Nature, being primitive when his action is minimal and natural when human action is decisive but maintains biological balance, physical stability and ecological dynamics.

The law considers landscape to be a human environmental component that human activity relies on and that should have specific protective measures aimed at improving quality of life at a national, regional and local level. A strong linkage between environment and spatial planning is considered crucial to protect specific landscapes (namely those with high cultural value).

A specific law concerning land use and urban planning system was published in 1989 (Law 48/98, 11th August). One of its main objectives is to improve quality of life while respecting cultural, environmental and landscape values. Land use and urban planning must safeguard natural values and landscape must be protected and improved. The law establishes three levels of planning (national, regional and local)

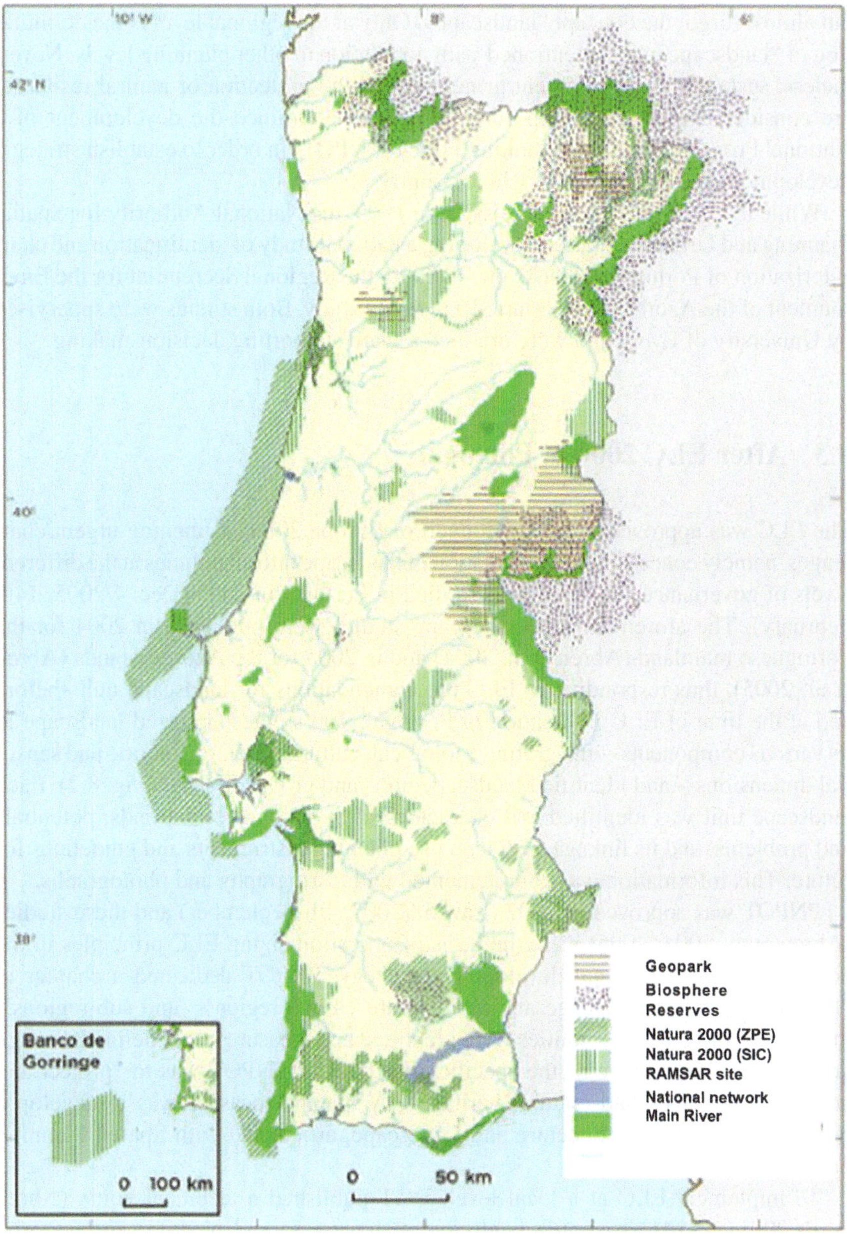

Fig. 8.1 Protected areas in Portugal. *Source* Institute for Nature Conservation and Forests (ICNF), 2017 (https://rea.apambiente.pt/content/sistema-nacional-de-%C3%A1reas-classificadas)

but almost forgot the concept "landscape". Only at the regional level is the delimitation of "landscape units" mentioned with no relation to other planning levels. Nevertheless, sustainable development principles and the protection of natural resources are considered to be very important. The law determined the development of a National Program of Spatial Planning Policy (PNPOT) in order to establish strategic development guidelines for the whole country.

While the ELC was under discussion in 1999, the National Authority for Spatial Planning and Urbanism started developing a national study of identification and characterization of Portuguese landscape. In 2000, the Regional Secretariat for the Environment of the Azores islands started a regional study. Both studies were supervised by University of Évora and were oriented toward supporting decision-making.

8.3 After ELC 2000 in Portugal

The ELC was approved in 2000 (Council of Europe 2000), launching urgent challenges, namely concerning the integration of landscape into all policies at the different levels of governance. The ELC was ratified in Portugal in 2005 (Dec. 4/2005, 14th February). The aforementioned landscape studies were published in 2004 for the Portuguese mainland (Abreu et al. 2004) and in 2005 for the Azores islands (Abreu et al. 2005), thus responding to ELC recommendations for landscape units before and at the time of ELC ratification in Portugal. The studies analyzed landscape in its various components—integrating ecological, cultural, socio-economic and sensorial dimensions—and identified landscape units and groups of units (Fig. 8.2). Each landscape unit was identified and characterized by its character, trends, potentials and problems, and its linkage with land use planning instruments and guidelines for future. This information was complemented with cartography and photographs.

PNPOT was approved in 2007 (Law 58/2007, 4th September) and these studies (Abreu et al. 2004, 2005) were the basis of operationalizing ELC principles in the development of a national development strategy. PNPOT dedicated a chapter to "Landscape, cultural heritage and architecture". Each region's (and sub-regions') main problems and opportunities were identified and measures were defined through an Action Program. One of the specific objectives of PNPOT was to "protect and enhance landscapes and cultural heritage". A specific measure was to develop a National Policy for Architecture and Landscape, articulated with Spatial Planning policies.

To implement ELC at a local level, DGT published a technical guide (Abreu et al. 2011) in 2011 to frame landscape management and planning in territorial municipalities, focusing on the strategic perspective of the Municipal Master Plans. The aim was to provide guidelines to integrate the study and management of landscape in the different planning model stages as well as in the implementation and evaluation phases, according to a methodology based on participatory processes. The guide explains the fundamental concepts and presents the ELC, highlighting its implementation as it directly relates to spatial planning, and how landscape is taken

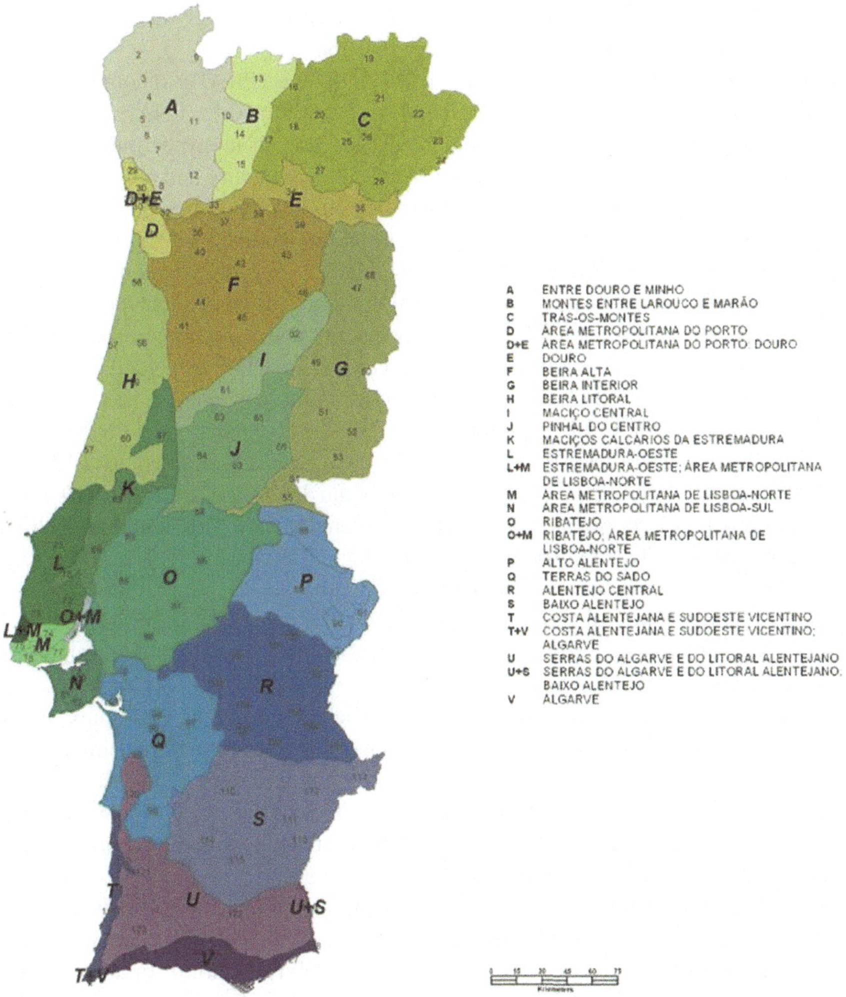

Fig. 8.2 Groups of landscape units. *Source* Abreu et al. (2004)

into account in the Portuguese Territorial Management System. Special attention is devoted to the definition of Landscape Quality Objectives through a participatory-based approach. In addition to the methodological aspects related to the study of the landscape, the document includes methodologies and techniques associated with governance, with a view to establishing consensus and commitments for landscape management. It also includes guidelines for assessing and monitoring the effects of the plan's implementation on landscape transformation and quality.

In 2014, the law concerning land use and urban planning (1989) was revised and major changes were made (Law 31/2014, 30th May). The law requires the promotion and enhancement of natural, cultural and landscape heritage. Based on this,

PNPOT2007 was reassessed and reviewed. Also, the 2011 municipal-level landscape guideline book (Abreu et al. 2011) is under revision.

A National Policy for Architecture and Landscape (PNAP), recommended in PNPOT2007, was approved in 2015 (RCM 45/2015, 7th July). PNAP is grounded on the recognition of architecture and landscape as patrimony of public interest and on the awareness that promoting built environment quality and architecture and landscape quality is a key factor in guaranteeing citizens' well-being and quality of life. Its main objectives are to (a) improve Portuguese quality of life and well-being; (b) achieve sustainable development and sustainable urban development; (c) protect and enhance Portuguese cultural and natural heritage; (d) promote and disseminate a territorial civic culture and (e) foster the competitiveness of the national economy and the affirmation of Portuguese culture and the country in Europe and the world.

As stated before, PNPOT2007 was reviewed and a new version was published in 2019 (Law 99/2019, 5th September). PNAP was integrated into PNPOT2019. Some measures have been defined to fulfill the guidelines defined by ELC and PNAP, namely "Valuing the territory through the landscape". PNOT2019 states that new territorial approaches are needed to promote rural, urban and peri Urban landscape quality. The reinforcement of ecological connectivity is a key factor (Fig. 8.3). Sustainable development of the country means integrating landscape quality objectives into all territorial plans as well as all financing programs (e.g., agriculture, forestry) and raising civic awareness of the cultural value of landscape and architecture.

In recent years, landscape observatories have been set up on the Portuguese mainland and Madeira Island. Those observatories have different structures, aims, activities and levels. In 2019, the Portuguese government determined the establishment of a Spatial planning and Urbanism Observatory to monitor and evaluate territorial dynamics and the management system, reinforce awareness and support strategic decision-making (Disp. 9315/2019, 15th October).

Last October, Portugal ratified the 2016 ELC amendments (Dec. 24/2019, 3rd October) "wishing to promote European co-operation with non-European States who wish to implement the provisions of the Convention" (preamble). In article 1, the title of the European Landscape Convention was changed to **Council of Europe Landscape Convention**.

8.4 Final Remarks

Despite all the changes and improvements concerning landscape in the reference documents and planning system, sectoral approaches continue to be applied to territorial frontiers defined in law (e.g., municipalities or regions). Instead, planning systems should be grounded on the concept of landscape as a unit with its own identity and landscape plan.

It is essential that there be comprehensive and transversal public policy on landscape planning and management based on the principles of sustainable development

Fig. 8.3 National ecological connectivity scheme. *Source* PNPOT2019 (Law 99/2019, 5th September, p. 47)

and the implementation of balanced and appropriate strategies that take into account the weaknesses and potential of landscape so as to respond to the aspirations of the people.

References

Abreu A, Correia T, Oliveira M (2004) Contributos para a Identificação e Caracterização das Paisagens de Portugal Continental. DGOTDU, Lisboa

Abreu A, Moreira J, Oliveira M (2005) Livro das paisagens dos Açores. DROTRH/SRAM, Açores

Abreu A, Botelho M, Oliveira M, Afonso M (2011) A paisagem na revisão dos PDM-Orientações para a implementação da Convenção Europeia da Paisagem no âmbito Municipal. DGOTDU, Lisboa

Council of Europe (2000) European landscape convention

Chapter 9
A Country Through Crisis and Beyond: The Case of the Greek Landscape

Theano S. Terkenli and Nikolaos Skoutelis

Abstract This chapter critically explores the relationship of Greeks to their landscape, attempting to make sense of the aspects of this relationship in its complex socio-cultural and political/economic context. It traces the factors and implications of the development of Greek landscape culture vis-à-vis the substantive landscape conceptualization and approach of the ELC, which Greece has not yet effectively implemented. The paper concludes with the presentation of the Rural Observatory of Crete, an initiative recently founded and funded by the local Prefecture authorities as a first-time attempt towards the instigation of (rural) landscape stewardship in Greece. The Observatory aims to contribute to a balanced development of the Cretan hinterland by promoting landscape sustainability, while acting as a counterpart to the excessive tourism development of Crete's northern coast.

Keywords Landscape · Greece · ELC · Crises · Regional planning · Sustainable development · Rural Observatory

9.1 Introduction. Context and Objectives

The Greek landscape, the setting of the birth of western civilization, has been extolled through the centuries by travelers, scholars and artists for its beauty and luminescence. The development of a landscape culture among Greeks, however, has been deficient as compared with other European societies.

This paper critically explores the relationship of Greeks to their landscape and attempts to make sense of the various, often conflicting, aspects of this relationship in its complex sociocultural and political/economic context. It traces the factors and

T. S. Terkenli (✉)
University of the Aegean, Lesvos, Greece
e-mail: t.terkenli@aegean.gr

N. Skoutelis
Technical University of Crete, Kounoupidiana, Greece
e-mail: nskoutelis@isc.tuc.gr

T. Matteini et al. (eds.), *Cultivating Continuity of the European Landscape*, Environmental History 15, https://doi.org/10.1007/978-3-031-25713-1_9

implications of the development of Greek landscape culture vis-à-vis the substantive landscape conceptualization and approach of the ELC, which Greece has not yet effectively implemented. The paper concludes with the presentation of the Rural Observatory of Crete, an initiative recently founded and funded by the local Prefecture authorities as a first-time attempt toward (rural) landscape stewardship in Greece, which finds itself in serious need of such public mobilization concerning participatory landscape governance and intervention.

9.2 The Construction of Greek Landscape. A Critical Overview

The myth of modern Greece has grown jointly out of the culture and history imprinted on, and reflected and depicted by, its landscapes. The symbolic physical Greek landscape features rich biodiversity, spectacular geomorphology and intricate insularity. The symbolic cultural Greek landscape reflects an ages-long tradition of urban development vs. harsh rural life, in contrast to the island and coastal leisure realms of domestic and international tourism.[1]

Classical 'antiquity', whether visible or subterranean, actual or representational, has always dominated the Greek legend and constituted the core of Greek (landscape) identity, omnipresent and heavily symbolic in Greeks' everyday life.[2] Since the founding of the modern Greek nation-state (1820s–1830s), Greek landscape identity/character has been constructed on the basis of two Western romantic ideals: (a) archaism and (b) orientalism.[3] These ideals were deeply embedded in Western conceptions of the Greek landscape and were only gradually replaced by indigenous depictions and local landscape ideals—in both formal and naive renditions of the Greek landscape—over the course of the twentieth century[4] (Fig. 9.1).

The need for economic recovery during the 1950s and 1960s through the Marshal Scheme led to utilitarian management of the land and its history, which would shape the country's industrial and tourism development. Ineffective State urban and regional

[1] The original ancient Greek word "τοπεῖον" or "τοπήιον", referred to the first human efforts to transform and incorporate nature into an urbanized space—connoting a fence/hedge delimiting property, but also serving aesthetic purposes. K. Moraitis, The art of landscape: cultural overview of modern landscape perspectives and interventions (in Greek), Department of Architecture Engineering, National Technical University of Athens, Kallipos—Greek Academic Electronic Books and Supplements:

Athens, 2015, and LSJ–Liddell-Scott-Jones, Greek-English Lexicon, 2020, https://lsj.gr/wiki/LSJ:GreekEnglishLexicon.

[2] Yalouri (2015).

[3] T. S. Terkenli, I. Tsalikidis, & F.-M. Tsigakou, "The physical landscape of Greece in nineteenth-century painting: An exploration of cultural images", in Ed. G. L., Anagnostopoulos, Art and Landscape.

[4] International Federation of Landscape Architects, Panhellenic Association of Landscape Architects—IFLA Symposium Athens, 1998, Panayotis and Effie Michelis Foundation, Athens, 2001, pp. 618–632.

Fig. 9.1 The ancient theater of Dodoni, in Epirus. *Source* N. Skoutelis

planning was never followed by monitoring, evaluation and control mechanisms, while planning policies conflicted with practice, which tended to be directed by local politicians.[5]

Diversions from the right path increased during the period of the military junta (1967–1974). The urbanization of the rural population continued; this led to the purchase of cheap land in city peripheries and, therefore, to illegal construction due to the absence of comprehensive urban plans. At the same time, uncontrolled industrial development and the growth of second-home construction in zones without urban regulations transformed rural land into a built-up continuum. In the countryside, the lack of planning in resource management and settlement growth led to linear small-scale industrial strips alongside country roads and the use of urban architecture as a model for development.

Buildings and building complexes are in some cases protected by the Ministry of Culture and Tourism, while rural landscapes are under the jurisdiction of the Ministry of Rural Development and Food. The landscape concept has appeared in national forest legislation since the 1920s but only in passing. Forests may be characterized as 'aesthetic' or as 'areas of remarkable natural beauty', whereas Natura2000 and Wildlife Areas are delineated as independent spaces without cohesion. The latter

[5] Philippides (1990).

were incorporated into national legislation and managed by the Ministry of the Environment, Energy and Climate Change but are widely considered to be inconvenient limitations, similar to those inflicted on archeological zones. Such a framework, in which natural-and-built-environmental value protection is characterized by fragmentation, almost always creates opportunities for arbitrary action (Figs. 9.2 and 9.3).

Fig. 9.2 Illegal out-of-plan construction next to archeological site, Crete. *Source* N. Skoutelis

Fig. 9.3 The urban continuum in the rural hinterland of Santorini. *Source* T. S. Terkenli

9.3 Current Advances and Challenges: Toward a Reconstruction of the Greek Landscape

In contrast to the top-down western scenic landscape ideal at the basis of development, at least in European landscapes and landscape cultures,[6] lie the original, bottom-up landscape notions and ideals represented by substantive local landscape polities and practices.[7] The ELC (Council of Europe 2000) advocates the latter substantive notion of landscape as it evolves through human-landscape interrelationships in space–time and the sociocultural context. It points to people's interrelations with landscape as central to who they are (Butler 2019) and to landscapes as the result and imprinting of such interactions on the land.

Greece readily signed the ELC in 2000 but only ratified it 10 years later (Law 3827/2010-Issue of Government Gazette 30/A/2010); at the same time, it established a 'Landscape Committee' within the Ministry of the Environment, Energy and Climate Change charged with its implementation. Unfortunately, very few and sporadic steps have been taken in the past decade in that direction,[8] mostly by the same Ministry, apart from the mention of 'landscape' in the environmental protection

[6] Cosgrove (1998).

[7] Olwig (2019)

[8] Terkenli et al. (2019).

clause of article 24 of the Constitution: namely, the nominal inclusion of landscape in national spatial sectoral plans for regional planning and sustainable development (for renewable energy, tourism, industry, etc.) and in the 2016–2017 regional plans for 7 Greek prefectures/regions. The most significant has been the inclusion of a landscape directive in the recent framework of the Regional Plans for Sustainable Development. Nonetheless, these laws and directives remain largely ineffectual and both public and private landscape matters tend to remain overwhelmingly dependent on circumstantial economic or political interests and claims, reflecting the problematic relationship of Greeks with landscape.[9]

Furthermore, during the era of its socioeconomic crisis (2008–2018), Greece witnessed a retrenchment vis-à-vis landscape matters as State as well as collective and private priorities were redefined due to a grave shortage of financial resources, and the 'Landscape Committee' became effectively obsolete. As Greece began to emerge from its crippling socioeconomic crisis, another crisis, the health/public safety crisis triggered by the COVID-19 pandemic in 2020 (exacerbated by various other urgent and often overwhelming crises, such as the refugee crisis and the Turkish geopolitical threat), further delayed the resurgence of interest in landscape.

9.4 The Rural Observatory of Crete

Crete is considered to be both an island and continental land by botanists, historians, geographers and ethnologists, not only due to its size difference relative to other Greek islands but also to the small scale of its morphological elements and climatic micro-regions, leading to great landscape diversity juxtaposed with frequently contrasting geomorphology[10] (Fig. 9.4). Its unique geography induced ages-long cultural exchanges and syncretism with peoples of the Eastern Mediterranean, expressed in Cretan traditional arts, built forms and land practices.[11] National policies in Crete, as elsewhere, led to the expansion of its three cities, Heraklion, Chania and Rethymno, and to the congestion of its northern coastal zone with big hotels and urban conglomerations for summer tourism. Inland areas were partially abandoned, revisited by locals only during weekends in order to keep some kind of relation with the native land and for agricultural production for nostalgic rather than functional purposes (Fig. 9.5).

In order to establish alternative policies for rural inland Crete and rationalize fund management, the region (Prefecture) of Crete created the Rural Crete Observatory (RCO), the establishment of which was announced on August 26, 2019. The RCO is the result of a partnership between the region of Crete, the Technical University of Crete and PLOIGOS, an educational-developmental NGO. It constitutes a prime

[9] Terkenli et al. (2011).

[10] Rackham and Moody (1996).

[11] Skoutelis (2003).

Fig. 9.4 The path linking Agios Ioannis with Agia Roumeli in Sfakia (SW Crete). *Source* N. Skoutelis

example of bottom-up action toward (rural) landscape stewardship in contemporary Greece.

The Observatory is currently going through its first year of pilot implementation. Planned as a flexible, interdisciplinary tool to record and support contemporary Cretan countryside multicenteredness and multifunctionality in its new development roles and goals, it proposes to organize the existing dynamics and potential of the multiplicity of different spatial units and transfer positive experiences from other European regions to create a flexible working scheme leading to resettlement of the countryside and consolidation of a new model for rural life in Crete. Specifically, a group of planners first assessed the Lassithi Plateau as a pilot project area for the Observatory's structure. By the end of June 2020, RCO will launch an ambitious project entitled "Register–Evaluation–Hierarchy of Traditional Settlements in Relation to European Path E4", which is a horizontal action highlighting strategies for a new attractiveness for Crete, adapted to the values and meanings of the place, its history and the human activity throughout the island (Fig. 9.6).

Fig. 9.5 Veni-Axos-Anogia, 3 villages in the rural hinterland of Crete. *Source* N. Skoutelis

9.5 Conclusion. Geography and Culture as Pillars of Landscape Sustainability

Modern Greece adopted western aspects of modernity by implanting and overlaying them on pre-existing cultural particularities and local ways of life. Specifically, the scenic landscape ideal was promulgated as the dominant landscape paradigm, especially in postwar Greece—which had severed its connections to its traditional substantive landscape polities and customary practices and instilled the normative landscape ideal through its westernized urban culture and planning/design disciplines. However, since 2010 when the ELC was adopted, Greek governments have not promoted further actions toward the formulation of tools to identify and manage cultural landscapes.

Addressing this gap in acknowledgment of the need to shape a contemporary dynamic narrative, tool and medium for the countryside, the Prefecture of Crete founded RCO as an institution for the production of development plans and instigation of actions based on Crete's geography and culture. The RCO aims to aid in the development of balanced socioeconomic systems in the Cretan hinterland by promoting landscape sustainability while acting as a counterpart to the hydrocephalus tourist development of its northern coast. It is certainly hoped that this undertaking

Fig. 9.6 Poster of the Rural Crete Observatory. *Source* Ploigos Εκπαιδευτική-Αναπτυξιακή Εταιρεία της Περιφέρειας Κρήτης

will further contribute to the renaissance of a landscape 'culture' and consequent appropriate landscape stewardship practices.

References

Butler A, Sarlöv-Herlin I (2019) Changing landscape identity—practice, plurality, and power, Landscape Research Vol. 44(3), pp 271–277

Cosgrove D (1998) Cultural landscapes. In: Unwin T (ed) A European geography. Longman, London, pp 65–81

Council of Europe (2000) European Landscape Convention, Council of Europe Treaty Series No. 176, Florence 2000. https://www.coe.int/en/web/landscape

Olwig KR (2019) The meanings of landscape. essays on space, place, environment and justice. Routledge, London

Philippides D (1990) Regarding the Greek city. Post-war development and future perspectives, (in Greek). Athens, Themelio

Rackham O, Moody J (1996) The making of the Cretan landscape. Manchester University Press, Manchester

Skoutelis N (2003) Crete. In: Philippides D (ed) Aegean Islands—architecture. Melissa, Athens, pp 160–178

Terkenli TS, Castiglioni B, Cisani M (2019) The challenge of tourism in terraced landscapes. In: Varotto M, Bonardi L, Tarolli P, Agnoletti M (eds) World terraced landscapes: history, environment, quality of life, environmental history, vol 9. Springer, Dordrecht, pp 295–309

Terkenli TS (2011) In search of the Greek Landscape: a cultural geography. In: Jones, Stenseke M (ed) The European landscape convention: challenges of participation. Springer, Dordrecht pp 121–144

Yalouri E (2015) The present past: classical antiquity and modern Greece. In: Aesopos Y (ed) Tourism landscapes: remaking Greece. Athens, Domes

Chapter 10
The ELC, a Game Changer in Scottish Policy Discourse

Sarah Govan

Abstract This paper explores how the study of institutional discourse can reveal the extent to which the concept of time in the form of past, present and future is reflected in how landscape is framed in Scotland.

Keywords Landscape · Time · Scotland · People · Place

10.1 Landscape in Time—Past, Present and Future

The Scottish landscape has long been subject to public debate, whether as countryside, natural beauty, scenery or as a landscape-scale ecosystem. It is often described as dramatic and picturesque, but there has also been a tendency to think of its people and their history as separate from nature and scenery.

The European Landscape Convention (ELC) discourages such a distinction, maintaining that "in practice, it is the interaction between the different elements that is more important than the elements themselves".[1] It has provided a vehicle for stimulating dialogue and debate across national, sectoral and disciplinary boundaries, bringing together people from different communities of place, interest, skills, knowledge and expertise.

This paper explores how the study of institutional discourse can reveal the extent to which the concept of time in the form of past, present and future is reflected in how landscape is framed in Scotland.

[1] Council of Europe (2013).

S. Govan (✉)
University of Edinburgh, Edinburgh, Scotland
e-mail: sarah.govan@ed.ac.uk

T. Matteini et al. (eds.), *Cultivating Continuity of the European Landscape*, Environmental History 15, https://doi.org/10.1007/978-3-031-25713-1_10

10.2 The European Landscape Convention—A Game Changer

The ELC definition of landscape presents a key starting point:

> An area, as perceived by people, whose character is the result of the action and interaction of natural and human factors[2]

The emphasis is on landscape as a holistic entity, encompassing multiple coexistent dimensions. All living things interact with their environment and with each other, impacting upon—and feeling the impact of—physical characteristics and processes in continuously dynamic change. For example, people dependent on a particular piece of land for their family's survival might extract lime to alter the pH of the soil, create lazy beds to grow crops and plant a shelter belt to protect their livestock, evidence for which can be detected in different ways in the landscape today (Fig. 10.1).

The act of perceiving relies on the daily interaction of people with their surroundings and the material and intangible processes that are part of how we experience being in the world. In focusing on the act of perception, there is a risk of landscape being fixed at the time it is perceived; but what is being perceived is the result of a complex process of multiple interactions that have continuously taken place over time.

A detailed study of the principles underlying the Convention was published in an accompanying Glossary, where landscape is described as "a system of interaction, both spatial and social, between its elements [...] When [...] elements or basic components of the landscape are studied or used in their own right, they cannot reflect the systemic, holistic dimension of landscape".[3] Emphasis is placed on all elements being taken together.

10.3 Institutions and Their Language

Society creates public institutions[4] to manage the selection of choices resulting from the continuity and change of daily life. Once an issue becomes a concern shared within a wider group, it prompts debate and becomes a legitimate subject for public discourse and policy-making. Legislation, guidance and advice are the principal mechanisms for driving action and any text published by a corporate body must be formally approved at a senior level prior to publication.

Institutions responsible for corporate governance will frame their understanding of a concept in their own way, based on their assigned objectives and corporate culture and in the context of contemporary politics. They are made up of a series of 'authors' and represent the considered perspective of those who identify with

---

[2] Council of Europe (2000a).

[3] See Footnote 1.

[4] Cairney (2016).

that institution. There are reasons behind all formal statements (whether definite or ambiguous) which are often the result of extended negotiation and which, once published, will not change for a certain period of time.

The research[5] that informed this paper examined outputs by formally incorporated public institutions as position statements of their policy on specific issues, with expert individuals employed to generate drafts for internal discussion, amendment and agreement.

Fig. 10.1 Grulin township, Eigg, including remains of the township, field-system, rig, lazy beds, dun and shepherd's bothy 2003, Crown Copyright HES

[5] Govan (2019).

10.4 The Language of Landscape

In Scottish policy discourse, landscape is characterised in several ways, particularly as scenery, natural beauty, countryside and an ecosystem, with evidence of past human societies recognised as individual features. More recent texts recognise an entity that is "perceived and experienced" by an individual but which also captures that which is being perceived (perhaps as scenery, countryside or natural beauty).[6]

Significant land use change in the period following the Second World War prompted serious concerns, resulting in a series of policy actions during the 1970s and early '80 s. This included the creation of 40 National Scenic Areas, largely in the more remote highland areas and described as "an accolade designation for areas of land which represent the very best of the landscapes for which Scotland is renowned".[7] There are no published criteria for their selection.

Scottish Natural Heritage (SNH) led public debate as the idea of the ELC emerged and in the decade following the re-establishment of the Scottish Parliament in 1999. This built on considerable investment and resulted in the completion of a national programme of Landscape Character Assessment (LCA) to provide a baseline understanding that, in turn, might inform land use planning decision-making. The aim was to capture the rather complex concept of character—"a distinct, recognisable pattern of elements in the landscape that makes one landscape different from another, rather than better or worse".[8] It was recognised that the idea of landscape could not be 'owned' by any one government agency alone, and SNH acknowledged a corporate "concern with the aesthetic and more natural qualities of the landscape and the enjoyment people may derive from this, always recognising the strong connections that exist with other *overlying values* which are primarily the responsibility of others".[9]

The use of the term 'overlying' is striking, echoing the traditional idea of landscape as palimpsest, made up of layers, first natural then cultural. It is true that well-placed monuments are a feature of the picturesque (Figure 10.2), but that often serves to hide the story of the generations of people who lived and worked in the place.

Discussion of the historic dimension of landscape has frequently concentrated on the mapping and analysis of specific landscape features, whether archaeological or monumental. The challenge is how to draw individual elements together into a series of patterns that present a coherent substantive material landscape entity surrounding us and that contain evidence—both tangible and intangible—of people who have previously interacted with this constantly changing environment.

---

[6] For example, Scottish Natural Heritage (2005) *Statement: SNH's Landscape Policy Framework* Policy Statement 05/01, 6.

[7] Historic Scotland (2012) Scotland's Historic Environment Audit 2012 Historic Scotland.

[8] Swanwick (2002).

[9] See Footnote 6.

Fig. 10.2 One rarely sees an image of Loch Alsh looking west without Eilean Donan Castle although generally without reference to the people who built it or who depended on it for their livelihood. In fact, there is still a thriving settlement nearby today

While predecessors of HES[10] were supportive of the principles of the LCA, they were concerned about the incomplete capture of the historic dimension, which overlooked "time-depth within the present-day landscape".[11] Scotland's Historic Land Use Assessment has recently been completed to capture the essence of this time-depth.[12]

10.5 The Different Discourses of Landscape

Although there are no clearly defined boundaries between the historic environment and landscape discourse, broadly speaking, the former conveys a worldview rooted in a temporal or historic dimension. In contrast, discourse rooted in issues related to countryside was found to be largely 'time-flat', where the focus has been on the challenges faced in the present time and how these can be met with direct remedial action. Discourse focused on natural heritage frequently links the concepts of landscape and ecosystems with reference to spatial scale, and rarely considers the idea of time-depth beyond circadian rhythms or geological epochs.

[10] Historic Environment Scotland—resulting from the merger of Historic Scotland and the Royal Commission for the Ancient and Historical Monuments of Scotland in.

[11] Millican et al. (2017).

[12] https://hlamap.org.uk.

Implicit throughout the discourse is a concern with human impact, particularly tied to the 'post-war period', generally in terms of loss, degradation or damage.[13] While there is no doubt that the second half of the twentieth century witnessed significant land use change, I found little recognition of earlier transformations which were equally dramatic; for example, the agricultural improvements of the eighteenth century or the industrial production of charcoal in the western highlands, the management of which is highly valued as ancient woodland today.

However, there is evidence that the ELC debate is bearing fruit, and in 2012 SNH published a guidance pack for local communities—*Talking about our Place* (SNH 2012). Taking a place-based approach, it argues that "landscape is more than just scenery; it is the interaction between people and place; the bedrock upon which our society is built".[14]

10.6 Two Scottish Landscapes

The town of Falkirk, incorporated as a Royal Burgh in 1600 is not formally recognised in the discourse as a coherent landscape area. Yet it is steeped in a rich natural and cultural heritage valued by the local community. It might be argued that the reason Falkirk is not considered a landscape is because it is not scenic in the traditional sense (Fig. 10.3). However, the Roman legions made a very deliberate choice to route the mass of the Antonine Wall (designated as part of the Frontiers of the Roman Empire World Heritage Site) through here and there is evidence of continuous settlement ever since. Today it is very clearly an area that can be perceived by people and clearly demonstrates complex interaction between human and natural factors over time. There are innumerable ways that the different dimensions of nature, culture, society, economy and ecosystems have interacted continuously through time.

On the other hand, Loch Leven (Fig. 10.4) has been designated as a special landscape area by the local authority (Perth and Kinross). It can be described as an area of natural beauty, and yet this land has been altered as much as that of Falkirk above. The level of the loch was lowered in the early nineteenth century for industrial power, reclaiming valuable agricultural land that continues to be farmed today. The surrounding fields contain buried evidence of prehistoric communities, and the hills in the distance are littered with settlement and ritual remains.

[13] For example, Scottish Government (2014, 47) National Planning Framework https://www.gov.scot/publications/national-planning-framework-3/.

[14] Scottish Natural Heritage (2012, C1).

Fig. 10.3 Characterised as 'Lowland River Valley', this area has no landscape designation. However, the woodland strip marks the line of the Antonine Wall (WHS), while the canals and railway connected 18th C Falkirk to markets/ports to the east and west, Crown Copyright HES

10.7 Conclusion

Scotland boasts spectacular landscape that reflects a complex story of continuity and change through the interaction of people with their environment through time. This process is dynamic and will evolve as natural and cultural forces continuously interact in the future. By not considering the temporal and cultural elements of landscape, a sense of continuity and change is lost, and while "the historic elements of the landscape are finite; they constitute at the same time a dynamic resource which grows and alters with time".[15]

[15] Macinnes (2002, 24).

Fig. 10.4 Loch Leven, Kinross—Designated as a special landscape area by the local authority, there is evidence of human interaction with the landscape over millennia. The only national designation however is for natural heritage, Crown Copyright HES

This continuous process from past through present to an unspecified future represents a dimension of landscape that enriches the understanding of the idea of *landscape* beyond that of a subject perceiving or experiencing an object.

With considerable commitment to the principles of the ELC, landscape discourse in Scotland has taken significant steps towards a holistic approach to a rich multi-dimensional totality. The recent global recession has had a considerable impact on already constrained resources delaying opportunities for innovation. This twentieth anniversary will hopefully spark renewed debate across a wide range of communities and I believe the ELC, its Glossary and Explanatory Statement[16] provide a firm foundation on which to build.

References

Cairney P (2016) The Politics of Evidence Based Policy Making. Palgrave MacMillan, UK

Council of Europe (2000) European Landscape Convention. http://convetions.coe.int/Treaty/en/Treaties/Html/176.htm

[16] Council of Europe (2000b).

Council of Europe (2000b) Explanatory report to the European landscape convention. https://rm.coe.int/16800cce47

Council of Europe (2013) European landscape convention glossary, 13

Govan S (2019) the past in the present: history, policy and the Scottish landscape. PhD Thesis http://hdl.handle.net/1842/35812

Macinnes L (2002, 24) The historic landscape in Scotland: towards a strategy for the future. In: Smout TC (Ed) Understanding the historical landscape in its environmental setting. Scottish cultural press, Dalkeith

Millican K, Dixon P, Macinnes L, Middleton M (2017) Mapping the historic landscape: historic land-use assessment in Scotland. Landscapes 18(1):71–87

Scottish Natural Heritage (2012, C1) Talking about our place. http://www.snh.gov.uk/protecting-scotlands-nature/looking-after-landscapes/communities/talking-about-our-place/

Swanwick C (2002) Topic Paper 1: recent practise and the evolution of landscape character assessment. Landscape Character Assessment: Guidance for England and Scotland, Cheltenham and Edinburgh: The Countryside Agency and Scottish Natural Heritage 8

Chapter 11
One Territory, One Plan, Many Authorities: The Future of Really Protecting the Landscape

Luca Di Giovanni

Abstract This paper deals with the theme of landscape plan and of his function of protection and value the territory. Because of this structure, landscape plan permits to safeguard the territory with a unified vision, which considers landscape and territory as two equivalent concepts, in according to the principles of European Landscape Convention. Under this profile, the tool of landscape plan seems to perfectly apply the European principle of integration of the landscape into the other territorial and sectorial politics. Although this good thing, in Italy exist many critical problems about a full application of the European Landscape Convention, for example, the question of participation and involvement of all public and private bodies into the administrative procedure for realising the landscape plan.

Keywords Landscape plan · Territory · Safeguard · Participation · Public authorities

11.1 Introduction

Landscape is protected and enhanced by a series of national and supranational regulatory documents which aim to preserve its primary and absolute value. At the European level, the main regulatory provisions are contained in the European Landscape Convention which was adopted by the European Council on 19 July 2000 and entered into force in Italy on 1 September 2006, with ratification law no. 14/2006. Nationally in Italy, reference must be made to the Cultural and Landscape Heritage Code which was drawn up after the European Convention and came into force with Italian Legislative Decree no. 42/2004.

The contents of the Convention and the Code bear certain similarities (considering the positive influence exercised by the former document on the latter); however,

L. Di Giovanni (✉)
University of Urbino, Urbino, Italy
e-mail: luca.digiovanni@uniurb.it

T. Matteini et al. (eds.), *Cultivating Continuity of the European Landscape*, Environmental History 15, https://doi.org/10.1007/978-3-031-25713-1_11

they are also divergent as is palpable in the contrasting notions of landscape which continue to be the subject of controversial case law and doctrinal debate.

Both regulatory documents recognise the need to use one or more instruments for landscape management and the coordination of territorial policies. In this respect, the Cultural and Landscape Heritage Code presents landscape plan as a suitable instrument for effective landscape protection.

This plan may be used not only to protect landscape but also to manage territory as a whole. The Italian National Code provides for two types of landscape plans: a "pure" landscape plan and an urban-territorial plan, with specific consideration of landscape values (art. 135). In addition to protecting the landscape, the latter type of plan presents certain territorial management characteristics, including aspects of urban planning and construction development with reference to specific territorial areas.

In consideration of the foregoing, this paper aims to examine the planning situation in Italy, identify any existing issues and suggest an instrument capable of resolving them, at least in part. This focus on scientific research is also accompanied by an awareness of how territories require suitable and effective protection which can only be achieved by simplifying the legislative framework in force and the planning tools provided for by law.

For some time now, legal experts have been bemoaning a legal framework characterised by regulatory glut and a patchy system of overlapping jurisdictions, resulting in a lack of linearity and unclear contents, which effectively undermines the complementary nature of the roles of public administrations involved in territorial protection. A solution has already been proposed: the simplification of planning instruments and their reduction into units in accordance with the golden rule "one territory, one plan, one authority".[1]

This paper aims to tweak this rule based on the belief that landscape plan, as configured in regulatory provisions like an urban-territorial plan, is the only instrument capable of responding to emergent territorial requirements. It is also argued that the constructive contribution of individual public administration authorities should not be shunned in favour of a single authority's role. In particular, the multitude of public and territorial bodies is an advantage and an added value which should be preserved and reinforced in accordance with regulations set forth by our Constitution.[2]

Recognition of landscape plan as the sole planning instrument capable of resolving or reducing existing critical issues stems from a consideration based on two factors. On the one hand, the subject of landscape plan (landscape) has been extended to include elements of an environmental nature alongside those of a territorial character in more general terms due to the influence of the European Landscape Convention on

---

[1] The most recent work in this sense is by Stella Richter (2018), 220. Indeed, the author underlines that "unification" is a problem whose resolution is "far from simple"; however, the "objectives of certainty and clarity must go hand in hand with the *reductio ad unitatem* of the multitude of plans which exist today". It is important to note that there are already inter-municipal and municipal structural plans, operational plans and planning regulations at a local level.

[2] Suffice to consider the theory of subjective administrative pluralism set forth in art. 5 of the Constitution. For more on this topic, see Esposito (1954), 74.

legislation. On the other hand, the Cultural and Landscape Heritage Code structural configuration of landscape plan is based on planning and prescriptive regulatory effects which prevail over all other urban and industrial planning instruments and also function as a means for coordinating economic planning.

11.2 Landscape and Territory: Two Equivalent Concepts

In order to affirm that territory can be managed by means of landscape plan alone, it is important that this planning instrument regulates something whose material extension coincides with the actual asset.

Pursuant to art. 9 of the Italian Constitution and art. 131 of the Cultural and Landscape Heritage Code, landscape plan focus on landscape of cultural-identity value. This means that the Italian legal system recognises protection for landscape exclusively with regard to "those aspects and characters which constitute a material and visible representation of national identity, as expressions of cultural values".[3]

In contrast, the European Landscape Convention states that safeguarding and protection must be extended to all types of landscapes, including those of no particular value or even those affected by degradation or abandonment processes.[4] Indeed, art. 2 of the European document grants protection to "all territories of the Parties and regards natural, rural, urban and suburban areas. It includes landscapes, internal and sea waters. It refers to landscapes which can be considered exceptional, landscapes of everyday life and even degraded landscapes". Therefore, landscapes subject to the discipline of European powers include those that are exceptional as well as those that are ordinary or even devastated.

Pursuant to art. 132 of the Landscape Heritage Code, supranational legislation prevails over national legislation.[5] Therefore in general terms, the contents of the European Landscape Convention should always prevail over those of the Cultural and Landscape Heritage Code.

Furthermore, the constitutional Court recognises that art. 9 of the Constitution regarding landscape heritage refers to the entire country and is therefore based on an integral landscape concept.[6] The Court found that the concept of landscape can identify the morphology of a territory, namely the visual aspect of the environment,

---

[3] Art. 131, paragraph 2 of the Cultural and Landscape Heritage Code. For a commentary on this provision, see Immordino and Cavallaro (2012), 983.

[4] For more on this topic, see Sorace (2007), 17.

[5] In paragraph 1, said article establishes that "the Republic complies with the obligations and principles of cooperation between States as set forth by international landscape conservation and enhancement conventions". The European Landscape Convention can be defined as an international document, adopted by the European Council on 19 July 2000, officially signed in Florence on 20 October in the same year and in force in Italy as of 1 September 2006, ratified with law no. 14/2006.

[6] See Italian Constitutional Court, 7 November 2007, no. 367 and Italian Constitutional Court, 23 November 2011, no. 309, in www.cortecostituzionale.it.

and that the constitutional value defined in art. 9 of the Constitution refers to the territory, acknowledged with reference to its territorial and cultural contents.

In short, according to the constitutional Court, landscape and territory are equivalent concepts,[7] transcending the abstract concept of "natural beauty" to embrace all things material, compositions thereof and interrelations with human beings.

In legal theory, the contrast between restrictive and more extensive orientations of the term "landscape" has also been transcended and has come to be of historic interest only.[8] Indeed, it is now recognised that with its indistinct reference to landscape, art. 9 of the Constitution is grounded on an integral concept of landscape. Nevertheless, in this integral concept of landscape is not included the environment (ecological elements of the territory): in fact, art. 9 of the Constitution has recently been reformed by constitutional law no. 1/2022, which has expressly introduced in this rule the concept of environment, separating it from the concept of landscape. Therefore now, the meaning of landscape coincides with the territory, without its ecological elements.

11.3 The Nature and Structure of Landscape Plan

If the subject of landscape plan can also be all the territory, it is important to verify the eligibility and suitability of this planning instrument in territorial management.

Pursuant to art. 135 of the Cultural and Landscape Heritage Code in its urban-territorial form and with specific consideration of landscape values, landscape plan serve four fundamental administrative functions in a logical sequence: "systemic knowledge of the territory, safeguarding, planning and management of landscape".[9]

The purpose of landscape plan is not only to safeguard the landscape characteristics of a certain territorial area, but also consists of "redeveloping compromised or degraded areas and identifying urban planning and construction development lines, based on compatibility with different landscape values that are recognised and protected".[10]

Furthermore, pursuant to art. 145 of the Code, this plan are superordinate with respect to other forms of planning (urban and industrial) and "set forth measures for coordination with territorial and industrial planning instruments as well as with national and regional plans, programmes and projects for economic development".[11]

In short, it can be affirmed that landscape plan, as outlined in the Cultural and Landscape Heritage Code, constitute "a general plan, vast in area, based on urban planning technique and regarding all possible ground use".[12] In particular, it is a

[7] See Italian Constitutional Court n. 641/1987, in www.giurcost.org.

[8] The restrictive orientation is represented by Sandulli (1967), 69. The extensive orientation is represented by Predieri (1969), 3.

[9] As stated in Amorosino (2010), 28.

[10] Art. 135, paragraph 4, letters b) and d) of the Cultural and Landscape Heritage Code.

[11] For more on this regulation, see Cartei and Ungari (2013a), 748.

[12] See Cartei and Ungari (2013b), 734.

territorial planning instrument in spatial terms and of economic significance as it affects the economic-social development of a specific landscape area; its provisions are for multiple recipients.

11.4 Multiple Authorities for Territorial Protection and Enhancement

The examination of the target and the relative instrument for its management should be followed by an analysis of the public entities required to apply landscape plans for the purpose of safeguarding and enhancing territory. The fact that there is more than one public entity raises two significant considerations: firstly, our Constitution values subjective administrative pluralism and promotes the decentralisation of public functions deployed on national territory[13]; secondly, the number and complexity of issues covered by any single plan require management by several entities with different competencies and specialisations.

Provisions set forth in the European Convention are also oriented towards favouring the participatory principle which, in accordance with the rules of the Aarhus Convention, are grounded on an attention towards the involvement of private entities as opposed to only public ones.[14]

The importance of the knowledge factor (deriving from a public and not a private entity) should not be underestimated as it is a necessary requirement in the pursuit of all human activities. Indeed knowledge means an awareness of past events, the collective understanding of which generates experience. In general, experience enables the improvement of existence by preventing the repetition of previous mistakes. In particular, knowledge of landscape develops knowledge of the contents of one of the fundamental components of Italian cultural and natural heritage, giving rise to human and ecological interrelations which characterise a specific territory.

[13] The regulatory framework consists of articles 5, 114, 117, 118, 119 and 120 of the Italian Constitution. In legal theory, for further information on subjective administrative pluralism in the Italian Constitution, in addition to the contribution by see Esposito (1954), see also De Marco (2015) and Amato and De Luca (2008), 907.

[14] As set forth in art. 5, lett. c) of the aforementioned convention and Part. I—General Principles, I.1, lett. G, of the Guidelines for the implementation of the European Landscape Convention (Recommendation CM/Rec (2008)3), which states that "all action taken to define, implement and monitor landscape policies should be preceded and accompanied by procedures for participation by members of the public and other relevant stakeholders, with the aim of enabling them to play an active role in formulating, implementing and monitoring landscape quality objectives".

Private entities may contribute knowledge of a technical or non-technical nature,[15] especially since the idea of true, objective and absolute knowledge of phenomena has come under discussion.[16]

11.5 Conclusion

This paper attempts to demonstrate the increasing importance of landscape plan in territorial management in the light of legislative and regulatory provisions.

In particular, supranational rules set forth in the European Convention, to be complied with pursuant to art. 117, paragraph 1 of the Italian Constitution, convince the jurist to formulate positive reflections regarding the use of this planning instrument as the sole means for territorial government.

The usefulness of landscape plan is ample and multifaceted insofar as they enable implementing public entities (the state and regions above all) to suitably tackle numerous emergent territorial issues and guarantee a high level of quality of life for local communities.

The only critical issue to emerge from legal analyses regards the contents of art. 144 of the Cultural and Landscape Heritage Code which establishes that the participation of local bodies is optional and subject to the regional legislator's explicit intervention.[17]

Therefore, in order to harmonise national legislation with provisions set forth in the European Landscape Convention, the Cultural and Landscape Heritage Code should be reformed to ensure the compulsory and effective involvement of autonomous local and civil society entities in the preliminary (and not only implementary) phases of the landscape plan creation process.

---

[15] The cognitive usefulness of the technical knowledge of a private entity belonging to the local community is highlighted by the European Landscape Convention (art. 6) and its Guidelines. The latter in particular recognises the importance of morphological, economic, archaeological, historic, cultural and natural knowledge of the landscape and extensively promotes the spread and accessibility of such knowledge so that it is available to all citizens (Recommendations CM/Rec (2008)3, Part. I- I.1, B e C). Non-technical knowledge means subjective information of a perceptive nature deriving from common people without specialised knowledge, which nevertheless may broaden the competent public administration's sphere of knowledge in safeguarding landscape heritage. On the significance of non-technical knowledge in landscape planning see Di Giovanni (2021), 189.

[16] The impossibility of achieving entirely truthful knowledge was highlighted by Wittegenstein (1971). In the field of science law, please refer to Orsi Battaglini (1991), 441–443.

[17] Pursuant to art. 144 paragraph 1, "in procedures for the approval of landscape planning, institutional cooperation, the participation of interested entities and stakeholder associations are ensured, identified pursuant to environmental and environmental damage provisions in force, along with ample forms of publicity. To this effect, by means of special legislation, regions discipline urban landscape planning procedures, also with reference to additional forms of participation, information and communication".

References

Amato C, De Luca G (2008) *I principi di sussidiarietà e leale collaborazione nell'esercizio del potere sostitutivo dello Stato.* Nuova rassegna di legislazione, dottrina e giurisprudenza 8

Amorosino S (2010) *Tutela e valorizzazione del paesaggio nella pianificazione regionale.* Istituzioni del federalismo 1

Cartei GF, Ungari P (2013a) *Rapporto dei piani paesaggistici con gli altri piani settoriali e con i piani urbanistici* (art. 145, Italian Legislative Decree 22.1.2004, no. 42). In: Battini S, Casini L, Vesperini G, Vitale G, a cura di. *Codice di edilizia e urbanistica.* UTET, Turin

Cartei GF, Ungari P (2013b) *Elaborazione, contenuti ed effetti del piano paesaggistico* (art. 143, Legislative Decree. 22.1.2004, no. 42). In: Battini S, Casini L, Vesperini G, Vitale C, a cura di. *Codice di edilizia e urbanistica.* UTET, Turin

De Marco E (2015) *Il regime costituzionale delle autonomie locali tra processi di trasformazione e prospettive di riforma.* Rivista AIC 2

Di Giovanni L (2021) *La pianificazione paesaggistica e la gestione integrale del territorio.* Editoriale Scientifica, Naple

Esposito C (1954) *Autonomie locali e decentramento amministrativo nell'art. 5 della Costituzione.* In: Esposito C, a cura di. *La Costituzione italiana: Saggi.* Cedam, Padua

Immordino M, Cavallaro MC (2012) *Art. 131.* In: Sandulli MA, a cura di. *Codice dei beni culturali e del paesaggio.* Giuffrè, Milan

Orsi Battaglini A (1991) *Libertà, legalità, mercato. Profili comparatistici del diritto della scienza.* In: AA.VV. *Scritti per Mario Nigro.* Giuffrè, Milan

Predieri A (1969) *Significato della norma costituzionale sulla tutela del paesaggio.* In: Predieri A, a cura di. *Urbanistica, tutela del paesaggio, espropriazione.* Giuffrè, Milan

Sandulli AM (1967) *La tutela del paesaggio nella Costituzione.* Rivista giuridica dell'edilizia 2

Sorace D (2007) *Paesaggio e paesaggi della Convenzione europea.* In: Cartei GF, a cura di. *Convenzione europea del paesaggio e governo del territorio.* Il Mulino, Bologna

Stella Richter P (2018) *I principi del diritto urbanistico.* Giuffrè, Milan

Wittegenstein L (1971) *Osservazioni sopra i fondamenti della matematica,* It. Translation. Turin

Luca Di Giovanni is Lecturer at University of Urbino. He has obtained a PhD in public, urban and environmental law at University of Florence. His research deals with landscape planning and cultural heritage, urban and environmental themes and public contracts.

Chapter 12
The Daily Governance of Landscape: The Contribution of Local Landscape Commissions

Raffaella Laviscio

Abstract The need of a widespread landscape quality has led, in Italy as abroad, to the establishment of Local Landscape Commissions (LLC) as a tool for a preliminary assessment of territorial transformations in relation to their landscape compatibility. These bodies are composed of subjects with long-term and qualified experience in the protection of the landscape and oversee the evaluation of the transformations subject to landscape authorization. In Italy, their widespread distribution throughout the territory makes them a potential effective tool for territorial transformation monitoring; their application, in many Italian regions, not only to listed areas, but to the whole territory, makes them an operational tool for the management and governance of the entire landscape, including degraded areas, as intended by the ELC. In Switzerland LLC are strictly connected to the Department of the Territory and contribute to the development of landscape local planning tools. In France the Conseils d'Architecture, d'Urbanisme et de l'Environment (CAUE) have a great role in landscape training and dissemination activities. In Netherlands the Architectuur Lokaal Foundation acts to create a link between landscape national policies and local practices. What can we learn from these experiences in a European perspective? The paper aims to reflect on how LLC could contribute to ELC implementation.

Keywords Landscape commissions · Landscape monitoring · Landscape compatibility · Landscape governance

---

R. Laviscio (✉)
Department of Architecture Built Environment and Construction Engineering, Politecnico di Milano, Via Bonardi 9, 20100 Milano, Italy
e-mail: raffaella.laviscio@polimi.it

T. Matteini et al. (eds.), *Cultivating Continuity of the European Landscape*, Environmental History 15, https://doi.org/10.1007/978-3-031-25713-1_12

12.1 Introduction: The Need for Widespread Landscape Quality

The European Landscape Convention (ELC, Florence 2000) states that to guarantee both outstanding and daily or degraded landscape quality (art. 2) "The effects of projects, whatever their scale, on landscape should therefore be evaluated and rules and instruments corresponding to those effects defined. Each planning action or project should not only match, but also be appropriate to the features of the places" (Council of Europe 2008).

An appropriate landscape project is, in fact, the only opportunity to highlight the specificities of places to be transformed whether through a conservative approach by adapting as much as possible to the contextual features or through an innovative approach by carefully reinterpreting contextual relations and character and proposing new significant references (Laviscio 2018a).

In Italy the entry into force of the Heritage and Landscape Code (Law no. 42/2004) after the signing of the ELC has resulted in the nation-wide introduction of "Local Landscape Commissions" (art. 148) which support local authorities in evaluating the effects of the projects on landscape in protected areas.

The Italian experience is only one of many similar experiences in other European countries. This paper proposes a comparison between several countries and the Italian case study.

Understanding the regulation, structure and functioning of Landscape Commissions could be useful to understand their role in ELC implementation.

Unfortunately, monitoring landscape practices at a project scale, despite the regional and national establishment of landscape observatories, is often still as rare in Italy as it is abroad (Peano 2011).

12.2 Local Landscape Commissions: Functions, Weaknesses and Strengths

So, what are Local Landscape Commissions? What are their functions? What skills are required? What tools are adopted? How effective is their work?

After the collection of various experiences (through interviews and web research) some qualitative considerations are possible.

In Italy in most cases, LLC have been established since 2008 under regional and municipal regulations. The Commissions are made up of technicians external to the Body (generally from 3 to 5 members), who have consolidated qualifications and experience in the field of landscape-environmental protection. Most experts belong to the technical disciplines (architecture, engineering, technical institutes) but to a lesser extent, they are also agronomists, foresters or naturalists.

Although the Code establishes that the scope of the evaluation is to be designated-area transformations, in some regions the scope is more extensive and affects the entire territory (as in Lombardy, Emilia Romagna, Lazio) (Fig. 12.1). In any case, they manage territory defined by institutional boundaries (in most cases, a municipality). This allows the Commissions to be widely disseminated but also involves the risk that different judgements will be expressed for neighbouring territories with homogeneous landscape characteristics. This expression of highly subjective judgements derives from the lack of adequate tools for reading the landscape at a local scale which results from the poor link to date between urban and landscape tools and a lack of landscape content that is still typical at the local scale.

Very often, judgements are limited to some minor aspects of the project (finishes, colours), while the choices with the greatest impact (such as the morphological layout, location or volumes) are imposed by urban plans without any consideration of landscape. LLC do not contribute to the definition of the landscape contents of local planning tools. LLC activities are not monitored and not produce effective monitoring of landscape changes (Laviscio 2018b).

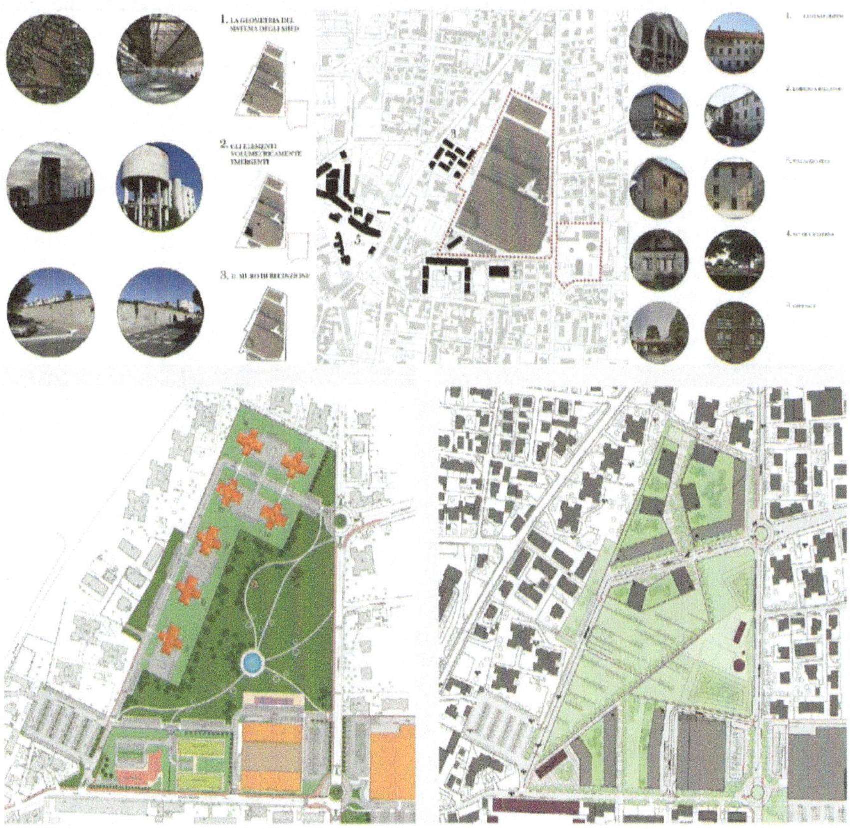

Fig. 12.1 Urban landscape transformation project in Italy: the study of the area, the initial project and the project after consulting the Landscape Commission (Laviscio 2018b)

In the Ticino Canton (Switzerland), the Landscape Commission is an advisory body established under cantonal law on territorial development which provides for "the orderly and harmonious insertion of new projects" (Art. 94 Lst. 2011). It is a multidisciplinary commission whose members (from seven to nine) are appointed by the State Council with the function of evaluating territorial plans, landscape projects and territorial projects of great significance. The consultancy therefore addresses both private and public subjects.

Its work is closely linked to that of the Department of the Territory; together they establish strategic guidelines for the development of landscape enhancement projects, participating in the drafting of guidelines and directives.

The definition of guidelines is one of the important tasks of the Commission as the guidelines define landscape reading criteria with which designers must comply and evaluation criteria adopted by the Commission (Fig. 12.2); this makes the expression of compatibility rulings clearer and more transparent (Gianoni 2018).

The guidelines specify the reading scales necessary for a landscape project as well as the knowledge already available. LLC organization responds to the need for interdisciplinary project evaluation based on a landscape culture of objective elements and not only of sensations or purely normative considerations that, in most cases, do not respond to the intrinsic complexity of landscape.

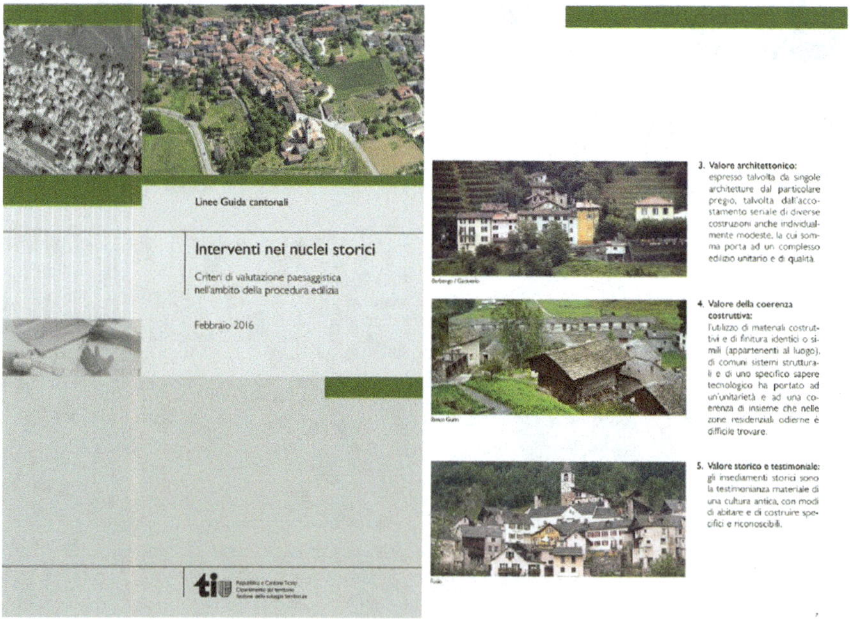

Fig. 12.2 Guidelines of the landscape commission in Switzerland: an example of indications related to historical centres (https://www4.ti.ch/fileadmin/DT/direttive/DT_DSTM_SST/Interventi_nei_nuclei_storici_022016.pdf)

The role of awareness is even more important in the experience of the Conseils d'Architecture, d'Urbanisme et de l'Environment (CAUE) in France; these are consultancy associations active since 1977 in each French department. In 2018, there were 93 CAUEs in 93 departments, 10 regional CAUE groupings and a national federation. The CAUE are aimed at both public and private subjects and operate at various scales to promote landscape quality. Their consulting activity is accompanied by robust training and dissemination work: public meetings, conferences, publications, databases, tools intended for a wide audience made up of professionals from different sectors (architects, agronomists, gardeners), but also teachers and citizens in general. Thus, they have a strong cultural role, focusing on training at all levels.

CAUE members are not tied to political changes; they operate at different scales, from research to professional practice; they transfer the results of research to the public and, vice versa, bring requests from the public to the attention of the State.

In 2018, the CAUE provided 67,168 consults to individuals as well as associations, companies, artisans, promoters and social landlords. They carried out 10,705 support actions with local authorities and their groups; 6,650 trainees were trained in architecture, town planning and landscape; and 62,805 school children benefited from educational activities (CAUE 2019).

Best practices are collected in a database published on the web (Figs. 12.3 and 12.4).

Their long-standing experience certainly represents a model for younger organizations born under the impetus of the ELC.

Similarly, the Architectuur Lokaal Foundation created in 1993 in the Netherlands is an independent centre of expertise and information devoted to commissioning building development in the Netherlands. This lightweight structure (10 people) is subsidized by four Ministries concerned with architecture (Culture, Town Planning, Environment and Transport) and has contact with both public and private clients which include local authorities as well as real estate developers and private individuals involved in building operations.

The Architectuur Lokaal's mission is to act as a link between national policies and local practices to help local agents apply national policies as well as incorporate local practices and experience into national decisions. It works towards building culture; therefore, it does not limit its work to professional consultancy, but addresses the whole of society by carrying out meetings, publications, field surveys and participatory events where projects are presented to citizens (Fig. 12.5).

12.3 What We Can Learn in a European Perspective

A comparison of these different experiences leads to some reflections.

Local Landscape Commissions have great potential to monitor landscape transformations at a local scale but only if the following points are carefully considered:

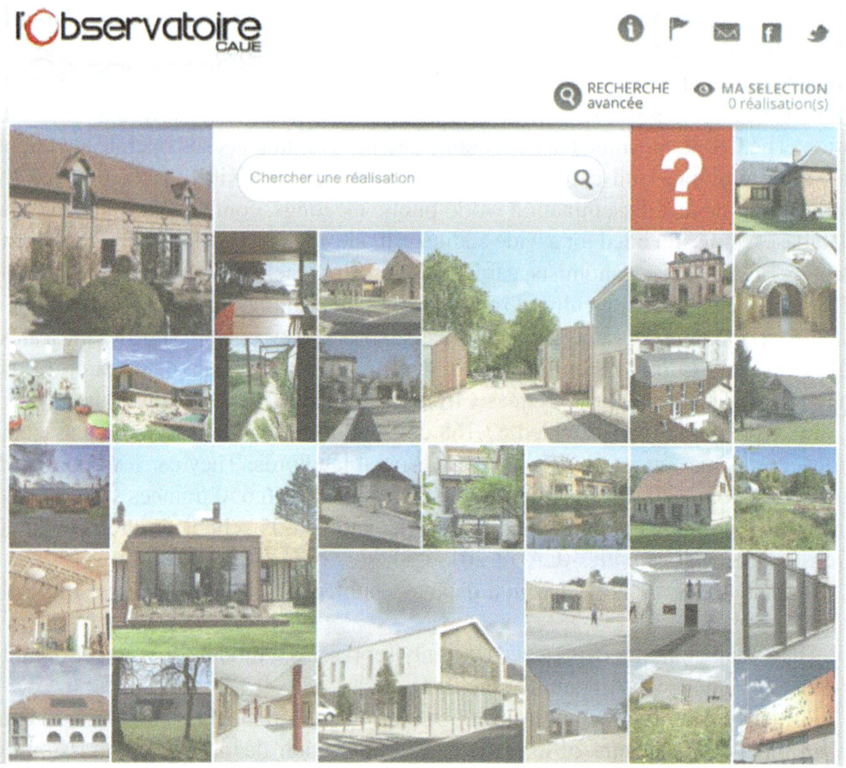

Fig. 12.3 CAUE observatory: the web database of best practices developed with CAUE consultation (http://www.caue-observatoire.fr/)

1. First of all, the link between landscape evaluation procedures and territorial and landscape planning tools: a landscape project even of limited extension is affected by its relationships to the wider context and by planning choices made at both the local scale and higher levels. In Italy, the effectiveness of Commission work has certainly been affected by the disconnection between landscape and territorial policies; the French and the Swiss experiences have been more positive in this sense, where Commissions actively participate in the definition of planning tools.
2. The independence of these bodies. If in Switzerland, France and the Netherlands, Local Landscape Commissions are a link between society and state, in Italy they are often conceived as an extension of local institutions, simply designed to support institutional decisions with a qualified point of view and thus fail to really contribute to the advancement of the discipline. The Commissions' work appears to be more effective and recognized if it is autonomous with respect to the institutions. Its role should not be to fulfil a simple procedural mechanism (within an institutional apparatus) but to mediate between practice and legislation, between society and institutions.

1 | 2 | 3 | 4 | 5 | 6 | 7 | 8 |

TRANSFORMATION DE LA GARE EN GÎTES ET ESPACES D'EXPOSITIONS - LEUGLAY

21290 Leuglay

HALLE | GARE | GITE RURAL | EXPOSITION | RESTAURATION | RECONVERSION DE BATIMENT | RECONVERSION DE SITE | REHABILITATION | ART CONTEMPORAIN | ARCHITECTURE CONTEMPORAINE

Transformation de la gare : nature du projet de

La volonté de l'équipe municipale, menée par François Maire du Poset (avec Françoise Spillmann et Françoise Naudet) pour **redonner vie au patrimoine paysager du site désaffecté de la gare**, en bordure du village de Leuglay (1), conjuguée à la rencontre avec Xavier Douroux, médiateur pour la **Fondation de France** dans le cadre du programme des **Nouveaux Commanditaires**, sont à la base de cette commande. L'idée du projet était, à partir de ce patrimoine, de constituer un **nouvel ensemble avec l'ancienne gare transformée en espaces collectifs** (cuisine, salle communes et appartement du gardien), la **grande halle aux machines devenant un lieu d'expositions** (2) et de **nouvelles constructions remplissant la fonction de gîtes**.

Organisation du projet

La restauration des constructions existantes (gare et halle technique) procède d'une remise en état ou d'une rénovation limitée afin de garder les ambiances d'origine. Les constructions nouvelles ont été, quant à elles, disposées entre les deux constructions existantes suivant des rapports de tensions et de distances fonctionnelles entre les volumes. Le sol minéral, constitué de ballast aux pourtours des constructions et d'un stabilisé uniforme pour les circulations, renforce la

PROGRAMME

Salle commune, gîtes individuels, halle d'expositions

CONCEPTEUR(S)

Architectures Adelfo Scaranello

MAÎTRE(S) D'OUVRAGE(S)

Commune de Leuglay

TYPES DE RÉALISATION

Equipement touristique et de loisirs

ANNÉE DE RÉALISATION

2014

SURFACE(S)

407 m² SU (réhabilitation : 87 m² - création : 320 m²)

COÛTS

Fig. 12.4 CAUE observatory: an example of a best practice inventory accessible to the public (http://www.caue-observatoire.fr/)

Fig. 12.5 The Architectuur Lokaal Foundation: a participatory event for the development of a new residential neighbourhood set up with the support of the Foundation (https://arch-lokaal.nl/wp-content/uploads/2020/03/Architectuur-Lokaal-96.pdf)

3. Adequate training of the actors involved. The multidisciplinary composition of a Commission guarantees interdisciplinarity in the expression of rulings, where the consideration of a multiplicity of aspects and scales is useful to formulate a landscape compatibility judgement that is as exhaustive as possible. This becomes especially essential in situations such as in Italy where there is inadequate training regarding landscape as evidenced by the poor quality of the projects submitted to the Commissions' judgement and the lack of a common and shared language between project designers and Commission experts. Involvement from the early stages of the project is essential for an awareness of all the factors involved.
4. The adoption of complementary tools such as guidelines, surveys and debates. These appear to be essential to increase the landscape awareness of the technicians involved in the process. A landscape project cannot arise from simple procedural mechanisms; it is not just a question of adequately fulfilling regulatory requirements, but of "transmitting a concept".
5. Raising landscape awareness outside and beyond project evaluations. Technical activity appears to be effective in associating activities aimed at a broad audience (debates, meetings, popular publications, courses for all age groups). This contributes not only to spreading landscape culture, but also to guaranteeing greater recognition of the role of the landscape architect.
6. Monitoring Commission activity. Commission activity is not always monitored. In Italy there is no monitoring (neither qualitative nor quantitative) of Commission work and, therefore, of its effectiveness. Much remains to be done in this

direction. The French example is useful, with its preparation of good practice databases (activities carried out, consultancies provided and consequently, projects carried out) and annual reports on the work of the committees.

The presence of Landscape Commissions throughout a territory has great potential to guarantee the widespread quality of extraordinary and daily landscapes. To date, there are few examples to be seen.

As usual, networking and knowledge can help create a basis for a common language in the spirit of the ELC.

References

Conseils d'Architecture, d'Urbanisme et de l'Environnement (2019) Reperes & Chiffres—Cles. Panorama de l'action des CAUE. https://www.fncaue.com/caue-reperes-chiffres-cles/

Council of Europe (2000) European Landscape convention. https://www.coe.int/en/web/conventions/full-list/-/conventions/treaty/176

Council of Europe (2008) Guidelines for the implementation of the European Landscape Convention. https://www.coe.int/en/web/landscape/guidelines-for-the-implementation-on-the-european-landscape-convention

Gianoni F (2018) Tutela proattiva codificata fuori d'Italia: la "Commissione del Paesaggio" del Canton Ticino (Svizzera). In: Ministero dei Beni e delle Attività Culturali e del Turismo, Osservatorio Nazionale per la qualità del paesaggio (a cura di), Rapporto sullo stato delle politiche per il paesaggio. Mibact, Roma, p 138, ISBN: 978-88-943177-1-8

Laviscio R (2018a) Landscape compatibility of wind energy: suggestions from Italian and foreign guidelines. In: Battisti L, Ricci M (eds) Wind energy exploitation in urban environment. TurbWind 2017 Colloquium. Green energy and technology. Springer, pp 221–235, ISBN: 978-3-319-749433, ISSN: 1865–3529. https://doi.org/10.1007/978-3-319-74944-0

Laviscio R (2018b) Le Commissioni locali per il paesaggio. Un ruolo di supporto da monitorare. In: Ministero dei Beni e delle Attività Culturali e del Turismo, Osservatorio Nazionale per la qualità del paesaggio (a cura di), Rapporto sullo stato delle politiche per il paesaggio. Mibact, Roma, pp 115–126, ISBN: 978-88-943177-1-8

Peano A (ed) (2011) Fare paesaggio: dalla pianificazione di area vasta all'operatività locale. Alinea editrice

Chapter 13
Enclosing the Horizon: Land Privileges and the Right to Landscape

Marius Fiskevold

Abstract The European Landscape Convention (ELC) represents a 20-year-old initiative which seeks to deal with these changes. Yet there is a mismatch between experts' management tools and people's perception of land as landscapes. This is evident when people talk about their landscapes in public decision-making processes. Thus, there is urgent to strengthen the argumentative and narrative aspect of landscape planning, motivated by the ELC.

Keywords Transformation · Rights · Landscape narrative · Common land · Visual horizon

13.1 Landscape Identity and the Properties of Land

Transformations of land have always provoked the prevailing human imagination of landscapes. The European Landscape Convention (ELC) represents a 20-year-old initiative which seeks to deal with these changes. As the Explanatory Report of the Convention states, Europeans are "no longer prepared to tolerate the alteration of their surroundings by technical and economic developments in which they have had no say (Council of Europe 2000:4)." The public reaction to the completed 2019 *National Wind Resource Strategy* of the Norwegian Ministry of Petroleum and Energy (OED) verifies this claim in an exemplary manner. The numerous protests showed that the affected population did not accept turning large parts of Norwegian territory into energy production sites and objects of investment. Drawing on policy documents and reports as well as research papers, media articles and scientifically developed inventories, I will pay special attention to the habitually ignored case that the *conception* of landscape differs fundamentally in the utterances of experts from those of the affected population. Although the experts' tools are apparently in line with the ELC's specific measures on identification and assessment, I will argue that public utterances clearly show that the tools of government do not encompass the

M. Fiskevold (✉)
The Norwegian University of Life Sciences, Ås, Norway
e-mail: marius.fiskevold@nmbu.no

T. Matteini et al. (eds.), *Cultivating Continuity of the European Landscape*, Environmental History 15, https://doi.org/10.1007/978-3-031-25713-1_13

ELC's emphasis on identity arising from everyday use, participation and reflection on any given piece of land. The mindset repeatedly reflected in standardized tools has effectively invalidated esthetic and experience-based knowledge in decision-making processes. Consequently, ignoring experiences obtained within the visual horizon promotes a self-fulfilling prophecy. When knowledge of landscape is isolated from the esthetic and identity-building experience of land, the natural and constitutive edge of the landscape, i.e., its visual horizon, is changed into ranks of standardized turbines decisively decreasing the vitality of human imagination.

13.2 Turning Common Land into an Object of Investment

In February 2017, the Ministry of Petroleum and Energy (OED) commissioned the Norwegian Water Resources and Energy Directorate (NVE) to make a proposal for future onshore wind power sites on the entire Norwegian territory. Wind resources and power distribution capacity were to be aligned with environmental and social interests. In this strategic document, the landscape concept emerges as a slightly rewritten version of the ELC definition: it "… is interdisciplinary, and embraces the natural, man-made and the landscape as an arena for interaction between people, and between people and the environment (Lundsbakken 2019:49)." Then the impact of the ELC's landscape concept falls silent. The overarching object, *landscape,* is effectively deconstructed into single topics like *bats*, *wild reindeer* and *nature types.* In the next move, *landscape* is reconstructed as a non-calibrated mixture of components as different as a database-based GIS-system (NiN Landskap), cultural historical landscape of national Interest (KULA) and selected cultural landscapes in agriculture (UKL) (Riksantikvaren 2019:23–26). The double action of deconstruction and reconstruction effectively hides the mismatch between the tools used and their ability to describe the relationship between humans and nature as a foundation of their identity.

Associating landscape with land, land with a map and the map with layers of themes is one way of excluding landscape as a holistic object of identity. Another way, as the licensing procedure effectuates, is to channel people's opinions to a fixed place in a system created by those with a bureaucratic-scientific knowledge formation. A recent study concludes that the right to land is addressed to the mindset and knowledge formation of commercial project developers, local landowners, municipalities, the NVE sectoral licensing body and the OED appeals department. As the study concludes, "[t]hese actors have come to dominate the licensing process to such a degree that NGOs, regional authorities, local private individuals and other stakeholders are marginalized (Inderberg et al. 2019:198)." The study clearly shows that the complexity of the licensing procedure, the large number of actors and the financial strength and strategic procedural advantages of the initiators reduce the impact of the people affected when decisions concerning existing and future landscapes are made. Any attempt to articulate the mixed use, memories and values attached to a piece of land is effectively blocked by the rules of play of the leading players. Even though the *National Wind Resource Strategy* was abandoned and the Government report to Parliament proposed changes in the licensing procedure, both the licensing authorities and the knowledge regime remain unaltered (Energidepartementet 2020:32).

In many ways, the transformation of Norwegian commons into energy production sites and objects of investment parallels the eighteenth century English commons enclosures. In Raymond William's version (Williams 1973:60–61):

> An estate passed from being regarded as an inheritance, carrying such and such income, to being calculated as an opportunity for investment, carrying greatly increased returns. In this development, an ideology of improvement—of a transformed and regulated land—became significant and directive.

The momentum of a transformation is always accompanied by the ideology and rhetoric of some grand narrative. In this case, narratives like the EU's Renewable Energy Directive and Norwegian energy policies link energy production to renewable natural resources like water, wind and the sun and label it "green." Despite their honest intentions, theses initiatives' heroic will to defeat every single individual problem makes them produce narratives which have successfully hidden or "greenwashed" the economic and financial exploitation of natural resources. Even when land privileges are given to faceless, ground-independent decision makers, the narrative's "green" plot stays the same.

The tools, procedures and legitimacy of the *National Wind Resource Strategy* reflect a modern mindset. According to the German scholar Albrecht Koschorke, this way of perceiving land as landscape does not primarily refer to our shared world of organic experience, but to signs and semantic representation (Koschorke 1990:318). The hegemonic belief in representation that characterizes the digital era, foreseen by Benjamin's narrative into information (Benjamin 1991:444) and Lyotard's humanity into performability (Lyotard 1984:46), not only geographically but also discursively displaces landscape as a basis for identity. The earth's visual horizon is effectively replaced by the virtual horizon of digital software. Huge land surface alterations and the orderly removal of territorial and visual commons are effectuated so rationally because they eloquently avoid confronting the experience of the landscapes they indirectly define.

13.3 The Liminal Border Line of the Visual Horizon

A multifaceted approach to landscape, as advocated by the ELC, did not enter the *National Wind Resource Strategy* scene through bureaucratic-scientific tools but through the voice of the people. In their experience-based narratives, they demonstrated that the right to landscape should be conceived as the privilege both to use land and to define the way land is seen as landscape. In order for it to become a privilege for all, the mutual interdependency of use and conception requires a non-human reference. The reference should not just inform human imagination but also be a source of reflection on our own possibilities within the limitations of earth's natural capabilities.

Acknowledging that any land surface is inscribed with one or many narratives of human imagination, the last threshold of common land seems to be at the borderline between earth and sky. Like Raymond William's analysis of the commons in eighteenth century England, the function of the commons for many Norwegians still resides in the pre-human and value neutral, visual horizon. As one of the neighbors affected complained, "On the neighboring island, they can never lift their eyes to the mountains and sky without getting rotating wind turbines "straight in the face" (Haram 2020). The fencing of the free horizon means the dissolution of the shared and generation-independent sight of human earthbound life.

In William's words, the commons was a place where everyone was free to take a step to the side and distance themselves from social obligations and everyday concerns (Williams 1973:107):

> When the pressure of a system is great and is increasing, it matters to find a breathing-space, a fortunate distance, from the immediate and visible controls. What was drastically reduced, by enclosures, was just such a breathing-space, a marginal day-to-day independence, for many thousands of people.

The German scholar Albrecht Koschorke emphasizes that the visual horizon is a reference for thought and is fundamental to our ability to establish a true framework for our experiences of reality. In his opinion, the horizon is neither a symbolic nor an empirical object, but a visual-material phenomenon, which frames and arranges everyday human practices and reflections (Koschorke 1990:7). Whether the horizon represents the mere limits of the world, the infinity of enlightened, rational life or a romantic utopia exceeding formal calculations or common-sense argumentation, it is, nevertheless, the horizon itself which provides the indispensable reference. As a matter of fact, the horizon is present to everyone on the globe at the same time. Yet it differs significantly in character, legibility and appearance with every step we take. Although the visual horizon frames our daily lives, it does not signal ownership or specific land-cover categories. The horizon shows both the finiteness of space and the infinity of time. The horizon connects our sensory experiences with our reflections on who we are and what we use our lives for. Thus the true value of the horizon lies in its ability to link the motion of nature to human comprehension of our shared existence on earth.

13.4 Conclusion: Expert Inventories into Landscape Narratives

As argued in this paper, the vast landscape changes implied by the *Norwegian Wind Resource Strategy* point to a fundamental discrepancy between the landscape intentions of the ELC and its operational reception by operational administrations. National parliamentarians and bureaucrats habitually separate facts about land from the holistic entity of landscape, thereby effectively dissolving the object at stake. In their experience-based narratives, however, the people affected prevail with their ability to perceive and articulate the complexity of these areas as meaningful landscapes. Accepting the decisive link between public decision-making processes, irreversible land transformations and the people's right to landscape, the now abandoned *Norwegian Wind Resource Strategy* nevertheless shows that there is an urgent need to strengthen the argumentative and narrative aspect of landscape planning, motivated by the ELC. As we have seen in this paper, inventories, maps and models should be paired with more experience-based narratives in order to meet the intentions of the ELC. Landscape is the commons of civilization and the visual horizon may be regarded as its last retreat.

References

Benjamin W (1991) Gesammelte Schriften, Frankfurt am Main, Suhrkamp

Council of Europe (2000) European landscape convention, florence, explanatory report. Council of Europe, Strasbourg

Energidepartementet O-O (2020) Vindkraft på land—Endringer i konsesjonsbehandlingen. 19. juni 2020 ed. Oslo: Olje- og Energidepartementet

Haram KA (2020) Dette er øya dei har tenkt å øydeleggje for all framtid. Aftenposten 24 June 2020

Inderberg TH, Rognstad H, Saglie I-L, GULBRANDSEN LH (2019) Who influences windpower licensing decisions in Norway? Formal requirements and informal practices

Koschorke A (1990) Die Geschichte des Horizonts: Grenze und Grenzüberschreitung in literarischen Landschaftsbildern. Frankfurt am Main: Suhrkamp

Lundsbakken M (2019) Forslag til Nasjonal ramme for vindkraft. NVE

Lyotard J-F (1984) The postmodern condition: a report on knowledge. Manchester University Press, Manchester

Riksantikvaren MO (2019) Nasjonal ramme for vindkraft. Innspill fra Miljødirektoratet og Riksantikvaren til eksklusjonsrunde 3. Oslo

Williams R (1973) The country and the city, London, Hogarth Press

Chapter 14
Landscape Observatory Västra Götaland & Balaton Ecomuseum—A Swedish–Hungarian Interaction for Sustainable Landscape Management

Bosse Lagerqvist, Sándor Némethy, and Anders Nilsson

14.1 Introduction

This paper addresses sustainable, inclusive approaches to landscape management, framed by the notions of 'heritage' and 'landscape' as instruments for democracy and sustainable development (Council of Europe 2000). The implications of 'heritage' in the CoE-convention carries the theories and standpoints of 'integrated conservation' formulated through the 1970s and 80s (ICOMOS 1975; Appleyard 1979; Feilden 1979; Rosvall and Aleby 1988; Engelbrektsson and Rosvall 2009). In spite of the time frame, integrated approaches based on this broad understanding of heritage and conservation, still needs to be developed. Research is required to improve the society's ability to work with landscape and heritage issues on all levels, from individual/local to national/international perspectives (Waterton et al. 2006; Ashworth 2011; Sully 2013; Luciani and Del Curto 2018).

In landscape related research the ecosystem approach is often used, and over time the paradigms of landscape and the ecosystem approach has merged (Morrison et al. 2018). Further research is require for understanding the alignment of landscape character, landscape perception and ecosystems functions (Morrison et al. 2018). Specifically the use of cultural ecosystem services is a necessary starting point, but operationalisation of such ecosystem services need to be improved and

B. Lagerqvist (✉) · S. Némethy
Department of Conservation, University of Gothenburg, Gothenburg, Sweden
e-mail: bosse.lagerqvist@conservation.gu.se

A. Nilsson
Region Västra Götaland, Gothenburg, Sweden

T. Matteini et al. (eds.), *Cultivating Continuity of the European Landscape*, Environmental History 15, https://doi.org/10.1007/978-3-031-25713-1_14

depends in turn on how conceptual and methodological assessment tools are developed (Plieninger et al. 2015). There are however, no studies linking landscape studies, character assessment, integrated heritage approaches and cultural ecosystem services from a theoretical as well as a practice-based perspective.

The hypothesis for this paper is that the principles for ecomuseums respectively landscape observatories might be merged into combined functionalities, creating a novel area of interactivity and providing models for sustainable heritage oriented landscape management or rather stewardship, with the Balaton Ecomuseum in Hungary, and the Landscape Observatory Västra Götaland in Sweden as cases.

14.2 Heritage as a Frame

Cultural heritage is both a product—the defined object or phenomena valued as heritage—and a process—how the definition and valuation have been managed. Traditionally the process is based on national formal organisations that top-down defines a national heritage—the Authorised Heritage Discourse/AHD (Smith 2016). The critique of the AHD is that it does not sufficiently employ a critical understanding of how the history might be interpreted and narrated, and that it not involves the life of the ordinary people that seldom have been visible in the big historical pictures.

According to Ashworth (2011) three different paradigms guiding heritage practices, has, since the 1980s, coexisted due to incomplete paradigm shifts. The *preservation paradigm,* late nineteenth century, represents a traditional perspective on the past with the objective to protect specific objects from change or other threats. In the 1960s it was followed by the *conservation paradigm* that included collections and environments together with specific objects addressing contemporary use in relation to an imagined future. Finally the *heritage paradigm,* in the 1980s, is based on how cultural values are constructed in the contemporary society and are not inherent in objects. It also implies a shift from experts to how (ordinary) people value historic remains.

The heritage processes in the frame of this paper are constantly performed in the realms of complex heritage dynamics (Fig. 14.1). The practice field can be divided into 'typical' cultural heritage areas or even societal systems understood in terms of different contexts, such as laws, regulations, professions and so on, exemplified by the boxes in Fig. 14.1. The activities within one box might have positive or negative effects in the other boxes. The success of the different sub-activities is dependent on more general attitudes to concepts such as preservation, or recycling, which in turn are the consequences of ongoing socio-cultural processes redefining the context for heritage work.

Rethinking heritage practice implies partly a combination of paradigms (Ashworth 2011), partly a development into a potentially integrated approach, where these boxes identified as interlinked sub-systems could have the strength to affect the attitude society as a whole takes towards history and preservation. In this way, the sub-systems might form one system that continually integrate historical properties and

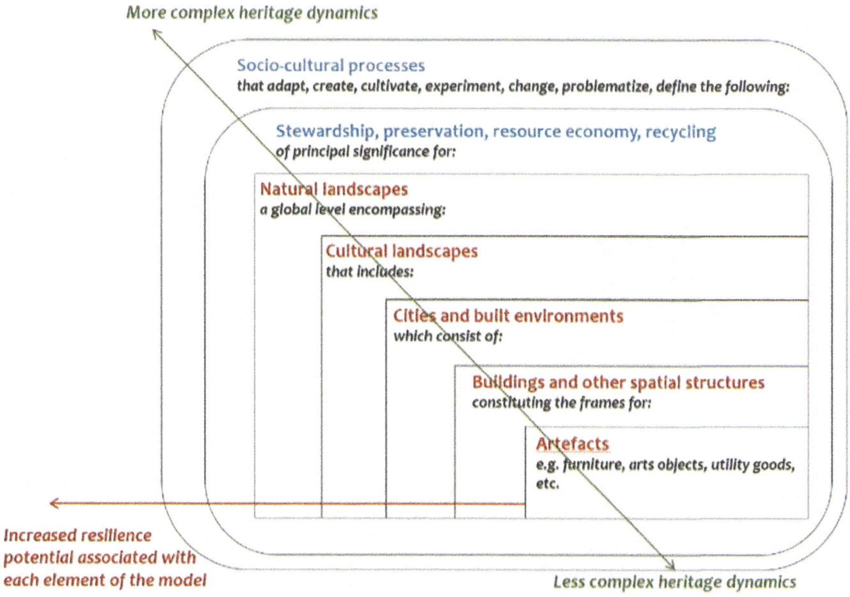

Fig. 14.1 Heritage dynamics of the practice field (Lagerqvist 2017)

qualities with contemporary needs and opportunities, underpinning a sustainable societal development. This figure of embedded boxes could also be given two further dimensions represented by, on one hand what we could call the *subject matter* and on the other hand *decision rationales*. In short, the former is composed of the data, information, knowledge, insights, skills, etc., on objects and phenomena within these boxes, and the latter describes the routines, procedures, processes, power structures, etc. for managing the *subject matter* in different societal applications. The two cases of this paper represents the two perspectives of this heritage dynamics illustration: The Balaton Ecomuseum has its aim towards the *subject matter*, while the Landscape Observatory Västra Götaland focuses the *decision rationales*. They of course overlaps each other but their centres of gravity can be characterised according to the two perspectives. The reasons for this paper to pick these two cases rests naturally in the fact that the authors are active within these organisations, and the organisations have initiated a cooperation.

Apart from *heritage*, ecosystem services represent an important framing perspective for this study. Ecosystem services are interconnected with natural and anthropogenic ecological cycle processes and the need of ecosystem approach is particularly pronounced in rural planning (Lagerqvist et al. 2016). Possibly the ecomuseum concept is possible frame for combining landscape studies, character assessment and cultural ecosystem services into integrated conservation approaches, supplying a landscape observatory with means for sustainable development. An ecomuseum is not an ordinary museum in the sense of a building with collections. Originally, the geographical territory was stated as the foundation for an ecomuseum, which has

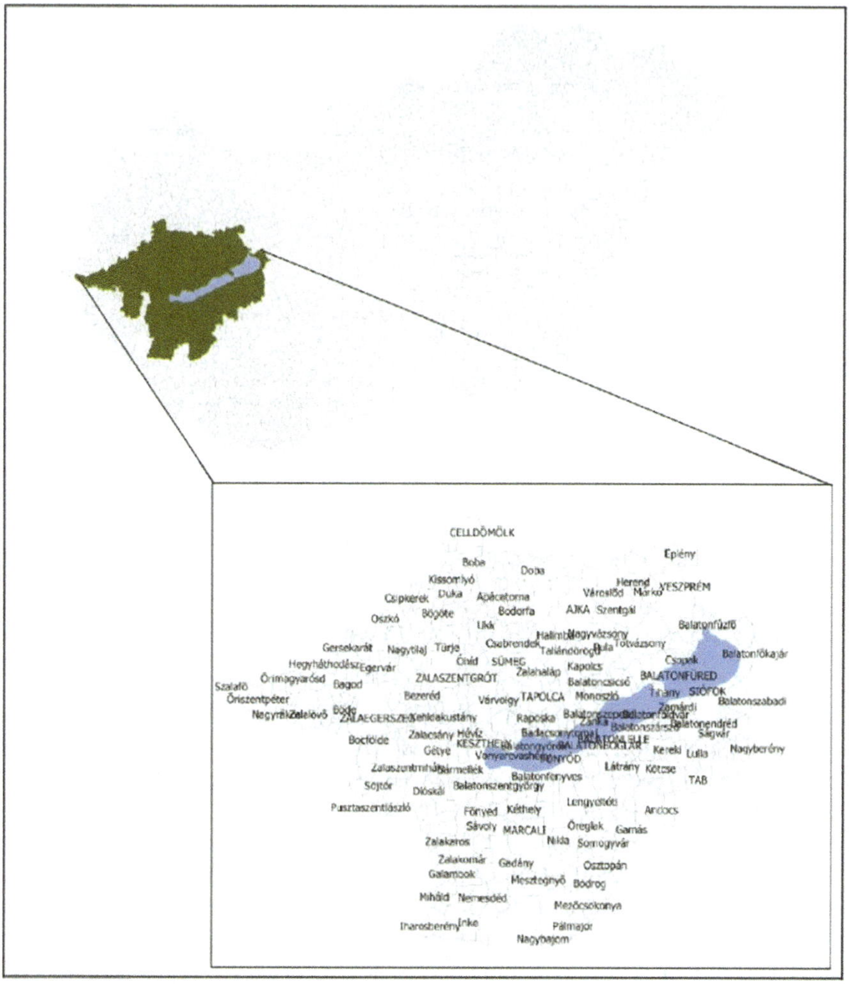

Fig. 14.2 The area of Balaton Ecomuseum (Némethy 2019)

been criticised by Davis favouring instead the concept of 'place' where 'sense of place' represent the connection between the community and the place (Davis 2011). This has been further discussed and the notion of 'space' is proposed by Bowden for the geographic context of ecomuseums, arguing for a non-rigid interpretation of a place but rather the space as the totality of all possible histories, stories, memories, emotions, etc., linked to a place (Bowden 2017). The ecomuseum thus might become an extended, or complementary, sensitive assessment resource for a landscape observatory. At least, this might be the hypothetical construct in the cooperation between Balaton Ecomuseum and Landscape Observatory Västra Götaland.

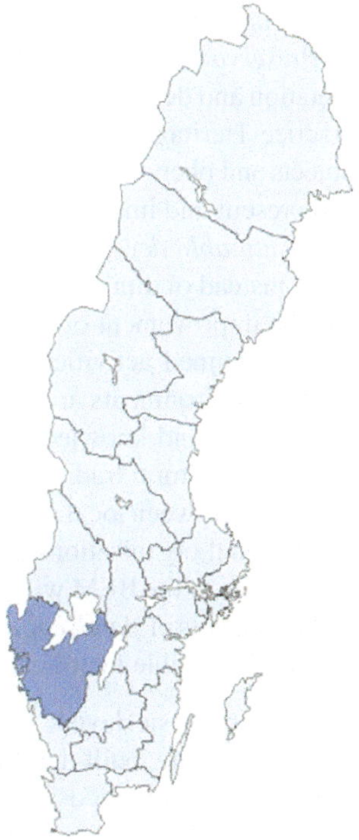

Fig. 14.3 Region Västra Götaland is one of 21 regions in Sweden (Wikimedia 2021)

14.3 The Balaton Ecomuseum

Through a cross-disciplinary project, the Balaton Ecomuseum, BEM, is under establishment, funded primarily from Interreg Central Europe Programme but also through Hungarian governmental funds. The area includes the catchment area and the surface of Lake Balaton, the area of the Balaton Wine Region, two national parks and a resort district, including as well geologically and hydrographically relevant areas. The organisation of BEM started as a project within the Lake Balaton Development and Coordination Agency that acts as a non-profit organisation to facilitate communication and cooperation of organisations interested in the development of the region. The BEM organisation will be more formalised during 2020.

The term *'éco'* refers primarily not only to ecology, but foremost to a holistic interpretation of cultural heritage made available with new, interactive methods of museum education, embracing the natural environment, the built heritage and the intangible heritage, contradicting traditional passive presentation (Corsane et al. 2007). Community involvement and active participation is essential for the functionality of ecomuseums.

An ecomuseum provides strategies and operations in real life that has the ability to change and improve the society and the landscape, through its community involvement, implying all possible persons within the concerned local society.

Community involvement implies engaging and involving people.

Preservation, interpretation and management means the constant on-going interpretation and development of heritage properties and qualities, as part of an everyday practice. Heritage is very close to the *Sense of Place*, including history of inhabitants, objects and phenomena, both tangibles and intangibles, but also memories of the past and present and imaginations of the future.

Sustainable development is central for ecomuseums and increase the value of a place instead of diminishing it, through two key elements: place-based development and the improvement of local networks.

The planned activities of the BEM will mirror the different geographical and cultural environments around Lake Balaton involving stakeholders with a wealth of experiences and knowledge derived from different geographic conditions, natural resources, cultural traditions and professional competences. A close cooperation is organised between local and regional authorities, businesses, university departments, professional organisations and scientific network.

The role of the BEM will gradually expand mirroring the evolution and a merging of the concepts of '*landscape*' and '*cultural heritage*', and is planned in several layers, with each tangible factor carrying a particular segment of the intangible heritage:

- A set of basic layers constitutes a general frame: Geodiversity and geomorphology; the built heritage with infrastructures; Biodiversity and protected species; Natural ecosystems and ecosystem services; and Lake Balaton surface waters including the catchment area.
- Within this frame a number of further layers has been identified providing meaning, participatory functions and development potentials: Description of intangible heritage assets; Agro-ecosystems and ecosystem services; Coordination of existing structures, i.e. geoparks and national parks; Tourism destinations' carrying capacity and products for the tourism economy.

14.4 Landscape Observatory Västra Götaland

The landscape of Västra Götaland is both a governmental administration form as *Västra Götaland County*, but also a local/regional administration form as *Region Västra Götaland*. The two administrations have different responsibilities in terms of the governmental top-down activity with regard to legislative effects, and the local/regional administration's bottom-up approach to primarily health service and regional development. The region is subdivided into 49 municipalities, which have urban- and land-use planning monopoly, through the Planning and Building Act although the County Board could start a review process of plans that has been decided on the local level.

The Region Västra Götaland is running the Landscape observatory Västra Götaland, LOVG, as a pilot-period of two years, to establish organisational structures, funding principles and operative and strategic routines. The LOVG Steering group with representatives from a local municipality, the County Board, the University of

Gothenburg, is headed by the Region and has as such a formal position in the public administrative system. The LOVG aims to find means to avoid 'silo planning' and instead out of a holistic landscape view, support increased cooperation and coordination for different sectors and administrative levels, such as *Regional development*, *Urban rural connection*, *Water management* and *Land use*.

Although the European Landscape Convention features a bottom-up perspective emphasising the social significance of the landscape, there is a natural top-down perspective in the landscape as a venue for national and regional exploitation interests, not necessarily coordinated with each other or with local interests. Not all planning matters are specifically subjects for a landscape observatory, but those with a high degree of complexity and opposing interests at several levels should be weighed against each other, and in the focus of a landscape observatory.

Water management issues are examples of such complex, and is becoming a growing concern from a landscape perspective, with effect on nature and cultural environments. Specifically the EU Water Framework Directive (Directive 2000/60/EC) has created conflicts between preserving cultural environments and the ensuring of good ecological status in streams. This is problematic when it comes to centuries old mill structures using the water as a power resource, and the human activity is partly the creator of the ecological status.

The platform for establishing the LOVG should engage in collaborative projects on national and regional level among government agencies through formal agreements. One example is a regional action plan for green economies and cooperation with other actors, where the government's material is available in various forms of protection through the Environmental Act and the Heritage Act, managed by the county administrative board. The civil society organisations, however, have massive information and knowledge on valuable qualities in the landscape. Not the least the landowner interests represents additional perspectives on values in the landscape. Green infrastructure and ecosystem services are increasing issues for regional action plans under way as an important measure following the government's proposal for a Swedish strategy for biodiversity and ecosystem services (Proposition 2013/14: 141). In short, there is a massive need for assessing the full totality of the subject matter of the landscape of Västra Götaland in order to table decision rationales applicable for integrating local, regional, governmental and other stakeholders activities and interests.

14.5 Concluding Remarks

Heritage is often treated as a side issue, regarded as not necessarily affecting the 'normal' everyday life of people and societies. This article rests on the value premise of 'heritage' being comparable with 'landscape' as defined in the European landscape Convention. The attention of BEM and LOVG is therefore not aimed at 'heritage landscapes' or 'cultural landscapes', but at landscape in the present, past and future where heritage is a fundamental resource for increasing the resilience of society (ICOMOS 2016).

Due to the geographical distance between the cases there will be no real life joint processes. However, through the cooperation between the two bodies there will be opportunities to operate model development in principle focusing on: Complex heritage dynamics assessments including ecosystem recording and development, and Integrated conservation approaches in planning and sustainable development.

Regarding this tentative model development, the *ecomuseum* comprises the subject matter of the landscape space, assessing, sampling and interacting with histories, memories, intentions linked to people, objects and sites. The inclusive approach reducing social gaps and promoting community feeling is important. Heritage processes preserves, maintains, develops and creates societal qualities and properties leading to employment opportunities and depopulation prevention. There is both an internal as well as external orientation in the activities of the ecomuseum, in the sense of addressing both the community and visitors to the area.

The *landscape observatory*, on the other hand represents support systems for decision rationales in all types of regional development and land-use planning, as receiver of subject matter made available through models applied for assessing the subject matter of the landscape space. The success factor rests on the ability to bring formal stakeholders, civil society organisations, the academia and school system, private businesses and the general public, into joint processes for understanding the landscape providing informed decisions. The landscape observatory is in this way more obviously oriented towards internal functions compared with the ecomuseum.

There are similarities between the European countries in terms of structures for urban and land-use planning, as well as how heritage is protected and incorporated in societal activities, while at the same time there are significant differencies (Pewtti et al. 2020). The establishment of landscape observatories across Europe shares the same incentive in terms of the ELC, however there are very different reasons, running periods, organisations and nature of activities represented by the observatories so far (Gotzman 2018). It is therefore a little early to try to compare results and effects from various activities in the subject field cross over the European continent. To date however, there has been no project or activity aiming at cooperation between the ecomuseum principle and the landscape observatory structure. The reason for embarking on this project, as outlined in this article, is the possibilities to achieve model development from two contrasting and at the same time parallel perspectives on landscape stewardship.

References

Appleyard D (1979) The conservation of European cities. Based on a conference held at the American Academy in Rome, June 1975. MIT P., cop., Cambridge, Mass

Ashworth G (2011) Preservation, conservation and heritage: approaches to the past in the present through the built environment. Asian Anthropology 10(1):1–18

Bowden A (2017) Pushing the boundary of territory: ecomuseums as dynamic, open spaces. Landscape Res. https://doi.org/10.1080/01426397.2017.1291921

Corsane G, Davis P, Elliott S, Maggi M, Murtas D, Rogers S (2007) Ecomuseum evaluation: experiences in Piemonte and Liguria, Italy. Int J Heritage Stud 13(2):101–116

Council of Europe. European landscape convention (2000) https://www.coe.int/en/web/conventions/full-list/-/conventions/rms/0900001680080621 Convention on the value of cultural heritage for society (Faro Convention, 2005) https://www.coe.int/en/web/culture-and-heritage/faro-convention

Davis P (2011) Ecomuseums: a sense of place, 2nd edn. Continuum, London

Directive 2000/60/EC of the European Parliament and of the Council establishing a framework for the Community action in the field of water policy. https://ec.europa.eu/environment/water/water-framework/index_en.html

Engelbrektsson N, Rosvall J (2009) Sustainable integrated and planned conservation of built environment and architectural heritage: principles of dynamic management of modern assets and their care. In: Canzani A (Ed) Conserving architecture. Planned conservation of XX century architectural heritage. Electa, Milano

Feilden B (1979) An introduction to conservation of cultural property. ICCROM, Rome

Gotzman D (2018) Oral presentation: overview of landscape observatories in Europe. In: Inaugural Seminar for the landscape Observatory Västra Götaland, Mariestad, April 19, 2018

ICOMOS (1975) The declaration of Amsterdam. Congress on the European Architectural Heritage, 21–25 Oct 1975. https://www.icomos.org/en/and/169-the-declaration-of-amsterdam

ICOMOS (2016) Cultural heritage, the UN sustainable development goals, and the new urban agenda. In: Hosagrahar J, Soule J, Girard LF, Potts A (Eds) ICOMOS concept note for the United Nations Agenda 2030 and the Third United Nations Conference on Housing and Sustainable Urban Development (HABITAT III)

Lagerqvist B, Némethy S, Dinya L, Huszár Z, Zádori I, Kőmíves T, Bujdosó Z, Nemeskéri Z (2016) Sustainable management of cultural landscapes: use of ecosystem services in rural tourism. In: Łabaj M, Sobczuk J, Beskidzka S (Eds) Ecological footprint in Central Europe. Multiauthorial monograph. The University College of Tourism and Ecology in Sucha Beskidzka (WSTiE), pp 91–106. ISBN: 978-83-926999-3-4

Lagerqvist B (2017) Conservation or reconciliation? Industrial heritage practices at a turning point. In: Walters D, Laven D, Davies P (Eds) Heritage and peacebuilding. Heritage Matters. Boydell & Brewer Ltd, Martlesham, UK. ISBN 978-1-78327-216-7

Luciani A, Del Curto D (2018) Towards a resilient perspective in building conservation. J Cult Heritage Manage Sustain Develop 8(3)

Morrison R, Barker A, Handley J (2018) Systems, habitats or places: evaluating the potential role of landscape character assessment in operationalising the ecosystem approach. Landscape Res 43(7):1000–1012

Némethy S (2019) The Balaton Ecomuseum. Conservation and management of cultural landscape and its intangible heritage. Lake Balaton Development Coordination Agency

Pewtti L, Trillo C, Makore BN (2020) Cultural heritage and sustainable development targets: a possible harmonisation? Insights from the European perspective. Sustainability 12:926

Plieninger T et al (2015) The role of cultural ecosystem services in landscape management and planning. Curr Opin Environ Sustain 14

Proposition 2013/14: 141. En svensk strategi för biologisk mångfald och ekosystemtjänster. (A Swedish strategy for biodiversity and ecosystem services). The Swedish Government. https://www.regeringen.se/49bb9c/contentassets/d11a7625086a4c3cb09fcf6322687aba/en-svensk-strategi-for-biologisk-mangfald-och-ekosystemtjanster-prop-201314141. Accessed 23 June 2020

Rosvall J, Aleby S (eds) (1988) Air pollution and conservation: safeguarding our architectural heritage. Published in cooperation with Swedish Institute of Classical Studies in Rome, Department of Conservation, University of Göteborg. Elsevier, Amsterdam

Smith L (2016) Uses of heritage. Routledge, New York

Sully D (2013) Conservation theory and practice. materials, values, and people in heritage conservation. In: McCarthy C (Ed) The international handbooks of museum studies: practice, vol IV. Wiley

Waterton E, Smith L, Campbell G (2006) The utility of discourse analysis to heritage studies: the Burra charter and social inclusion. Int J Heritage Stud 12(4)

Wikimedia (2021) https://commons.wikimedia.org/wiki/File:SverigesL%C3%A4n2007V%C3%A4stra_G%C3%B6taland.svg. Accessed 7 Dec 2021

Chapter 15
Landscape and Public Participation in Norwegian Local Planning—An Example of Implementing the European Landscape Convention

Marte Lange Vik

Abstract The ambition of the European Landscape Convention to create landscape democracy is oft quoted and discussed, but few studies have evaluated the local effects of the convention. This study follows previous studies of the Norwegian pilot project for implementation of the ELC, focussing on a local spatial planning process. The ambition of the paper is to analyse public participation in this process and to shed light on how landscape and public participation are integrated in Norwegian planning, with particular interest in the manifestation of the ELC's «democratic promise». The analysis distinguishes between approaches to landscape form and function in planning. A main finding is that experts and local actors have differing approaches to landscape, complicating effective participation. The concluding argument is that successful implementation of the ELC, relating both to landscape and democracy, relies on the planning authorities' ability to make both substantial and procedural changes.

Keywords Public participation · Local spatial planning · Landscape geography

15.1 Introduction

The European Landscape Convention (ELC) emphasises the importance of landscapes for human activities and well-being, understanding landscape as a carrier of democratic potential (CoE 2000a). Public participation is seen as the prime means to achieve landscape democracy (Jones 2018). In Norway, the Planning and Building Act (PBA) serves an important role in implementing the ELC. Increased focus on both landscape and public participation in the last revision of PBA (PBA 2008) can be understood as efforts to meet the objectives of the ELC (Eiter and Vik 2015).

M. L. Vik (✉)
Department of Environmental Sciences, Western Norway University of Applied Sciences, Sogndal, Norway
e-mail: marte.lange.vik@hvl.no

T. Matteini et al. (eds.), *Cultivating Continuity of the European Landscape*, Environmental History 15, https://doi.org/10.1007/978-3-031-25713-1_15

Following previous evaluations of a Norwegian pilot project for implementation of the convention (Conrad et al. 2011; Eiter and Vik 2015), this study has two aims: assessing the effectiveness of a participatory planning process and discussing how the democratic promise of the ELC is manifest in practice.

15.2 Landscape, Planning and Public Participation

Landscape was actualised in Norwegian planning and management after adoption of the ELC. Landscape considerations and focus on public participation in spatial planning were both strengthened in the last revision of the PBA (2008). Although the participation development happened independently of the ELC, it clearly resonates with the convention.

Public participation and landscape imply different focuses in planning studies. Whereas landscape considerations require focus on planning substance (Skjeggedal 2013), public participation is firmly rooted in process-oriented planning approaches (Amdam 2012).

Eiter and Vik (2015) argue that Norwegian spatial planning may be characterised by a functional landscape approach due to the role of zoning in spatial master plans. This is contrasted to emphasis on form and visual aspects, often associated with studies of landscape paintings and scenic landscape. There are indications that landscape function and form is not fully integrated in Norwegian planning practice and that a dichotomous relationship between the two still exists (Eiter and Vik 2015).

Studies on barriers to effective participation in Norwegian spatial planning conclude that diverging discourses and approaches in planning often limit inclusion, and that dominant discourses tend to marginalise alternative voices in planning processes (Hanssen and Saglie 2010). Kristiansen (2012) finds that alternative voices often take experienced reality as their departure point, relating to everyday practices rather than distant ideals. Vik and Refstie (2014) describe how resistance and exhaustion develop when participatory planning results do not reflect local inputs. These studies pose fundamental questions to the relation between public participation and democracy. According to Jones (2018), the ELC does not discuss possible shortcomings of participation in creating a "true 'landscape democracy'" (CoE 2000b, Art 8).

15.3 Sund and the Skogsskiftet Plan[1]

Sund municipality[2] participated in the Norwegian pilot project[3] for implementation of the ELC with a plan for the local centre, Skogsskiftet (Fig. 15.1a). The plan combined spatial and societal objectives, aiming to create a vital centre based on broad local demand for a renewal and facelift of the centre. At start-up, Sund municipality hired an external consultancy responsible for preparing the plan.[4]

Sund is located on the southern part of an island outside Bergen, on the western coast of Norway. The settlement structure is decentralised, with small communities along the shore. Skogsskiftet was defined as a municipal centre in the early 1980s when land-based transport increased and is characterised by random developments without a unified plan. Stores and services are mainly located in the small commercial centre collocated near the town hall (Fig. 15.1b). The topography is challenging with altitude differences restraining overview and movement (Fig. 15.1c).

When Sund entered the pilot project,[5] they got funding for engaging external landscape consultants. A landscape character assessment which was meant to provide input on landscape qualities and ease safeguarding of these in the plan, was added to the process. The previously mentioned private consultancy which was hired for preparing the plan, was also engaged to undertake the landscape assessment.

The process had four phases with separate focuses (Table 15.1). An initial phase provided analyses and assessments that served as a knowledge base for the preparation of a proposed plan in the second phase, and adoption of the plan in the last phase. Two key elements in the process were the centre relocation discussion (Asplan Viak 2007a; Sund kommune 2007a, b) and the definition of and emphasis on managing undesirable elements[6] in order to plan a good centre (Asplan Viak 2007a).

Special consultancy reports enhanced focus on landscape (Asplan Viak 2007a, b), emphasising landscape characteristics such as sunlight conditions, slope, view, and waterfront proximity. Waterfront proximity especially gained momentum early

[1] The study is based on qualitative data from a combination of observations, document studies, and semi-structured interviews. The interviews were carried out among actors who participated actively in the pilot project process, covering a total of 14 interviewees. The interviews are supplemented with observations in and of the area, and thorough document studies (c. 140 documents) of public maps as well as written material from the process.

[2] From 2020, Sund is part of Øygarden municipality after the Norwegian local government reform.

[3] A national pilot project (2006–2009) was carried out for implementation of the ELC in local planning. Hordaland County and four related municipalities participated with different plans. The pilot project aimed to improve management of landscape qualities in municipal spatial planning. Defined project goals included identification and analysis of landscape qualities; development and implementation of tools for integration of landscape in local planning; awareness-raising on qualities and safeguarding of landscape (Fylkesmannen i Hordaland and Hordaland fylkeskommune 2007).

[4] When hiring an external planning consultancy, their responsibilities are to carry out the planning process and developing a planning proposal. The municipal planner still is responsible for the quality of the outcome, and the elected municipal council is responsible for adopting the plan.

[5] The planning process was initiated (2005) before the pilot project started (2006).

[6] These elements were a grocery store, bus terminal, gas stations, fire station, a field for motor-cross, and an area with activities generating heavy transport through Skogsskiftet (Asplan Viak 2007a).

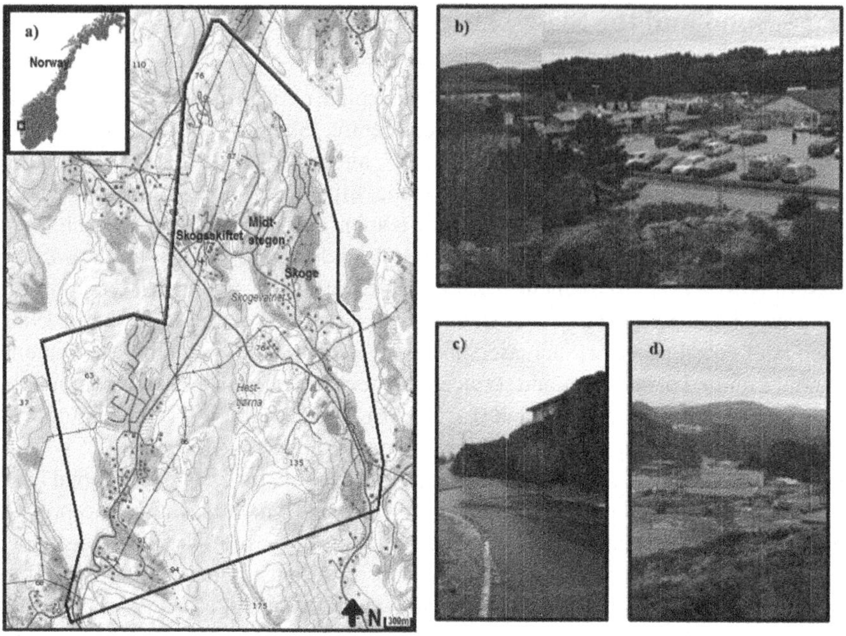

Fig. 15.1 (**a**) Map over Skogsskiftet, the planning area and localisation in Norway (**b**) The existing centre (**c**) Large differences in altitudes (**d**) Midtstegen industrial area (Map and photos: Marte L. Vik)

Table 15.1 Overview over phases in the planning process

Period	Label	Content/Focus	Public participation
March 2005–March 2007	Pre-phase	Initiation, definition of scope and progress	Start-up announcement First round of inputs Public meeting
March 2007–April 2008	Initial phase	Mapping and analysis, landscape character assessment, evaluation of alternatives for centre localisation	Engagement of schools and kindergartens Public meeting Second round of inputs
June 2008–February 2009	Second phase	Preparation of planning map with regulations, plan description and impact assessment	Meetings with land owners and commercial interests
March 2009–May 2011	Last phase	Ensure political support, revisions, adoption of plan	Hearing and public inspection Third round of inputs

on and guided the process based on the argument that a seafront location is characteristic for local communities in western Norway[7] (Asplan Viak 2007a). When the option to relocate the centre by lake Hesttjørna was rejected, lake Skogevatnet, close to Skogsskiftet, gained importance (Fig. 15.1a). Midtstegen, an industrial area generating heavy vehicle transport through the centre, slopes south towards the lake (Fig. 15.1d). Favourable landscape features, combined with the desire to (re)move the cause of heavy vehicle transport, made Midtstegen the focal area for centre development, despite the existing industrial area.

Several public participation measures were carried out beyond the mandatory minimum of the start-up announcement, hearing, and public inspection (PBA 2008, §§ 11–12, 11–14). Two public meetings were arranged, along with a school visualisation project, a survey for kindergartens and school parents, and two open office days,[8] intending to give people who could not otherwise forward suggestions an opportunity to dialogue with the planning experts (Eiter and Vik 2015). Moreover, three rounds of written inputs generated a total of 29 inputs from 15 private actors. These concerned delimitation of area and target issues for the plan and suggestions for development of private land for housing, school, and sports arena, or for cultural and commercial activity (round 1, 10 inputs); reactions to the centre relocation discussion as well as suggestions for developments in the existing centre (round 2, 6 inputs); and objections to the proposed plan and suggestions for alternatives and additional elements (round 3, 13 inputs).

Despite a wide range of public participation measures and broad engagement, accounts from participants in the process revealed a dissatisfaction with how public participation was managed. Key points of criticism were lack of realism, failure to comply with local inputs, and lack of actual influence on the plan.

15.4 Concluding Discussion

The local discontent with public participation parallels other studies. Vik and Refstie (2014) stress that effective participation needs to focus not only on participation measures, but on how inputs influence the resulting plan. This corresponds with Skjeggedal's (2013) criticism of the emphasis on process over substance in Norwegian planning discourse and practice.

Landscape considerations imply focus on planning substance, but different approaches to landscape limit the impact of local inputs. Hanssen and Saglie (2010) find that established planning discourses are led by overall aims. In the Skogsskiftet plan, these overall aims were represented by an ideal landscape, focussing on visual

---

[7] Whereas historical seafront location had a *functional* logic based on the importance of the sea for livelihood, transport, communication, and trade, today, transport and livelihood strategies have changed drastically, and the arguments are now related to *form* more than function, emphasising visual and aesthetic qualities.

[8] There are no official records from these meeting days; therefore, it is difficult to date them exactly. According to accounts from the municipal planner, these were arranged after the public meetings.

and aesthetic qualities presented by the consultancy. This contrasted the landscape approach in local inputs, making specific and place-bound suggestions for developments of the existing landscape, corresponding to Kristiansen's (2012) finding that alternative voices in planning often use the experienced world as their departure point.

On a more theoretical level, the consultants' landscape approach was characterised by a focus on landscape form, whereas local inputs were centred on the uses of landscape, everyday life and economic interests, implying a focus on landscape functions and practices[9] (Eiter and Vik 2015). The local demand for renewal and a facelift of the centre related to landscape arguments used by the consultancy may have strengthened the focus on form, favouring landscape ideals over the existing landscape and disturbing communication between the planners and the public (Hanssen and Saglie 2010; Kristiansen 2012).

This implies that landscape and public participation are disintegrated in Norwegian planning, rather than being treated as integrated issues, as promoted by the ELC (CoE 2000a). Public participation, on one hand, relates to planning process, while landscape, on the other hand, relates to planning substance. Successful implementation of the ELC requires efforts from Norwegian authorities to make both substantial and processual changes.

Furthermore, landscape assessments carried out by external consultants may challenge the participatory process if experts approach landscape without integrating perspectives of local actors. The Sund case has shown that engagement of external consultants may lead to heavier influence of scholarly landscape knowledge and less influence of local landscape knowledge in plans. This opposes the aim of heterogeneity in planning which participatory processes are supposed to promote (Jones 2018), threatening the democratic promise in the European Landscape Convention (CoE 2000b).

Acknowledgements This study was part of my Ph.D. project, hosted and financed by the Department of Geography, Norwegian University of Science and Technology. I want to thank all interviewees for sharing their experiences. Thanks also to anonymous reviewers, as well as late Eirin Hongslo, and Karoline Daugstad and Michael Jones, for discussions and inputs on earlier versions on the manuscript.

References

Amdam R (2012) Fokus på prosess: Kommunikativ og samarbeidande planlegging. In: Aarsæther N, Falleth EI, Nyseth T, Kristiansen R (eds) Utfordringer for norsk planlegging: kunnskap, bærekraft, demokrati. Cappelen Damm Høyskoleforlaget, Kristiansand, pp 273–290

Asplan Viak (2007a) Fortetting Av Skogsskiftet Eller Nytt Senter. Asplan Viak, Bergen

Asplan Viak (2007b) Landskapsanalyse. KDP Skogsskiftet - Sund Kommune. Asplan Viak, Bergen

[9] Only documents from the consultancy refer explicitly to landscape. By locals, landscape contents are implicit rather than explicit.

CoE, Council of Europe (2000a) European Landscape Convention. 2000. http://conventions.coe.int/treaty/en/treaties/html/176.htm
CoE, Council of Europe (2000b) European Landscape Convention. Explanatory Report. http://conventions.coe.int/Treaty/EN/Reports/Html/176.htm
Conrad E, Cassar LF, Jones M, Eiter S, Izaovičová Z, Barankova Z, Christie M, Fazey I (2011) Rhetoric and reporting of public participation in landscape policy. J Environ Planning Policy Manage 13(1):23–47. https://doi.org/10.1080/1523908X.2011.560449
Eiter S, Vik ML (2015) Public participation in landscape planning: effective methods for implementing the European Landscape Convention in Norway. Land Use Policy 44(March):44–53. https://doi.org/10.1016/j.landusepol.2014.11.012
Fylkesmannen i Hordaland, and Hordaland fylkeskommune (2007) Prosjektplan for Landskap i Kommunal Planlegging. Fylkesmannen i Hordaland & Hordaland fylkeskommune, Bergen
Hanssen GS, Saglie I-L (2010) Cognitive closure in urban planning. Plan Theory Pract 11(4):499–521. https://doi.org/10.1080/14649357.2010.525373
Jones M (2018) Landscape democracy: more than public participation? In: Egoz S, Jørgensen K, Ruggeri D (eds). Defining landscape democracy. A path to spatial justice. Edward Elgar, Cheltenam, UK, pp 16–28. https://doi.org/10.4337/9781786438348.00011
Kristiansen R (2012) Kritisk-Alternative Initiativ i Norsk Byplanlegging - En Studie Av Byutviklingens År i Tromsø Og Nyhavna- Regjeringen i Trondheim. Avhandling for Graden Philosophiae Doctor. Institutt for sosiologi, statsvitenskap og samfunnsplanlegging, Fakultet for humaniora, samfunnsvitenskap og lærerutdanning, Universitetet i Tromsø, Tromsø
Planning and Building Act (2008) Lov Om Planlegging Og Byggesaksbehandling (Plan- Og Bygningsloven). https://lovdata.no/dokument/NL/lov/2008-06-27-71?q=plan-%20og%20bygningslova
Skjeggedal T (2013) Prateplanleggingas Triumf Og Tragedie. Plan 45(01). http://www.idunn.no/ts/plan/2013/01/prateplanleggingas_triumf_og_tragedie
Sund kommune (2007a) Referat Frå 13. Møte i Prosjektgruppa for Kommunedelplan Skogsskiftet, 12.09.2007. Sund kommune
Sund kommune (2007b) Referat Frå 14. Møte i Prosjektgruppa for Kommunedelplan Skogsskiftet, 05.11.2007. Sund kommune
Vik ML, Refstie H (2014) Medvirkning, Makt Og Avmakt i Planlegging - Norge Og Malawi. Kart Og Plan 74(4):280–290

Chapter 16
Implementing Landscape Policies by Promoting Sustainability, Democracy, and Participation. Awareness Raising and Landscape Education in Trentino, Italy

Gianluca Cepollaro and Bruno Zanon

Abstract The chapter analyses the experience of a "Landscape School" activated in the Province of Trento (Trentino), Italy. The European Landscape Convention (ELC) introduced the concept of landscape as a space perceived and lived in by the local community. The emphasis is on the relationships established between people and the space they live in, or the places they experience. Following the ELC orientation, a new institutional system has been activated in Trentino, which revolves around a School for spatial planning and landscape management (step), responsible for implementing landscape education activities, supporting the Landscape Observatory, as well as carrying out other actions on outstanding landscape values. The School operates within a network of local actors, operators, and stakeholders who express an educational demand while bringing in experiences and knowledge. It adopts an experiential method, making use of a variety of "learning devices" to actively involve recipients in activities based on the sense of landscape as a common value.

Keywords European Landscape Convention · Landscape education · Trentino

G. Cepollaro (✉)
Scuola per il Governo del Territorio e del Paesaggio (step), Trento, Italy
e-mail: gianluca.cepollaro@tsm.tn.it

B. Zanon
Università degli Studi di Trento and Scuola per il Governo del Territorio e del Paesaggio (step), Trento, Italy
e-mail: bruno.zanon@tsm.tn.it

T. Matteini et al. (eds.), *Cultivating Continuity of the European Landscape*, Environmental History 15, https://doi.org/10.1007/978-3-031-25713-1_16

16.1 The Educational Challenge of the European Landscape Convention

The European Landscape Convention (ELC) introduced the concept of landscape as a space perceived and lived in by the local community (Council of Europe 2000a, b). The emphasis is not on the features or qualities—natural or anthropic—of the places involved, but on the relationships established between people and the space they live in, or the places they experience (Cepollaro and Zanon 2021).

Landscape, being the sum of features and opportunities—as well as of meanings, symbols, and imagery that are inherited, but also lived, used, and transformed over time by people and communities—is a common good. It is, therefore, a socio-political space of responsibility that requires participation (Debarbieux 2010).

The ELC affirms the need to implement awareness-raising actions and educational and training processes, while the guidelines for the implementation of the Convention solicit the active participation of stakeholders in the definition of landscape policies (Council of Europe 2008). This educational challenge implies disseminating the knowledge of landscape values and consolidating the people's sense of belonging to their place. Since landscape is constantly evolving, it is also necessary to strengthen the technical skills of professionals and improve the political and administrative capacity to manage conservation procedures and transformation processes.

16.2 The Experience of a "Landscape School"

In the past decade, an innovative experience has been activated in Trentino (Province of Trento, Italy), as a result of the approval, in 2008, of the "Autonomous Province of Trento Spatial Plan", based on the ELC concepts. This new orientation has stimulated the creation of a new institutional system, which revolves around a *School for spatial planning and landscape management (step)*, responsible for implementing landscape education activities, supporting the Landscape Observatory, as well as carrying out other actions on outstanding landscape values, namely the UNESCO Dolomites site.

The School and the Observatory constitute an "integrated system of actions" supporting the implementation of the ELC. *Step* is engaged in awareness-raising, education, and training activities, and the Observatory—supported by a Forum that gathers the most representative stakeholders from both the public and the private sector—is in charge of monitoring landscape dynamics and taking initiatives aimed at improving architecture, open space projects, and heritage preservation. The main instruments, in this regard, are public involvement and participation.

The organisation of the "School" is simple: a director is responsible for a staff of four people; a Scientific Committee, nominated by the Province, and made up by five members—experts and representatives of professional boards of architects and engineers—is responsible for the elaboration of the activity programme. *Step* takes action on a wide range of topics and reaches different target groups, operating within a cooperation network composed of a plurality of local institutions and actors: Schools of different levels, museums, eco-museums, associations, professional orders. They all express an educational demand while bringing in experiences and knowledge. The educational activities, therefore, are developed by leveraging their skills and involving high-level experts.

Special attention is devoted to the younger generations, considering landscape as an "educational space" to actively address environmental, historical, and planning issues, and to take responsibility towards the common living environment (Cepollaro and Zanon 2019, 2021). And landscape, as a common good, is an arena for debate on diverse—and often controversial—values and visions, offering an extraordinary opportunity for citizenship education.

16.3 Awareness Raising and Educational Activities

The ELC challenge requires the development of new approaches, methodologies, and tools. On the one hand, awareness-raising activities on the landscape must involve the entire population; on the other, the individual target groups, characterised by different age ranges, personal experiences, knowledge bases, and motivations, require tailor-made educational and training activities (Fig. 16.1).

The initiatives of the School are based on an experiential method that shifts the focus from things to relationships, from landscapes with distinctive features

Fig. 16.1 Workshop "attuned landscapes" with professionals—2017

to perception processes, from peculiar environmental features to people's "everyday living space". In this endeavour, a number of "learning devices" have been developed and used (Cepollaro and Zanon 2021). The idea is to integrate consolidated teaching tools with experiences that give the participants an active role, since they are not passive recipients of knowledge transfer.

Landscape is a mosaic which requires sectoral knowledge to be coherently composed to form a synthesis, and direct experience is pivotal. In this regard, experts can provide disciplinary foundations; people with different cultural backgrounds and skills must be encouraged to exchange their ideas, knowledge and know-how; the elderly can pass on experiences, points of view, and values to the younger generations; professionals can re-orientate their practices towards shared quality objectives. The learning activities are based on an interdisciplinary, intercultural, and intergenerational approach.

Activities with Schools combine disciplinary knowledge and personal engagement, highlighting the connections between different fields: ecology and biodiversity; cultural heritage and territorialization processes; agricultural practices, food production, fertile soil preservation; aesthetic values and cultural heritage; quality of the "living space"; and sense of belonging (Cepollaro and Zanon 2019; Zanon 2019). The expected effect is a synthesis that goes beyond specific points of view or peculiar interests, towards a shared vision and commitment that are at the centre of an educational path to citizenship (Fig. 16.2).

As concerns professionals, architects, engineers, and agronomists play an important role in shaping the landscape and designing a sustainable future. In general, they

Fig. 16.2 Field trip to a protected area—2015

Fig. 16.3 European landscape day 2019—performance "journey towards utopian landscapes"

must be urged to overcome the usual small scale of their interventions by taking into account environmental processes, visual aspects, the inhabitants' role and identity. Technical skills must be integrated considering that dimension, form, and materials of the objects matter, in the landscape; that not only buildings must be considered, but also soil modification and the use of vegetation. Practical activities that mix different professional profiles allow for the exchange between diverse points of view, skills, and methods, against the background of citizens' needs.

As concerns the general public, a number of initiatives have been developed, generally in cooperation with local communities, associations, eco-museums, etc., aimed at raising awareness on the multiple values of landscape and on the need to protect the quality of outstanding sites as well that of the urbanised space. The approach must be positive, not based on a multiplication of limitations and constraints, but rather on the opportunities to improve the landscape by developing concerted projects (Fig. 16.3).

16.4 Results Achieved by the School

After twelve years of activity, the School has become the hub of a cooperation network operating on landscape culture (Cepollaro and Zanon 2019). The figures of the events organised and of the people involved demonstrate the effectiveness of the awareness-raising activities and highlight the educational demand on landscape. The quality of the initiatives has been demonstrated by the special mention assigned to *step* by the Italian Ministry of Cultural Heritage in 2018 as part of the Landscape Award of the Council of Europe (Italian session).

Step manages a website which is a window on the landscape of Trentino and the events going on in the province, thus acting as a meeting point for landscape actors and as a communication tool for different targets. An important aspect is the cultural production regarding publications and books on the landscapes characters of Trentino (Zanon 2019), the elaboration of educational materials such as exhibitions, booklets, a "travel diary" for excursions, etc. Other materials are available on the web, in particular video interviews and documentary films.

A critical analysis of the School's activity is continuously carried out based on the participants' feedback on events, while a self-assessment process is developed together with the partners (Cepollaro and Zanon 2019).

16.5 Lessons Learnt

The School's activities go beyond the task solicited by ELC to raise awareness of landscape values. Landscape is considered an "educational space" that supports disciplinary learning and citizenship education. In this endeavour, a variety of "learning devices" based on personal experience are used, from on-site visits to role-playing games, the elaboration of plans and projects, and hands-on experiences. Some principles are at the basis of the educational methods, namely the ecological dimension of landscape, the personal role—starting from the corporeal and sensory experience, then personal identity—, and consequently the political commitment (Cepollaro and Zanon 2021). A connected perspective regards the consideration of landscape as an arena where to meet, exchange different points of view, and debate controversial issues. There are no recipes for a sustainable future, and landscape, as a common good, requires sharing values and visions, which means democratic participation. What is needed is not simply a trade among the stakeholders based on personal/group interests, but informed participation and political engagement. And the School has taken on this task by supporting the local community in promoting sustainability, democracy, and participation (Fig. 16.4).

Fig. 16.4 "Utopian landscapes": role-play with young people—2018

References

Cepollaro G, Zanon B (2019) Il paesaggio, spazio dell'educazione. Una giornata studio sull'educazione al paesaggio. Atti del convegno tsm-step Scuola per il governo del territorio e del paesaggio, Trento 15 giugno 2018, ETS, Pisa

Cepollaro G, Zanon B (2021) The landscape as a learning space. The experiential approach of a 'landscape school' in Trentino, Italy, Landscape Res. https://doi.org/10.1080/01426397.2021.1942442

Council of Europe (2000a) European landscape convention. Council of Europe, Florence, Strasbourg, CETS No. 176. https://rm.coe.int/1680080621. Accessed May 2020

Council of Europe (2000b) European landscape convention: explanatory report. Council of Europe, Florence, Strasbourg, CETS No. 176, https://rm.coe.int/16800cce47. Accessed May 2020

Debarbieux B (2010) The political meaning of landscape. In: Malpas J (ed) The place of landscape. MIT Press, Cambridge, pp 131–149

Zanon B (ed) (2019) Il paesaggio del Trentino. Caratteri, percezioni e vissuto di un territorio alpino ETS, Pisa

Website

https://www.tsm.tn.it/step
https://www.paesaggiotrentino.it

Gianluca Cepollaro is director of tsm—*step—Scuola per il Governo del Territorio e del Paesaggio* of the Province of Trento, Italy. His area of expertise concerns adult education and professional training, topics on which he has published extensively. He has developed inquiries and research on landscape perception, psychology of landscape, and approaches and methods for landscape education.

Bruno Zanon taught Urban and Spatial Planning at the University of Trento, Italy, until 2018. His research interests focus on sustainable development of mountain areas, regional and urban planning, and landscape protection and management. He has been a member of the Scientific Committee of *step—Scuola per il Governo del Territorio e del Paesaggio* of the Province of Trento, involved in awareness-raising and educational activities on landscape values.

Chapter 17
Common Property in Europe: Landscape Design Evolution Over Time

Monica Pantaloni, Giovanni Marinelli, and Davide Neri

Abstract The term "commons" is used today to indicate a set of land property models, use or management of resources based on a fundamental element: namely allowing, regulating and ensuring land access and use over time by a variously identified community. In Europe, common property plays a significant role in many countries for managing extensive agricultural surfaces. In Italy, state-owned land and common property extend over approximately 1.5 million hectares involving about 5% of the entire national territory. The collective nature of assets that communities have always managed necessarily leads to a greater awareness of sustainable development to preserve the common good for future generations. In 2018, the Italian National Landscape Award of the Italian Ministry of Cultural Heritage focused on the interaction between common property and evolving landscape and awarded first prize to a particular collective property with ancient origins dating back to the Middle Ages located in the Delta Po Region. The winning project: "Between Land and Water, Another Way of Owning" offers a holistic vision of "evolving landscape" and multiple values: public awareness-raising, applied research work, innovative agricultural practices and social cohesion and identifies criteria for a "new generation" agricultural-environmental model capable of combining top-quality agriculture with "ecology" and biodiversity demands. In 2019, the project received an award from the Council of Europe for its great value, thus becoming a part of the Landscape Award Alliance of the Council of Europe Committee of Ministers as a source of inspiration.

Keywords Landscape making · Care · Transition

M. Pantaloni · D. Neri
Department of Agricultural, Food, and Environmental Sciences, Polytechnic University of Marche, Ancona, Italy
e-mail: m.pantaloni@staff.univpm.it

D. Neri
e-mail: d.neri@staff.univpm.it

G. Marinelli (✉)
Department of Materials, Environmental Sciences and Urban Planning, Polytechnic University of Marche, Ancona, Italy
e-mail: g.marinelli@staff.univpm.it

T. Matteini et al. (eds.), *Cultivating Continuity of the European Landscape*,
Environmental History 15, https://doi.org/10.1007/978-3-031-25713-1_17

17.1 Common Property in a European Perspective

The term "commons" is used today to indicate a set of land property models, use or management of resources based on a fundamental element: namely allowing, regulating and ensuring land access and use over time by a variously identified community (Gatto 2017).

These ownership models are implemented with different connotations and widely represented in Europe (Bravo and de Moor 2008).

In many European countries, common property plays a significant role in the management of extensive agricultural surfaces (Table 17.1). These areas, located mainly in the mountains as in the Italian case, are important from an environmental and landscape point of view. Maintaining these surfaces as public property may ensure land protection, biodiversity and sustainable animal husbandry.

The 2010 EU Member State agricultural census allowed more light to be shed on the concept of agricultural common land (CL).[1]

The areas pertaining to CL, in countries where this form of property exists, are generally quite large. They amount to about 10 million hectares (9.3 million, in addition to those relating to Germany, which cannot be quantified) and, on average, they account for 7.6% of the Total Utilizable Agricultural Area (UAA) of the Member States considered and 18.9% of permanent meadows and pastures.

Table 17.1 European countries where various forms of common property were surveyed

Common property	Surveyed/estimated in each surveyed as farms farm having right of use	Surveyed at the nuts 3 level (municipalities)
Also surveyed in the past	Germany, Cyprus, Spain, Croatia, Italy, *Norway*	Germany, Portugal, Spain, Italy, Romania, Austria
Surveyed for the first time during the 2010 agricultural census	Slovenia, *Montenegro*	France, Bulgaria, Hungary Greece, United Kingdom, Ireland
Not present	Czech Republic, Estonia, Latvia, Lithuania, Luxembourg, Malta, Netherlands, Poland, Slovakia, Finland, Sweden, Belgium, Denmark	

In italics, non-EU countries for which Eurostat data is available (Greco 2014)

[1] In this respect, Eurostat's endeavour was critical to bring out the statistical debate about significant agricultural areas, but often neglected or not reported in the official statistics because not directly relating to the farms being surveyed (Greco, 2014). Not less than 18 Member States declared that they had forms of common property within their respective territories. In many countries, cases of "common property" (also known as common land or collective ownership) were detected considering them as real farms or attributing their acreage to individual holdings which have the right of use. (Greco, 2014).

The countries where common property has a greater impact are Greece (32.8%), Bulgaria (19.2%) and Romania (11.7%). In Italy, it accounts for about 610.000 ha, 4.7% of the total UAA and 17.8% of the total of permanent meadows and pastures (Greco 2014).

In 2013, common property amounted to about 9 million hectares (8.8 million) and, on average, accounted for 7% of the total UAA of the Member States considered. While in countries where CL areas had a greater impact in 2010, the values remain more or less the same (Table 17.2), in Italy they account for about 285.000 ha (corresponding to 2.3% of the total UAA, compared to a mean value of 8,1% of the CL/UAA of the countries considered, Fig. 17.1). Considering the updated data of the UAA (2018), the mean percentage of CL (2013)/UAA (2018) is quite similar (Table 17.2).

Table 17.2 % Incidence of common land (CL) on the total of utilizable agricultural area (UAA)

	Utilizable agricultural area UAA (thousand ha) 2013	Utilizable agricultural area UAA (thousand ha) 2018	Common land (ha) 2013	% CL (2013)/ UAA (2013)	% CL (2013)/ UAA (2018)
Bulgaria	4995.11	5030.28	856	17.1	17.02
Ireland	4477.64	4516.04	423.02	9.4	9.37
Greece	5213.04	5288.05	1475.268	28.3	27.90
Spain	23,494.57	24,201.91	1605.369	6.8	6.63
France	28,975.97	29,020.16	675,128	2.3	2.33
Italy	12,426	12,908.75	285,266	2.3	2.21
Hungary	5339.98	5343.78	67,228	1.3	1.26
Austria	2862.44	2653.84	202,133	7.1	7.62
Portugal	3718.16	3591.42	102,239	2.7	2.85
Romania	13,904.64	13,413.74	1,514,634	10.9	11.29
Slovenia	478.89	477.93	8733	1.8	1.83
United Kingdom	17,259	17,357	1,195,246	6.9	6.89

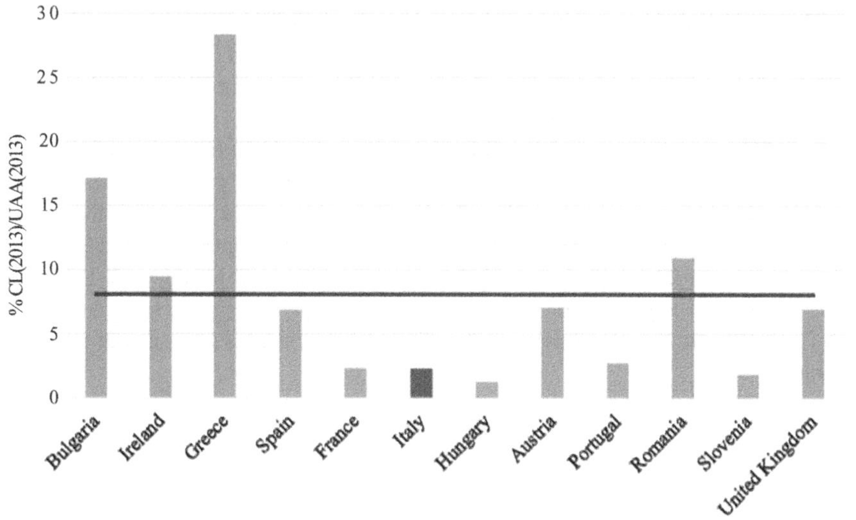

Fig. 17.1 % Incidence of common land (CL) on the total of Utilizable agricultural area (UAA) for each country. Mean of the % CL (2013)/UAA (2013) of the considered countries (red line). *Data source* Eurostat

17.2 Role and Opportunities of Common Property in Italy

In Italy, state-owned land and common property extend over approximately 1.5 million hectares,[2] (ISTAT 2010, Fig. 17.2) involving about 5% of the entire national territory.

The collective nature of the assets that communities have always managed necessarily leads to a greater awareness of sustainable development, to "protect" and enhance the common good in its entire value for future generations.

For this reason, common property, much more than other forms of property, can and must contribute specifically to land development and enhancement, especially following its recognition as a core legal entity for the protection, preservation and sustainable economic growth of the environment and landscape by the Italian Republic, as stipulated by the recent approval of Law no. 168/2017.[3]

[2] The 2233 number of collective properties with agricultural areas detected in the 2010 census consider 610 thousand hectares of Sau plus more than one million non-agricultural hectares, calculated in the Total Area, and mainly represented by woods (Greco, 2014).

[3] Common property is protected by current legislation according to the same requirements applying to state-owned land, namely inalienability, non-applicability of statutory limitations and acquisitive prescription procedures and the principle of immutability (Law n.168/2017, art.3). These are the four principles that have safeguarded the integrity of these properties from any form of "land pathology", such as "privatization" or "public property acquisition and ownership".

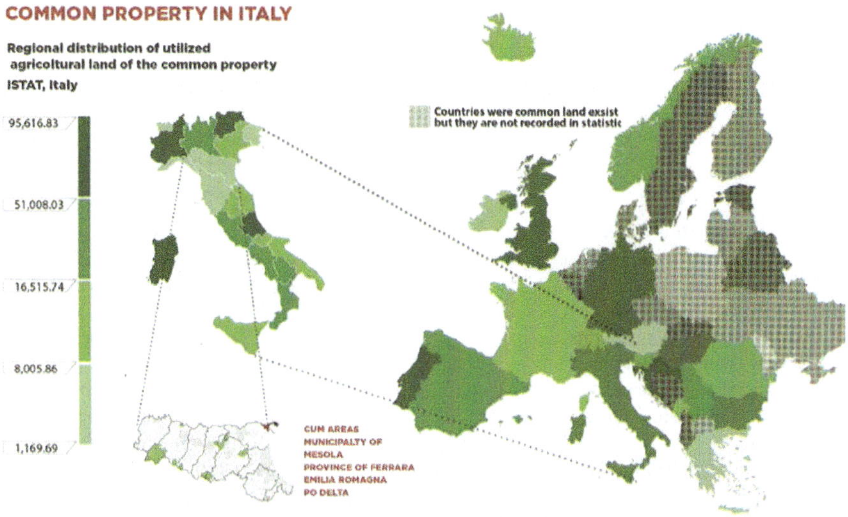

Fig. 17.2 Common property in Italy: an initial Italian geographical mapping. *Source* https://www.coe.int/en/web/landscape/italy

These two "statutes" show unity of purpose in the sharing of their founding values such as "community", "environment" and "identity". They define the desire not to passively take on these components, but to transform their values into active tools with which to act through the landscape project, for the conservation, enhancement and planning of these natural resources and their related social practices, in an ever-evolving relationship between permanence and change, between past and future and between tradition and innovation. Indeed, Italian Law No. 168/2017 "Rules on collective domains" and the treaty open to the Member States through the European Council "European Landscape Convention" represent two cornerstones on which to base the actions and programmes needed to relaunch Italy's fragile areas (Marinelli and Pantaloni 2021) (Table 17.3).

Table 17.3 Comparison between Italian Law n.168/2017 "Regulations on common property" (art. 2) and the European Landscape Convention (art.1)

LAW no. 168, dated 20 November 2017 "Regulations on common property"	European Landscape Convention—Florence, 20/10/2000 *Treaty open for signing by Council of Europe member states and for access by the European Union and European non-member States*
Article 2. Jurisdiction of the State The Italian Republic protects and enhances the collective use of common property, as (a) a fundamental factor in the life and development of local communities (b) a primary tool to ensure the conservation and enhancement of national natural heritage (c) a stable component of the environmental system (d) a local community historical institution intended to safeguard cultural and natural heritage (e) a landscape and environmental system protecting national agriculture, forestry and livestock grazing landscape (f) a source of renewable resources to be used and enhanced by locally entitled communities for their well-being	*Article 1. Definitions* For the purposes of the Convention (a) "Landscape" means an area, as perceived by people, whose character is the result of the action and interaction of natural and/or human factors (b) "Landscape policy" means an expression by the competent public authorities of general principles, strategies and guidelines that permit the taking of specific measures aimed at the protection, management and planning of landscapes (c) "Landscape quality objective" means, for a specific landscape, the formulation by the competent public authorities of the aspirations of the public with regard to the landscape features of their surroundings (d) "Landscape protection" means actions to conserve and maintain the significant or characteristic features of a landscape, justified by its heritage value derived from its natural configuration and/or from human activity (e) "Landscape management" means action, from a perspective of sustainable development, to ensure the regular upkeep of a landscape, so as to guide and harmonize changes which are brought about by social, economic and environmental processes (f) "Landscape planning" means strong forward-looking action to enhance, restore or create landscapes

Source Between land and water, "another way of owning" ISBN 978-88-946213-0-3

17.3 Between Land and Water, "Another Way of Owning" Special Mention for the Council of Europe Landscape Award—European Landscape Convention

In 2018, the Italian National Landscape Award of the Italian Ministry of Cultural Heritage and Activities and Tourism (6th Session of the Council of Europe Landscape Award) was awarded to the "*Uomini di Massenzatica*" Consortium (CUM), for their self-financed and developed project: *between land and water, "another way of owning"*.

In 2019, the project received an award from the Council of Europe for its great value, thus becoming a part of the Landscape Award Alliance of the Council of Europe Committee of Ministers, as a source of inspiration for the Governments of Europe.

The successful experience of the model of the Consortium opens up important future scenarios for the Italian collective land ownership structures for their role as territorial bases of historical institutions to safeguard the cultural and natural heritage, as custodians of the Italian and European landscape.

17.4 Landscape: A Continually Evolving Collective Project

CUM is a collective property of 353 ha in Massenzatica, Mesola, Ferrara Province, Italy.

Its origins date to the Middle Ages when the abbot of the Pomposa Abbey granted local people pasture rights in these marginal lands, which were regarded as unsuitable for farming (Gandini 2009). These lands and the delta in the south of Ferrara province are a "landscape in transition" characterized by changeable borders between land and water generated by the Po river and sea sedimentation processes, subsidence and climate change in a continually evolving land, where man, labour and environment are indissolubly melded and bound together by a shared destiny.

Thanks to their constant labour in these once marginal and "unhealthy" areas, the *Uomini di Massenzatica* community gradually won their independence (Fig. 17.3).

During the last 50 years, the aim has been to supply food market commodities and fruit trees from nurseries, but the present tendency is to enhance the territorial specificity of local produce by establishing a territorial brand known as "Terre Pomposiane".

Since 2010, CUM has been working on an integrated project to promote awareness, applied research work, innovative agricultural practices and social cohesion.

In the project submitted, between land and water, "another way of owning", CUM identified the criteria for a "new generation" agricultural-environmental model combining top-quality agriculture with "ecology" and biodiversity demands (Fig. 17.4).

These actions are based on a "landscape pact" establishing a management model by selecting twenty regularly certified agricultural businesses who each manage 80/100 ha within the Delta Po Biosphere Reserve[4] (Fig. 17.5).

[4] Considering the high environmental value of this area (the protected area of the Po river delta is one of the most important parks on a river delta in Europe), the recognition of "Biosphere Reserve", by UNESCO's explicit statement began in 2011 when the institution of the Regional Park of the Po River

Fig. 17.3 Winning project of the National Landscape Award and Italian entry for the VI edition of the Council of Europe Landscape Award. *Source* https://www.coe.int/en/web/landscape/italy

The expected outcome of the project is to extend its highly positive social capital entrepreneurship model and economic and landscape values across the whole farming network (to over 1500 ha of land) to generate mutual growth and positively stimulate the fragile landscape areas of the Po Delta Biosphere in order to hand down a harmonious landscape and a multifunctional vision of agriculture to future generations.

According to the three sustainability pillars (Our Common Future 1987), the expected results of the project actions are

1. Enhanced environmental sustainability by working on
   - the water and land resource system: improving water use efficiency in specific saline land conditions, especially to promote high quality crops, such as radicchio and plant nurseries;
   - the residual green area and productive landscape system: rebuilding the environmental network via planned landscape investments in the crop plans of the consortium and its partner businesses (estimating 11% re-naturalization of the total farming area);

Delta started to interact with several local entities concerned with the possibility of incorporating this area into a process of candidacy to become a Biosphere Reserve.

Fig. 17.4 Between land and water, "another way of owning" Identical special mentions at the Council of Europe Landscape Award—*6th session—2018–2019* at Council of Europe Landscape award, European Landscape Convention. *Source* https://www.coe.int/en/web/landscape/italy

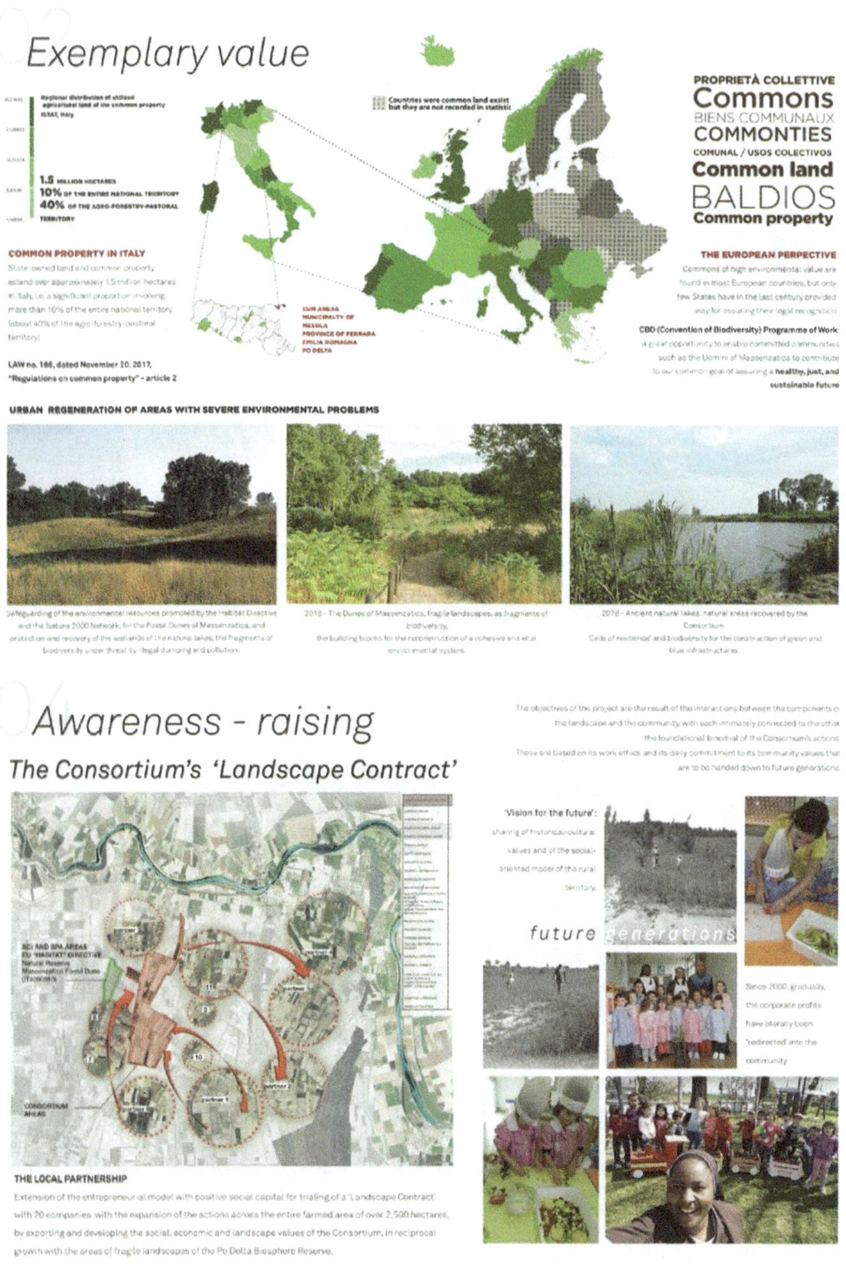

Fig. 17.5 Between land and water, "another way of owning" Identical special mentions at the Council of Europe Landscape Award—*6th session—2018–2019* at council of Europe Landscape award, European Landscape Convention. *Source* https://www.coe.int/en/web/landscape/italy

2. Enhancing economic and social sustainability by adopting a "social oriented"[5] economic model whose revenues and profits have grown over the last 15 years as has investment in the community in response to the collective's external social needs.

   - In fact, since 2000 the Consortium's profits have gradually been re-invested in the community in the following ways (which can be considered good practice benchmarks):
   - Supporting work targeting the community's most vulnerable individuals;
   - Financing educational projects with social value;
   - Refurbishing rural buildings with historical and cultural value, preserving agricultural landscape and fragile environmental resources for future generations;
   - Combating deterioration and promoting the regeneration of abandoned areas, such as the "Terre Pomposiane multi-function area", an educational-experimental landmark for agricultural produce transformation, classroom teaching and exhibitions on history and local traditions.

17.5 Common Property and Biodiversity

There is growing awareness of the key role of local communities in conserving the biological diversity that sustains life on Earth. With the adoption of the Aichi Biodiversity Targets[6] in 2010, this is now fully incorporated into the Work Programme of the Convention on Biodiversity (CBD), a binding international treaty subscribed to by all European countries and the European Union.

Experts have acknowledged that both wild and agricultural biodiversity can be preserved especially where local and indigenous communities have been able to exercise common rights over their land and natural resources, often on the basis of their own ancient rules and governance models (Bassi 2016).

Commons of high environmental value are found in most European countries, but in the last century only few states have provided a way for ensuring their legal recognition, especially at a sub-national level as in the case of Spain and Italy. Countries like Scotland and Romania have recently engaged in providing laws and policies as a means to reconstitute their commons (Bassi 2021).

In 2017, Italy adopted Law no. 168 on "Regulations on common property", linking rural commons to environmental conservation and creating a single national category

---

[5] Thus, the CUM experience fits into the "social oriented" social capitalism experience, as opposed to the more traditional "profit-oriented" capitalism (Gandini, 2009); a different approach which attaches a special importance to the way in which profit is distributed and ensuring the living and working conditions of employees and the poor present in one's community, or in the community where one lives (Muhammad Yunus, author of "A World Without Poverty" (2008), the Nobel Peace Prize winner in 2006 and founder of the Grameen Bank in 1977.

[6] Strategic Plan for Biodiversity 2011–2020, https://www.cbd.int/sp/targets/.

out of the previous heterogeneity (Di Genio and De Vita 2005; Postiglione 2007; Nervi 2008; Bassi 2012).

This is in line with the Programme of Work of the CBD, a great opportunity to enable committed communities such as the *Uomini di Massenzatica* Consortium to contribute to our common goal of ensuring a healthy, just and sustainable future.

Its particular legal history, which has linked common property regimes to landscape and environmental protection, and sizable existing literature has put Italy in a privileged position to play a pioneering role in initiatives to support community heritage within the European Union[7] (Ostrom 2006).

Throughout their history, common property institutions have played a unique role in preserving natural heritage by preventing its destruction and maintaining its ecological, economic and cultural value.

Today, with a significant part of the agroforestry resources of our country suffering from problems related to non-management, abandonment and the loss of a functioning ecosystem, community assets are still current models offering management solutions that provide society with opportunities for widespread access to natural resources.

Acknowledgements The research project was developed by the multidisciplinary group of the Polytechnic University of Marche through interdepartmental collaboration between the Department of Agricultural, Food and Environmental Sciences *(D3A)* and the Urban Planning and Landscape Area of the Department of Materials, Environmental and Environmental Sciences and Engineering. Urban Planning, SIMAU, Faculty of Engineering, entitled: "Preliminary studies for the sustainable development of agri-environmental, horticultural and socio-economic resources of the Men of Massenzatica Consortium aimed at drafting a Landscape Pact in the rural areas of the Lower Ferrara province".

References

Bassi M (2012) Recognition and Support of ICCAs in Italy. In: Kothari A with Corrigan C, Jonas H, Neumann A, Shrumm H (eds). Recognising and supporting territories and areas conserved by indigenous peoples and local communities: global overview and national case studies. Secretariat of the Convention on Biological Diversity, ICCA Consortium, Kalpavriksh, and Natural Justice, Montreal, Canada. Technical Series no. 64

Bassi M (2016) Nuove frontiere nella conservazione della biodiversit€: Patrimoni di comunit€ e assetti fondiari collettivi. Archivio Scialoja-Bolla. Annali di studi sulla proprietà collettiva. 1.2016. Giuffrè Editore, pp 111–136

Bassi M (2021) The environmental value of rural commons. A global overview. In: "Between land and water, "another way of owning". CUM, Italy, pp 194–205, ISBN 978-88-946213-0-3

[7] The sale or division of land does not make sense, because the fragmentation of land into many smaller individual plots cannot produce the same wealth as through an integrated management scheme held by the community as a whole. This model was thoroughly studied by Elinor Ostrom, winner of the Nobel Prize for Economics in 2009, who showed that the best strategy to support man's life in the most challenging habitats is common property and management of common pool resources.

Bravo G, De Moor T (2008) The commons in Europe: from past to future. Int J Commons 2(2)
Di Genio G, De Vita L (2005) Promozione degli usi civici e tutela ambientale. Archivio Scialoja–Bolla (1):147–156
European Landscape Convention (2000) Florence. https://www.coe.int/en/web/conventions/full-list/-/conventions/rms/0900001680080621
Eurostat (2013), Common land statistics – backgrounds, modified in 19 Sept.2013
Eurostat (2013) Data coherence and quality Fss 2013, documento Cpsa/SB/718 presented at Working Group "Structure of Agricultural Holdings", 17 and 18 June 2013
Eurostat (2017) T2 Common land (recorded using Method B), 2000–2013. https://ec.europa.eu/eurostat/statistics-explained/index.php?title=File:T2_Common_land_(recorded_using_Method_B),_2000-2013_(hectares).png&oldid=344803
Eurostat (2018) Utilised agricultural area by categories. https://ec.europa.eu/eurostat/tgm/refreshTableAction.do?tab=table&plugin=1&pcode=tag00025&language=en
Gandini A (2009) Vers un nouveau capitalisme? Il caso del Consorzio Uomini di Massenzatica, Archivio Scialoja-Bolla. Annali Di Studi Sulla propriet'a Collettiva 1:2009
Gatto M (2017) Accesso alle terre e assetti fondiari collettivi: uno sguardo alla situazione internazionale e italiana. Agriregionieuropa anno 13(49)
Greco M (2014) Le statistiche sulle Common Land nell'Unione Europea e in Italia. Agriregionieuropa anno 10(36)
Grossi P (1977) Un altro modo di possedere: l'emersione di forme alternative di proprietà alla coscienza giuridica postunitaria. Giuffrè
Istat (2010) 6° Censimento Generale dell'Agricoltura, Istruzioni per la rilevazione. Rome, pp 163–166
Marinelli G, Pantaloni M (2021) Common property and the European Landscape Convention. In: Between land and water, "another way of owning". CUM, Italy, pp 182–189, ISBN 978-88-946213-0-3
Nervi P (2008) Istituti regolieri e protezione della natura. Archivio Scialoja-Bolla 1:59–80
Ostrom E (1990) Governing the commons: the evolution of institutions for collective action. Cambridge University Press—Italian version: Governare i beni collettivi, Marsilio, Venezia
Postiglione A (2007) Demani civici e proprietà collettive: una risorsa per la biodiversità, Archivio Scialoja-Bolla. Annali Di Studi Sulla propriet'a Collettiva 1(2007):69–77
Postiglione A (2009) La proprietà collettiva: una proposta contro la residualità di suolo e campagna, Archivio Scialoja-Bolla. Annali Di Studi sSulla propriet'a Collettiva 1(2009):27–36
Postiglione A, Maglia S (2012) Natura e biodiversità. Irnerio Editore
Regione Emilia-Romagna, Comitato tecnico Aree Interne (2015) Rapporto di istruttoria per la selezione delle Aree interne. SNAI, Strategia Nazionale Aree Interne
Regulations on common property, Law no. 168, Nov. 20, 2017. https://www.gazzettaufficiale.it/eli/id/2017/11/28/17G00181/sg
World Commission on Environment and Development (1987) Our common future. Oxford University Press, Oxford. https://www.coe.int/en/web/landscape/italy

Chapter 18
Collaborative Governance for Reviving Tiber Landscapes Downstream of Rome

Anna Laura Palazzo and Romina D'Ascanio

Abstract The regeneration of river landscapes in urban and peri-urban areas is increasingly relevant on local agendas to promote integrated policies framing within landscape quality objectives both environmental protection and communities' well-being. Centuries old relationships between the Tiber and Rome, at the basis of the city's development, have gradually faded. Despite crossing a long stretch of diverse human and natural landscapes within the municipalities of Rome and Fiumicino, the river is far from a green blue infrastructure, nor is it conceived as a *Common.* At close distance from the riverbanks, suburbs interweave with traditional countryside, informal settlements, and landfills. Moreover, the administrative complexity and the overlapping of competences from different levels make the establishment of a river governance particularly difficult. New expectations have recently been raised by the adoption of the Tiber River Contract, that is, a collaborative and participative tool to enhance and innovate the effectiveness of decision making process for the protection and management of water resources and mitigation of hydraulic risk, taking into account landscape and local development issues as well. This contribution points out synergies and conflicting logics between traditional planning tools and bottom-up initiatives in reuse and re-appropriating processes, focusing on the south-western area downstream of Rome.

Keywords Tiber River Contract · Peri-urban landscape · River policies · Multilevel governance

A. L. Palazzo · R. D'Ascanio (✉)
Department of Architecture, Roma Tre University, Rome, Italy
e-mail: romina.dascanio@uniroma3.it

A. L. Palazzo
e-mail: annalaura.palazzo@uniroma3.it

T. Matteini et al. (eds.), *Cultivating Continuity of the European Landscape*, Environmental History 15, https://doi.org/10.1007/978-3-031-25713-1_18

18.1 Introduction

Waterways as a settlement location factor and tremendous exchange and civilization vector since the earliest antiquity have developed ever-changing relationships with their riverfront cities. This was the case for Rome, whose rise and economy largely depended on the exploitation of the Tiber River.

Connected to the River Port of Fiumicino, Ripa Grande was the main urban terminal and sorting center for a great variety of goods subject to heavy import duties—food, raw materials, building materials, and luxury goods from overseas—while upstream from Rome, Ripetta received consumer products such as firewood, wine, wheat, and other foodstuff from the hinterland of Lazio, Umbria, and the Marches (D'Errico and Palazzo 2008).

However, after unification and the establishment of Rome as the Capital City of the Kingdom of Italy (1870), the Tiber started to lose its role. Engineering works harnessed the river within high walls to prevent flood risk, cutting off previous relationships to the city and conveying emerging values of appropriateness and representation elsewhere. Navigability fell into decline as goods were carried out by the railroad system (Avarello et al. 2004; Neri 1988).

At the same time, a huge commitment by the State would support reclamation works in the large wedge between Rome and the coastline for the next 70 years up to the Second World War. The coastal area of Rome and the rural south of the city had been privileged targets for interventions such as the reclamation works of Ostia Antica, started in November 1884 with the settlement of 500 laborers from northern regions' hinterland (Orlando 1941; D'Errico 2007; Palazzo 2014). Accordingly, permanent settlement in the countryside (*Agro Romano*) would be encouraged after the First World War in the form of agricultural centers and urban–rural villages that accepted new, intense migrations waves.

Still, from the very beginning, such agriculture regeneration policy critically overlapped with a linear industrial city along the Tiber, jointly nurtured and funded by the State and the Municipality: a railway line to the seaside neighborhood of Ostia, the Via Ostiense, the Via del Mare, and a brand new channel for commercial needs were expected as main facilities.

Under the Fascist regime (1922–1943), *Rome by the Sea* (Ferraris 1907), concept first proposed decades before to functionally regain the lost connection between the Capital to the coastline, was the emphasized slogan tied to the rhetoric on reminders of the past glories of the Roman Empire; concurrently, the half-planned 1942 World's Exhibition (E42) became a main opportunity to boost urban transformation in the whole area (Cuccia 1991; Quilici 1996) (Fig. 18.1).

In the wake of massive urban development in the second half of the twentieth century due to planning policies and illegal settlement dynamics, the whole area underwent remarkable diffusion, relocation, and restructuring phenomena that can be read as the result of an array of individual choices (Insolera 1962; Clementi and Perego 1983) (Fig. 18.2).

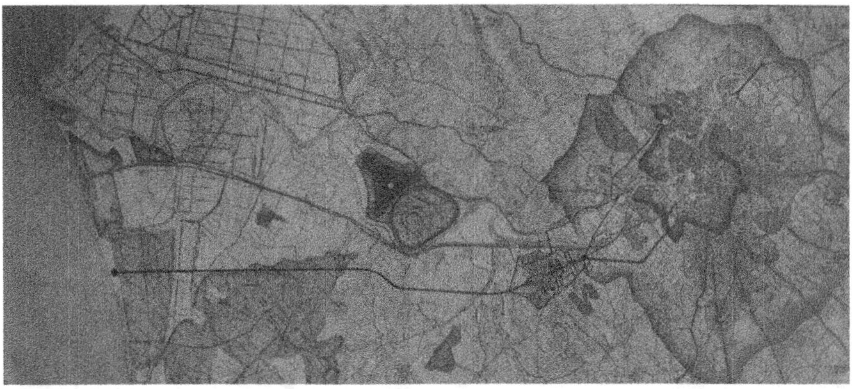

Fig. 18.1 *Tiber River between Rome and the seaside*. The scheme of Via Imperiale connecting Rome to the coastline crosses the pentagon of the E42. Provisions in the 1920s and '30s paved the way for a huge transformation called *Rome by the Sea*. At that time, the new coastal town of Ostia was a garden city provided with facilities: Via Ostiense, Via del Mare and the Roma-Lido railway running parallel to the Tiber. Some improvements were made alongside it: the seaplane base at Fiumara Grande and the rectification of the riverbed at Magliana. *Source* Master Plan by E. Del Bufalo, C. Roccatelli, G. Nicolosi, A. Del Bufalo, 1938, Archivio Marconi

Around 338,000 people currently live along the left bank of the Tiber River between E42 and the seaside, (Borough IX and Borough X), whose access to the Tiber is forbidding even at short distance. On the other side of the Tiber, the Municipality of Fiumicino alone (that separated from Rome in 1993) comprises 78,000 residents, while Borough XI contributes another 232,000 inhabitants (Fig. 18.3); a linear medium-size city in its own, lacking public facilities and social infrastructure and featuring poor relationships among local politicians and municipalities' administrative bodies.

Altogether, the pace and scale of urban growth has hastened the involvement of ever larger portions of countryside, merging dense and scattered settlement and production patterns (Palazzo 2005). The structure and trends of such hybrid space (urban and peri-urban areas) surrounding the consolidated city play a major role due to citywide facilities (the Fiumicino international airport, two harbors, a major exhibition center, several main roadways, a subway line and two regional railways). These assets, along with big retail boxes, shopping centers and multiplex cinemas, merge into or stand opposite the few still untouched stretches of the river, major landmarks and minor traces of ancient settlements, enduring farming features and reclamation landscapes dating back to the first half of the twentieth century (Fig. 18.4).

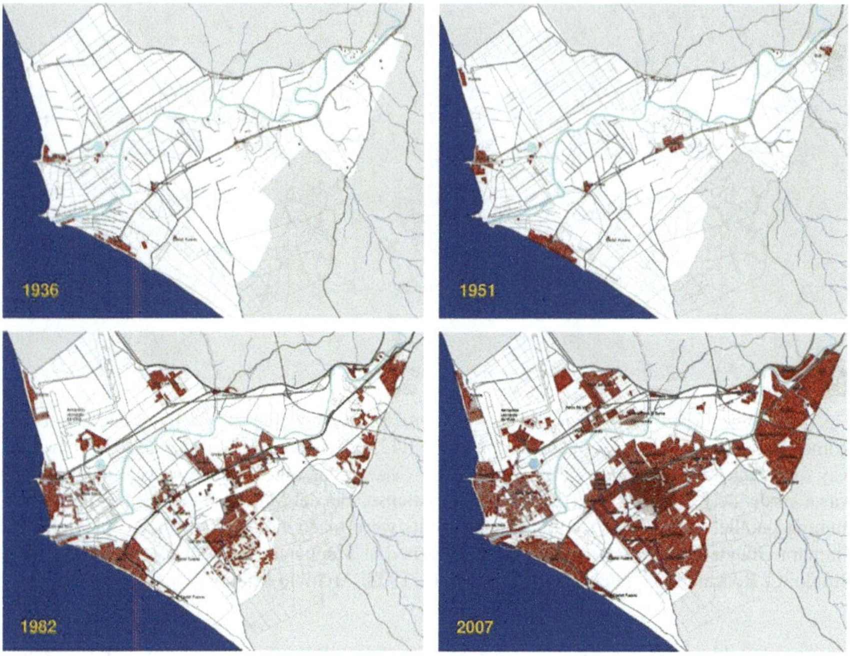

Fig. 18.2 In the first half of the nineteenth century, huge reclamation works coupled agriculture patterns with permanent settlements. Since 1914, the estate of *Monti San Paolo* in an area next to the Tiber has hosted the first rural nucleus close to the railroad (*Borgo Acilio*) for a few settlers coming from the Marches and subsequently a low density district (Acilia) which gathered some 150 workers' families from Lazio and other regions. From the '50s onward, the hinterland has been occupied by middle class residential areas (districts of *Casal Palocco, Ostia, AXA*) and social housing estates (*Spinaceto, Tor dei Cenci, Laurentino 38, Ostia*) adjacent to informal settlements (*Acilia, Madonnetta, Infernetto*). *Source* Interreg Europe STAR CITIES, Regional Assessment of Lazio Region and Roma Capitale Borough IX *EUR*, 2019

18.2 Ongoing Dynamics

Unlike many European cities committed to key urban and landscape projects and partnerships for the management and use of riverbanks, Rome has not yet been keen to exploit opportunities to enhance the role of the Tiber as a common good within the urban fabric or outside the urban walls. As result, ever larger communities settling down in the Tiber Plain are exposed to high environmental risk. In fact, despite being acknowledged as the principal backbone of the Ecological Network set up by the General Plan of the Capital City of Rome (2008), the Tiber continues to be neglected; urban sprawl and land fragmentation are still at work, notably in the vulnerable area of the Delta (water table, flooding phenomena, etc.).

The increasing demand on the part of citizens to access and enjoy the Tiber River is in tune with claims throughout Europe and North America for an extension of all kinds of *Commons*, according to seminal work by Ostrom (2009) spotlighting

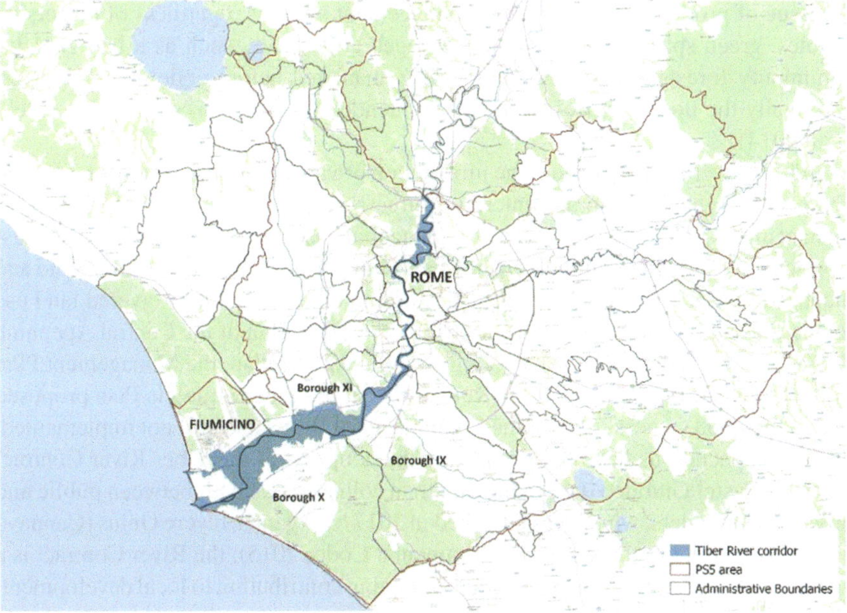

Fig. 18.3 Territorial overview of the Municipality of Fiumicino and the Borough XI *Arvalia* on the right bank of the Tiber and the Boroughs IX *EUR* and X *Ostia* on the left bank. In red the perimeter of the Metropolitan Tiber River basin and in blue the river corridor provided by the Metropolitan Tiber Plan of the Central Apennine Basin District Authority (PS5)

Fig. 18.4 Typical rural landscape of the Agro Romano on the left bank of the Tiber River, Corsetti Estate, Borough X, Rome. *Source* Romina D'Ascanio, October 2020

the issue of production and common management systems regardless of ownership statutes: green spaces organized and managed collectively such as urban gardens, community forests, green areas temporarily managed before undergoing changes, and finally the floodplains of rivers, often intended for river parks (Goni Mazzitelli et al. 2014).

In Italy, river areas fall under the jurisdiction of various public bodies for different competences that overlap and interfere, often resulting in immobility and lack of management. The Tiber represents a very peculiar case due to manifold land concession schemes along the publicly owned banks under the jurisdiction of the State and managed by Lazio Region from 2001 onward, in addition to multi-layered land use regulations and planning tools: the Metropolitan Tiber Plan of the Central Apennine Basin District Authority (PS5), the Regional Landscape Plan, the Management Plan of the State Natural Reserve of the Roman Coast, the Tiber Strategic Plan proposed within the 2008 General Plan of the Capital City of Rome and still not implemented.

New expectations have been raised by the adoption of the Tiber River Contract from the Castel Giubileo dam to the mouth, a collaborative path between public and private stakeholders, which was launched in 2017 by Agenda Tevere Onlus (Cannavò 2018). According to the Italian Environmental Code (2015), the River Contract is a voluntary tool of strategic and negotiated planning contributing to local development, protection and management of water resources, as well as to mitigation of flood risk.

To date, the Tiber River Contract has collected about 90 subscribers from public bodies, universities, research centers, private companies and associations (Fig. 18.5).

Framed by the Tiber River Contract, the management of the riverbanks between the Castel Giubileo dam and the mouth (some 70 km) entails both spatial planning and public–private agreements, a multilevel governance process encompassing public bodies, associations, private stakeholders, and the community at large.

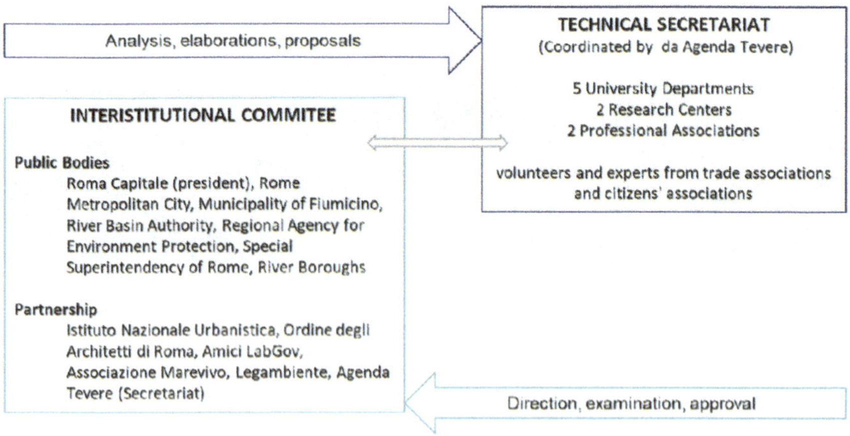

Fig. 18.5 Two pillars of the Tiber River Contract governance system. *Source* processed from the Tiber River Contract website www.contrattodifiumetevere.it

In 2018, the Strategic Document was drawn up, then the first Action Plan was constructed, including about 40 actions for 79 million euros, and finally the River Contract was signed in February 2022.

The Lazio Regional Landscape Plan states the importance of the agricultural system in the construction (in past times) and in the protection (today) of historical landscape. Thus, also in compliance with landscape quality objectives European Landscape Convention (Council of Europe 2000), and European Green Infrastructure Strategy objectives (EC 2013), manifold issues are expected to be dealt with, such as water management, river bank recovery, and recreational activities.

Downstream of Rome, the Tiber River could act to all effects as a backbone connecting the city center to the seaside. Grassroots movements claim the river and its values as part of the public realm, such as the cycling paths that run along the Tiber River on both banks entirely within the State Natural Reserve of the Roman Coast.

Here, for about fifteen years, associations and volunteers have opened up long sections of bike routes by constantly hacking back the vegetation to keep the trail open (Fig. 18.6). However, these justifiable demands find it hard to meet counterparts due to the disinterest of public bodies and legal uncertainty. Communities complain about illegal occupation preventing accessibility.

Since the Corona Virus Pandemic outbreak, the municipality of Rome has been more committed to enhancing slow mobility (notably walkability and cycling policies) both for citizens' daily commutes and for leisure. Some minor links still need to be singled out on the initiative of the boroughs in a bottom-up logic to increase ecological connectivity within the districts and enhance the current fruition of cycle path networks along some reclamation channels.

Important historical and natural assets remain a major challenge for biodiversity conservation and maintenance and for developing leisure and cultural initiatives along the river. The latter could be framed as green infrastructure in light of the EU Green Infrastructure Strategy (EC 2013). Over time, these places and some biodiversity

Fig. 18.6 Cycle path along the right bank of the Tiber River. *Source* www.reginaciclarum.com

hotspots rich in oak forests, wetlands, pastures, and agricultural landscapes have been singled out and protected after long-standing citizens' claims and fights.

Ongoing modernization of the Roma-Lido railway and the purchase of new trains by the Lazio Region are targeting by mid-2024 an increase of the frequency of the service with trains running every six minutes. Besides freeing citizens from car dependency, public transport is expected to better support everyday outdoor recreational practices along the Tiber, its feeders and the system of parks and reserves that make for primary components of the ecological connectivity in Rome, complemented by slow mobility paths.

18.3 Final Remarks

In recent decades, new environmental and ecological concerns have arisen in the general public, together with claims for sustainability and landscape protection policies.

In order to revive the riverbanks as *Commons*, conflicts between top-down and bottom-up initiatives in reuse processes resulting in trade-offs between inclusionary paths and market-driven issues in re-appropriating river spaces can be overcome by framing a collaborative governance approach such as the River Contract. Collaborative governance brings public and private stakeholders together in collective forums to engage in consensus-oriented decision making (Ansell and Gash 2007). New projects and partnerships such as the ones provided by the Action Plan of the River Contract may contribute to refining regional and municipal tools dedicated to enhancing the Tiber River as a primary component of ecological connectivity and as a green infrastructure (Bastiani 2011; Scaduto 2016).

Under this respect, a further step will be to prioritize needs and expectation alongside different threads in order to build a shared sustainable Agenda within the River Contract. Several issues and scales are at stake: urban agriculture along with nature-based solutions able to enhance urban quality and human well-being and ultimately improve community resilience (EC 2020).

A main chapter of such Agenda is to be devoted to a broad open-air museum system, diffuse and centered upon the historical core of Rome (UNESCO World Heritage), sites such Ostia Antica, Portus, and several minor settlements and ruins from Antiquity, offering many opportunities in the fields of restoration, tourism, cultural, and professional training (Palazzo and Pugliano 2015).

The Tiber River, in its capacity as a major historical driver in the development of the area and of the city itself, will carry out a main role in unfolding different narratives:

- Myths, memories, and iconographies (D'Onofrio 1980).
- Nature through history: river landscapes and ecologies in the past and present day (Calzolari 1997, 2012).

- The navigated Tiber: urban and rural economies in the modern age (Pascoli 1740; Paola Scavizzi 1991).
- The navigable Tiber: technical knowledge and water regimes (Rasi 1827; Cialdi 1845; Enzi 2006). *Rome by the Sea*: schemes and master plans, images and imageries for the city and the river (Orlando 1941; Polci 1996).

Clearly, the main criticalities inherent to multilevel governance are rooted in different styles and traditions in planning and decision making. Hopefully, the landscape project embedded in the River Contract (D'Ascanio et al. 2018) is likely to become permanent home for dialogue and cooperation between players endowed with different powers and/or knowledge, reaffirming the role of the Tiber River as a main component of the ecological network and of urban life as well.

References

Ansell C, Gash A (2007) Collaborative governance in theory and practice. J Public Adm Res Theory 18:543–571

Avarello P, D'Errico R, Palazzo AL, Travaglini C (eds) (2004) Il Quadrante Ostiense tra Otto e Novecento. Roma moderna e contemporanea, XII, n. 1–2

Bastiani M (2011) Contratti di Fiume. Pianificazione strategica e partecipata di bacini idrografici. Flaccovio Editore, Palermo

Calzolari V (1997) Storia e Natura come sistema. Argos

Calzolari V (2012) Il Sistema storico-ambientale dell'area romana come fondamento del suo piano direttore. In: Calzolari V (ed) Paesistica = Paisaje, Istituto Universitario de Urbanismo

Cannavò P (2018) Verso il parco urbano del Tevere. Tracce Urbane, 4

Cialdi A (1845) Delle barche a vapore, della navigazione del Tevere e della foce di Fiumicino. Tip. delle Belle Arti, Roma

Clementi A, Perego F (1983) La metropoli spontanea. Il caso di Roma. Dedalo, Roma

Council of Europe (2000) European Landscape Convention, Florence 20.10.2000. European Treaty Series, No. 176. 7p

Cuccia G (1991) Urbanistica, edilizia, infrastrutture di Roma Capitale 1870–1990. Laterza, Roma-Bari

D'Ascanio R, Andreucci MB, Palazzo AL (2018) River management: an opportunity to implement landscape policies. In: Healthy landscapes: green, regeneration, safety. Book of extended abstracts, UNISCAPE conference, Bologna, Italy, 6–8 June 2018, Lorenzo de' Medici Press, pp 298–303

D'Errico R, Palazzo AL (2008) Il Tevere navigato e navigabile. Note sul trasporto fluviale tra Restaurazione e Unità d'Italia. In: Travaglini C (ed) La città e il fiume. Secc. XIII-XIX, Collection de l'Ecole française de Rome, n. 394, Atti del Convegno internazionale, Roma, pp 265–282

D'Errico R (2007) Ostiense. Assetti proprietari e trasformazioni economico-sociali di un settore dell'Agro romano (secoli XVIII-XX). Croma, Roma, p. 166 ss

D'Onofrio C (1980) Il Tevere. Romana Soc. Ed., Roma

Enzi S (2006) Le inondazioni del Tevere a Roma tra il XVI e XVIII secolo nelle fonti bibliotecarie del tempo. In: Mélanges de l'École française de Rome. Italie et Méditerranée, tome 118, n°1, pp. 1320

European Commission Communication from the Commission to the European Parliament, the Council, the European Economic and Social Committee and the Committee of the Regions (2013) Green Infrastructure (GI)—Enhancing Europe's Natural Capital, Bruxelles, Belgium

European Commission, Communication from the Commission to the European Parliament, the Council, the European Economic and Social Committee and the Committee of the Regions (2020) EU Biodiversity Strategy for 2030. Bringing nature back into our lives, Bruxelles, Belgium

Ferraris M (1907) Roma al mare. Nuova Antologia 860:529–575

Goni Mazzitelli A, Palazzo AL, Rizzo B (2014) Urban green commons. Prospettive nel caso romano, Atti della Conferenza SIU L'urbanistica italiana nel mondo. Prospettive internazionali, contributi e debiti culturali, Planum, pp 1736–1741

Insolera I (1962) Roma moderna. Einaudi, Torino

Neri ML (1988) Note sull'espansione di Roma verso il mare. Storia Urbana 82–83(1998):31–48

Orlando P (1941) Alla conquista del mare di Roma. Tipografia della Camera dei fasci e delle corporazioni, p 406

Ostrom E (2009) Governare i beni collettivi. Marsilio, Venezia

Palazzo AL, Pugliano A (2015) The burden of history: living heritage and everyday life in Rome. In: Auclair E, Fairclough G (eds) Theory and practice in heritage and sustainability. Taylor and Francis, pp 54–69

Palazzo AL (2005) Campagne urbane. Paesaggi in trasformazione nell'area romana. Gangemi, Roma

Palazzo AL (2014) Politiche urbane e processi decisionali dall'Unità d'Italia al Governatorato. Roma Moderna e Contemporanea, Anno XX, fasc. (1):37–64

Paola Scavizzi C (1991) Navigazione e regolazione fluviale nello Stato della Chiesa fra XVI e XVIII secolo, Edilstampa, Roma

Pascoli L (1740) Il Tevere navigato e navigabile, Antonio de' Rossi, Roma

Polci S (ed) (1996) L'ingegno del Tevere: attraverso vicende storiche, valori ambientali, progetti e risorse inespresse, Mediocredito, Roma.

Quilici V (1996) (a cura di) E42 Eur. Un centro per la metropoli. Olmo Edizioni, Roma

Rasi GB (1827) Sul Tevere e sua navigazione da Fiumicino a Roma, Tipografia Perego Salvioni, Roma

Scaduto ML (2016) River Contracts and integrated water management in Europe. Springer International Publishing AG Switzerland

Chapter 19
Planning the Adaptation of European Landscapes to Climate Change: The Experience of the EIT Climate-KIC AELCLIC Project

Daniele Torreggiani, Juanjo Galan, Francisco Galiana, Bas Pedroli, and Emilio Servera-Martínez

Abstract Despite growing awareness about climate change (CC) issues, European landscapes are facing increasingly severe challenges related to CC for which a common understanding, vision and effective integrated solutions are not yet available or implemented on a large scale. Therefore, planning the adaptation of landscapes to CC is crucial for our well-being and socio-economic development. From this concern stemmed the EIT Climate-KIC AELCLIC-Pathfinder project (aelclic-pathfinder.com), aimed at defining and testing transferable and scalable models in different climatic, socio-economic, cultural and bio-geographical European conditions for the creation of local networks able to define their own Landscape CC Adaptation Plans (LACAPs). The paper presents some of the main findings of a comparative analysis of the 15 pilot landscapes included in the project. The following were found to be key AELCLIC principles: 1. The landscape concept functions as a systemic and integrated framework to overcome sectoral approaches and to foster a new approach to CC adaptation, deeply rooted in participatory learning and co-creation processes. 2. The construction of multi-stakeholder local networks helps assure the implementability, legitimacy and feasibility of the outcomes. 3. The community-based

Thematic session: Landscape Policy and Governance
Topic: TOPIC 2 | Landscape democracy and rights

D. Torreggiani (✉)
University of Bologna, viale G. Fanin 48, 40127 Bologna, Italy
e-mail: daniele.torreggiani@unibo.it

J. Galan
Department of Architecture, Aalto University, Espoo, Finland

J. Galan · F. Galiana · E. Servera-Martínez
Department of Urban Planning, Universitat Politècnica de València, Valencia, Spain

B. Pedroli
Wageningen University, Wageningen, The Netherlands

T. Matteini et al. (eds.), *Cultivating Continuity of the European Landscape*, Environmental History 15, https://doi.org/10.1007/978-3-031-25713-1_19

identification of landscape values, CC impacts, goals and topics for CC adaptation enhances local support and the solidity of the adaptation strategy or LACAP. Moreover, by following this process, opportunities for innovative actions, potential solutions for foreseeable problems and options to overcome barriers can be discussed in a more productive and visionary way. In all its pilot landscapes, AELCLIC led to the co-definition of programmatic inputs for LACAPs, whose scope and form varied based on the specific context (strategic plans, thematic or specific plans, pilot actions, etc.). Their potential integration with other policies and planning tools were taken into account. AELCLIC has created new spaces and expanded existing ones for international and intercultural dialogue on an issue of crucial importance for both local communities and Europe. This shows that CC adaptation can become an opportunity for promoting landscape quality and strengthening landscape identities.

Keywords Climate change adaptation · Landscape planning · Participatory planning · Co-creation process

19.1 Introduction and Goals

Despite growing awareness about climate change (CC) issues, European landscapes are facing increasingly severe challenges related to CC for which a common understanding, vision and effective integrated solutions are not yet available or implemented on a large scale. Therefore, planning the adaptation of landscapes to CC is crucial for our well-being and socio-economic development. From this concern stemmed, the EIT Climate-KIC AELCLIC-Pathfinder project (aelclicpathfinder.com) aimed at defining and testing transferable and scalable models in different climatic, socio-economic, cultural and bio-geographical European conditions for the creation of local networks able to define their own Landscape CC Adaptation Plans (LACAPs). This paper presents some of the main findings of a comparative analysis of the 15 pilot landscapes included in the project.

19.2 Methodology, Process and the Participatory Approach

AELCLIC addressed CC adaptation from a landscape perspective, which meant taking landscape as a framework for CC adaptation, rather than as an object, and using the landscape concept as an ideal platform for seeing the whole integrated picture instead of individual parts. Among the main implications of this approach is the adoption of a systemic, cross-disciplinary and participative perspective to interpret the relationships between environmental, cultural, economic and social factors and to co-define holistic interventions on them based on co-construction and design thinking processes. The stakeholder-driven process characterizing the activities of the AELCLIC project was based on a sound consideration of CC impacts, risks and vulnerabilities derived from state-of-the-art knowledge and previous experiences at a local level, tightly integrated with an analysis of the public's perception of them. Since European landscapes, expressing our history, diversity and identity and supporting our economic development and well-being, are the result of our collective creation and of the action and interaction of natural and human factors (ELC), their CC adaptation must be site-specific and based on the contribution of all the people and groups that live, work and enjoy those landscapes. Therefore, AELCLIC was firmly based on a bottom-up approach aimed at including—more or less directly and consciously—all the relevant stakeholders involved in landscape transformation from the initial phases of co-defining the workplan and goals to the co-identification of impacts and the co-design of different visions and aspirations.

This process was developed and tested in 15 pilot landscapes distributed in rural, urban and fringe areas, selected to cover Europe's diversity and to enable the production of transferable and scalable models (Fig. 19.1). The project was carried out by 20 European organizations (of which 9 Climate-KIC partners), including universities, local and regional administrations, NGOs, social and economic actors.

The AELCLIC process was started in a set of leading pilot landscapes, one in each of the four European macro-regions of the project (Northern Europe, Atlantic-Alpine Europe, South-Western Europe, South-Eastern Europe), where tandems of Universities and Regional/Local Authorities (all Climate-KIC Partners) initiated the definition of different models for the generation of local networks of stakeholders. The AELCLIC process was then propagated to a set of multiplier pilot landscapes with the support of academic, regional/local authorities, business and non-governmental third parties. A total number of more than 100 public and private organizations and 500 participants covering societal, economic, institutional and academic sectors were

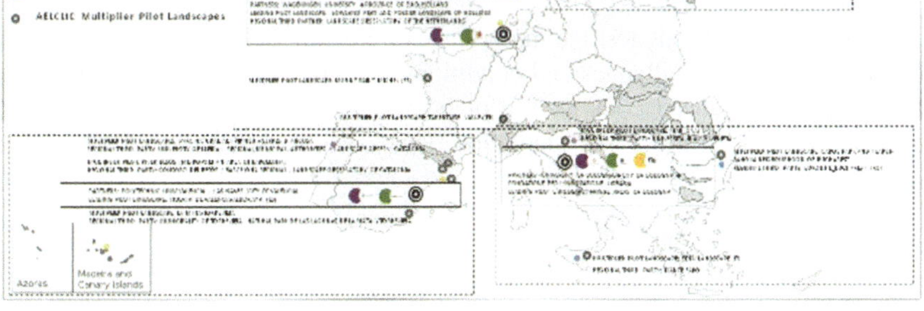

Fig. 19.1 AELCLIC's European macro-regions and pilot landscapes

involved in all the pilot landscape network processes. In addition, the participation of the UNISCAPE and CIVILSCAPE networks provided a solid platform for dissemination and scalability. Regular meetings between the universities leading the implementation of the project in the European macro-regions and an international workshop in which representatives of each pilot landscape participated resulted in combining a site-specific and local approach with a trans-national vision and exchange of experiences that, indirectly, promoted the co-creation of feasible and sustainable follow-up actions.

A circular and iterative process allowed the pilot landscapes to fruitfully exchange experiences with cross-fertilization and mutual learning of both methods and results. In each pilot, the project partners promoted and led the identification and creation of open and inclusive networks of stakeholders. These networks started their work by defining a workplan to be developed throughout the year 2019 that would lead them to the co-definition of a basic framework for future preparation of their own landscape CC adaptation plans (LACAPs). In particular, each group went through a series of hands-on workshops under the facilitation and the knowledge-based guidance of the project partners and other experts, where they co-identified the impacts that CC might have on their landscapes and their own ways of living.

After this diagnosis, the project activities focused on the co-definition of objectives, structure and basic contents of future LACAPs whose scope and form varied depending on the specific context of each pilot landscape, ranging from strategic plans, to thematic or specific plans, pilot actions, etc. Moreover, these different typologies of LACAPs were taken as clustering criteria for the international workshop activities where representatives of the various local stakeholder networks had the opportunity to work together, even across the European macro-regions, to launch follow-up phases. In this way, the project was conducted on a local scale but always referring to a European vision and to the generation of new alliances among landscapes with similar expectations and challenges.

The project was based on an interdisciplinary and participatory approach as well as on the use of strategies to promote active learning, team building and planning/design thinking through a wide range of complementary tools and methods that were gradually calibrated by the partners to better fit the specificities of the various pilot landscapes and local networks. The project involved the combined use of multi-criteria analysis with literature, modelling and stakeholder or expert inputs to achieve robust outcomes.

The project results suggest the importance of creating flexible workplans adaptable to the different challenges and conditions that affect different landscapes and of aiming at a balanced combination of bottom-up and transdisciplinary expert-supported processes in the CC adaptation field, covering science, technology and policy issues.

19.3 Results

As a general criterion, LACAPs can include regional/local policies, strategies, pilot actions and initiatives to promote CC adaptation. Consistently with the principles of the ELC, the programmatic inputs for future LACAPs in all the pilot landscapes were always related to landscape quality objectives and varied depending on the specific nature of each landscape. Due to the variety of topics, objectives, challenges, barriers, values and resources of each local landscape, LACAPs cannot be referred to a standard set of instruments ex ante, but must adapt to local needs in order to guarantee flexibility. Among the most important general findings are the following:

- LACAPs may be translated into specific codified plans or instruments, or into specific layers of climate adaptation policies and planning, encouraging land use, spatial and urban planning policy adaptation, integration with existing or under-development plans and pilot action coordination;
- Integrated landscape projects, including systemic actions on several themes and sectors, can facilitate the implementation of a LACAP and increase its internal consistency and alignment with EU, national and regional/local strategies;
- Climate risk or vulnerability assessments, landscape characterization and the analysis of perceived CC impacts, opportunities, solutions, actions and barriers are fundamental contents of each LACAP, which should also include an implementation and a monitoring plan.

Various factors influenced the development of AELCLIC activities such as the sociocultural and governance environment, the existence of a clear landscape or community identity and its urban or rural character, the type of connections between local inhabitants and their landscapes, existing knowledge of CC and mitigation/adaptation strategies and the level of involvement and role of local and regional authorities. The combination of all these factors led to very different results and the understanding of this diversity constitutes in itself an important result of the AELCLIC project that can also inform the transferability of the AELCLIC methods to other landscapes.

From a methodological point of view, the main lessons learned include the importance of

- guaranteeing openness and representativeness in local networks so as to recognize the different landscape identities democratically, permit a holistic vision and increase the implementability, legitimacy and feasibility of their decisions;
- engaging representatives of supra-regional organizations, key economic actors and administrations in charge of territorial and environmental planning and landscape protection;
- prioritizing options using multi-criteria analysis and stakeholder inputs;
- developing joint workshops or sessions with other projects;
- integrating CC interventions with existing or future policies and planning tools;

- considering the environmental and social effects of CC, from global to landscape-specific scenarios.

19.4 Conclusions

AELCLIC fostered a participatory process in which CC adaptation represented a challenge to be addressed collectively. Global threats and trends were addressed in a local and site-specific perspective. This permitted reflection on CC as a boundary condition for landscape protection, management and planning, as promoted by the ELC.

The following were found to be key AELCLIC principles: (1) The landscape concept functions as a systemic and integrated framework to overcome sectoral approaches and to foster a new approach to CC adaptation, deeply rooted in participatory learning and co-creation processes. (2) The construction of multi-stakeholder local networks helps assure the implementability, legitimacy and feasibility of the outcomes. (3) The community-based identification of landscape values, CC impacts, goals and topics for CC adaptation enhances local support and the solidity of the adaptation strategy or LACAP.

Moreover, by following this process, opportunities for innovative actions, potential solutions for foreseeable problems and options to overcome barriers can be discussed in a more productive and visionary way. In all its pilot landscapes, AELCLIC led to the co-definition of programmatic inputs for LACAPs, whose scope and form varied based on the specific context (strategic plans, thematic or specific plans, pilot actions, etc.). Their potential integration with other policies and planning tools were taken into account. With the creation of local networks connected at a European level, AELCLIC has created new spaces and expanded existing ones for international and intercultural dialogue on an issue of crucial importance for both local communities and Europe. This shows that CC adaptation can become an opportunity for promoting landscape quality and strengthening landscape identities.

Acknowledgements The AECLIC pathfinder project was funded by EIT Climate-KIC.

Chapter 20
The European Baykal—Regional Development Inspired by the European Landscape Convention

Nataliya Pozhidaeva, Alexandra Kruse, and Victor Pozhidaev

Abstract The main idea of the NLSF project is to organize a territory of comprehensive sustainable development, based on tourism in Przhevalskoye of Smolensk region. The project was inspired by the European Landscape Convention and additionally motivated after an onsite working meeting of experts, local actors, and politicians from Russia and Germany, among others the authors of this article. The location was chosen due to the high-quality landscape and human potential, sufficient proximity to Moscow and because of the will of the local actors, to overcome the depression, which escalated these last decades—a typical phenomenon in rural areas. Even if areas are considered by foreigners as beautiful or even outstanding, depression can be striking due to missing employment opportunities. The base of the project is a package of innovations by authors, partly approved by the Council of Europe. The article will present the concept for valorizing the landscape, protecting the natural environment by at the same time creating a sustainable local economy based on innovative landscape approaches, like layouts of Europe and Russia, landscape passports of birthdays, script-game scenario, etc. The active involvement of residents, the use of best foreign practices and direct participation, primarily of Germany, will be further developed.

Keywords Agricultural landscape types · European database · Landscape as heritage

N. Pozhidaeva (✉) · V. Pozhidaev
The Interregional Public National Landscape Salvation Fund—NLSF, 12/7 Tverskaya str., Moscow 125009, Russia
e-mail: civilland@mail.ru

V. Pozhidaev
e-mail: civilland@mail.ru

A. Kruse
insitu World Heritage Consulting, 10bis, Rue du Haras, Buc 78530, France
e-mail: akruse@whconsult.eu

T. Matteini et al. (eds.), *Cultivating Continuity of the European Landscape*, Environmental History 15, https://doi.org/10.1007/978-3-031-25713-1_20

20.1 Introduction

In 2018, the Interregional Public National Landscape Salvation Fund (NLSF) organized the international meeting of experts, "European Baykal" in the urban village Przhevalskoye at Lake Sapsho, situated in the National Park and UNESCO Biosphere Reserve "Smolenskoye Poozerie" (Fig. 20.1). The Smolensk region is located in Western Russia, close to the boarder with Belarus. The lake Sapsho was called Mini Baykal by the world famous traveler Nikolay Przhevalsky, honorary member and academician of many European academies. Przhevalsky was leader of five expeditions to Mongolia, China, and Tibet, discoverer of many geographical phenomena and also eponymous of the wild Przhevalsky horses. The meeting's aim was to evaluate the potential of this place as model landscape for sustainable regional development, based on the valorization of the landscape itself. Foreign experts had been asked to present and to inspire with foreign best practice examples. Earlier, in 2006, the Parliamentary Assembly of the Council of Europe in Strasbourg unanimously approved the PACE report «Protection and use of European landscapes» by the reporter Mr V. Sudarenkov, Russia (Rec № 1752). Among other things, the reporter offered the program of the NLSF and the National Trust in Russia, based on the research of their leader N. Pozhidaeva about "The formation of national socio-natural landscapes system is the real way for sustainable development", approved by the Russian government—Ministry of Economic Development of Russia (1996)

Pozhidaeva (1995, 1997, 2000, 2001, 2002). Under the special item—11.2—PACE recommended to set up a European program to establish a "pan-European system of national socio-natural landscapes as a genuine mechanism for sustainable development" (http1). After a careful survey of possible territories for the future project in eight Russian rural regions, Przhevalskoye was elected by N. & V. Pozhidaevs in 2016. In the meanwhile, both of them have achieved many steps and presented their ideas to many different actors on local, national, and international level. In order to convince, they lean on elements from the Russia's traditions and culture, e.g., handmade prepared models (layouts) of the Planet, including Europe and Russia as well as Przhevalskoye (Fig. 20.5), the presentation of the European Baykal project in the Cultural center of the Ministry of foreign affairs in Moscow for the Embassies in Moscow, etc. (http2). Several exhibitions had been developed, e.g., architectural projects, the youth landscape village for sustainable development, the water parks/expositions; the "Tree of harmony", and the "Landscape House".

Sustainable development and a landscape approach are not yet well known among the ordinary people in Russia. Therefore, new approaches to raise awareness have been created, e.g., a script-game scenario, based on the N. Pozhidaeva book-fantasy "The soul of landscape" (Pozhidaeva 2013) and its map «A vision of Russia" (as the "traveling in Russia from Kaliningrad to Vladivostok in search of sustainable development"). Furthermore, 365 landscape birthdays passports and other unique and unexpected forms of awareness raising often with a script-game scenario have been created in order to promote the landscape approach (Fig. 20.5).

(a) Lake Sapsho, called the European Baykal is known for its beauty. (V.Pozhidaev, 2020)

(b) The Biosphere Preserve gives home to many species, here we see a heavy beaver's construction. (A. Kruse 2018)

(c) One of four Sapsho islands– the calmness is what attracts tourists. (V.Pozhidaev, 2020)

(d) Birches are the dominating species of the local forests on moor and peat soils. (A. Kruse 2018)

Fig. 20.1 Impressions of the Biosphere Reserve Smolenskoye Poozeriein Smolensk Region

The following article explains the specific approach to revitalize a rural area, which is facing depression and rural exodus after the collapse of the former Soviet regime. The complex project of the NLSF is essentially important for the local community. The NLSF was inspired by the European Landscape Convention (CoE 2000). Its leaders N & V. Pozhidaevs repeatedly spoke at official parliamentary meetings about the need for Russia to sign the European Convention (Bulletin of the Council of Federation No. 3 2008). Last but not least, the project will help the local inhabitants to gain a biophilic view of life after the pandemic of COVID-19 and to believe in the reality of sustainable development prospects, based on the European Landscape Convention.

20.2 The Smolensk Region and Its Lake Sapsho, Called the European Baykal

Smolensk region has a significant tourist and recreational potential and has good prospects for becoming one of the leading tourist regions of Russia and creating competitive tourist products in foreign markets, especially addressing Europe.

Tourist traffic is formed along the route Moscow-Minsk, which partially repeats the historical path "The Old Smolensk Road". Today, the most popular format of visits are one-or two-day trips. Smolensk is a transit city for tourists traveling from Moscow to the Republic of Belarus or to Europe and in the opposite direction.

Many famous persons are connected to the Smolensk region, which is why tour operators often complement tourist programs by visiting memorable places associated with famous Smolensk citizens:

- In Gagarin: Museum of Yu.A. Gagarin (town of Gagarin and the village of Klushino), Local history museum, art gallery, children's museum "Games of Yuri Gagarin," Museum of the First Flight.
- Museum M.I. Glinka in the village Novospasskoye (Elninsky District);
- Museum AS Griboyedov in the village Khmelita (Vyazemsky district);
- Museum of the great naval commander P. Nakhimov (Smolensk and Khmelit);
- Museums of the poet A.T. Tvardovsky (Smolensk and the farm "Zagorje" in Pochinkovsky district);
- N.M. Przhevalsky House Museum (urban village Przhevalskoye, Przhevalsky settlement in the Demidov district);
- Museum of S.T. Konenkova's Sculpture (Smolensk);
- Places associated with Y. Nikulin (Demidov);
- Places connected with A. Papanov and L. Kasatkina (Vyazma).

In 2012, more than 180 thousand people visited Smolensk region, in 2015 even 300 thousand. Smolensk itself was in the top ten of the most popular tourist cities in Russia, due to the relatively small distance from the capital and the low cost of tours for trips to the New Year and May holidays. On the territory of the Smolensk region are located 4225 objects of cultural heritage, as well as 1754 objects of archaeological significance. Some of the most famous Russian tourist brands are located here:

- Gagarin's Motherland,
- Smolensk fortress wall,
- The Old Smolensk Road,
- Smolensk Lakeland.

However, as of 2018, the Smolensk Region is not yet present among the constituent entities of the Russian Federation for which regional goods are registered (appellations of origin of goods (AO)).

20.3 Tourism as Driving Factor and Source of Local Income and Development

The UN WTO (http3) report about international tourism highlights from 2019 points out a total of 1.4 billion international tourist arrivals per year and which represents 1.7 trillion USD. This makes the sector a true global force for economic growth and development, driving the creation of more and better jobs and serving as catalyst for innovation and entrepreneurship. For the seventh year in a row, tourism exports grew faster than merchandise exports. Tourism is an important component of export diversification both for emerging and advanced economies, with a strong capacity to reduce trade deficits and to compensate for weaker export revenues from other goods and services. Digitalization, innovation, greater accessibility, and societal changes are expected to continue shaping tourist sector.

Destinations in Central and Eastern Europe grew in line with the region's average. Visa facilitation led to an increase of Chinese tourists in some destinations. The Russian Federation hosted the 2018 FIFA World Cup and recorded double-digit increase in international tourism receipts. Positive performance in Western Europe was led by Belgium and the Netherlands, who consolidated their rebound. Four out of five tourists travel within their own region. Certainly, in 2020 the COVID pandemic reinforced this trend, which might even have a longer lasting effect as many people discovered the beauty of their home countries and of destinations nearby.

According to the UN WTO study, the following services are still considered important: provision of basic tourist services such as accommodation and meals, transport services, and equipment of tours. Sightseeing services are also an important element in the configuration of the tour company. The United Nations (UN) General Assembly proclaimed 2017 as the International Year of Sustainable Tourism in order to develop this sphere, to improve mutual understanding between peoples, spread knowledge about the rich heritage of various civilizations, and ensure respect for the eternal values of different cultures and thereby strengthen peace throughout the world. The tourism industry is the main source of employment, currency, and tax revenues for developing countries. The foreign exchange flow of tourist income in developing countries is much larger than all aid flows from foreign donors (Fig. 20.2).

Before corona pandemic, an increasing number of countries around the world were becoming easily reachable for tourists. States invest in tourism development, turning it into the main driving force of social and economic progress through the creation of new jobs and enterprises, the generation of export earnings, and the development of infrastructure. Over the past six decades, tourism has experienced

(a) Ecohotel "Spring" (V. Pozhidaev 2020)

(b) New holiday houses grouped around a common spa at Lake Sapsho. (A. Kruse, 2018)

(c) Valorisation and recognition of local (national) cultural heritage and at the same time handcraft traditions by local artists. (A. Kruse 2018)

(d) Local handy craft traditions, here artistic objects from birch trees and master classes are offered: Unfortunately not very well known. Even the neighbour was not aware of the capacities. Communication and cooperation is needed. (A. Kruse 2018)

(e) Many personal initiatives to establish a rural agri-tourism with different standards of accommodation and direct marketing of agricultural products. The biggest problem is the missing internet connection – the houses had been "forgotten" and administration needs time to get things in order. (A. Kruse 2018)

Fig. 20.2 Tourism activities in Smolensk

long-term development and diversification and has become one of the largest and most dynamic sectors of the world's economy.

The study "Tourism 2020 Vision", conducted by the UN World Tourism Organization (UNWTO) (http3 2021), identified the most promising directions and types of tourism of the XXI century (http4 2021). Experts believed that by 2020, recreational tourism will become the most popular type of tourism. Based on the results of this study, the participants of the UNWTO conference in Lisbon identified five most promising tourist destinations of the twenty-first century, among which four directions will be developed as part of the proposed project European Baykal. This tourist trends must be significantly expanded.

It was supposed that agri-tourism will be a sector of the tourism industry, which, according to various expert estimations, in particular the UN WTO will become one of the five main strategic directions of tourism development in the world until 2020. The great demand for rural tourism in Europe is confirmed by the fact that this tourism sector brings about 20–30% of the total income of the tourism industry. Ecotourism is a natural tourism, which includes the study of the natural environment. This direction includes both thematic educational tours for ecotourism lovers and tours for holidaymakers at resorts with visits to specially protected natural areas (PAs). This kind of tourism is very popular in Northern and Central Europe. In some regions of Russia, the process of developing ecological tourism is also quite active. According to the Federal Agency for Tourism (http5), since 2011 the flow of environmental tourists to the protected areas of Russia (national parks and reserves) has increased significantly. Since 2014, Russia has been developing the popular international tourist brand "The Most Beautiful Villages" (http6), which was first established in France. This tourist model brings together rural communities with exceptional heritage and has many opportunities for development through the receipt of various grants and subsidies.

Next to agri-tourism, also adventure tourism is considered as a still growing sector. This type of tourism combines all travels associated with active modes of movement, with the aim of obtaining new sensations, impressions, and the study of the most remote corners of the Earth. The geography and theme of adventure tours are quite extensive and diverse; new tourist needs require the development of an appropriate tourism product.

The third category with growing potential is cultural and educational tourism. The market of informative tourism is wide. The largest flows of tourists traveling for educational purposes go to Europe, Asia, and the Middle East. Diverse informative tourism is divided into two main types: stationary tours (including city tours) and itinerary tours (most of the proposed comprehensive tourist projects and travels). For small groups of tourists, informative tours are organized in the form of day trips with visits to cultural monuments (http7, http8).

What does this mean for the European Baikal region? In the course of the research, it turned out that the proposed new brands of Smolensk and, in particular, "European Baikal—Republic of Brothers of Months"[1] and "Smolensk Region—Landstandard*****[2]" will be able to successfully compete with such leading brands in other countries.

In the sphere of tourism, this concerns not only the more advantageous transport logistics of the Smolensk project (450 km from Moscow—the megacity № 9 in the world) and the more valuable potential of the natural and cultural heritage of the model territory. It was searched for a brand, which attracts people not only at one date or season.

The chosen brand name Landstandard*****offered by N&V Pozhidaev's (Neverov 2018), which can be assigned to several names of relevant agricultural products and tourist services at once, does combine "land" as rural area with its positive associations like calmness, health, nature, with "standard" as a term of quality, referring to a five star hotel. At the same time, it is a wordplay, also referring to "art". Generally speaking, the brand means the products and services are of high quality due to the outstanding landscape.

20.3.1 Analysis of the General Situation in the Russian Market of Tourist Services

In Russia in 2019, the number of inbound trips to the country decreased slightly compared with 2018. The indicator fell by 0.9%. As for outbound tourism, on the contrary, the indicator has increased. In the first nine months of 2019, Russians made more than 35 million trips abroad. This figure is 7% higher than in the corresponding 9 months of the previous year 2018. The most frequently visited Russian cities are Moscow, St. Petersburg, Novosibirsk, Sochi, as well as the entire Krasnodar territory, Kazan, Yekaterinburg, Vladivostok, and Kaliningrad. If we consider outbound tourism, the following tourist destinations are most popular among Russians: Turkey, Finland, China, Italy, and Germany.

The number of travel agencies has increased over the past 8 years by more than 30%, but the total number of tourist packages that they sell annually does not change in total. From this, we can conclude that part of the population began to travel independently or the number of trips is not so much related to recreation as, for example, business or medical tourism. The number of incoming tourists from 2014 to 2019 decreased from 25,438,000 foreign tourists to 24,419,000. This can be interpreted as

[1] In Russia, there is a very famous fairy tale by S. Marshak "Twelve months", where the heroes are namely 12 brothers named after the months (January, February, etc.). They represent the forces of nature. Walt Disney was interested in adapting it as a cartoon. However, it was Japan where similar films already exist, who became the award. In Pozhidaev's book "The soul of landscape" this fairy tale inspired the key story "The republic of brothers of Months" deals with sustainable development.

[2] Protected designation.

a change in the political situation or an insufficient attractive level of service. From 2017 to 2019, outbound tourism increased by 30%—from 39,29,000 overnight stays to 4,533,000. (http9).

20.4 The Inspiring Character of the European Landscape Convention

Landscape is often only connected to nature and therefore to nature protection. However, it is so much more: It is cultural heritage and habits (Kruse and Pungetti 2007). As stated in the preamble of the European Landscape Convention (ELC), "landscape is what is perceived by the people!" (CoE 2000: preamble). Landscape is very diverse and what people perceive as "nice" or "beautiful" or "ugly" depends on their background—especially childhood—, education, expectations, culture, well-being, and often the moment itself: Do we experience a landscape during our holidays or as our daily environment? (Kruse and Pungetti 2007).

Even if landscape can be considered as a common good, we are missing clear responsibilities, conventions, guidelines and laws. Furthermore, in Europe, we miss a common understanding about heritage and language. Most approaches are from a regional respectively a national point of view. This fact explains why the ELC is so important—and also so successful! It is still the only convention, which looks from an upper level to all landscapes as equal. Cultural landscape is the result of interdependencies and feedbacks between natural conditions and human impacts in history, present, and future (Kruse et al. 2010). The target is not only to document the peculiarities of the historical landscapes for conservation purpose. Rather, the task is to design and lead the future change and dynamics of landscapes for new possibilities of identification for the inhabitants in their homeland. One important step forward to the valorization, recognition, and protection of landscape was the setting-up of the European Landscape Convention in 2000 in Florence. It was ratified in 2004. Today, it is signed by 41 and ratified 39 by States—among those, neither Russia, nor Germany. Ten countries have included the definitions of ELC in their national legislation (especially regarding spatial planning, natural, cultural environment). Ireland, Belgium, Spain, and France have copied or integrated the definitions in their national laws (BHU 2016).

In the preamble is also stated that the "ELC of the CoE promotes the protection, management, and planning of the landscapes and organizes international cooperation on related issues". Even if already cited many times, §1 of the ELC is still and always valid: "Landscape" means an area, as perceived by people, whose character is the result of the action and interaction of natural and/or human factors. Noting that the landscape has an important public interest role in the cultural, ecological, environmental, and social fields, and constitutes a resource favorable to economic activity and whose protection, management, and planning can contribute to job creation. Aware that the landscape contributes to the formation of local cultures and that it is a basic component of the European natural and cultural heritage, contributing to human well-being and consolidation of the European identity" (CoE 2000).

There are many good reasons for its national implementation: *Landscape actions* can be a combination of protection, management, and planning. Knowledge about landscape is needed, because good decisions can only be taken on sound knowledge basis. Therefore, landscape identification, description, and assessment are needed. In order to improve acceptance but also European cooperation, this should be done harmoniously in a more or less standardized and comparable way. A definition of (national/regional) landscape quality objectives can help to define local, regional planning goals and will provide a framework to the attainment of these objectives by protection, management, and planning over a period of time. Mapping and assessment is needed to localize these landscape-based quality objectives. A constant monitoring of changes as well as the evaluation of the effects of policies is needed, accompanied by a possible redefinition of choices.

Alarm signals have to be fixed as well as punishments in mayor cases. The ELC gives advice for all these tasks. Last but not least: ELC fosters the safeguarding of cultural heritage without excluding the use! However, many countries have already understood that ELC matters. The following list gives a few reasons why landscape protection is so important (BHU 2016: 16): Attention that we do not lose our cultural and regional specialties, due to globalization ("McDonaldization"). Demographic changes lead to rural exodus which leads to a loss of cultural landscapes and related heritage systems including the related knowledge. Agricultural landscapes are energy landscapes in both meanings: As a production ground for renewable energies but also because their force provides personal energy to the people. The destruction of home landscapes lead often to the loss of regional identity and sense of place. However, people, even living in the cities, do need their connection to "their" landscapes (Kruse and Pungetti 2007). Landscape is transvers on political level. Therefore, it cannot be treated only locally. Research has clearly proven that people long for diverse, "nice/harmonious", qualitative attractive, high value landscapes (see divers research conducted by Schwarz-von Raumer et al. (2019), Renes et al. (2019), Roth and Buchecker (2010), Buchecker et al. (2007). In order to answer to the formulated demands, raising awareness, sharing responsibility, and liabilities must become the driving factors in local planning and policy. The ELC helps with ideas and networks. In this sense, expertise gets stronger when linked with ELC (and other conventions) and is enriched by international comparison, especially as often people's meaning does not count locally. ELC, with its workshops and publications brings examples, arguments, and experts to the people so that they can make use of it and can discuss with politicians and planners on equal level. Last but not least, a challenge might be new in one location, but it might be already solved in another. Therefore, international exchange can help to speed-up processes or to provide solutions.

Democratic principles, human rights, equality, and freedom have always been connected to landscape and the need for living space, if one analyze peoples' movements in history but also today. Also the use of the land, e.g., in forms of transhumance, which is still practices in many regions around the world has been a driving, landscape-related factor (BHU 2016: 30). Last but not least traditional rights of land use, like « the right of way», hunting, collecting berries, mushroom, fruits, medicinal herbs, or wood cannot be left behind.

There is a new research branch about landscape justice and landscape democracy (Dalgish http10, PECSRL 2018, http11).

Most often landscape is considered as a common good, belonging to all of us. It is connected to many intangible values (and believes). Therefore, landscape can be considered as a human right, as it is place and object of personal spatial identification (sense of place/place attachment, Heimat). It is also object of society involvement and productivity and public discourses. On the other hand, this fact of being a common good makes it so difficult to define responsibilities and management authorities. There is the will and the need to maintain landscape in a good shape so that it can fulfill "its" tasks but who is in charge? In this sense, the ELC principles (CoE 2000) play an important role, because local actors can lean on the following principles.

- Principle of personality
- Principle of benefit for all
- Principle of subsidiarity
- Principle of solidarity—and taking responsibility
- Principle of justice
- Principle of sustainability.

20.4.1 …and Beyond—How ELC Inspires Other Countries

As mentioned above, many countries have ratified the ELC. However, also in countries where the government has not adopted the ELC, there are many citizen approaches but also administrative implementations existing, e.g., Germany, which shows the importance but also the power the ELC has. Although Germany hast still not ratified the convention, the following tools and approaches are very common.

- Landscape plan/planning includes the mapping and assessment of natural and cultural resources as well as historic landscape analysis and definition of the "Eigenart" (character)
- Definition of "Leitbild" (guiding principle) and quality principles of cultural landscape development
- Compilation of protection and development measures of cultural landscapes
- Moderation, communication, stakeholder participation
- Monitoring
- Regional development concepts and (overall) plans based on the research results of cultural landscapes (Fig. 20.3)
- Cultural landscape educational itineraries
- Local building codes, local/regional development concepts; renewable energy atlases
- Cultural landscape management concepts—often on a voluntary basis.

Fig. 20.3 Examples for the valorization of landscape and landscape participative awareness rising approaches at Klingenberg/Main in Germany (All photos by Kruse 2018)

One example for countries which adopted the ELC already early is Poland that joined ELC in 2004 as one of the first countries and shows many inspiring ideas for national implementation:

The ELC is located at the ministry for environment due to the analysis that after the falling of the iron curtain, there had been many changes in politics and planning, which led to economic progress. Enforcement of the regions and a liberal property market was needed. It became clear that tools and regulations for the protection and management of landscapes were missing. On the other hand, there was already a long tradition in landscape and settlement research. Landscape assessment did already exist since the 1950ties. In 1992, a Landscape Atlas was published, based on a national Landscape Map. However, the ELC was considered as a good tool to strengthen landscape and to provide necessary rules and approaches. Since 2015, Poland has a Landscape Law, which includes landscape audit (including mapping and assessment). The aim is to define (and protect) priority landscapes, their protection and management. The so called «Woiwodschaften» is reviewed every 20 years. Lewandowski and Szewczyk (2008) invented the idea of the Red Book of landscapes (without legal consequences)—similar to the Red Book of plants. As far as the authors of this article know, this is the only case of its kind, which should serve as example.

Furthermore, Poland has a Spatial Planning + Monument—and Nature Protection Act. UVP, SUP, etc., are planning instruments which also include the assessment of landscape image and the definition of landscape quality aims, as well as mapping and description of nature elements (= opracowanie ekofizjograficze). Pilot (projects) for the valorization of cultural landscape are all Culture Parks (e.g., Wilanowski Park Kultury), historic monuments (e.g., Grünwald, a revitalized battlefield), landscape protection parks (e.g., Nadbuzanski, Dezydery Chlapowski) and some National Parks (BHU 2016).

20.5 The Living Landscape Concept of the NFLS: Tourism as a Trigger for Regional Development—Rio+

As stated above, tourism is one of the most profitable and rapidly developing sector of the world's economy. In addition, with regard to profitability, tourism is one of the most powerful factors of strengthening the country's prestige and identity. Analysis of UNWTO (2019) (http3 2021) has pointed out that ecotourism is increasingly popular in the world, as well as the agro-tourism, which has also developed strongly in several European countries. Residents of big cities and megacities of developed countries show an increasing interest in visiting natural and cultural attractions, situated in their natal country. We are talking primarily about Western Europe and the USA; however, as mentioned before, the COVID pandemic has even strengthened this development.

Therefore, it seems very promising for Russia, to organize a fundamentally new competitive national tourism sector, which can be described as "landscape tourism", where the high quality of the tourism product is primarily provided by the high quality of the landscape, not only its natural, but also socioeconomic and traditional-cultural components. The general agreement about international and landscape tourism between the Russian State Committee of Physical Culture and Tourism and the NLSF was signed by June 5, 1996 (No. 170), based "on the conception of landscape tourism", offered by NLSF. Later, this definition was included in the PACE report by Mr. Sudarenkov (Sudarenkov 2006). We are talking about tourism in the areas, where the most balanced ratio can be provided: balance between the natural complex, the economy, and the social sphere; growth of the quality of local people's life in limits of the economic capacity of the host socio-natural landscape.

Such places mean areas of sustainable development. In fact, this meets one of the key issues of the UNESCO 2015 sustainable development goals (http12). For this purpose, it is planned to create the network of International Landscape Centers

(ILC) in the Russian pilot regions, mentioned as ILC network in the document of the Committee of Ministries of Council of Europe CM/AS(2007)Rec1752 (CoE 2007). In this sense, it is proposed by NLSF to form a national tourist program "Trip to the landscape country" and, at the same time, the first network of territories of sustainable development. The project ILC "Rio-i-" has been developed for more than a decade and has received pan-European, international, and national recognition (Pojidaeva 2007a, b). ILC "Rio-i-" is an image and lighthouse project: It forms the image of Russia in the world as the landscape country for sustainable development (Fig. 20.5) (Pozhidaeva, 2010).

The concept was approved by the Federal tourism Agency, the Ministry of agriculture, and the Administration of Smolensk region. The current version is based on the case study of the village Przhevalskoyea and takes into account proposals and comments of participants of the Russian-German meeting. This meeting was organized together with the third meeting of the International club of territories of sustainable development (the club was organized by the NLSF and the International Information Academy in General ECOSOS UN status, Moscow, which took place in Przhevalskoye, Demidovsky district of Smolensk region in the national park "Smolenskoye Poozerye" from 3 to 5 November 2018). Based on the example of Przhevalskoye and its surroundings, the main indicators of necessary expenses and profits were calculated, and the substantiation of an economic efficiency of the project was carried out.

The project includes all four vectors of complex development of tourism infrastructure (CDTI) in accordance with the classification of the Foreign Economy Bank of Russia/Vnesheconombank (Внешэкономбанк).

1. Development of objects of commercial tourist infrastructure,
2. Development of regional and municipal infrastructure,
3. Organization of "anchor" mass events,
4. Development of a regional brand.

ILC "Rio+" -agropark of sustainable development is a new generation of parks. An innovative area of sustainable development will be formed: an increase in the quality of life of the local population within the economic capacity of the enclosing landscape. The park will be a multifunctional system object, clearly demonstrating the balance of the economy, the natural complex, the traditional culture, and the social sphere. It offers like the first step in Russia of innovative landscape economy, which determination as well as landscape tourism, offered by NLSF, was also included then in PACE report.

This tourism product, due its complexity, covers several touristic industries at the same time, which, according to the UNWTO study «Tourism 2020 Vision», are among the most promising tourism directions of the twenty-first century: ecotourism, agri-tourism, adventure, and cultural and educational tourism. All four directions address to a very large segment of consumers, both, domestic and foreign people of different age, financial opportunities and life convictions. At the same time, it is proposed to organize a new tourism sector—“Landscape tourism” in forming sustainable development areas, where the economic and social development is proceeding, while the high quality of surrounding landscape is preserved.

The project’s economy provides the development of a synergistic effect from two main interrelated directions: the tourism and the agro-economy complex. While tourism gives a quick return of investment costs, the farming sector has a longer payback period.

The development of the agro-tourism sector within the framework of the Project includes the creation and promotion of own agro-tourism products (almost not present at the moment) as the component of the landscape tourism, as well as the production of the own agricultural products. Therefore, the Landstandard***** (Ландстандарт *****) brand was created. It determines, high quality, advanced technologies in combination with the high quality of the landscape, where the products are produced. It is planned that the creation of new tourist brands “European Baikal—the Republic of Brothers-Months”, “Smolensk region—Landstandard*****” on the basis of the project will make it possible to establish production and services that are competitive in national and foreign markets. It is assumed, that these new brands will be formed and developed thanks to the active participation of the local population. In detail, the Landstandard***** concept is composed by the following six pillars:

1. Products and services with a high landscape standard. Food and other products of agricultural landscapes, rural landscape tourism, forestry and building complex products—wooden house-building, wild-growing plants (collection only in the territories allowed for this type of activity) and others. At the beginning the dominant marketing will be direct selling on the local fair etc;
2. Innovative architecture “12+ 7”. House of landscapes, village of harmony, tree of harmony in local style and tradition (Fig. 20.4);
3. Innovative infrastructure and tourist attractions. Landscape House and the Golden Age exposition, landscape module/settlement of sustainable development, a complex of eco-farms and other agro facilities of green economy, a tourist village; regional, national and international park complexes of the series “Soul of Landscape” and “Landscape society”
4. Innovative symbols and images of sustainable development: Tree of harmony, landscape module;
5. Innovative recreational, tourist, excursion programs for sustainable development;
6. Landscape economy: protection of the natural complex, and cultural heritage + direct mass participation of the local population (improvement of human settlements, tour programs), employment and growth of the educational level.

While the brand “European Baykal” is a local one, only for the Smolensk regions, the “Landstandard*****” is a brand which can be used in any Russian territory, which corresponds to it and where local actors would like to develop their region accordingly.

Next to Russians, Germans are considered as main tourism target group as they are known to be a traveller’s nation. There is already an existing cooperation with German partners for many years, including their participation in parliamentary events

(a) Private houses are often very colorful and traditionally with many decorative elements. This house represent the month July. (A. Kruse, 2018)

(b) N.M. Przhevalsky House. The Museum counts around 20000 visitors per year and combines nature, culture and local art exhibitions.(A. Kruse, 2018)

(c) The House in traditional architecture can be named March on European Baykal (N. Pozhidaeva, 2020)

(d) The post office (A. Kruse, 2018)

(e) This village street gives a good impression of the settlement layout: colorful wooden houses with accurate handy craft, gardens and traditional wooden fences combined with metal structures. (A. Kruse 2018)

Fig. 20.4 Examples for the colourful wooden traditional architecture in Przhevalskoye, Smolensk Region of the Council of Federation on the NLSF program, meetings of the NLSF International Sustainable Development Territories Club, international conferences, NLSF Landscape Days in Moscow schools and others.

In the following, a short overview on the landscape related, innovative component of the NLSF project will be given.

20.5.1 *European Baikal: The Republic of Brothers-Months*

Distributed ILC Rio+, its basic component “European Baikal: The Republic of brothers-months” (Fig. 20.5) in the Przhevalskiy of Smolensk region has six priority features of a breakthrough investment project, namely.

1. Birthday: According to the project, the birthday is not only a holiday and the main date of the year for many people. The innovative leitmotif of the project—“Life under the flag of birthday” fills by sense (and the corresponding motivation for tourist trips) not only holiday, but—that is especially important—other grey everyday life. In this sense, Przhevalskoye will offer very personal birthday celebrations in harmony with the circle of nature at that day.
2. People have birthday all around the year. In this respect, there is no “high” and “low” tourism season. People will come to celebrate their day, no matter if this is in November, December, or February. This means for the local people that their service is needed all year long.
3. The project of the international landscape center ILC “Rio+” is based on the key issue of the UN sustainable development goals (see the final documents of UN summits: 1992 (Rio), 2002 (Rio+ 10 in Johannesburg), 2012 (Rio+ 20 in Rio), 2015 (anniversary summit of UN adopted the 2030 agenda and 17 goals for sustainable development). The Chairman of the 75th anniversary UN session New York 2020 Volkan Bozkir promised to step up the fight against the pandemic, promote sustainable development, and reform the UN.

   To achieve this goal, the authors have created: an innovative Glossary, conception, key indicators/indicators, which were then used in the unanimously approved report of PACE “Preservation and use of the landscape potential of Europe”. For the first time models and layouts of the territories and other objects of the exhibition of ILC Rio+, the architectural projects on the principles, game, and educational scenarios of the ILC Park and tourist programs had been presented. The following websites may give a first impression:

   - https://iiaun.ru/activity/directions/european-baikal/
   - https://iiaun.ru/news/iia/2018/11/05/club-of-the-territories-sustainable-development-3
   - https://iiaun.ru/news/iia/2020/06/15/20th-anniversary-of-the-florentine-convention
   - https://iiaun.ru/news/iia/2013/05/20/european-baikal-republic-of-brothers-months

4. Seeing that various different kind of accommodation does already exist and are ready to receive tourists, there are only low start-up costs which implies a short payback period in comparison with similar tourist investment projects (high level of attraction and maximum use of the existing infrastructure on the model territory at the 1st stage). The former sanatorium does also provide accommodation, as well as the brand new high level hotel of Koniy dvor, which was opened beginning of 2017 and landscape hotel La Taiga, (the first stage) opened in this year together with the Golden Age museum. The museum was specially builded by N. & V. Pozhidaev's for their authors exhibition and was opened in summer, 2023. It has to be added that the few local actors are all very innovative, creative, and ready for investment. The only problem was the missing cooperation and communication, which was finally overcome thanks to the NLSF initiatives.
5. The high level of international and national recognition of the NLSF program has achieved already first steps in implementation, including the innovative project of the ILC Rio+, a package of scientific developments of the project ILC Rio+ (p. 3.): PACE, the UN Information center in Russia, an international conference of National Trusts and others, the conclusion of the relevant Federal ministries, and the recommendations of the parliamentary events.
6. Luckily, the NLSF authors had been able to lean on foreign experience. International expert visits gave a lot of awareness and motivation locally—people started to see their own landscape through the eyes of the visiting experts and started to be proud—and therefore convincingly also on local level. Book, film and other tourist attractions, which has become the main base of Disneyland, fairy-tale parks in Germany, the village of hobbits in New Zealand and other international tourist centers will not be needed here. Instead, the stakeholders act very carefully, e.g., the author's exhibition of mobile ILC Rio+, which was firstly presented in the UN information center in Moscow in 2013, the book "Soul of the landscape" (Pozhidaeva 2013) with an corresponding map that shows the names and the screenplay; layouts of the territories, parade of dolls, personalized landscape passports Roden[3]. They are the basis of a single game scenario "The Patron of technocrats of the planet against the Knight of landscapes" (Fig. 20.5b). Its components "Russia fantasy" and "planet fantasy" (Fig. 20.5f, g.) are a prototype of the exhibition on the water and the amusement park of the investment project as well as innovative tourist programs.

20.5.2 The International Landscape Center (ILC) Rio+

The goal of the pilot project ILC Rio+ in the Smolensk region is to ensure a breakthrough of socioeconomic development. As shown above, the model area has a high-quality natural complex, good transport logistics both, nationally and internationally, and other necessary characteristics, e.g., highly motivated private entrepreneurship,

[3] Birthday in Russian is composed by two words: **ro**zhdenie is the Russian word for birth, **den** is the Russian term for day.

(a) With the building of museum

(b) Handmade model of the *Patron of technocrats against the Knight of landscapes*. Plastic, 30 cm. With twelve months, they are the chief fairytale heroes of N Pozhidaeva book "The soul of landscape"(Pozhidaeva N.V. (2013) . Together with odinary people and other fairytale heroes, they form two hostile camps - the forces of landscape creation (land-filias) and the forces of landscape destruction (land-phobias). (V. Pozhidaev, 2020)

(c) *November* is welcoming guest. This is part of the tourism concept: personalized birthday parties (A. Kruse 2018)

(d) new total vision of Exposition with new layouts «the Planet»

(e) The birthday pass is a little book with information about the day from different sectors, like art, nature etc. Also, birthday pass videos are produced and personalised. (A. Kruse, 2020)

(f) «European Baykal—Republic of brothers-mothers»/reception

- (g) The working layout of N. Pozhidaeva "Ideal Europe"- as territory of sustainable development" - the part of the global model, based on the author's research and the world map "Present day landscapes",hand made by N. Pozhidaeva (V. Pozhidaev, 2020). A mixture of plasticine and glue; sequins, beads. 0, 33x 0,26 m for global model 1,9x1,3 (prepared)

Fig. 20.5 Innovative tourism concept for the revitalization of the Przhevalskoye region bases on the landscape experience local food and direct marketing, local handicraft, accommodation, and meeting locations to create a sustainable market in agri-tourism. This will allow to organize a sustainable development area "national socio-natural landscape" by achieving a balance between the development of a highly efficient economy and social sphere, on the one hand, and a productive natural complex and traditional culture, on the other (Figs. 20.2 and 20.6). It is expected that along with the model ILC, other territories of the Smolensk region will be gradually involved in the project, in order to create a regional landscape network of territories of sustainable development.

During the implementation of the investment project in the economic sphere, the following steps are necessary:

- Approval of the tourist program "Travel to Landscape Country:"—"European Baikal—Republic of Brothers-Months)" by the Russian Ministry of economic development;
- Introduction of the eponymous regional brand "Landstandard***** of Smolensk region which will be applied for agricultural production and touristic activities as well as for other component parts of the landscape green economy, specially focusing on youth by means of education and art;
- Setting-up of a year-round landscape festival "Meeting of the month", every first Sunday of a month—integration with local Sunday fair and market (food, handicrafts, etc.) in Przhevalskoye.
- Furthermore, a feature film "The Soul of the Landscape. Republic of Brothers-Months" is planned.

20.5.3 Available Tangible and Intangible Assets of the Project

All developed component parts and ideas are originally invented for ILC Rio+ project. The UN Center in Russia rated the ILC Rio+ exhibition as a "truly innovative project for a green economy and sustainable development." (official letter of the director A. Gorelik of the UN Information Centre in Moscow by 11.02.2013). The project is based on the strategic development of N.V. Pozhidaeva, "Forming a System of National Socio-Natural Landscapes—the Real Way of Sustainable Development", which became the basis for preparing the Russian report "Preserving and Using the Landscape Potential of Europe"—reporter V.V. Sudarenkov (doc. CE 9450, 2002), unanimously approved by the Council of Europe. A special paragraph of the Recommendation on giving it the status of a pan-European program was adopted by the decision of the Committee of Ministers of the Council of Europe (2007, Rec 1752). Based on the development of an international team of experts, the separate report was prepared—the rationale for the UN Global Landscape Network program (Pozhidaeva et al. 2012; Mazur and Pozhidaeva 2002). "The strategy of protection and use of Russian landscape potential for sustainable development" in the book "Strategy and Problems of Sustainable Development of Russia in the XXI Century", mirrors a "slightly different slice of all the problems of sustainable development, since natural landscapes can be viewed as a reserve of sustainability of the biosphere (territories

(a) Reproduction center of the European Wisent with 16 animals of which 14 stravaig through the National Park Smolenskoe Poozerie (stand 2018). (A. Kruse, 2018)

(b) An exhibition informs about the European Wisent and the project itself. A translation into English is planned. (A. Kruse, 2018)

(c) Several nature and visitor trails have been installed in the Biosphere Preserver and the National Park including observation facilities. (A. Kruse, 2018)

(d) Close to the International Landscape Center (ILC), the nature trail is adapted for people with reduced mobility. (A. Kruse, 2018)

(e) Ecotrail to Russian Troy - ancient settlement Verzhavsk (8 century) (V. Pozhidaev, 2020)

(f) Reconstruction of a watch tower of the former settlement. (A. Kruse, 2018)

Fig. 20.6 Leading principle for all visitor and information of facilities in National Park and Biosphere Preserve is sustainability and user friendliness unaffected by human economic activity), and socio-natural landscapes are the main reservoir of human economic activities and at the same time represent the areas of its settlement and daily habitat." The book was published for the World Summit in Johannesburg, Chap. 6 bases on the author's participating in the interdepartmental working group under the Ministry of Economic Development (1996). Thus, in international practice, about 140 indicators of sustainable development were used in that time, often contradicting each other. The author's strategic development focusses on the ratio of two indicators: a new glossary which provides new definitions of socio-natural landscape, national socio-natural landscape, landscape tourism, etc., and the integral classification of landscapes for sustainable development (Pozhidaeva 2000, 2001a, b, 2002).

20.5.4 The Main Impact of the ILC Rio+ on the Local Development

The International Landscape Center (ILC) will have impact on the following three levels:

1. Creation of a constant year-round tourist flow and production of high-quality agricultural products addressing both at local people and visitors. This will create and maintain jobs as the main source of income and therefore be the basis for funding all other sectors of development, including infrastructure elements.
2. Creating a healthy socioeconomic environment of the settlements within the model area of the ILC, which will hopefully turn back the rural exodus:
   - reducing unemployment, increasing the financial well-being of the local population through the creation of jobs on a permanent basis;
   - reducing the rural exodus of population toward large cities, especially among young people.
   - creating an incentive for schoolchildren to return after graduating from higher education: searching for a twin-town and twin-school among the countries of Europe, organizing student exchanges; increasing the interest of schoolchildren in the work of the ILC—the creation of a testing ground for their creative, intellectual, physical development;
   - reducing the disunity of the local population by involving them into the work of the ILC, creating collective goals, thematic entertainment events "for one's own" (without the participation of tourists);
   - gradual transition to a maximum of self-sufficiency in basic consumer goods, development of local food production based on already existing enterprises and facilities, components which can also serve as tourist attractions (bakery, dairy farm, bee apiary, etc.), creation of new workplaces in this sphere of production. In the long run—the expansion of agricultural production shall allow access to the markets of Russia with farm products of the brand Landstandard*****.

3. Implementation of the environmental mission and improvement of the environmental situation:

   - Based on the results of the implementation of aims 1 and 2, reduction of cases of violation of the regime of special protection of the national park by tourists and local residents, reduction of the randomness of the existing recreational activities, the creation of a regulated and distributed tourist flow.
   - Elimination of unregulated waste disposal: a significant reduction of waste by moving from an unregulated tourist flow (where the main purpose is beach holiday, picnics, etc., camping, tourist facilities). One aim is to accommodate tourists in tourist hotels with an organized catering network. Introduction of waste separation and implementation of recycling; reducing plastic consumption—replacing plastic bags by paper bags, selling Landstandard*****'s own products only in paper packaging.
   - Collaboration with the national park: environmental education and ecotourism. Several lighthouse projects have already been undertaken in the last years (Fig. 20.6). In order to address international tourists, more emphasis must be laid on translations—at least into English, followed by German, Polish, and French.

20.6 Discussion

The Landstandard***** is a very nice example, how inspiring ELC can be even in countries which have not signed the convention. It also shows how important international exchange is: Local stakeholders get ideas and knowledge from other initiatives where local actors have been in similar situations. The case of Smolensk region shows on the other side, that private initiatives, no matter how engaged people are, will be stopped at one point due to political decisions at national and international level. Due to the recent sanctions Russia is facing, it is very difficult for the local actors to participate in European programs—even in exchange programs like, e.g., COST or Erasmus. Access to funding is very limited. The support of the Administration of the Smolensk region is a prerequisite for the implementation of the envisaged project. In this regard, it is assumed that, along with the funds planned to attract German as well as other foreign and also domestic investors and in connection with the interest of the Federal Agency of Tourism and the Ministry of Agriculture of Russia, the project will be implemented by attracting funds from two federal targeted programs "Development of domestic and inbound tourism in Russia 2025" and "Sustainable Development of Rural Areas".

In order to start the international cooperation, the World student orienteering championship was supposed to be held in Smolensk and Przhevalsky in 2020 (Fig. 20.7). The NLSF offered to the administration of the Smolensk region to include in the cultural program of the championship the grand opening and the first stone laying of the International Youth Tourist Village "Twelve months". Due to the pandemic, the championship was postponed and then was planned to be again organised in 2022, but then rejected.

Foreign participants will be invited to participate in the ceremony of laying of the first stone for starting the joint project.

The International landscape Center ILC Rio+ is designed for a multidirectional cooperation primarily with European countries. Thanks to Russian parliamentary initiatives, the author's concept of a landscape approach to the sustainable development was unanimously approved by all 46 European member countries of the Council of Europe, including Germany, as a pan-European program. First of all, a joint investment project of the ILC Rio+ could be created with Germany. This is due to several reasons: The "landscape" perception of the environment, the problems of

(a) Biathlon stade, facilities for the athletes. Newly erected. (A. Kruse 2018)

(b) The biathlon stadium looks already back to a long history of national championships. (A. Kruse 2018)

(c) The administration wing was newly constructed in in the traditional wooden style. (A. Kruse 2018)

(d) The beach of Lake Sapsho is adapted to organize Volleyball championships. The former sanatorium, today a rehabilitation center and hotel does provide the necessary accommodation facilities. (A. Kruse 2018)

Fig. 20.7 Smolensk region possesses sports facilities, which are adapted for international and national championships—a further mean for local, sustainable development sustainable development, economic activity, natural, and cultural values are characterized primarily by the mentality of the population of European countries, namely Germany introduced the concept of the *philosophy of landscape* which inspired the Russian local actors.

The landscape approach in Germany has become an effective tool for spatial planning. Joint Russian-German projects in this area have been implemented in the regions of Irkutsk and Kaliningrad. Germany is one of the leaders in the field of green economy, with one of the main priorities creating technologies for mass consumption by the population (energy saving, separate waste collection, etc.), including the creation of energy-efficient residential buildings, e.g., the German Passive-House Institute of Darmstadt (http13). New technologies are especially relevant for the cold winter of the Smolensk region, which includes the entire model territory. The German participants of the outreach session proposed a number of promising directions of cooperation on the Rio+ ILC project, where the best practices of Germany could be used—right up to role-playing games with the local population in order to actively participate in the project.

20.7 Conclusion

The implementation of the pilot investment project ILC Rio+ in the pilot region of Smolensk will provide the formation of a Russian agri-tourism product that is competitive in the domestic and foreign markets and new brands of Smolensk Landstandard***** and the European Baykal—Republic of Brothers-Months. The tourist attraction of the same name has good chances to become recognizable and attractive far beyond the borders of Russia.

Taking into account the assessment of the Federal Agency for Tourism, the pilot investment project “European Baikal—Republic of Brothers-Months” will launch the entire program “Journey to the Landscape Country” aimed at further shaping the positive tourist vision of Russia in the world. A series of projects distributed by the ILC Rio+ will be created in key Russian regions, where the experience of the Smolensk pilot project will be replicated, and the Smolensk region itself will become the basic testing ground for the entire project.

Along with the planned participation in the Federal Target Program “Development of domestic and inbound tourism 2019–2025” and in the Federal Target Program “Sustainable Development of Rural Territories” the first sustainable development territory in Russia will be established in the Smolensk region. The question was previously worked out with the Ministry of Agriculture of Russia. All this will give a new impetus for the transition of the Smolensk region to a model of sustainable development—the main priority for the development of the country, declared by the President of Russia in his inauguration speech on May 7, 2018, and will also help further international cooperation of the region—primarily with Germany and other European countries.

References

Buchecker, M. (2010): Social functions and benefits of agricultural landscapes. In: Gloria Pungetti and Alexandra Kruse, European culture expressed in agricultural landscapes, Palombi Editori, Roma. S. 60–65.

Buchecker M, Kianicka S, Junker B (2007) Value systems: drivers of human-landscape interactions. In: Kienast F, Ghosh S, Wildi O (eds) A changing world: challenges for landscape research. Landscape series. Springer, pp 7–27

Bulletin of the Council of Federation No. 3 (2008) (Вестник Совета Федерации 2008), Москва; С.46–49

Council of Europe (2000) The European landscape convention. ETS, No. 176. https://www.coe.int/en/web/conventions/full-list/-/conventions/treaty/176

Council of Europe (2007) Conservation and use of the landscape potential of Europe Parliamentary Assembly Recommendation 1752 (2006-i.i.112, 11.3, Committee of Ministries of Council of Europe CM/AS (2007)Rec1752, final 19 Jan 2007. http://www.assembly.coe.int/nw/xml/XRef/Xref-XML2HTML-EN.asp?fileid=17447&lang=en

Dalgish, Ch. (without year) Landscape justice (http10)

Diakonov KI (2006) Дьяконов КИ (2006) Ландшафтоведение в современном обществе и актуальные задачи ландшафтных исследований. В: Ландшафтоведение: Теория, методы, региональные исследования, практика. Материалы 11 Международной ландшафтной конференции. Географический ф-т МГУ, Москва, С. 8–13

French Ministry of the Environment. State Committee of the Russian Federation for Housing and Construction Policy (1997) Министерство окружающей среды Франции, Госкомитет РФ по жилищной и строительной политике. Москва - Париж. Природа и градостроительство. Инокомбук, Москва, 1997. С.123

HRH the Prince of Wales: "A vision of Britain. A personal view of architecture". The edition in Russian language (2005). Его королевское высочество Принц Уэльский «Взгляд на Британию» (2005) Личное видение архитектуры». – пер. с англ. Элима, Москва, 2005. 185 с.

Kalinkin LA, Pozhidaev VA (1995) Калинкин ЛА, Пожидаев ВА(1995) Спортклуб на экостадионе. В: Здоровье и физическое состояние населения России на рубеже 21 века . Материалы Всероссийской научно-практической конференции . Государственный Комитет РФ по физической культуре и туризму, ВНИИ физкультуры и спорта, Москва, 1995. С.38–39

Kruse A (ed), Centeri C, Renes H, Roth M, Printsmann A, Palang H, Benito Jordá L, Velarde M-D, Kruckenberg H (2010) Glossary on agricultural landscapes. J Landscape Ecol (Hungary) Special issue:136 p

Kruse A, Pungetti G (2007) EucaLand, European culture expressed in Agricultural Landscapes. In: Landscape and rural heritage, European spatial planning and landscape, No. 88 Council of Europe Publishing, pp 219–222

Lewandowski W, Szewczyk R (2008) Poland—Disappearing landscapes (Polska Ginacekrajobrazy). Sport ITurystyka – MUZA SA Warsaw

Mazur II, Sudarenkov VV, Pozhidaeva NV (2003) Мазур ИИ, Сударенков ВВ, Пожидаева НВ. Ландшафтная политика и механизм Национальных Трастов - перспективное направление международного сотрудничества. В кн..Дипломатический ежегодник -2003г. Научная книга, Москва, 2004.С 122–131

Mazur II, Pozhidaeva NV, Sudarenkov VV (2003) Мазур И.И Пожидаева НВ, Сударенков ВВ (2003) Как сформировать и обустроить национальный ландшафт России? В: Журнал Президиума Российской Академии Наук «Энергия», № 10, Москва. С 16–24

Mazur II, Pozhidaeva NV (2002) Мазур ИИ, Пожидаева НВ (2002) Стратегия сохранения и использования ландшафтного потенциала России для устойчивого развития. В: «Стратегия и проблемы устойчивого развития России в XXI веке. Под редакцией

А.Г.Гранберга и другие. – Экономика, Москва, 2002. Глава 6, с. 177–184, предисловие с.17

Minister of Regional Development of the Russian Federation V. Basargin (2009). Министр регионального развития Российской Федерации В.Басаргин (2009) Письмо Главе представительства Европейской Комиссии в России М.Франко № 12.05.2009 № 14022-В.Б./14

Ministry of Economic Development of Russia (1996) Министерство экономического развития России (1996) - состав рабочей группы по подготовке проекта государственной стратегии устойчивого развития Российской Федерации; приложение к приказу № 114 14.08.1996

Neverov SI (2018) Неверов С. И.Руководитель Фракции Всероссийской политической партии «Единая Россия» в Государственной Думе Федерального Собрания Российской Федерации, епутат от Смоленской области. Письмо ФСНЛ № 5.1–30/ 221 от 14.05.2018

Pozhidaeva NV (1995) Пожидаева НВ. Устойчивое развитие: этноразнообразие и ациональный ландшафт. В : Стенограмма Международной конференции «Устойчивое развитие: этноразнообразие и национальный ландшафт». Организатор - Комитет Совета Федерации Федерального Собрания Российской Федерации по социальной политике. Парламентский Центр, Москва, 1995

Pozhidaeva NV (1997) Пожидаева НВ. Формирование системы национальных социо-природных ландшафтов - реальный путь устойчивого развития. Авторское свидетельство. РА0, Москва, 1997; 2000

Pozhidaeva NV (2001a) Пожидаева НВ Национальный ландшафт как компромисс интересов экологии, экономики и социальной сферы. Материалы 3-го совещания Общенационального экологического форума. Ред: В.А.Черешнев, В.В Аникеев. г.Волгоград – г.Пермь, 2001. S.181–186

Pozhidaeva NV (2001b) Пожидаева Н.В. Национальный социо-природный ландшафт-интегрирующий путь преодоления социальных, экономических и экологических проблем и онфликтов территорий. В кн.: Озеро Байкал - мировое наследие. Экономика, экология, инфраструктура, туризм: Материалы Всероссийской конференции . 23.08.2001г., Улан-Удэ, ыпуск 1. ВИНИТИ Российская Академия Наук и Минпромнауки России. Москва S. 37–40.

Pojidaeva N (2007a) The international landscape network. A vision of Russia. In: «Heritage Souvenior» INTACH, New-Delphi. S.100

Pojidaeva N (2007b) The international landscape network and the international national trusts organization. In: Heritage and development. Papers and recommendations of the 12 international conference of national trusts. INTACH, New-Delphi. S 60–68

Pojidaeva N (2005) Key strategies and next steps: building an international advocacy network. In: The report on the 11th international conference of national trusts, Washington

Pozhidaeva NV (2009) Пожидаева Н.В. Национальные ландшафты – сценарий безопасного бескризисного развития. В журнале «Женщина и политика» № 9, Москва, 2009. С. 71–78.

Pozhidaeva NV (2010) Пожидаева Н.В (2010) Цивилизация ландшафтов - с чего начинается родина в России? В журнале «Женщина и политика» № 1–2. Москва С.122–123

Pozhidaeva NV (2013) Пожидаева НВ (2013) Душа ландшафта. Вече, Москва. 320 с

Pozhidaeva NV, Mazur II, Fokin YE (2005) Пожидаева НВ, Мазур ИИ, Фокин ЮЕ (2005) Предисловие к изданию на русском языке книги Его королевского высочества принца Уэльского «Взгляд на Британию». Личное видение архитектуры». – перевод с англ., Элима, Москва, 2005. С. 14–29

Pozhidaeva NV, Pozhidaev VA, Fokin YE (2009) Пожидаева НВ, Пожидаев ВА, Фокин ЮЕ (2009) Формирование национальных социо-природных ландшафтов России – стратегия бескризисного развития. В: Стенограмма и Рекомендация выездного заседания Комиссии Совета Федерации по жилищной политике и жилищно-коммунальному хозяйству. 16.02. 2009 Деденево Московская область

Pozhidaeva NV, Pozhidaev VA, Fokin YE (2008) Пожидаева НВ, Пожидаев ВА, Фокин ЮЕ (2008) Формирование системы национальных социо-природных ландшафтов России – реальный путь долгосрочного развития и первая очередь международной ландшафтной сети: доклад на парламентском круглом столе «Экологическая инфраструктура городов» 14.02.2008г.. В: стенограмма и журнал Совета Федерации Федерального собрания Российской Федерации «Вестник Совета Федерации» № 3 2008, Москва, с.46–49

Pozhidaeva NV, Ugland K, Pozhidaev VA, Drevatne D, Gatsulaeva K, Nebolsin R (2012) Пожидаева Н.В., Угланд К, Пожидаев В.А., Древатне Д., Гацулаева К., Небольсин А. (2012): Обоснование Международной программы к Рио+20 ООН «Формирование национальных социо-природных ландшафтов – реальный путь устойчивого развития». Письмо советника Президента Российской Федерации, специального представителя Президента Российской Федерации по вопросам климата Бедрицкого А. ФСНЛ № А-60 -18–102; 07.03.2012

Renes H, Centeri, C, Kruse A, Kučera Z (2019) The future of traditional landscapes: discussions and visions. Land 8(6):98. https://doi.org/10.3390/land8060098

Sudarenkov V (2006) PACE report "Protection and use of European landscapes" at Parliamentary Assembly of the Council of Europe (Rec № 1752, - i.i.112, 11.3) (http15)

Sudarenkov VV, Mazur II, Pozhidaeva NV (2003) Сударенков ВВ, Мазур ИИ, Пожидаева НВ. Сохранение и эффективное использование ландшафтного потенциала – путь к повышению качества жизни, устойчивому развитию и процветанию России. В журнале :Нефть. Газ. Строительство - Апрель. 2003. Роснефтегазстрой. Москва С.64–69

The Council of Federation of the Federal Assembly of Russian Federation and others (2007) Совет Федерации Федерального Собрания Российской Федерации, Министерство сельского хозяйства Российской Федерации и Государственный университет по землеустройству. Сохранение и использование ландшафтного потенциала Европы (2007) Сборник документов Парламентской Ассамблеи Совета Европы., Москва, 2007. 100 с

UNWTO (2019) International tourism highlights. https://tourlib.net/wto/WTO_highlights_2019.pdf

UNWTO (UNWorld Tourism Organisation) (2020) Tourism 2020 vision vol 7. Global forecast and profiles of market segments, eISBN 978-92-844-0466-7 (http3)

Literature

BHU (2016) Konventionen zur Kulturlandschaft, Bonn [In German] (http14)

BHU (2012) Beispiele und Methoden zur Kulturlandschaftsvermittlung, Bonn [In German]

BHU (2011) Vermittlung von Kulturlandschaft an Kinder und Jugendliche, Bonn [In German]

Http References

http1: http://www.assembly.coe.int/nw/xml/XRef/Xref-XML2HTML-EN.asp?fileid=17447&lang=en

http2: www.iiaun.ru

http3: https://tourlib.net/wto/UNWTO_highlights_2019.pdf. Accessed by 24 May 2021

http4: https://www.e-unwto.org/doi/book/10.18111/9789284404667. Accessed by 24 May 2021 http5: https://tourism.gov.ru/en/. Accessed by 24 May 2021

http6: https://eng.krasaderevni.ru/. Accessed by 24 May 2021

http7: http://intjournal.ru/mirovoj-rynok-turistskih-uslug-tendentsii-i-perspektivy/. Accessed by 24 May 2021

http8: https://www.marketing.spb.ru/mr/travel/tourism_industry.htm. Accessed by 24 May 2021

http9:https://rosinfostat.ru/turizm/. Accessed by 24 May 2021

http10: https://www.communitylandscotland.org.uk/wp-content/uploads/2017/09/Landscape-justice.pdf. Accessed by 24 May 2021

http11: Closed session about landscape justice by Kenneth Olwig and Laurence Le Du, Landscape Research Group: https://pecsrl2018.sciencesconf.org/resource/page/id/48. Accessed by 24 May 2021 http12: https://en.unesco.org/sustainabledevelopmentgoals. Accessed by 24 May 2021

http12: https://passivehouse.com/01_passivehouseinstitute/01_passivehouseinstitute.htm. Accessed by 24 Apr 2021

http13: BHU = Association for Heimat and Environment (Bund für Heimat und Umwelt): https://bhu.de/. Accessed by 24 May 2021

http14: http://www.assembly.coe.int/nw/xml/XRef/Xref-XML2HTML-EN.asp?fileid=17447&lang=en. Accessed by 29 Apr 2021

Chapter 21
Twenty years of Landscape Policy and Governance in Europe, and the Way Ahead

Claudia Cassatella

Abstract The European Landscape Convention (ELC), which came into effect in 2004, has been ratified by 40 countries since its signing in 2000. For the first time in history, all these countries with rich and diverse landscape cultures as well as different institutional frameworks agree on a legal definition of landscape and common goals for landscape policies. The implementation of the ELC has ushered in innovations in their legislation, governance and administrative processes, and spatial planning systems. A systematic comparison is lacking and, while landscape heritage protection has internationally accepted protocols, landscape planning has not. This paper proposes a framework for a comparative analysis and, based on the collection of contributions to this section, drafts some initial considerations. After more than twenty years of implementation of the ELC, the most challenging issues seem to be participation, conflict management and strategy making.

Keywords Landscape policies · Landscape governance · Juridical issues in implementing the ELC · Landscape planning · Landscape democracy and participation

21.1 A New Baseline for 40 Countries

The European Landscape Convention (ELC), which came into effect in 2004, has been ratified by 40 Countries since its signing in 2000. For the first time in history, all these countries with rich and diverse landscape cultures as well as different institutional frameworks agree on a legal definition of landscape and common goals for landscape policies. The implementation of the ELC has ushered in innovations in their legislation, governance and administrative processes, and spatial planning systems.

C. Cassatella (✉)
Interuniversity Department of Regional and Urban Studies and Planning, Politecnico di Torino, Turin, Italy
e-mail: claudia.cassatella@polito.it

T. Matteini et al. (eds.), *Cultivating Continuity of the European Landscape*, Environmental History 15, https://doi.org/10.1007/978-3-031-25713-1_21

To mention one example of the effects of the ELC, Italy, a country where landscape protection as a principle has been enshrined in Constitutional Law since 1948,[1] shifted its legislation focus from landscape protection to planning; introduced new forms of collaborative planning between State and regions and innovated administrative procedures for landscape interventions; and set up new educational programmes.[2] In addition, many bottom-up civil society initiatives, such as local landscape observatories, have arisen.[3]

Although this book is devoted to exploring future challenges for the ELC, the "Landscape Policy and Governance" Section also celebrates the progress made since it came into effect. The diversity of concepts, cultural backgrounds, academic traditions, and social practices related to landscape in Europe is well known and documented by the preparatory documents of the Convention (see Seguin 2017). Even the official English and French versions of the text refer to two different semiospheres when making use of the words landscape and *paysage*. Nevertheless, these words now have the same definition in law and have been translated into dozens of languages.

In some countries, "landscape" simply was not a legal matter before the ELC; in others, it was an issue only for conservationists or, very occasionally, an ordinary matter for spatial planners or landscape architects. Even today, the ministry tasked with the ELC reveals much of a country's diversity and traditions. In fact, landscape is a matter for a variety of Ministries, such as Culture, Agriculture, Land use, or others.[4] Interestingly, a search for the "landscape" concept in European countries' constitutional laws reveals that a few countries do not explicitly mention landscape (nor environment); a small number of them address both landscape and environment (separately or jointly); and the rest refer to the sustainable development concept (Cerrina Feroni 2019).

21.2 Towards a Comparative Perspective

A comparative perspective is always challenging and in the case of the ELC, an investigation of its implementation should take into account a wide array of policies (Civitarrese Matteucci, *infra*). Nevertheless, the Convention facilitates much easier understanding, comparison and, in the end, collaboration at international level. In

[1] "The Italian Republic protects the landscape and the historic and artistic heritage of the Nation", Italian Republic Constitution, 1948 (Article 9), translated by the author.

[2] In particular, in 2001 two new educational programs were introduced into the Italian university degrees' system: Landscape Architecture and Regional, Urban, Environmental, and Landscape Planning.

[3] The ferment induced by the ELC is well testified by the notable presence of Italian scholars in this book (and section) as well as in the literature and networks regarding the Convention.

[4] See the Information System on the Council of Europe Landscape Convention, https://www.coe.int/en/web/landscape/landscape-observatory. Last Access: July 2021.

fact, it would be possible to review legislation and regulation across European Countries through a shared international effort using themes identified in the ELC (definitions, general measures): the legal definition of landscape; implications for institutions and related competences at a national level; landscape protection systems and related categories of designation; the role of landscape planning in the spatial planning system, and established procedures for active participation.

Recognition of landscape in law can be taken for granted, as already mentioned. Nevertheless, the translation of the concept and the context where the concept is applied open interesting issues. In particular, the definition of landscape as an area "as perceived by people" may refer both to experiential perception and to conception. "In the Danish, Norwegian, and Swedish versions, "perceived" is translated to a word ("opfatte") that is closer in meaning to "conceived"" (Primdahl 2018: 1473). On the contrary, in my country, Italy, perception is sometimes understood as "visual or sensorial" instead of "social" by the many who have not read the Convention's explanatory texts. Moreover, the main Italian legislative text of ELC implementation (Italian Republic, Code of Cultural Heritage and Landscape, 2004 and subsequent amendments) introduces the notion of national identity which may be interpreted as the opposite of the plurality of "people's perceptions".

So we are far from a shared understanding and intention when implementing the ELC, but reaching homogeneity might not be so desirable. "It is clear that this linguistic and semantic diversity has not been, and is still not, a barrier to implementing the European Landscape Convention", according to Seguin, who analysed landscape in the languages of Europe, "[...] this definition provides an aid to communication and discussion, which all Europeans can adopt and use" (Seguin 2017: 222, 225).

The text of the ELC provides a baseline for dialogue as the annual "Council of Europe Meetings of the Workshops for the implementation of the Council of Europe Landscape Convention" demonstrates. What is still missing is a systematic study which goes beyond the cumulation of individual case studies on national policies.

A starting point might be a comparison of landscape preservation since it was the first public policy relating to landscape and, consequently, some countries have had a well-established preservation system for over a century. The result is a variety of designation categories which address cultural assets, natural features and environments, scenic resources, places of memory, and so on, not always labelled as "landscape" but related to the actual conceptualization of landscape in various ways. In addition, the action of international organizations and agencies, such as UNESCO, ICOMOS, IUCN, has contributed to the circulation of concepts such as "cultural/natural landscape", or "protected landscapes" with their related analytical frameworks, facilitating international dialogue and comparisons among conservation experts.

Nothing similar can be said about planning and management. Landscape is linked to a variety of separate or integrated sectoral and spatial policies and European spatial planning systems are too diverse. The diverse institutional frameworks obviously influence land regimes, spatial planning systems and territorial governance in terms of ownership rights, the role of the State, the distribution of powers, subsidiarity, and participation. A distinction can be made, for instance, between planning systems derived from the Napoleonic Code, those based on common law, or those influenced by the Soviet period.[5] Not surprisingly, there are few comparative studies on landscape planning systems and measures.[6]

Let me use my country once again as a case study: when commenting on the implementation of ELC "landscape planning" policy, the majority of Italian scholars automatically refer to the Italian "landscape plan", a specific instrument designed by the State to match the ELC guidelines. But landscape planning, defined and intended as "strong-forward looking actions" (art. 1), can take many other forms: a sectoral plan or study, the integration of landscape concerns into ordinary measures, the assessment of interventions on landscape, or even financial measures (see the CM Rec 3/2008). Above all, landscape planning should include creation and future envisioning (which is far from the Italian approach, more focused on conservation). Interestingly, commenting on Denmark, a country belonging to a very different spatial planning system family, Primdahl (*infra*) complains about the same absence of strategy-making in its landscape planning practices.

Last but not least, a significant aspect for comparing ELC implementation is participation. From an institutional perspective, it would be interesting to see how and to what extent states respond to the call for "active" participation with ad hoc procedures. In Italy, for instance, landscape plans are subject to the same mandatory procedures of any spatial plan (meaning interested parties can submit comments in a given phase of the plans' approval process). In addition, Regional Landscape Observatories have been set up which have no specific mission by law but carry out a variety of activities.

21.3 Initial Evidence and Gaps

The present section presents the feedback to our call for papers (2020 UNISCAPE Conference, Florence, 16–17 October 2020). A dozen contributions have been collected which deliver evidence from many countries, from Portugal to Norway. Da Costa and Curtado boldly face the challenge of comparing different countries and autonomous regions (Andorra, Catalonia, Hungary, Ireland, Mallorca, Portugal, Switzerland, and The Netherlands). They point to the lack of monitoring systems and financial provisions as major weaknesses of national and regional policies. A group of authors present in-depth analysis of ELC implementation in each of their countries, offering expert assessments and descriptions of related policies and systems

[5] EU country families or models have been identified with specific reference to spatial planning systems and territorial governance by the ESPON TANGO, Territorial Approaches for New Governance (2013) project, and further investigated by ESPON COMPASS (2018).

[6] Sala et al. (2014) is one of the few studies to systematize landscape planning systems in different countries. In the collaborative framework of the EU ESPON Programme, Targeted analyses, see LP3LP—Landscape Policy for the 3 Countries Park and LIVELAND—Liveable Landscapes: a Key Value for Sustainable Territorial Development.

in Italy (Di Giovanni), Scotland (Govan), Portugal (Ramos and Freire), and Greece (Terkenli).

Interestingly, the authors from Italy (Di Giovanni, Laviscio) focus on landscape protection and landscape controls, revealing how the traditional apparatus (derived from heritage preservation policies) still influences discourse on the ELC. In this context, participation and education practices are presented (Cepollaro and Zanon, Marinelli et al.) or called for (Palazzo and D'Ascanio) as needed innovations. On the opposite side, democracy and participation, with particular attention to the local scale and community level, seem to be the main discourse in Northern Countries (Lagerqvist, Némethy and Nilsson; Vik). Conflicts between local and regional authorities as well as between authorities and the general public are documented in both contexts.

The idea of "commons" is also proposed as a key concept. In particular, Marinelli et al. provide data on public land in the EU, with specific reference to farmland and its management. Once again, "commons" doesn't have the same meaning in different legal systems (sometimes it is nearly untranslatable) and is still debated as a legal topic but apparently, it has brought new insights and inspiration.

Managing transformations and creating landscapes, or the role of projects, also find space in this collection of papers. The conceptual framework and process proposed by the ELC (assessing multiple values through active participation, integrating sectoral perspectives, capturing aspiration for the future, and managing change) offer a clear path for designing both large-scale policies (Torreggiani &C) as well as site-specific visions and territorial development strategies (Marinelli et al., Palazzo and D'Ascanio).

21.4 Conclusion

The present section dealt with landscape policy and governance from various perspectives: juridical issues, government and governance systems, planning and management practices.

All the contributors, despite the variety of topics, disciplines and geographical contexts, testify to a struggle to improve landscape democracy by calling into question spatial justice and governance issues such as relationships between authorities, government levels, the general public, local communities, interest groups. They reason on the limits and possibilities of improving transparency, participation and knowledge, pillars of good governance. The ELC is an initiative of the CoE Congress of Local and Regional Authorities, managed by the General Directorate of Democracy. Since its inception, democracy and participation have been key principles of the ELC (CoE 2018) and yet they remain the most challenging to implement. As a consequence, conflict management's theories and practices offer a promising field of research (Calderon and Butler 2020), far from the illusion of landscape holism as a guarantee for reaching mediation and consensus. The present collection of papers demonstrates that the ELC has contributed to creating a baseline and a common language, has brought innovations into national frameworks and changed the rules with a rippling effect, challenging regional and local levels. Still, comparative studies are lacking, particularly on landscape planning systems and measures.

In conclusion, landscape policy and governance is a fundamental topic that needs international collaboration in order to be explored from both a comparative perspective and an interdisciplinary point of view. UNISCAPE is the network that can carry on this programme.

Abbreviations

CoE Council of Europe
ELC European Landscape Convention

References

Calderon C, Butler A (2020) Politicising the landscape: a theoretical contribution towards the development of participation in landscape planning. Landsc Res 45(2):152–163. https://doi.org/10.1080/01426397.2019.1594739

Cerrina Feroni G (2019) Il paesaggio nel costituzionalismo contemporaneo. Profili comparati europei. In: Federalismi.it, n. 8/2019, pp 1–25

Council of Europe, Committee of Ministers, Recommendation CM/Rec (2008) 3 on the Guidelines for the implementation of the European Landscape Convention, 6 February 2008

Council of Europe (2018) Council of Europe landscape convention contribution to human rights, democracy and sustainable development. Council of Europe Publishing

ESPON TANGO (2013) Territorial approaches for new governance. Final report, Luxembourg: ESPON

ESPON COMPASS (2018) Comparative analysis of territorial governance and spatial planning systems in Europe. Final report. Luxembourg: ESPON

Primdhal J (2018) The contested nature of farmed landscape. In: Marsden T (ed) The SAGE handbook of nature. SAGE Reference, London, pp 1468–1487

Seguin J-F (2017) Wealth and variety of terms, instruments and approaches to landscape in Europe. In: Council of Europe (ed) Landscape dimensions, reflections and proposals for the implementation of the European Landscape Convention, Council of Europe Publishing, pp 221–233

Sala P, Puigbert L, Bretcha G (eds) (2014) Landscape planning at a local level in Europe. Observatori del Paisatge de Catalunya; Andorra la Vella: Govern d'Andorra, Olot (Documents; 2)

Websites

CoE (2018) Information System on the Council of Europe Landscape Convention. https://www.coe.int/en/web/landscape/landscape-observatoryLast Access July 2021.

EU ESPON programme targeted analyses. https://www.espon.eu/targeted-analyses. Last Access July 2021.

Part III
Landscape Design and Time

Chapter 22
Landscape Design and Time

Rita Occhiuto and Margherita Vanore

Abstract Design tools have been prefigured over the last 20 years to protect, transform and adapt landscape in relation to changing environmental, economic and cultural conditions. Making landscape is here intended in the sense of its production and its continuous transformation depending on a project, that is to say on a complex prefiguration of places, resources, spaces, nature, constructions, physical and cultural relationships, elaborated by design experts able to integrate competences on the basis of a shared vision of the future from which the actions, processes and forms of the territory derive their sense. The Landscape Design and Time section focuses on the characters of landscape projects that infrastructure, conform and produce sustainable places; cure wounds and take care of the territory over its long time of transformation; cultivate visions and new imagination; define themselves on, and incorporate relationships between different scales; are inclusive and never finished because they are always open to new relationships, new processes and progressive adaptation.

Keywords Open project · Active infrastructure · Landscape making · Taking care · Feeding imaginations

Landscape as a continuous project assigns a decisive role to the integrated prefiguration of territory. The conformation of living spaces as well of the formal process and integration of the various ecosystems gives rise to ever wider, even global, responsibilities with regard to achieving a balance between the different determining factors.

In the design, implementation and protection of landscape, including policies, good practices and perspectives, it is evident how design tools have been prefigured

R. Occhiuto
Department of Architecture, University of Liège, Liège, Belgium
e-mail: r.occhiuto@uliege.be

M. Vanore (✉)
Department of Architecture and Arts, Università Iuav di Venezia, Venice, Italy
e-mail: vanorem@iuav.it

T. Matteini et al. (eds.), *Cultivating Continuity of the European Landscape*, Environmental History 15, https://doi.org/10.1007/978-3-031-25713-1_22

over the last 20 years to protect, transform and adapt landscape in relation to changing environmental, economic and cultural conditions.

Making landscape is here intended in the sense of its production and its continuous transformation depending on a project, that is to say on a complex prefiguration of places, resources, spaces, nature, constructions, physical and cultural relationships, elaborated by design experts able to integrate competences on the basis of a shared vision of the future from which the actions, processes and forms of the territory derive their sense.

Interest in the Landscape Design and Time section focuses on landscape projects that infrastructure, conform and produce sustainable places; cure wounds and take care of the territory over its long time of transformation; cultivate visions and new imagination; define themselves on and incorporate relationships between different scales; are inclusive and never finished because they are always open to new relationships, new processes and progressive adaptation.

Infrastructural Landscapes involve **active infrastructure**, able to **take care** of as well as conform and produce sustainable places by enhancing resources and production systems to feed coordinated actions between interventions of transformation of the territory, irrespective of administrative boundaries. An **open project** does not aim to complete an arrangement, but create an articulated system capable of being inclusive and of welcoming progressive adaptation to continuously changing conditions.

Landscape Time Design involves the **making** of **landscape** over a long period of time, where **time** is considered capable of taking on the processes that qualify landscape. A project unavoidably articulates its intervention across different scales of action that activate multiple relationships and an integrated vision. In this case, the ability of a project to overcome a one-dimensional vision of territories to formulate possible relationships between innovation and tradition in order to implant and cultivate integrated visions, **feeding imaginations.**

Taking up these themes allows us to re-examine the natural processes of transformation of places. And at the same time, we find we can read human action as a force acting on soil that imposes constant change. These actions respond to human intentions to "use" land or to "orient" its becoming until it is transformed into a new artefact of "cities or countryside" corresponding to a "Third Nature" that we no longer recognize today. But the question that interests us most is understanding whether, in new landscape "reconstruction or regeneration" processes, the time plays an active role and how this is inserted in the design process.

The two keynote speakers for the Conference Session "Landscape Time and Design", Navarra and Hossler, contribute on a path that links projects to the times and character of the earth's crust and the stratification of soil, but also to the ability to adapt designing to the processes of shared transformation, aspects that appear to be underlined by what Delbaere describes in his paper as the capacity for alteration.

Marco Navarra presents the idea of **geologic architecture** in relation to the concept of "Fragile Earths". This suggests an architecture montage, a practice that takes complexity as an indissoluble interweaving of space and time with geography and history as its foundation. The word "Earth" brings the earth's crust back to the centre of project concerns in a complex vision, while the adjective "Fragile" expresses multiple levels of criticality, but at the same time identifies a plastic force of a dynamic and operational concept. This idea leads us to imagine "inactual" architecture capable of reconstructing a state of hazard in which the ruins and fragments of the past can re-interview life and open up to the future. The task of architecture for "Fragile Earths" must be to generate a new conjunction between living and building, a new way of working with time and matter.

Michel Hössler retraces the Agence Ter project for Vilaine Aval in Rennes as a time-tested design and work. There is the long time of the geohistory, but also the time of the project, of its elaboration and its spatialization, which can take place over several years. The question becomes how to maintain the course of a project while accepting its enrichment by programmatic or usage inflections. The governance of a project integrates time management and development strategy with collaborative work in a process changing direction from top-down action to bottom-up work. Making territory is thus built up over time through work associating large and small actions, carried by a strong and shared political will, redirecting the classic work of design into a shared and co-constructed project approach. This work over time is a paradigmatic shift for constantly working on spaces, re-interrogating and sometimes weakening a landscape designer's work, but with the capacity to manage the inevitable contradictions.

Denis Delbaere emphasises how the relationship between the landscape architecture project and time is above all a relationship with the alteration of what was defined by the project. It is thus important to be familiar with the terms of the alteration formula: on the one hand, the place as we see and experience it several years after its construction, and on the other hand, the original project in its circumstances, ambitions, renunciations and pride factors. Delbaere specifically identifies seven concepts that form the basis of an *alterative* grammar: the ruin, the island, the support, the contour, the wood, the seed and the pole. These *alterative* dimensions are not exclusive but are part of a global dynamic whose effects are always temporary. There is no "climax of the project" any more than there is a biological climax. Evolution never ends. The need to accept the unpredictable becomes unavoidable for places subject to the continuous action of natural agents and an "altered natural grammar". What might therefore be perceived as one's own landscape project is actually a collective creation. And it can undoubtedly be said that a landscape project, in its re-composition of social and political legitimacy, is a public space.

The proposed sequence of presentation of the selected papers follows a narrative path that seeks continuity. Starting from the contributions of keynote speakers, integrated by a paper that has well interpreted the themes of the entire session, **five areas of reflection** are proposed to recompose the infrastructural role and the temporal component in the making of landscape. These are w*ater spaces and times, parallel places narratives, fragility in perspective, regenerative infrastructures and revealing itineraries.*

Water spaces and times. Attention to water areas as liquid landscapes holding together multiplicity and diversity (Colafranceschi) fuels a design vision based on water spaces and times. The specificity of the waters dividing and connoting archipelagos, lands that are close to each other but on opposite shores, take on the value of public space.

In the same way, inhabited riversides highlight the **aggregating and relational power of water,** (Anastasia) becoming opportunities to generate/regenerate and promote the local potential of urban areas.

These are lands conditioned by water where new possibilities arise for proposing strategies of adaptation and sustainable intervention to rediscover new ecosystem balances.

A systemic approach makes it possible to coordinate the anthropic processes related to urban and territorial transformation phenomena (Mencarini et alii) and ecosystem protection, admitting reciprocal continuity and functionality. In fact, the experience of building a wetland at the edge of a port area teaches us that the protected area and port environments must be considered as an integrated infrastructure.

In the meantime, wind and tides mark out a visible rhythm (Tornieri) on the coasts: this is demonstrated by intertidal vegetation in the case of a resistant, mineral coast or wide intertidal plains, sandy or muddy, in the case of a free coast.

The reading of time requires a project to identify and understand variability and "movement" as conditions which are also independent of places shaped by water.

Parallel places narratives. Landscape has always changed through biological and social times (Di Lauro), and today it reveals an interweaving between the fast events of society and the slower evolution of places, due to nature and human work. The garden is however a "microcosm" in which the short and long times of landscape—the immediacy of everyday life, the slower transformations of territories, the perpetual renewal of nature—all converge and are amplified in a narrative of symbols and metaphors. Here, more than elsewhere, "contemporaneity"—as the time span of human life—is expressed as landscape, in complex relationships with the past, present and future.

Similarities can be found in other detached contexts, but with strong identity ties (Alici et alii) that connote parallel and underground places with the local community. The heritage dimension is the basis for visions that interface with forms of transition management, also to integrate accessible and sustainable tourism, technological innovation, as elements requiring the development of initiatives at a local and regional level.

Fragilities in perspective. Various fragilities permeate intermediate spaces, marginal areas or areas with no apparent function. These have considerable design relevance (Angi, Soci) and represent one of the fields for any credible urban regeneration plan.

Fragility and wounds produced by disastrous events and seismic phenomena show how in the face of the slow time of landscape construction which follows a process of sedimentation (D'Agostino, Vannelli), extraordinary events can stop time and determine new beginnings and new processes. The acceleration of these processes can, moreover, generate risk for the quality of life as well as that of landscape. More than ever, then, a project needs to be a process, strategic, adaptive and co-evolutionary, in short, an open project that accepts the unexpected and changes. In these cases, the architect aims to provide tools and not objects for a project made up of multiple temporalities.

The abandonment phenomena re-proposes the need for new mapping processes (Ruggeri) able to recompose available urban heritage into an overall view in order to make a critical classification of buildings and urban spaces to be recovered, modified, replaced or demolished to integrate green spaces in the dense body of an ancient city.

In the meantime, as military and industrial complexes are abandoned, nature operates freely as a designer (Fabris, Camerin), creating new transitional landscapes, "self-regenerating" landscapes that preserve a strong historical/anthropic identity as well as the potential to increase biodiversity.

Moreover, abandonment is not a phase of immutability (Caravello, Baldin), but rather one of regeneration. A continuous capacity for self-repair is in fact manifested in abandoned industrial sites through processes of spontaneous re-appropriation. The chance to return to time cultivation fosters interrelationships between the various natural components in the recovery of time signs as memory to be translated into new planning, but also as indicators of the reasons for the use of the land, its transformation and reinvention.

Regenerative infrastructures. If we still see the lack of effective integration of infrastructures into landscape (Miralles) despite numerous studies and various procedures for environmental impact assessment, Green Infrastructures have considerable potential for intervention through systemic and punctual projects (Caprari, Gulic, Trusiani). A network of slow mobility, made up of greenways and urban parkways of an educational and naturalistic nature, can become connections between neighbourhoods, centres and intermodal exchange nodes. The vision of landscape as an "open project" also proposes scenarios that take care of the territory in both the short and long term, while the territorial framework is configured through an ecological infrastructure that influences urban planning by providing ecosystem goods and services.

Revealing itineraries. Some sporting events highlight landscape features by following itineraries capable of narrating and revealing its richness, as well as highlighting particular infrastructures for the well-being of the urban area and its inhabitants. This is the case of the *Venicemarathon* (Triches) whose route aims to present Venice in its metropolitan articulation, promoting the idea of city-landscape. Over time, the event has triggered numerous actions aiming to improve the well-being of local communities, raise public awareness of environmental issues, educate children and families about physical activity and promote and support sharing initiatives.

Faced with cities full of impenetrable borders, barriers and frontiers, the availability at *km0* (Battle, Rivera) of productive spaces equipped for sharing can promote biodiversity, sustainable energy production, water control, the cultivation of local foods and leisure time for citizens.

With a *km0* strategy promoting proximity, boundaries become doors opening in both directions, neighbouring spaces are able to unite and connect public spaces with the agricultural, rural or natural environment that was previously only seen, or perceived, in the distance above a grey wall of infrastructure. Public space and nature thus merge into new infrastructure for slow and sustainable mobility. Slow time connotes the use of intermediate areas and infrastructural fringes, restoring continuity to territories inside and outside the city, allowing unexpected spaces and systems of new public spaces to be rediscovered.

In the making of landscape, time is an unavoidable part of its designing, its quality and awareness, as well as its capacity for integration and alteration. Alteration introduces vital fragility and instability, like figures reflected on a variable water surface, while new infrastructural regeneration perspectives, the unveiling of potential and other narratives are being outlined.

Chapter 23
Geologic Architecture

Marco Navarra

Nature is the maker. It is the giver of presences. You can have a thought, but a thought has not presence until you call on Nature to exercise its powers of order to make it manifest (...) Material is spent light you're spent light, the trees are spent light, the atmosphere is spent light, water is spent light. All material is spent light.
Louis I. Kahn

Abstract The idea of Geologic Architecture experienced in these occasions suggests a procedure that not only goes beyond the idea of work autonomous from its context, since it is composed of stratification and the juxtaposition of materials and sections, but also eliminates the boundary between background and object. Each transformation is a node connecting different materials and levels, stratifications of the Earth's crust and human communities, long durations and short times, and cyclic and unique, unrepeatable events. The opposition between nature and artifice disappears and new hybrid matter appears in which boundaries are no longer identifiable, and we are immersed in a hyper-natural environment. The forms of architecture, rather than mere technology, become a response to environmental issues. Even the idea of circularity is bent and reinterpreted within a temporal framework in which instants and long durations are intertwined.

Keywords Geologic architecture · Landscape · Earthling architecture · Atmosphere · Post-human · Post-colonial

---

M. Navarra (✉)
Department of Civil Engineering and Architecture, University of Catania, Catania, Italy
e-mail: marco.navarra@unict.it

T. Matteini et al. (eds.), *Cultivating Continuity of the European Landscape*, Environmental History 15, https://doi.org/10.1007/978-3-031-25713-1_23

23.1 In the Landscape Folds

In 1963 Aldo Sestini edited the book "Il paesaggio" (Landscape) for the Italian Touring Club, the seventh volume in the series "Conosci l'Italia".[1] It is a journey through Italy from North to South, presenting its variety of landscapes and showing and explaining differences, characters, and the often invisible reasons for the outward forms. The volume describes Italy's geographical landscapes as "objective elements manifest to direct observation, in themselves and in their reciprocal spatial relationships".[2] The author clarifies a number of points in the foreword, starting from the idea that the earth's surface, not as a whole but in its individual features, is the basis of landscape.[3] He explores the subtle dividing line between "actual views" and possible views, perception and objective data that the word landscape weaves together. In this way the landscape concept is not bound to the idea of a specific view and becomes "a complex combination of objects and phenomena linked together through mutual functional relationships (as well as position) so as to constitute an organic unity".[4] The text is accompanied by black and white or colour photographs, analytical and illustrative drawings, and maps. These are sequenced and flank the writing to clarify descriptions and illuminate words with clear and incisive images (Figs. 23.1 and 23.2).

The varying didactic illustrations consist of drawings ranging from analytical views to sequential evolutionary schemes, sections with perspective views, geological sections with vegetal cover, morphological representations (outlined with the most representative profiles), or axonometric cutaways with geological stratifications. The drawings appear as small, finely etched miniatures at a very high level of definition and precision. The maps are a reworking of those of the Istituto Geografico Militare with the most important parts highlighted by superimposing colours and fields. The book is formed by interweaving and juxtaposing writing, photos, schematic illustrations, sections, and axonometric drawings. Often the photos are juxtaposed with drawings or maps in order to make the sensitive aspects immediately comprehensible through an analytical illustrated explanation. The space occupied by photos, maps, and drawings conveys the atmosphere and dense exchange of energy that builds landscapes and architecture.

Among the graphic representations, the axonometric cross-sections are significant and original. They are pure fragments of the earth's surface that clearly show the strong relationships linking the visible and invisible, current events and those of ancient times. It is very clear how activities on the surface of the earth's crust are affected by both deeper geological layers and atmospheric movement. Under the earth's crust, the nature and consistency of rock can be understood (through slow sedimentation), while erosion and movements on its sensitive surface show the actions of the atmosphere. Often, to make the process even clearer, the axonometric

[1] Richard Saul (1986, p. 29).

[2] Aldo (1963, p. 11).

[3] Ibidem, p. 9.

[4] Ivi, p. 10.

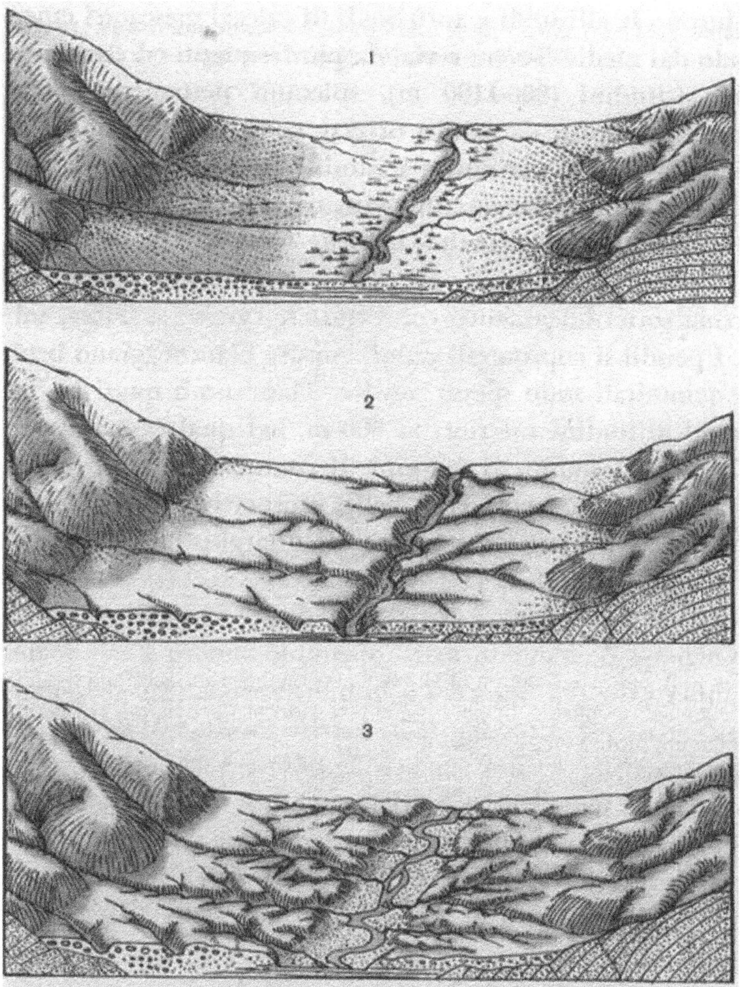

Fig. 23.1 Drawing taken from Aldo Sestini's book 'Il paesaggio' (1963). Sequence of geological sections showing the evolution of an Apennine intermountain basin

cutaways are placed in sequence like pictures in a figurative narrative to describe and narrate the slow changes that take place over very long periods of time. These drawings are an invention, an isometric axonometric projection with two geological sections on each side, as an expedient to unite the thickness of the earth's crust with the shape of its surface. We can immediately read the relationships between geological processes and the visible elements observable in a landscape. The shape of these axonometric clumps also suggests the invisible to our eyes. In fact, looking at them,

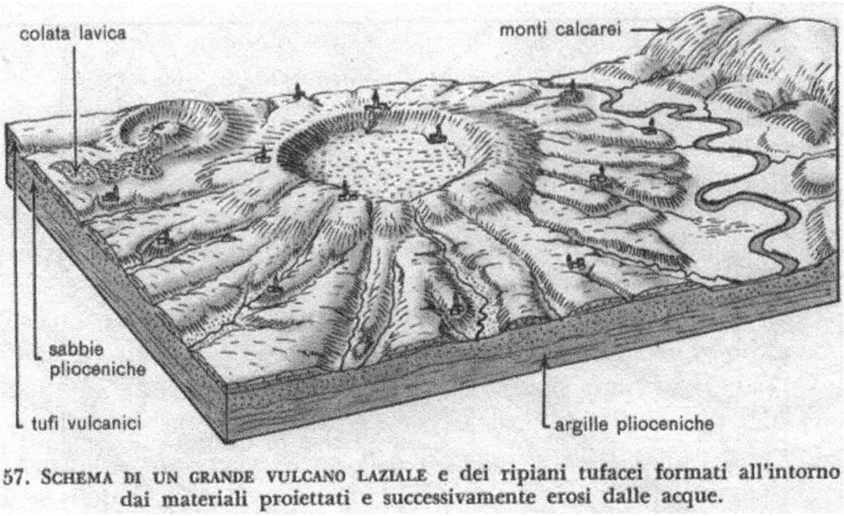

Fig. 23.2 Drawing taken from Aldo Sestini's book 'Il paesaggio' (1963). Axonometric cross-section of a large volcano in Latium and the tuffaceous shelves formed on its inside by materials projected and subsequently eroded by water

it is easy to imagine the section of sky that acts on these surfaces and models them. Their sequences alongside the photos and text show us the dynamics of geology and let us imagine further.

"Geological strata have been formed through (at least) two distinct operations. On close observation, it is striking that each layer of rock contains others and that each of these is composed of pebbles almost homogeneous in size, shape, and chemical composition. Since pebbles do not all have the same shape and size in nature, there must be a sorting mechanism, a device that takes pebbles of a heterogeneous nature and distributes them in roughly uniform layers.

Geologists have identified a similar mechanism in rivers, which function as hydraulic computers (or at least as sorting machines). (…) When raw materials have been sorted into more or less homogeneous groupings on the sea floor (i.e. when they have settled), in order to transform these free collections of pebbles into a higher scale entity, sedimentary rock, a second operation is required which consists of cementing the selected components into something new with its own characteristics, such as overall strength or permeability, which cannot be attributed to the sum of the individual parts".[5]

Manuel De Landa's descriptions (2003) could well be illustrated by Sestini's drawings, and not only in their geological but also for the theoretical aspects. Architectures and landscapes result from articulations that allow not only the formation of well-defined configurations but also a change of scale. After the operations of selection and accumulation, assemblages and sedimentation occur slowly and are

[5] De Manuel (2003, pp. 57–58).

completed through the incorporation of other matter. This palimpsest of superimposed layers is subsequently broken up and corrupted by other forces, generating new conformations.

How do you move from one layer to another? How do you work with the interfaces defining that thin surface of transition between different materials where movement, slippage, and folding take place?

Sestini's drawings are not simply illustrative and didactic but suggest a working method that uses drawing to hold different worlds together by composing them. The first useful indication is that which shows how to construct a drawing using different techniques of representation, for example, an axonometric projection with geological sections and qualitative and expressive representations of the elements occupying the surface (from the vegetal mantle to the orographic and hydrographic morphology, anthropic signs, or traces of events). We discover a tool that succeeds in proffering a synthetic vision with representations that are both rigorous and measurable. By presenting different timescales together through the repetition and juxtaposition of narrative frames, it can recount and describe cyclical and long-time variations. The role of a section is central both as a technique whose vivisection of the living body of landscape shows the invisible relationships between the inside and out, the visible and invisible, the comprehensible and mysterious, and as an investigative tool able to examine fragments and clues to allow us to discover unsuspected features. A cutaway makes it possible to assemble elements lying on different planes (strata) into a single configuration. A section goes against the flow, against time, to reveal complexity. By moving in the opposite direction to stratification, dissection discovers and puts diverse materials and natures, and different times and histories, into tension. Through transversal cuts, axonometric drawings present the architecture of not necessarily horizontal, superimposed planes to show the complex articulation of their relationship to the surface. These drawings suggest an idea of architecture as a tensor, a uniting ring, a leap transposing and transfiguring a material through metamorphosis into an open form towards other configurations.[6] Just as sedimented layers are broken into faults and wrinkled into folds by the slow processes of orogenesis, so a design, like a scalpel, breaks down and reassembles different materials and worlds. A section pushes a project to grow and conform places and materials that come into play in the evolutionary process of transformation "from within".

[6] "Zeno does not pose here the question of the transit, but the question of the figure. The latter is always still unfaithful. The journey here does not go from one point to another, but down the steps of a ladder. It piles the cards on top of each other in a space of overlapping sheets. He searches for the limit of what can be represented, he searches for the real in the recesses of a fragment. (...) Let us suppose that, of a region of space, we possess several maps of different scales; let us arrange them one on top of the other in a volume of several sheets. Zeno does not follow several directions and senses along one of these maps, but descends normally to the successive sheets, pierces a piece in the thickness of the representations. The resumption, the reiteration of his turn, actually concatenates the succession, he draws an exact series. And he descends infinitely towards the room. His path connects the stairs". Michel (1984, pp. 120–121).

23.2 In the Field. Some Experiences

1994. Linear Park between Caltagirone and Piazza Armerina

This park reinforces its former railway line incisions thus intensifying the vivisection of the land to bring to the surface all the faults and discontinuities of the landscape. The landslides, upheavals, and tampering of recent years have produced deformations, adjustments, new configurations, suggesting the possibility that a project does not build closed objects but rather forms capable of defining themselves and adapting to subsequent dynamics. Interpreting the infrastructure as a graft and a cut in the landscape, the project used these features to organize and design movement and visions along a line defining a field where figures unfold, generated by energy vectors in tension with each other (Fig. 23.3).

1998. Garden-Arena at the Temple

A reclaimed landfill has taken on the form of a raised landscape clod that blends in with fields at the bottom of the valley and on the opposite hills. A new suspended horizon line delineates nearby objects, such as the water tower, blending them with distant presences. This new object in the landscape thus appears like the petrification of a telluric movement that has disrupted the continuous line of the slope.

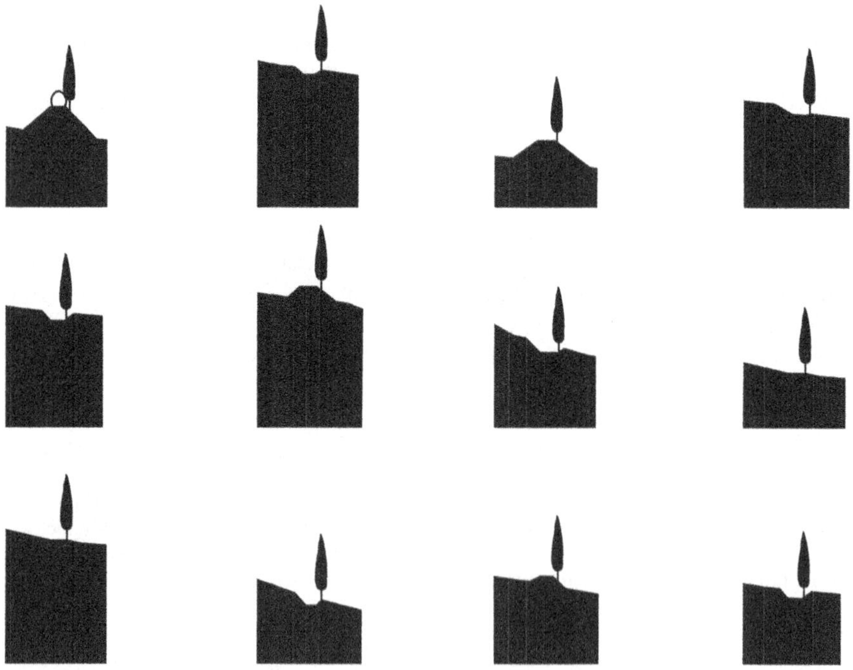

Fig. 23.3 1999–2000 Linear Park between Caltagirone, San Michele di Ganzaria, and Piazza Armerina Tomographic sections

The Nicosia stone skin-cladding recounts the stages of a geological metamorphosis, fixing the final outcome. Starting from the ground, the stone surfaces present an isodomum construction, but at a certain point something happens: internal energy springs forth, disrupting the original musical score, producing a laceration, a displacement, which disarticulates the ordered sequence of horizontal planes and bends them into unpredictable diagonals.

The Garden-Arena is a clod of soil raised by slow movements along a fault line that separates and unites times, landscapes, and architecture. Its surfaces intercept seeds and collect rainwater, select sights and cut out fragments from the objects present, reassembling them into a new configuration that melds with human events. The Garden-Arena transposes crumpled rubbish sedimented by abandonment and neglect into a new geographical form.

2010. Giampilieri

Since all the torrential river courses from upstream to downstream were involved, reconstruction projects drawn up after the 1 October 2009 flood had to modulate instruments adapted to different contexts, from landscapes to urban centres, while seeking to develop internal consistency. Geological layers are reproduced by tomographic sectioning that generates faults and breaks where different densities, directions, and inclinations intersect. The dissection reads new levels of movement in space where signs of times and events distant from each other appear on the same surface. The juxtapositions produce gaps and dilations. Reinterpreting these territories' characters and features through sectioning opens up new possibilities for investigating the most interstitial and hidden parts of their interior (Fig. 23.4).

The initial project for the hydraulic canal in the Vallone Puntale area, drawn up by the Civil Engineers of Messina, was based on a number of rigid choices that offered minimal intervention margins to mitigate the impact of the work, reduced to burying the reinforced concrete box and covering the side walls with local stone. The alternative suggested by the "Riparare Fiumare" group radically re-considered the purely technical solution to the problem, proposing a complex urban project attentive to places and their relationship to the community, which the previous hypothesis would have irreparably damaged.

Fig. 23.4 January 2010/December 2011. Scaletta Superiore and Marina. Tomographic sections with panoramic view

The particularity of the interventions along the torrent required the refinement of some project tools such as sections and topographic models. In particular, repeated Tomographic Section (TS) sequences enabled management of the relationship between intervention insertions in the landscape and single details linked to the micro-physical contingencies of the place (Fig. 23.5).

Thus the sections became a two-pronged instrument able to extract actual physical forms and transform them into elementary profiles so as to simplify reading the territory and connect the project to the geological history of the place and, at the same time, allow the generation of abstract profiles from the elements linked to the location—orography, hydrography, climate, matter, history, experience—to make the objects identified unique and encourage a site specific project. Water works were also studied by means of Tomographic Sections that were then contextualized differently within the larger territorial sections considering their original position along the river.

Fig. 23.5 January 2010/December 2011. Giampilieri Superiore. Topographical study sections of the variant project for the new security water canal in c.da Loco Grande. Study model of the variant project for the new security water canal on via Puntale

The Tomographic Section tool enables the generation of inventions and displacements starting from the manipulation of objective data. To transform the sequence into a narrative, some elements are isolated through the use of single sections according to the geological stratification method in order to describe and control the space around the lines of movement. Tomography is a cognitive tool able to describe and recount the surfaces crossed, becoming an opportunity for unexpected readings of the territory obtained by manipulating objective data through precise procedures and constraints. The obsessive exercise of repetition and superimposition opens up the possibility of reconsidering space in a nomadic key, concentrating not only on traces that disappear and move along the route but also on those archaeological fragments that permanently emerge after a traumatic event (Fig. 23.6).

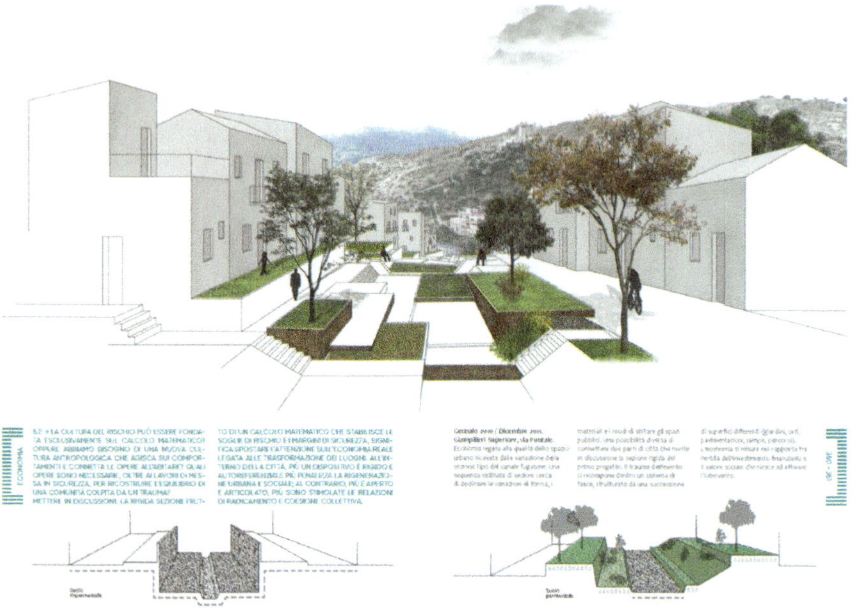

Fig. 23.6 January 2010/December 2011. Giampilieri Superiore, via Puntale. New urban space derived from the variation of the standard section of the new security water canal

23.3 Conclusions. For a Geologic Architecture

The idea of Geologic Architecture experienced in these occasions suggests a procedure that not only goes beyond the idea of work autonomous from its context, since it is composed of stratification and the juxtaposition of materials and sections, but also eliminates the boundary between background and object. Each transformation is a node connecting different materials and levels, stratifications of the Earth's crust and human communities, long durations and short times, cyclic and unique, unrepeatable events. The opposition between nature and artifice disappears and new hybrid matter appears in which boundaries are no longer identifiable and we are immersed in a hyper-natural environment. The forms of architecture, rather than mere technology, become a response to environmental issues. Even the idea of circularity is bent and reinterpreted within a temporal framework in which instants and long durations are intertwined.

Architecture is part of the erosion and sedimentation process of the earth's crust. The materials we use are Eroded Mountains, Water, and Light. The earth's crust is the millennial petrification of a slow and violent battle. An architect participates in the construction of the earth's crust; he uses the materials of this battle. If he uses breccia or granite, he works with either millennial erosion or sedimentation, the violence of an abrupt solidification, or the trauma of a sudden jump in temperature. We work on the sedimentation and erosion of visions, movements, and sensations.

We move on the faults that open up, making space for the atmosphere; the air that we inhale and exhale defines the rhythm of our participation in the world. Sedimentary rocks tirelessly tell us how fragments can re-compose themselves into even more wonderful formations than before. Architecture recomposes and invents perceptions and effects.

Many cities, as Francesco Rodolico (1976) reminds us, have modified their urban terrain by transforming the ruins of the past or the accumulation of waste and debris into new topographical configurations on which to erect new architecture and new spaces. "Antiquity offers us the most curious artificial relief that has ever been inserted into an urban landscape: Testaccio (Mons Testaceus) in Rome, next the left bank of the Tiber downstream from the city near the Porta San Paolo. Testaccio because it was formed from the shards (testa) of amphorae, mostly wine or oil, which for over a millennium were unloaded on this site by the cargo ships that had sailed up the river to Emporium. It rises 35 m above the surrounding plain and the perimeter at its base is about 850 m. (…) This Testaccio is an exceptional example; generally, artificial reliefs present in cities derive from the accumulation of earthworks or rubble, perhaps left in place after the collapse of buildings". Rome is an emblematic example of the geological modifications carried out by architecture in a slow process of stratification, cutting, erosion, and excavation, which has ceaselessly woven the palimpsest of the city. It is this forgotten material that should still be placed at the centre of projects for contemporary cities and territories. This material should not be considered inert, but rather some-thing alive, only temporarily dormant and absent.

The most recent stratifications have often erased and removed those natural elements that for centuries were considered the founding principles of cities. Geological architecture has the task of rediscovering them, digging them up, and reconnecting them to contemporary life.

Geologic Architecture activates a process of correspondence: not imposing form by a preconceived strategy on a raw material substance, or an inert territory, but rather drawing out or actualizing the potential immanent in a transfiguring world. Life is espied and discovered from within matter—in a territory—thus allowing it to collaborate. Within this paradigm, a project must "see", through its design, what matter—a territory, places—can do.

References

Sestini Aldo (1963) Il Paesaggio. Touring club, Milano

De Landa Manuel (2003) Mille anni di storia non lineare. Rocce, germi e parole, Instar libri, Torino

Serres Michel (1984) Passaggio a Nord-Ovest. Hermès V, Pratiche editrice, Parma

Wurman Richard Saul (1986) What will be has always been. The words of Louis I. Kahn, Access and Rizzoli, New York

Chapter 24
Time-Tested Design and Implementation: Differing Temporalities in Landscape Design

Michel Hössler

Agence Ter is a landscape and urban design practice, working on a wide range of project types and scales, from designing pocket, urban or ecological parks, to the redevelopment of river banks, urban projects for new districts and the definition of major territorial strategies.

During more than 30 years that have passed since the agency's founding in 1986, we have observed the evolution of our profession and the emergence of new social and environmental concerns, particularly concerning the effects of climate change.

Time has always been an important factor in the work of landscape design, where the imperatives and rhythms of nature offer both opportunities and constraints. This is even more true in French, where the word '*temps*' is used to designate time, in the sense of duration and also climate, in the sense of the weather.

I have chosen three examples to illustrate the importance of this notion in our work, each reflecting a different temporality. Time can mean the long timeline of the geohistory of the project site, where landscape functions as a time recorder; historical time can be read in the visible traces and ruins left by past inhabitants, and geological time in the landforms of events from previous millennia, such as ice ages. There is also the shorter time of the project, of its conception, elaboration and then its spatialization.

M. Hössler (✉)
Agence Ter, Paris, France
e-mail: mhoessler@Agenceter.com

T. Matteini et al. (eds.), *Cultivating Continuity of the European Landscape*, Environmental History 15, https://doi.org/10.1007/978-3-031-25713-1_24

24.1 Geological Time: Nantes Saint-Nazaire, an Estuary in Motion

The upheaval of conditions due to climate change demonstrates almost daily the tremendous acceleration of time taking place in our era. In previous geological periods, a change of 5 °C would take place over 100,000 years, whereas now the climatologists of the Intergovernmental Panel on Climate Change (IPCC) predict the same 5 °C change over a period of less than 500 years.

We therefore need to deconstruct our 'fixist' vision of the landscape, frozen in the sepia images of old postcards, to project ourselves into the future that is fast approaching dangerously. For example, we now need to invent new resilient palettes of plants, capable of reversing the inevitable fall in biodiversity, in anticipation of the rapid changes to come.

In the Nantes Saint-Nazaire estuary, we have designed a forward-looking metropolitan strategy '*Bords à Bords*' (Edge to Edge) which anticipates rising water levels and is exemplified through pilot sites representative of the diversity of the landscapes. The proposed territorial grid is based on the geography and history of the territory. These landscapes are marked by the presence of water in all its forms. Where water meets land, the edges, true spaces of interface in all their diversity, become the privileged places of the project. Inaccessible and wild today, these spaces have very limited contact with the urban world. The project is based upon reconnections with these ignored and neglected spaces through leisure activities, while maintaining ecosystems. The proposed developments combine tourist interests with the maintenance of agriculture. Visitors are thus made aware of these territories, of their history, their heritage and their uses.

Two main project objectives were articulated, taking into account the future spatial conditions created by rising water levels:

- to find continuities of new routes based on existing paths and
- to protect certain wild spaces by concentrating visitors on well-identified routes.

An extensive work of interpretation, education and signage has been proposed to create awareness of this territory and help make it a special place within the metropolis of Nantes-St Nazaire: a new metropolitan public space accessible by multiple slow transport modes: bicycle, train and on foot.

24.2 A Time of Healing: A Master Plan for the Renewal of Liege Steelworks Sites.

As part of the development of a masterplan for 300 hectares of derelict land from four abandoned steelworks sites, we are working on the integration of time through a process which we assimilate to healing. This concept links time management to the urban or landscape actions essential to the economic life of the project and, most importantly, facilitates a dynamic of renewal over time.

The healing of brownfield sites is first of all biological; it encourages the return of biodiversity and of human activity to a site that people have previously degraded and abandoned. As in healing in the human body, where scars attest to the process, certain elements that make up the ecosystem will permanently retain traces of past degradation even after renaturation.

The scarring is also urban. The process reintegrates the four sites into the surrounding urban fabric that hosts them, remodeling them in continuity with the city, rejoining the traffic that today stops at the edges, to traverse the sites, encouraging and easing access for pedestrians and cyclists.

The healing is architectural because the reconversion of old disused industrial sites raises the question of the preservation or removal of the built traces of a flourishing past, witnesses to a complex industrial process. These traces are inseparable from the economic, political, social and cultural history of a city or a region, an integral part of European cultural heritage.

Finally, the healing process involves the place of the site in the civic consciousness. The future of the sites no longer rests with the large steel companies; it belongs to citizens now, and the healing aims to stimulate other imaginations, other fields of possibilities and a sense of wonder and optimism for the future.

24.3 The Time of the Project: The '*Vilaine Aval*' Project in Rennes

Sometimes the time of the project, of its conception, elaboration and then its spatialization, takes several years, or even decades. This poses the question of maintaining the course of the project, its internal logic and overall sense while accepting its enrichment by programmatic or usage inflections. The '*Vilaine Aval*' (Vilaine Downstream) project led by Agence Ter illustrates our response to the challenges of a project being realized over a relatively long timeline with an innovative approach to participatory conception.

The Vilaine valley is a segment of the '*Manche-Oc6an*' link connecting Rennes to the Atlantic coast. This area, characteristic of the Rennes metropolis, is home to occasional industrial activities involving the extraction of aggregates in the heart of a high-quality natural and agricultural landscape. This landscape is a protected area with high biodiversity potential; it is ecologically dense, complex and sensitive, particularly regarding the omnipresent water features (ponds, flood zones, wetlands).

The project began in 2013 with a competitive dialogue process bringing together three multidisciplinary teams led by landscape designers, who over a period of six months refined their territorial development strategy through regular exchanges with a committee of elected representatives and technicians from '*Rennes M6tropole*', the intercommunal municipal structure centered around the city of Rennes.

The site was not precisely defined, it was vaguely indicated by an outline that the candidates could question. The initial brief was for all elements of the project to emerge at the same time: a place for leisure, sports and outdoor events and an area of development and innovation based on agricultural and aquatic components with strong tourist appeal.

The context, new for us in 2013, was complex and led to an adaptation of our usual methods of analysis and conception: the site was huge, (nearly 3500 hectares, over 25 km in length) and the timeframe envisaged was more than ten years.

The project management was complex: municipal actors included the city of Rennes (200,000 inhabitants) but also six other smaller municipalities (from five thousand to twenty thousand inhabitants).

One of the difficulties was to take into account the needs of the different groups of people who would be using this area. Pedestrians alone or in groups, runners, cyclists and horse-riders, but also farmers and fishermen; different groups with occasionally contrasting visions and desires.

This is when the question of time became critical, with a choice between imposing a project with very limited consultation on a shorter timescale, or conceiving a broad but lengthy process to seek the views of citizens and stakeholders, with again two possibilities. The first, and more conventional process, was to seek a consensus, which seemed improbable and difficult. The preferred second method was to bring together all stakeholders in a work of co-construction, exploring the contrasts between different people, different groups and the elected representatives of the different municipalities.

This work was carried out in parallel with our work as landscape designers on the development of certain public spaces on a shorter timescale. This part of the commission was urgent because of political temporalities—in France, municipal elections take place every 6 years. This political time is sometimes incompatible with other project imperatives.

The project development process changed from a 'top-down' approach to 'bottom-up' work, in a partnership based on dialogue and exchange, developed through participatory workshops and exchanges as well as occasional prefiguration and pilot actions. A co-construction approach has been put in place, in a logic of contextualization through testing in the field. The Cuesta cooperative accompanied Agence Ter through this process, co-organizing events with cultural or sports associations and artists and bringing an innovative cultural approach to project conception.

These prefiguration actions feed into the work of the landscape designers and serve to fuel, or even redirect, a classic work of design into a shared and co-constructed project. A paradigm shift in the way of working on spaces, this approach constantly reinterrogates and sometimes can be thought to weaken the landscape designer's work. But we feel that this complementary approach enriches our studies, and by opening the process to local actors and inhabitants, a territorial project emerges with strong and enduring support. In addition, ecological, environmental and economic issues are brought together on this scale, and the project benefits all aspects of local society.

24.4 Crossings and Stopovers

Carried out in 2015 and 2016, '*Traversées et Escales*' (Crossings and Stopovers) were among the earlier methods of meeting and communication with the inhabitants and stakeholders of the territory. Designed and developed in collaboration with Cuesta, the Crossings and Stopovers took the form of exploratory hikes of the territory with each of the municipalities concerned and involved elected representatives, members of associations, economic and artistic players and inhabitants. They involved collecting content for innovative communication and enhancement tools (reference points, digital and contributory tools for presenting the project, a creative walking guide enriched with artistic content and feedback from former users).

24.5 The Year of the Vilaine

Running in parallel with project design, participatory work in the territory continued with '*l'Ann6e de la Vilaine*' (the Year of the Vilaine) between spring 2018 and summer 2019. This was a cultural experience of this specific territory between town, and country, and river, by navigating and finding new recreational, sporting or economic uses for the water, through artistic, cultural and sporting proposals. '*l'Ann6e de la Vilaine*' was a collective proposal, in which everyone was a stakeholder—sports and environmental associations, farmers and craftsmen, companies and institutional partners, artists and citizens. They contributed to making a kind of open-air metropolitan facility and to testing its governance and animation in the long term.

24.6 Summary

The success of such an approach lies both in the capacity to manage the inevitable contrasts between the actors of the territory and the elected representatives, but also in the difficulty of reconciling project aspirations with a restricted budget. The projects result from collective work involving actors who have something useful to propose for the territory and its inhabitants and who wish to co-construct an innovative common proposal in a perspective of perpetuation. As the process continues over several years, the various participants build a network of support points for the territorial strategy.

Above all, the key to success for such a complex project on a territorial scale is to respect the need for the **time of the project.**

Chapter 25
The Art of Failing Urban Planting Precariousness as Landscape Project Ecology

Denis Delbaere

The relation of a landscape architecture project to time is mostly a relation to its own *alteration*. This acknowledgment, which clearly appears to each landscape designer honestly thinking of his own work, is nevertheless hidden although, as I will try to explain in this paper, it should be thought of as the most interesting aspect of a landscape architect's work because it relates it to the social and biological will for greater adaptability in public spaces (Fig. 25.1).

25.1 The Contradictions of Landscape Architecture Regarding Time

That a landscape architecture project mainly deals with time is a commonplace. But the way landscape architects speak about time in their speeches, papers or conferences assimilate it to maturation. The maturation logic comprises the following statements:

1. A landscape architecture project is by nature incomplete when built
2. To achieve conformity with its designer's will takes several years
3. Plant growth and, in general, the use of living materials illustrate this necessity
4. When the project has been well designed, time makes it perfect and brings it to its ideal and terminal state.

According to this idea, the landscape architect would plant a little seed and only he would know how it will grow. And if he has done his work well, time will prove he was right. If this is not the case, if the project fails, if plantings don't grow as was planned, if benches or street lights are destroyed, if areas are used for different goals than the

D. Delbaere (✉)
Laboratoire Conception Territoire Histoire (LaCTH), Ecole Nationale Supérieured'Architecture et de Paysage de Lille (ENSAPL), Villeneuve-d'Ascq, France
e-mail: d-delbaere@lille.archi.fr

T. Matteini et al. (eds.), *Cultivating Continuity of the European Landscape*, Environmental History 15, https://doi.org/10.1007/978-3-031-25713-1_25

Fig. 25.1 Bois Leurent Park's observatory, in the North of Lille, in 2003 and in 2015

ones they were designed to support, if paths are diverted by a dense spontaneous path network completely independent of their geometry, it is less because of the project's errors than because of the failure of maintenance services or of the local authorities who have forgotten the initial aims of the project. From this point of view, the usual landscape architecture discourse about time is contradictory. On the one hand, it claims that time is completely a part of and the most important condition of its art, but on the other hand, it only considers those effects that confirm the project and its ideal achievement.

25.2 The Precariousness of Ordinary Public Spaces

The truth, as the facts show, is very different. It can be seen if we drop the beautiful professional reviews showing beautiful photographs on glossy paper of nearly always prestigious projects signed by the most famous designers, involving large budgets and included in ambitious urban masterplans in the centers of big urban metropolises all over the world. These kinds of projects are, as we know, weakly exposed to the effects of time because they involve important maintenance contracts and because their designers have often favored architectural hardware rather than landscape software: large pavements and elegant and expansive benches, stone borders and sculptural fountains, deep foundations and major infrastructure, in order to reduce the risk of alteration.

But this only concerns a small part of the landscape production of public spaces. On the contrary, it ordinarily concerns cheap budget projects involving large areas to be managed at very low prices and in less preserved urban contexts, mainly in suburban environments where social practice control is weak, and conflicts often major. Making a landscape project in such conditions is difficult but also stimulating. We have to plant a lot and build spaces with slopes and topography rather than with expensive sustaining walls and infrastructures. We have to find the precise balance between excavated and embankment ground, even if it involves very heterogeneous and unstable artificial grounds. We have to minimize the street furniture and assimilate it to the opportunities offered by topographical declivities. We have to plant small in order to produce mass effects. Public spaces designed in such a way are characterized by their fragility and, moreover, by their profound precariousness. And then, in most cases, as soon as building is achieved, a slow but irreversible erosion of the project starts which, in a few years, can disfigure it—and then reconfigure it—in order to make it become *something else.*

25.3 A Critique to Understand the Landscape Temporality of Public Spaces

For the last fifteen years, I have been systematically engaged in returning to landscape projects, mine as much as those of my colleagues.

This last case leads me to critical activity based on a retrospective evaluation of projects, considering not only what designers had wanted to do but also, and in the same proportion, what their design really induced. Since 2013, the *banC public* review has regularly and freely published the results of such enquiries. My "landscape searcher" activity extends this critical work to projects built during the sixties and the eighties. In my second book, "Table rase et Paysage", I published the results of several surveys in French social housing districts, showing how in such areas, buildings increasingly deteriorate while landscape grows and often produces quality public spaces although nobody has designed them.

I have called this kind of landscape project, "metaproject", that we can only discover through a retrospective survey of the urban landscape. It really is a project in the sense that it would not have existed if no architect, landscape designer or engineer had worked there. But nevertheless, none of those designers can claim such spaces as their own creation because they are mostly shaped by the progressive and obviously unpredictable alteration of initial projects (Fig. 25.2).

In recent years, I and my colleague Frederic Pousin have extended this critical approach to the first regional plans drawn up during the sixties and the seventies by OREAM. Those public planner teams had to imagine new French metropolises to reduce the Parisian area's territorial weight, and they included several young French landscape architects such as Jacques Sgard, Michel Viollet, Jean Challet and Pierre Dauvergne. All these designers are the authors of sometimes very interesting landscape situations. But if we consider those spaces in relation to the plans they came from, we soon discover that *they are less produced than induced by them.*

25.4 Altered Public Space as an Open Project

But, if these metaprojects often seem very disappointing to their designers, who look at them as a disfigured version of their own goals or are even unsure whether to consider them as their own creation at all, they nevertheless generally constitute high-quality places according to the other social actors.

On the environmental side, those places are often partly abandoned and become precious refuges of urban biodiversity, to the point that they have now sometimes become completely part of regional and metropolitan green networks. This is what I have shown for the last ten years by exploring the Likoto spontaneous green network, whose 1300 km of almost continuous green spaces are the result of the alteration of the great infrastructure projects built between Lille, Tournai and Kortrijk in France and in Belgium since the sixties (Fig. 25.3).

Fig. 25.2 Example of metaproject, the Blosne district, in the south of Rennes (aerial view up). Built since the beginning of the 70' and planned by Louis Arretche (see the redrawing of the plan under the aerial view), the urban landscape has not followed the very regular grid of the project and has been built in several steps during 15 years and without real order. But each lot has involved some peripheral trees plantations, and little by little all those plantations have placed the all district in the middle of what could be perceived as a real urban bocage (see our sketches). The first elements of this bocage come from the conservation of some preexisting linear of hedges and corpses. The aerial photographs' analysis reconstitutes the progressive growth of the Blosne's bocage. (all drawings by Denis Delbaere)

Fig. 25.3 Likoto Eurometropolis' spontaneous green network appears through a cartographic survey of all the green corridors growing beside the great infrastructures between Lille, Tournai and Kortrijk (on this map, lawns in yellow, bushes in light green and woods in dark green). Those plantings and green wastelands now create a green landscape everywhere in this eurometropolis which particularity is to have been built into a wide European infrastructural crossroad

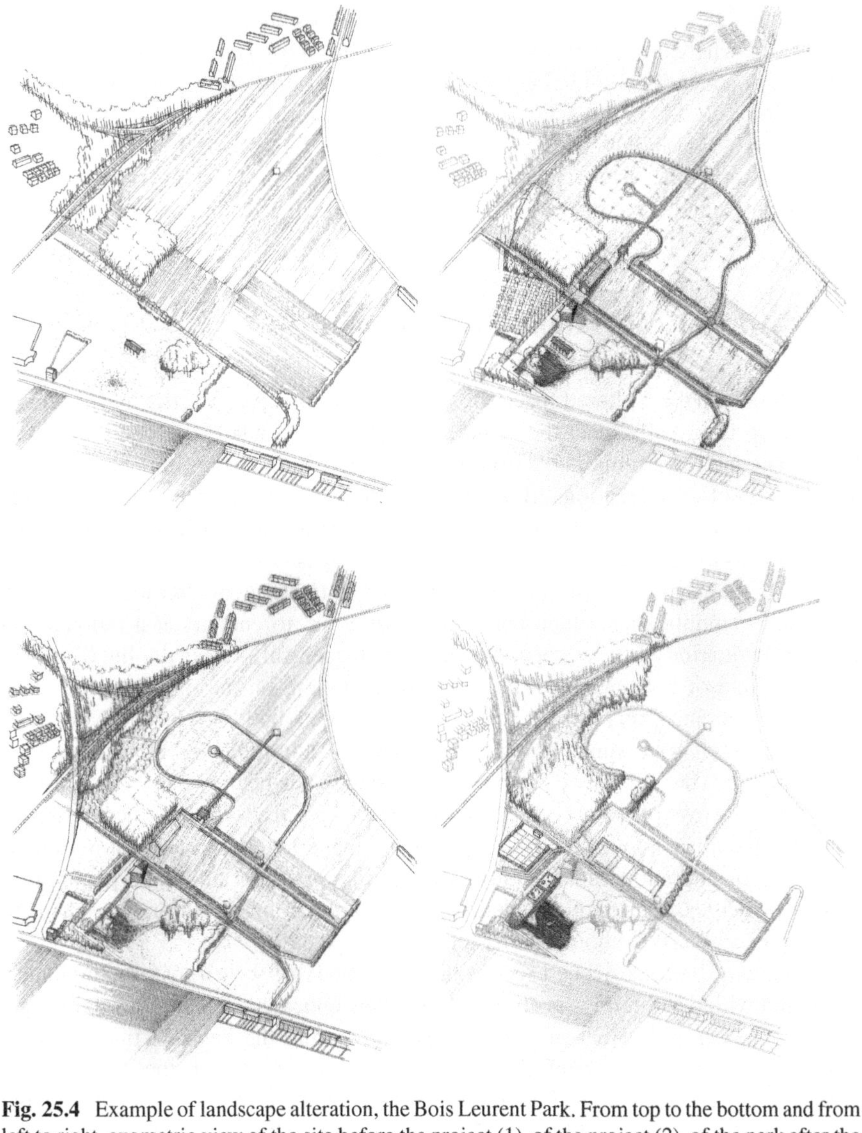

Fig. 25.4 Example of landscape alteration, the Bois Leurent Park. From top to the bottom and from left to right, axometric view of the site before the project (1), of the project (2), of the park after the building (3) and of the park 15 years later (4). (all drawings by Denis Delbaere)

On the social side, these altered places are intensively used by their neighborhoods who often reallocate and, if necessary, share them when their goals diverge. These huge grasslands, copses and slopes welcome different social activities, excluded from the city, and as diverse as picnicking, sledging, walking with dogs, small daily walks, mountain biking, jogging, as well as temporary habitats and camping. Claire Simon has perfectly shown the compensative social function of the Bois de Vincennes' margins, where the landscape design drawing vanishes under spontaneous trees and bushes, and where the sexual, spiritual and artistic needs of a large part of the Parisian population, generally confined in a too regulated and constraining environment, can be liberated. Weakly or not at all ruled, beyond the control of the police, those places are territories that living social groups share and contest, generally reaching a kind of pacific, sometimes friendly, status quo, even if it takes a shape that our ideal of an open society condemns: the reign of fences, walls, hedges and ditches.

25.5 A Landscape Alteration Grammar

Reinvestment in these altered spaces by landscape architecture practice and discourse would help it to really meet the actual needs of society, instead of confining itself to brilliant and often remarkable exercises of composition, more similar to architectural than landscape creations. But we first need to change the way we consider the alteration of our own projects, so that, behind the appearing disorder of disfiguration, we can guess the growing reconfigurations.

That is the purpose of my last book. It aims to imagine an alteration grammar, a physics of public space landscape reconfiguration, to encourage a rational and positive evaluation of these phenomena. To be successful, we need a huge amount of knowledge of both poles of the alteration equation: on the one side, the place as we can see and experience it several years after its building; on the other side, the project itself—we must know its circumstances, ambitions, renouncements and strong points. The best way to gain such knowledge has been to carry out a critical analysis of my own projects because I have at my disposal a complete documentation of them. So in my book, I analyze eight of my projects, chosen according to their high level of alteration and diversity: a large urban park, a landscape planting on motorway verges, a little square, an urban walk, a plaza, an agricultural park and the green spaces of a social housing district.

For each of these projects, I reconstituted the diary of their transformation and I wondered which common aspects and differences appeared on their way to change. I then came up with a list of seven concepts forming the basis of the *alterative* grammar: the ruin, the island, the stand, the outline, the wood, the seed and the pole. These *alterative* dimensions are not exclusive but are parts of a global dynamic whose effects are always temporary. There is no more a "project climax" than there is a biological climax. Evolution never ends.

As evidence, among these different "landscape autocritiques", those which most inspire me deal with the most ancient and malleable projects, for which I can observe complete reconfigurations twenty years later. For example, the Bois Leurent Park (Fig. 25.4), whose alteration was so cruel to me during the first years because it devastated numerous aspects of the place that I had worked hard on and that I was quite proud of, is now moving toward another spatial organization whose sense is slowly appearing and makes me glad. Reality, after the first disappointments, now seems full of promise. I would even be tempted to present some aspects of this park as the effects of my will though I know perfectly well they are the result of the action of some inhabitant, user or public gardener who was completely ignorant of the original project.

How better to say that what I can perceive as my own project is, in fact, a collective creation? And that is undoubtedly why it can be claimed that a landscape project, more than ever and in a very pertinent way in our period of recomposing social and political legitimacy, *is a public space.*

Chapter 26
Seascape Design. A Limit for the European Landscape Convention

Daniela Colafranceschi

Abstract Talking about lands is also talking about the sea, of the sea that belongs to it and the territories that limit its shores. Because the sea is part of our perception up to the horizon and because from the horizon, we are perceived as part of it. Like a liquid landscape that holds together the geography of heterogeneous cities and multiple places, the sea is a 'full' and not a 'void'. Water is not all the same, the sea is not all the same.

Article 2 of the European Landscape Convention says: "(…) this Convention applies to the entire territory of the Parties and covers natural, rural, urban and peri-urban areas. It includes land, inland water and marine areas. (…)" In addition, the guidelines—Recommendation CM/Rec(2008)—mention: "(…) the landscape forms a whole whose constituent parts are considered simultaneously in their interrelations".

It is precisely the interrelation between land and sea that makes the coastal border a fringe, a sensitive swath possessed of intrinsic value due to its sense of identity and the fact of belonging to a territory that is liquid but still 'landscape'.

The water of rivers and lakes is an integral part of our territories; however, landscape stops there, along the coastline of our regions.

The dynamics of views and perceptions from land to sea and back to land again convey meanings that are related to use, to passage, to all the technology and infrastructure dedicated to moving, travelling, transporting and passing between coasts. Today they encompass profoundly changing cultural values that landscape design is called to interpret.

Water, as well as landscape, is a cultural product.

The ELC has been a revolution; we are all aware of this.

It has completely overturned the way of conceiving the dimension of landscape in terms of quality, identity and entity. It establishes a set of articles that affirm a concept

Thematic Session B. Landscape Design and Time—Topic 1 | Infra-Structural Landscapes Key words: Sea, cultural device, liquid public space

D. Colafranceschi (✉)
University of Naples Federico II, Napoli, Italy
e-mail: daniela.colafranceschi@unina.it

T. Matteini et al. (eds.), *Cultivating Continuity of the European Landscape*, Environmental History 15, https://doi.org/10.1007/978-3-031-25713-1_26

of landscape which is dynamic and non-static, hybrid and not homogeneous; it directs design not towards 'landscapes of quality' but towards 'quality of landscape' in its entirety; it has consolidated its autonomy in terms of strategies, planning, design and launched a sustainable development directive in perspective.

But everything refers to territories. Everything seems to end along that line separating land from sea, even if it does not interrupt the landscape, and instead, as a hybrid entity, binds the one to the other. Water is a meeting space because it belongs to the same system.

The Mediterranean as a land and not as a sea, as a liquid landscape that holds together the geography of heterogeneous cities and multiple places is a 'full' space and not a 'void' one.

However, the sea is not all the same.

The water is special between neighbouring shores, between islands of the same archipelago, or when it is treated and transformed into land. Here, the landscape takes on the value of a liquid public space. Here, the sea is not an interruption; it is not a silence.

John Cage loved to teach us that even when sounds cease and silence comes, the music continues, but we stop listening to it.

I would like to present some examples I consider emblematic in this sense.

26.1 Straits

The closer coasts are to each other, the stronger the tension and the relationships that water weaves between its lands. This is the case of sea Straits, as a condition of mobility *infrastructure*, as a *device,* as a space for translation and modification of values, as a *system* for the conceptual, as well as the physical scale of the identification of its landscape, to be *places* in the different levels of identity inhabited by people.

For the Gibraltar Strait, critical frontier of our European continent with North Africa, the sea takes on an asymmetry between two worlds and measures a reversal of the economic, social, political and ideological balances between the two shores.

There is the Strait of Messina in which, through the Strait Park—for which I have been fighting for years—and its 'liquid square', two Italian regions would find the cultural system of a landscape centrality that substantiates the reasons for a 'Metropolitan Area of the Strait' in the objectives of its draft law.[1]

An even more striking case of special water is İstanbul Boğazı, a narrow strip of sea where just 700 m separate Asia from Europe. The densely trafficked Bosporus, privileged theatre of flows and currents, merges two continents and more (Fig. 26.1).

[1] Draft law no.142. The 'Metropolitan Area of the Strait' draft law aims to place the territory overlooking the Strait of Messina of the provinces of Reggio Calabria and Messina in a context of economic and social integration that enhances their respective potentialities. (http://www.consiglio regionale.calabria.it/pl8/142.1.htm.

Fig. 26.1 L'acqua non è tutta uguale. Collage 1, by Daniela Colafranceschi, july 2020

Straits are landscapes of passages. Trajectories, flows, networks identify the fabric sewn between the two sides: stratifications of identity, of ways of living, of ways of relating and moving. They are liquid spaces of translation and innovation of non-traditional landscapes. A type of sea where the infrastructure of displacement, economic, merchant or fishing activities, and the daily life of work and human relationships mark the speed, scales and times of its transit, supporting reasons, meanings and feelings.

Water is an element of interaction, an 'exchange' area of immaterial and tangible values, which merge sea and land into a unique and unitary landscape.

26.2 Archipelagos

In 2003, with resolution n.99/2003, City Council established 'Venice Lagoon Park'[2] then, by extension, the regional environmental and anthropological park of local interest of the Northern Lagoon of Venice was raised in 2014. Sea and islands become 'territory', even before becoming a Park, as the resolution says "the Institution must promote and disseminate knowledge and information relating to the

[2] "The main task of the 'Lagoon Park' Institution is the environmental and socio-economic enhancement of the Northern Lagoon area, through the definition and promotion of uses compatible with the protection of the naturalistic, archaeological, historical and cultural values of the places" (http://www.agendavenezia.org/it/organizzatore-194.htm).

Fig. 26.2 L'acqua non è tutta uguale. Collage 2, by Daniela Colafranceschi, july 2020

territory of the Northern Lagoon also through the involvement and active participation of local communities, both in the planning and management of initiatives, in order to strengthen awareness, identity and responsibility of the same".[3]

A lagoon ecosystem composed of cities, monuments, islands, architectures, water and sea canals. The value of a liquid in the quality of the space it conforms to is inseparable and essential. Venice is the sea of the lagoon (Fig. 26.2).

In 2010, King Juan Carlos I issued Law 44 on the Canary Islands waters, published in the Boletín Oficial del Estado n.318. This is the only case of inclusion and attribution of the sea to a Spanish Autonomous Community, that of the Canary Islands.[4] Thus, the spatial context pertaining to the Canaries is modified by attaching the territory of the islands to that portion of the sea inside the polygonal figure obtained by intercepting the extreme points of each island of the archipelago. The new perimeter defines a single mixed surface of water and land as a special maritime area: a unique 'territory'. The sea of this archipelago simply does not exist (Fig. 26.3).

[3] Ibidem.

[4] "Among the most salient extreme points of the islands and islets integrated within the Canary archipelago—according to art. 2 of its Statute of Autonomy—a perimeter will be traced that follows the general configuration of the archipelago, as established in the annex to this law. The waters within this perimeter will receive the denomination of 'Canary Waters' and will constitute a special maritime area of the Autonomous Community of the Canaries. State and autonomous community authority over the Canarian waters and, in its case, over the remaining maritime spaces surrounding the Canaries over which the Spanish state exercises sovereignty or jurisdiction, will take place taking into account the material distribution of constitutionally established competences and statutory as much for the aforementioned spaces as for the terrestrial ones" (https://www.boe.es/diario_boe/txt.php?id=BOE-A-2010-20140).

Fig. 26.3 L'acqua non è tutta uguale. Collage 3, by Daniela Colafranceschi, july 2020

26.3 Land from the Sea

The Dutch Rijkswaterstaat (Directorate-General for Public Works and Water Management of the Ministry of Infrastructure and Environment)[5] is responsible for monitoring an area of 90,278 km^2 of water, 214 km of dams and banks and 2969 km of rivers and canals and recognizes water as a characterizing and strongly identifying element of the natural and urban landscape. By managing infrastructure that reclaims land from the sea and promotes the sustainable development of the territory of the Netherlands, it fosters the configuration of new agricultural and pastoral landscapes as an engine for the regeneration of the public space of cities.

The land wrestled from the North Sea becomes a 'common good' for a plural society. Water that is both an economic, urban and landscape resource. Surfaces, which are now liquid, now wet but still places, new relational spaces, new territory for communities. A country for which liquid space assumes a central strategic, social and economic role, which has both landscape and environmental sustainability values in continuous transformation. A country where water draws the dynamics of collective living by letting flows and fluids meet (Fig. 26.4).

These are examples that testify to much more than protection value; it is a clear desire to attribute a design value to the sea water of their latitudes, as a heritage that determines their future.

[5] www.rijkswaterstaat.nl.

Fig. 26.4 L'acqua non è tutta uguale. Collage 4, by Daniela Colafranceschi, july 2020

They are landscapes defined by the journeys, habits, visits and sight of those who live or belong to them. Not because they are 'beautiful' but rather because they are made by people.

Even the migratory phenomenon of crossing the Mediterranean from North Africa to our countries has changed its identity, in its reasoning but above all in its feelings. Water that is full of worries, pain, expectations, death, uncertainties, abandonment. Water made of people, once living or survivors that change the value of the sea, in the dramatic landscape of transit. How do the eyes and souls of these people perceive our coasts from the sea? Where do we start to be part of their landscape?

The culture of contemporary landscape design has the difficult task of giving profound shape and meaning to the architecture of these geographies. We are called to define that margin of 'respect', that essential swath that allows us to maintain a relationship between the sea and its territory; we must be able to 'measure', based on values of quality and identity, the distance—but at the same time the relationship—that water is still able to weave between the sea and its land.

Organic strategies of thinking and fulfilling—and activating—effective, complex and broad responses cohesively, welding areas that are only apparently distinct and composed of similar sounds and pauses into a score of the same music.

In International Law, territorial waters are recognized within a range of 12 miles from the coast, then there are the 'exclusive economic zones' which go out to 200 miles, and also 'international waters', 'navigable' ones, 'forbidden' ones, 'pertinent' ones and so on.

But what is the right distance? How many metres from the coast is the sea still landscape? How many metres from the coast are we still able to support and perceive the tensions that, like wires, attach us and anchor us to the mainland? We must be able to read and interpret these threads.

Just as infrastructure for mobility has profoundly expanded and transformed the concept of public space and enhanced the urban quality of the relationship and exchange areas in contemporary cities, so it is with the large liquid square that is the sea, when limited portions of water combine with neighbouring lands.

The sea belongs to the territory; it is a square. It is a complex space, a landscape as a cultural device.

The European Landscape Convention could—and should—extend its important directives to this liquid territorial condition.

From here, we must activate a more widespread process of recognition of a 'liquid landscape' as a cultural product, a complex and hybrid space on the scale of the territories that belong to it.

(Fig. 26.5. As composition of1 + 2 + 3 + 4).

Daniela Colafranceschi, *L'acqua non è tutta uguale*. Collage, July 2020.

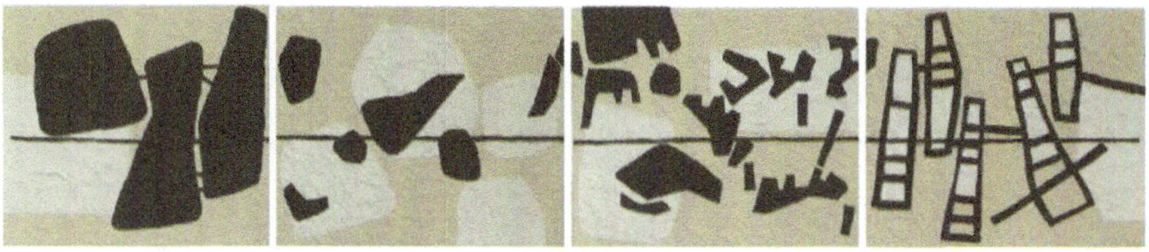

Fig. 26.5 L'acqua non è tutta uguale. Collages in sequence, by Daniela Colafranceschi, july 2020

Chapter 27
Water-Related Projects for the Urban Environment. The Case Studies of Lisbon Metropolitan Area

Caterina Anastasia

Abstract This work highlights how water is becoming a support for landscape and urban projects that tend towards a better liveability of inhabited areas. While analysing how in densely urbanised areas public spaces along the watercourses become terrain where urban policy aspires to build a greener and habitable metropolitan future, the intent is to emphasize the potential of urban regeneration projects bound to water. The case studies of this research are regeneration projects located in the city settled along the Tagus Estuary hydrographic system—dubbed the City of the Tagus Estuary (CTE, Lisbon Metropolitan Area, and Portugal). The article investigates the design action implemented by the public realised interventions and, at the same time, the context-based design processes that support them. The work emphasises that, regardless of whether they are natural or artificial / visible or invisible (because currently canalised and buried), estuarine water spaces—potentially the place par excellence of the sensitive territorial project—are today a resource for the design of the CTE, above all as generators of new urban facades and triggers of territorial greenery.

Keywords City and water · Water landscape · Water-related regeneration project · Lisbon metropolitan area · City of the Tagus Estuary

27.1 Introduction

This paper focuses on the relationship between the city and the hydrographic network that passes through it. Focusing on the case study of the huge Tagus Estuary water space and its humid system located in the Lisbon Metropolitan Area—LMA (Portugal)—the work highlights how watercourses today are potential areas for urban regeneration projects.

C. Anastasia (✉)
CIAUD, Research Centre for Architecture, Urbanism and Design, Lisbon School of Architecture, Universidade de Lisboa, Lisbon, Portugal
e-mail: caterinaanastasia@fa.ulisboa.pt

T. Matteini et al. (eds.), *Cultivating Continuity of the European Landscape*, Environmental History 15, https://doi.org/10.1007/978-3-031-25713-1_27

The Tagus Estuary banks and humid system are currently becoming the way and place to support an appealing urban regeneration formula. Accordingly, we will illustrate realised public regeneration projects along the main estuarine riverbed and its drainage watercourses (with surface or underground water flow) aiming to mark watercourses as a primary resource and tool for city projects.

The work starts from the hypothesis that the river corridor is firstly a natural area that supports an ecological network. According to current urban planning and landscape urbanism trends, a river is an element that can introduce a project into a city to entrust the environmental network with the regeneration of the territory at its different scales. This means that in urban transformation processes and policies, the water network is a key player in the promotion, habitability, and adaptation of a city (Anastasia, 2023).

Relating to this, the paper identifies realised interventions and organises the analysis according to where they are located: along the Tagus main riverbed and along the estuarine drainage corridors. The proposed division intends to address two great potentials of the water edge projects in relation with their geographical location. On one hand, the estuary as a 'generator of new urban facades' and, on the other hand, the 'water corridors as triggers of territorial greenery'. In both senses, corridors defined by water represent a land of urban opportunity.

27.2 Regeneration Projects Bound to Watercourses

27.2.1 The Case Studies: Water and Built-Up Space

The Tagus River defines the geographical opposition of Portugal's north and south. Close to its mouth, the Tagus incorporates an internal transition sea (*Mar da Palha*) around which the main Lisbon metropolitan conurbations have settled and connected. Of the 18 metropolitan municipalities belonging to the provinces of Lisbon and Setubal, 10 are located along the main Tagus riverbed.

From a geographical and environmental point of view, and as a clearly legible natural infrastructural corridor, the Tagus Estuary is the matrix and core of the LMA, the physical element that structures both banks at ecosystemic, formal, and socio-economic levels. Despite the marked differences between the north and south banks, this work identifies the Tagus Estuary as the key structural element for the ecological and socio-territorial connectivity of what we dubbed 'the City of the Tagus Estuary' (CET). As highlighted by Viganò, Secchi, and Fabian regarding the empty space of today's Venice Lagoon, we can say that in our day, the Tagus Estuary also remains the centre of the metropolitan territory, 'logically and conceptually' (Viganò et al. 2016, p. 15).

The drawing of the Tagus Estuary hydrographic network (Fig. 27.1) shows the 'fundamental ecological structure' (Ribeiro Telles 1999) of the LMA, where inhabited riverfronts and drainage corridors highlight the aggregating and relational power

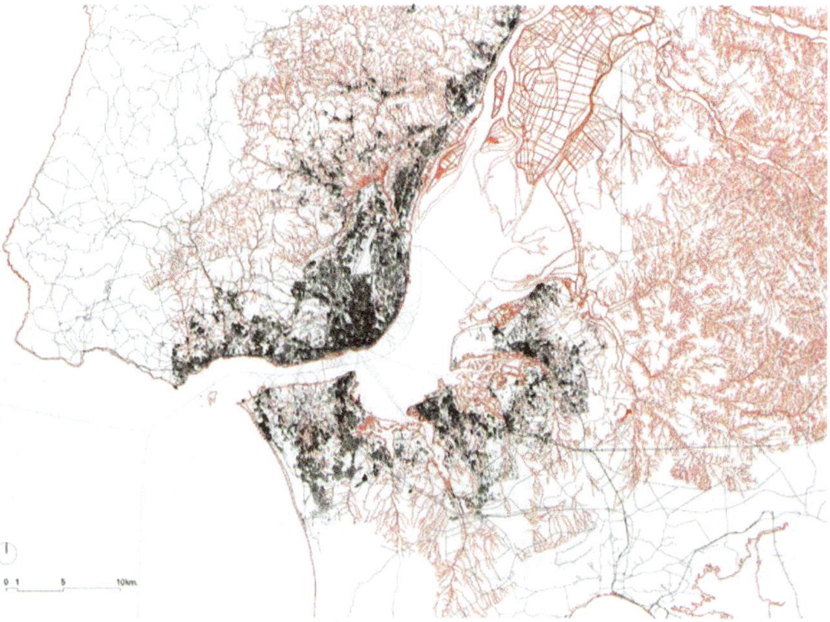

Fig. 27.1 Tagus Estuary—the hydrographic network, settlements along the main riverbed, and main mobility infrastructures (by land and water)—Source: author's drawing

of water by becoming opportunities to generate/regenerate and promote the local potential of its urbanised surroundings.

27.2.2 The New Urban Estuarine Facade

During the last thirty years, the banks of the Tagus Estuary have undergone several transformations. On both side of the river, territories gather urban areas, brownfields, and natural reserve areas with different economic, commercial, and industrial activities, have become terrains for public regeneration projects (Fig. 27.2).

The late 1980s (post-industrial era) saw the beginning of a paradigm shift. New attention was given to the river by Lisbon's urban cultural environment and the relationship between the city and the Tagus was pushed to enhance, as expressed at the time, 'returning the city to the river' (Bruno Soares and Trinidade Ferreira 2017).

The renowned Lisbon riverfronts were activated in 1993–94 with the POZOR (*Plano de Ordenamento da Zona Ribeirinha*) and continued with the PGIFR 2008 (*Plano Geral de Intervenções da Frente Ribeirinha*), which aimed to establish new urban continuities by extending the Expo '98 model from the west part of the city to the east (Salgado 2012). At a metropolitan level, from the idea of 'the city of two banks', stated by the LMA Regional Plan for the Territory Management (PROT AML)

Fig. 27.2 Tagus Estuary riverfronts: Almada (right), Seixal (top right), and Barreiro (bottom right)—*Source* Author's photographs

in 2002, the objective of achieving a metropolitan area centred on the Tagus Estuary seems to be a shared manifesto also at the municipal and district level. In fact, from then on, numerous public institutions of riverside conurbations promote the symbolic centrality of the estuary by linking urban renewal programs and promotions with the Tagus' waters. A part of the Lisbon riverfronts, Table 27.1 focuses on metropolitan renewal examples that use the main Tagus riverbed as a support for urban projects.

Based on direct observation, a literature review and information provided by the riverside municipalities, Table 27.1 presents 11 groups of variables through which this work aim to understand the design action set up by the 16 regeneration of public spaces facing water. The group of variables take into consideration features of the interventions—namely: the projects' materials and characteristics—and the new relationship created with the surroundings areas—links with river's identities. The focus is more on new promenades rather than 'greener' interventions (parks or beaches) or the individual restoration/revalorisation of old buildings linked to the estuary's activities and identity (i.e. renovation of tidal mills).

As shown in Table 27.1, in all cases the riverside interventions aim to create new promenades/public spaces for pedestrians and bikes, reduce car circulation, and increase green plots. Almost all the projects seek to revaluate and adapt local heritage for cultural activities, with a few examples of public space art exhibiting the Tagus' identity (e.g. statue of a salt worker). New views towards the Tagus—fewer barriers and increased pavement surface continuity—mark numerous projects. In a few cases, colourful paths signal the presence of cycling routes favouring an idea of continuity

Table 27.1 Estuary riverfront regeneration projects, north bank (NB) and south bank (SB). Rising Water Level (RWL); Cultural Heritage (CH)—Source: author's elaboration

Intervention	Location	Completion	New pavement, street furniture, lighting	Green plots	Reduction of car circulation / street pedestrianisation	Parking lots	Colourful paths	Increased continuity of pavement surfaces	Less barrier towards the Tagus and its view	Measures to mitigate RWL of the estuary	Valorisation of tangible CH / appropriation of local CH	Tagus' identity in public space	Link with coastal protected areas
Passeio do Tejo – Requalificação da frente ribeirinha	Alcochete SB	2014	•	•	•	•				Extension of existing containment wall to mitigate swell / RWL	•	•	•
Requalificação Miradouro Amália Rodrigues	Alcochete SB	2021	•	•	•	Ordering parking areas					•		
Frente ribeirinha Esteiro da Quebrada - Cais dos Vapores	Montijo SB	2011	•	•	•				•	Nnew dyke to retain river water / to drain rainwater from the city at critical times	•		•
Frente ribeirinha da Caldeira	Moita SB	2006	•	•	•		•	•	•		•		•

(continued)

Table 27.1 (continued)

Intervention	Location	Completion	New pavement, street furniture, lighting	Green plots	Reduction of car circulation / street pedestrianisation	Parking lots	Colourful paths	Increased continuity of pavement surfaces	Less barrier towards the Tagus and its view	Measures to mitigate RWL of the estuary	Valorisation of tangible CH / appropriation of local CH	Tagus' identity in public space	Link with coastal protected areas
Passeio ribeirinho Augusto Cabrita	Barreiro SB	2009	•	•	•		•	•	•		•		•
Passeio ribeirinho	Seixal SB	2017	•	•	•	Ordering parking areas		•	•		•	•	
Tratamento passeio ribeirinho Cacilhas - Olho de Boi	Almada SB	2012	•								•		
Miradouro/ elevador da Boca do Vento; Jardim do Rio	Almada SB	2000	•	•				•	•		•		
Passeio marítimo Santo Amaro - Paço de Arcos	Oeiras NB	2009	•	+ Beach	•	•		•	New view				
Passeio marítimo Caxias - Cruz Quebrada	Oeiras NB	2017	•	+ Beach	New pedestrian / cycling pathway		•	•	New view				

(continued)

Table 27.1 (continued)

Intervention	Location	Completion	New pavement, street furniture, lighting	Green plots	Reduction of car circulation / street pedestrianisation	Parking lots	Colourful paths	Increased continuity of pavement surfaces	Less barrier towards the Tagus and its view	Measures to mitigate RWL of the estuary	Valorisation of tangible CH / appropriation of local CH	Tagus' identity in public space	Link with coastal protected areas
Percurso ribeirinho Santa Iria de Azóia - pontão da BP	Loures NB	2017	•		New pedestrian / cycling pathway				New view		•		•
Percurso ribeirino Moinhos da Póvoa / Ciclovia do Tejo	Vila Franca de Xira NB	2018	•	•	New pedestrian / cycling pathway	•		•	New view	Tidal tanks	•		•
Parque urbano Póvoa de Santa Iria	Vila Franca de Xira NB	2013	•	•	•						•		
Parque linear ribeirinho estuário do Tejo	Vila Franca de Xira NB	2013	•	•	New pedestrian / cycling pathway			•	New view		•		•
Caminho pedonal ribeirinho Alhandra - Vila Franca de Xira	Vila Franca de Xira NB	2008	•		New pedestrian / cycling pathway		•		•		•		•
Recalificação cais Vila Franca de Xira	Vila Franca de Xira NB	2013	•		•						•	•	

capable of linking territories belonging to the same river beyond their administrative boundaries. In addition, we note that few interventions (in Montijo, Vila Franca de Xira and Loures) implement measures to mitigate the effects of rising water level in the estuary by contemplating the phenomenon as part of the project. Links with coastal protected areas are especially improved in project areas located within the limits of the Tagus Estuary Nature Reserve (*Reserva Natural do Estuário do Tejo*) to the east.

27.2.3 Visible and Invisible Water Corridors as Triggers of Territorial Greenery

According to the main idea argued by the *Plano Verde de Lisboa* (Ribeiro Telles 1999)—to establish continuity and sustainability of both the ecological and built systems, to ensure biodiversity, the presence of nature in urban areas, protection, and cultural enhancement of the architectural, natural, landscape, and historical heritage—we need to construct and maintain metropolitan corridors by defining links between natural and agricultural areas.

As clearly proven by recent urban regeneration trends, in addition to riverfronts, also inland waters and their adjacent open spaces can be terrains for local regeneration. To design green corridors supported by watercourses, surface or underground water flow appears to be an appealing formula to intervene in consolidated and often degraded urban areas.

Among the planned structural and non-structural green corridors implemented by the Lisbon Municipal Master Plan now in force (*Plano Diretor Municipal*—PDM, 2012), the Structural Green Corridor of the Alcântara Valley is currently under construction. The unfinished intervention of the Alcântara Valley Corridor—formed by an already canalised and buried Tagus tributary—aims to create a new pedestrian and cycling route and to recover the memory of the old Alcântara River by introducing water from a nearby wastewater treatment plant (Anastasia 2019a) into the project area for scenic effect and irrigation (Fig. 27.3).

By transforming the current infrastructure corridor used purely for mobility, the implemented route along the Alcântara Valley intends to minimise the isolation of some nearby neighbourhoods and (re)qualify them, and to valorise the local cultural heritage (the *Águas Livres* Aqueduct). In addition, we note that the Alcântara intervention goes beyond the PDM and the current Lisbon General Drainage Master Plan (*Plano Geral de Drenagem de Lisboa 2016–2030*—PGDL) by choosing a more resilient solution: a method for using water already present in the area.[1]

In the Lisbon Metropolitan Area, we find other examples of 'greener' intervention corridors drawn by the estuarine drainage waterlines. The urban regeneration action

[1] In order to control the risk of flooding and contribute to the adaptation of challenges posed by an increase in exceptional climate events, the DGDL suggests constructing the diversion tunnels to lead waters from the upper drainage basins towards the main Tagus river bed.

Fig. 27.3 Structural Green Corridor of the Alcântara Valley. *Source* Author's photographs

for the conservation and promotion of natural heritage integrated in the *Estrutura Verde Principal da Cidade de Montijo*[2] (*Sítio das Nascentes*, south bank) proposes the landscape recovery of two rainwater streams. Here, the goal is to ensure interconnection between the Tagus Estuary and land in the interior, and aggregation of the city centre with urban expansion to the east. Also in Montijo, along a watercourse flowing into the Tagus, we find the *Corredor Verde da Mundet* (Green Corridor of Mundet) that networks different municipal green spaces, including the *Jardim do Vale Salgueiro* (Salgueiro Valley Garden). In Loures (north bank), the *Plano Estratégico para a Rede Hidrográfica*[3] aims to improve the riverside habitat and accessibility to inhabitants all along the 250 kms of the municipality's existing inland waterlines (headed). In all cases of greener interventions along inland waters, we note fewer design variables than in riverside projects. The main project materials and features are: new pavement, street furniture and lighting for new cycling-pedestrian paths, and, in some cases, valorisation of tangible cultural heritage and/or appropriation of local cultural heritage.

27.3 Final Considerations: Addressing the Tagus Estuary Humid System Potential

In times of change, both in terms of use and the attention given to water, the Tagus Estuary hydrographic network—natural courses, buried drainage canals, green/blue corridors—requires a comprehensive project. Its watercourses and the open spaces bound to them can become an occasion to propose adaptive strategies to mitigate the effects of climate emergency, to counteract the consequences of increasing urbanisation, and to regenerate degraded urban and natural areas. All this in an atmosphere of renewed vicinity to water and of consensus on the restoration of ecological balances.

[2] TN main green structure of Montijo city.

[3] TN hydrographic network strategic plan.

In projects along the main estuarine riverbed and its hydrographic network, we note that all the estuarine **riverside projects contribute considerably to renewing the metropolitan city's image** and, consequently, **to rapidly boosting the real estate value of new and existing buildings** (Anastasia 2019b).

Unfortunately, except for very few interventions, **the new facades of the estuarine cities do not foresee a common and shared plan addressing the consequences of climate emergency** (especially in terms of adapting to rising estuarine water levels). In fact, only the *Frente ribeirinha entre Esteiro da Quebrada e Cais dos Vapores* (Montijo, south bank) and the *Parque ribeirinho Moinhos da Póvoa* (Vila Franca de Xira, north bank) adopt measures clearly addressing rising estuary levels, by providing new spaces that accommodate water. In other projects, elevated pedestrian and cycling paths are planned to cohabit with changes due to rising estuary levels and a fragile environment: *Percurso ribeirinho* (Loures, north bank) and *Parque linear ribeirinho estuário do Tejo* (Vila Franca de Xira, north bank).

We also note that, apart from the idea of physical connection—cycling-pedestrian paths able to connect riverfronts belonging to different municipalities, new bridges across drainage watercourses (north bank), and planned restoration of the existing bridges linking different urban centres (e.g. Seixal with Barreiro, south bank)—**there is a weak common and shared idea of the entire north/south bank, and even less of the entire estuary**, which goes beyond administrative boundaries.

A common idea of *arco riberinho sul / arco riberinho norte* (southern river bank / northern river bank) and a whole estuarine identity could be achieved with an awareness of the estuary's landscape peculiarities underlying its primary essence: its changing/shifting character. And it is precisely this essence that should be enhanced by riverside regeneration projects.

In summary, the regeneration project along the water towards a 'shared estuarine identity' should leverage two main issues: building a common CTE adaptation strategy that starts from its core (the Tagus Estuary)—climate change and ecological emergencies as an opportunity for a common estuarine project; managing LMA water(s) recourses more efficiently and sensitively—drought, and an increase in exceptional events, as an opportunity to improve water drainage and the use/reuse of water already present in the intervention areas.

'As always the water(s) and the energy(ies) will be the key to all 'success'' (Gaspar 2018, p. 117).

Acknowledgements This work is financed by national funds through FCT—Fundação para a Ciência e a Tecnologia, I.P., under the Strategic Project with the references UIDB/04008/2020 and UIDP/04008/2020.

This research was funded by the Portuguese Foundation for Science and Technology (*Fundação para a Ciência e a Tecnologia*—FCT)—'*Orçamento de Estado do Ministério da Ciência, Tecnologia e Ensino Superior*'—and the European Commission / European Social Fund (ESF)—'*ao abrigo do Quadro Estratégico Comum (2014–2020), através, nomeadamente, do Programa Operacional do Capital Humano*', under the Postdoctoral Research Fellowship, individual postdoctoral grant (SFRH/BPD/116331/2016), carried out at CIAUD (URBinLAB); Lisbon School of Architecture; Universidade de Lisboa, Lisbon (Portugal).

References

Anastasia C (2019a) Urban regeneration projects bound to water, along and towards the Tagus Estuary (Portugal). Sustainability 11:6578. Available online https://doi.org/10.3390/su11236578

Anastasia C (2019b) More tourism? mass tourism? Mars tourism? Esempi Di Archit Int J 6:9–18

Anastasia C (2023) Designing a captivating proximity to water: two case studies of the Iberian Peninsula. In: Giudice B, Novarina G, Voghera A (eds) Green infrastructure. The urban book series. Springer, Cham. https://doi.org/10.1007/978-3-031-28772-5_10

Bruno Soares L, Trinidade Ferreira P (2017) Reabilitação do espaço público na Frente Ribeirinha da Baixa Pombalina: Terreiro do Paço e Cais do Sodré/Corpo Santo. PORTUS: the online magazine of RETE, 33, RETE Publisher, Venice. Available online https://portusonline.org/pt/reabilitacao-do-espaco-publico-na-frente-ribeirinha-da-baixapombalina-terreiro-do-paco-e-cais-do-sodrecorpo-santo/

Gaspar J (2018) Geographical entries for a vision of Lisbon's futures. Futures of Lisbon. Museu de Lisboa, Lisbon, pp 112–123

Ribeiro Telles G (Coord) (1999) Plano verde de Lisboa. Colibri, Lisbon

Salgado M (2012) Reconquista da Frente Ribeirinha de Lisboa. *Portusplus*, 3. Available online http://retedigital.com/wp-content/themes/rete/pdfs/portus_plus/3_2012/Contribuciones/Manuel_Salgado.pdf

Viganò P, Secchi B, Fabian L (eds) (2016) Water and asphalt: the project of isotropy. Park Books, Zurich

Chapter 28
Soil Displacement. Landscape Project as an Infrastructure Across Building Geography and Grounding Metabolism. The Case Study of Pialassa Piomboni Constructed Wetland in Ravenna

Vittoria Mencarini, Gianni Lobosco, Luca Emanueli, and Massimo Tondello

Abstract In current circumstances, man has the great responsibility of better managing the effects and products of those intense anthropic processes—such as waste disposal or excavation materials—which cannot be assimilated in a reasonable time by spontaneous renewal cycles. A systemic approach is needed for coordinating man-made metabolic procedures concerning the inevitable phenomena of urban and territorial transformation and ecosystem protection, admitting that one can be consequent and functional to the other. By being open to a strongly interdisciplinary approach, landscape design can offer precise and contextualized design solutions which may involve large temporal and spatial scales in reference to the extension of the problem. The proposed case study of the Ravenna commercial docks shows how the progressive expansion and continuous maintenance of the seabed through dredging has led to the creation of newly constructed wetlands in continuity and balance with existing ones. In particular, the excavated material was transferred into Pialassa Piomboni (protected wetland pSCI and SPA) through topographic modeling suitable for improving the ecological state of the humid area. After the intervention, the Pialassa area is expected to increase its ecological potential. Such an approach has turned waste treatment into an opportunity, allowing the wetland system to grow stronger and reducing costs compared to transporting the material to

V. Mencarini (✉)
IDAUP International Doctorate in Architecture and Urban Planning, Department of Architecture, University of Ferrara, Ferrara, Italy
e-mail: vittoria.mencarini@unife.it

G. Lobosco · L. Emanueli · M. Tondello
Sealine Research Center, University of Ferrara Department of Architecture, Ferrara, Italy
e-mail: gianni.lobosco@unife.it

L. Emanueli
e-mail: luca.emanueli@unife.it

M. Tondello
e-mail: massimo.tondello@unife.it

T. Matteini et al. (eds.), *Cultivating Continuity of the European Landscape*, Environmental History 15, https://doi.org/10.1007/978-3-031-25713-1_28

a remoter site. Several years late, building on the site is still ongoing in 2020. The project time schedule was not completely adhered to, with a strong impact on the expected outcome. In light of this experience, the contribution discusses different solutions for integrating and observing the regenerative time of nature in the design process and soil handling in spatial planning by activating new metabolic processes in its replacement in the environment.

Keywords Landscape making · Transition · Forms of time

28.1 Introduction

Nowadays, man has the great responsibility of better managing the effects of those intense anthropic processes—such as waste disposal and excavation materials—which cannot be assimilated in a reasonable time by spontaneous renewal cycles (Sijmons et al. 2014; Tilly et al. 2014; Belanger 2016). Moving materials and debris from one place to another to maintain the status of human habitat means reshaping topography and creating new processes that lead to new patterns and topology (Ellis 2018; Clement 2015). The time for regenerating and metabolizing the artificial switches is longer than the time to achieve this change, undermining the ability of self-regulation and autopoiesis present in natural systems (Naveh 2000). Environmental conditions that generate biodiversity have been greatly altered and disrupted by anthropic dynamics and the immanence of these sites has become a condition strictly dependent on human responsibility. The maintenance of several protected areas and the preservation of their ecological peculiarities is only possible through artificial interventions that reproduce hydro-morphological situations, in which spontaneous environmental processes can be developed, especially in transitional water lagoon sites. Using a metaphor, man is called to become a *planetary* gardener (Clement 2013), taking care of the spatial organization and maintenance over time, turning his gaze to a worldwide scale and looking forward. An attitude that a landscape designer has to explore in contemporary praxis.

The proposed case study of "Pialassa Piomboni" restoration in the Ravenna commercial docks shows how the expansion and continuous maintenance of the seabed through dredging should have led to the creation of a novel constructed wetland in continuity and balance with the existing context. In particular, a large amount of excavated material was transferred into Pialassa Piomboni through topographic modeling (Fig. 28.1) suitable for improving the ecological state of the humid area, and also by damming the water exchanges between it and the harbor channel. After the intervention, the Pialassa area was supposed to increase its ecological potential, preserving the protected habitats as pSCI and SPA. Such an approach should have turned the waste treatment into an opportunity allowing the wetland system to grow stronger and reducing costs compared to disposal of the material farther away. The case study is a paradigmatic example for highlighting the opportunities, limits and constraints in translating such kinds of projects into reality, in which the concurrence

Fig. 28.1 Soil displacement, building geography grounded metabolism. Pialassa Piombone construction site. *Source* Piacentini group website

of systematic operations has to follow an accurate time-phasing program related to the fragile equilibrium, in which the construction site is inserted.

28.2 Territorial Framework and Landscape Evolution of the Ravenna Harbor

Through the centuries, the Ravenna territory—located in Italy on the Adriatic coast of the Emilia Romagna region—has been shaped by important land reclamation and infrastructural works (Fig. 28.2). Its current configuration reflects the interaction between the natural tendency of the area to be a wetland and the artificial transformations carried out (Fig. 28.3). Especially during the twentieth century, major industrialization centered mainly along the Candiano channel connecting the city center to the sea involved works exploiting underground resources such as gas deposits and aquifers, thus increasing subsidence in several coastal areas. In particular, the so-called Pialasse and their hydrodynamic system have been deeply affected by this phenomena. These special wetlands are important elements of Ravenna's territory, occupying vast depressed areas parallel to the coast. Originally saltwater lagoons, since the sixteenth century, they have been reshaped by man in order to create a network of channels allowing the regulation of tidal flows and improving accessibility to the harbor. The name "Pialassa" comes from the local dialect and literally means "take and leave" (i.e., "pia and lassa") referring to its dynamic function of accumulating and returning water to the sea.

These systems were fully operational until major changes in the harbor layout were made from the 1960s onward. The 2.5 km extension of the harbor breakwaters in 1968 has resulted in a constant reduction in tidal flow which has affected their hydrodynamics and water exchange with the sea. This situation has had an especially strong impact on "Pialassa Piomboni", located south of the port. In 2000, the site was declared a Preservation Zone (Fig. 28.4), while the commercial docks extension has been considerably decreasing its wetlands extension (Fig. 28.5). The presence of such a major industrial site close to the project area prompted a reframing of the relationship between them in order to preserve the ecological function of the wetland and its habitats. The Harbor Authority promoted a project for the ecological restoration of the site involving the physical separation of the wetland from the harbor channel extension and its bathymetric reorganization through materials from the dredging of new inner channels.

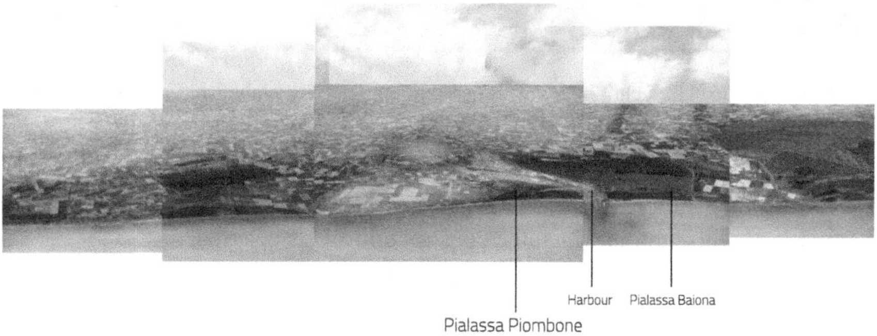

Fig. 28.2 Visualizing Ravenna: territorial framework overview. *Source* author reprocessing of Google Earth

Fig. 28.3 Pialassa panoramas. Left: Industrial Site in Candiano Channel, non-protected area, December 2018; right: Pialassa Baiona protected area, March 2019. *Source* Vittoria Mencarini

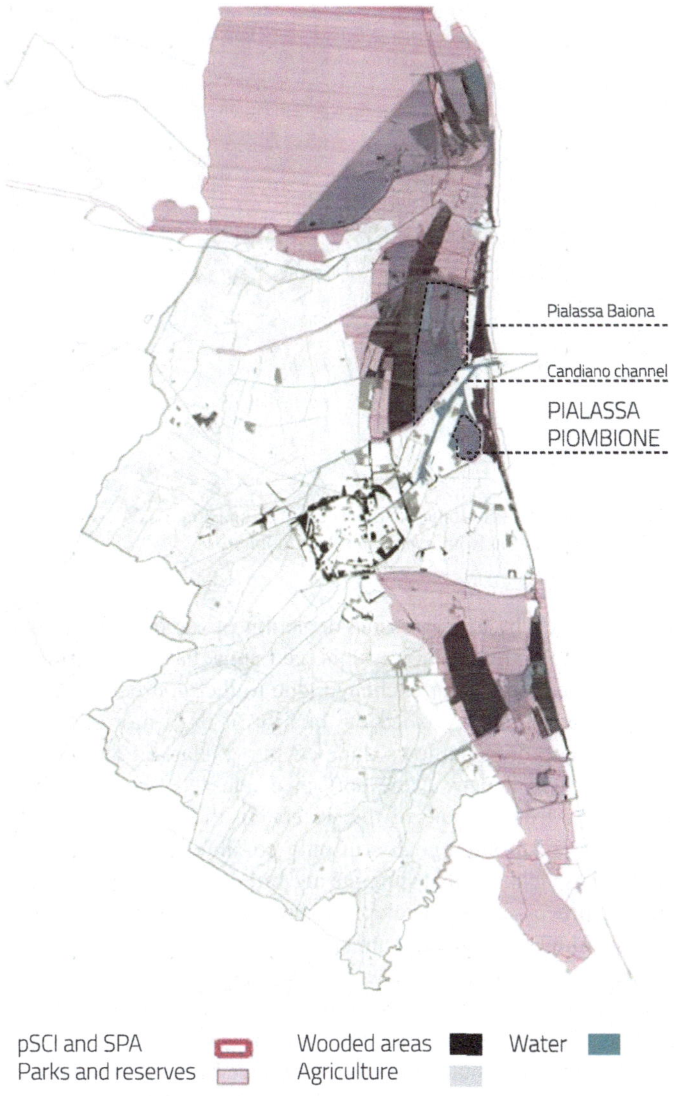

Fig. 28.4 Territorial framework, cartography of Ravenna with protected areas highlighted. *Source* Author

28.3 The Project: Ambitions and Constraints of a Constructed Wetland

The Pialassa Piomboni joins the Ravenna port canal and comprised semi-submerged areas and low beds bordered by artificial banks and a dense network of major and minor canals. During the last 50 years, the morphological modification of the Pialassa

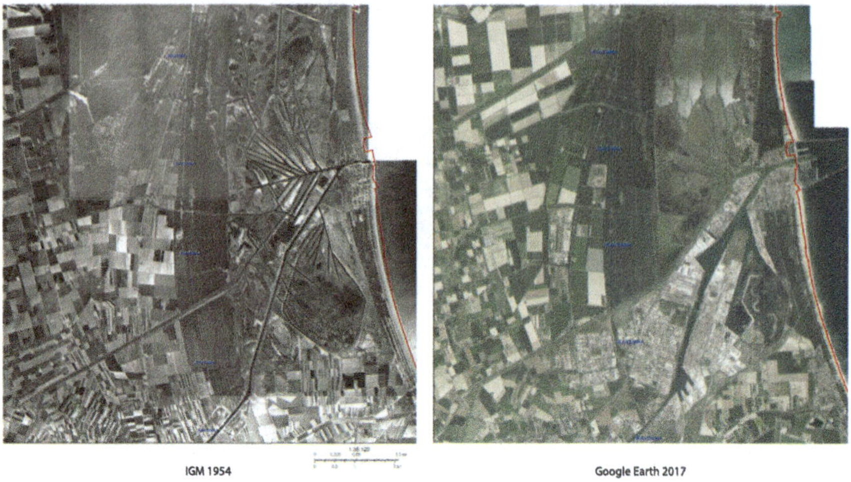

Fig. 28.5 Urbanization process along Candiano channel and Pialassa's evolution from 1954 to 2017. *Source* Moka GIS Emilia Romagna, Satellite orthophotos

has been accompanied by a progressive depletion of the quality of environmental conditions (Fig. 28.6). The attrition of emerged areas has reduced the presence of poultry, while the reduced water exchange, due to the modified bathymetric structure, has progressively impoverished the quality of the water environment. Over time, agricultural and civil outflows dumped into Pialassa Piomboni from the S. Vitale drainage system and from the port have released large quantities of nutrients, increasing the trophic state of the waters. In this framework, achieving the project's ecological restoration goals was only possible through the artificial recreation of some precise features that simulate and reproduce a morphological situation, in which natural processes can develop. This operation has meant addressing three main topics: providing tide water dynamism; improving water quality; implementing bathymetric reorganization (Fig. 28.7).

Concerning the bathymetric reorganization, new sea canals needed to be dredged and debris materials used to construct inner islands and rises, together with material from outside the environmental zone. Different interventions were designed in relation to the density and shape of the baronial and emerged areas; in order to create the best conditions for poultry repopulation in the area, the project provides for:

- a perimeter bank that creates a separation between the port and natural areas;
- internal tide channels to facilitate tidal wave entry, connected to three inlets and two outlets;
- four "venetian doors" which allow water outflow in one direction only, creating a forced recirculation of water in the Pialassa and hydraulically isolating the site from the port area especially when conditions in the latter present high pollution levels;
- a navigable Vincian door;

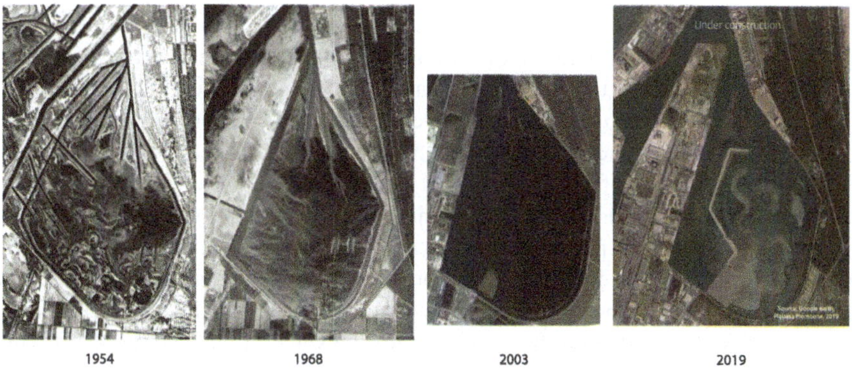

Fig. 28.6 Pialassa's evolution, from 1954 to 2019. *Source* Pialassa's Restoration General report—Definitive unified project and environmental impact study and Google Earth

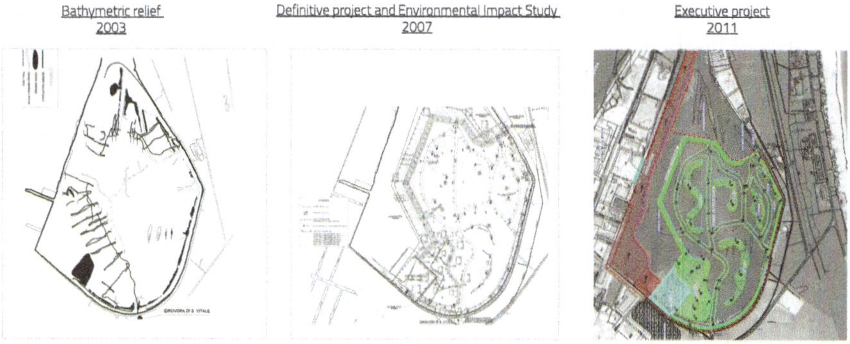

Fig. 28.7 Design path. In the middle the in-depth project, subject of the environmental impact study and basis of the integrated tender of assignment. *Source* General report—Definitive unified project and environmental impact study—Apr 2007—Design team: Med-Ingegneria

- some islands and internal sandbanks with a shape suitable for favoring the resumption of poultry population;
- resetting the district canal;
- a 12-hectare phyto-depuration area located next to the outlet of the S. Vitale drain pump in order to limit the supply of nutrients.

All these interventions have been planned to maximize the reuse of on-site dredging materials and to accommodate a portion of the dredging volumes from port channel maintenance. The emerged areas designed inside the Pialassa require 240,500 m^3 of soil provided entirely by the 247.750 m^3 available from dredging of the inner channel. The other interventions concerning the perimeter bank and the areas of re-naturalization were supposed to use the remaining 7250 m^3 and the dredging material of the port channel. In this way, the project would have acted as a metabolic landscape capable of creating a virtuous relationship between industrial and protected areas. The interventions were organized in a time phased layout that investigated compliance between diachronic and synchronic construction steps.

In order to establish the best-performing topological reshaping of soil displacement, an analysis of project impacts was carried out by comparing speed and flow range patterns over time in three alternative scenarios (Fig. 28.8). It concerns as many different levels of intervention that simulate the behavior of bathymetry in relationship with tidal flow. The first consists of non-intervention scenarios both on the naturalistic and the port area. The second scenario depicts the situation of the Pialassa in the case of dredging works on only the harbor canal. Finally, the last one shows the effectiveness of perimeter embankment combined with internal channels in terms of tidal dynamism, necessary for the value of the whole system. This hypothesis corresponds to the project's full implementation according to the time schedule and the specific operations proposed (Fig. 28.9).

Eventually, none of these scenarios came to be. In 2014, the construction of the separation embankment between the Pialassa and the harbor was interrupted because of a change in ministerial parameters concerning the re-collocation of dredging materials (becoming Law n.116/2014 in August 2014). Most of the building site works continued, but the physical separation of the areas has not yet been completed. The construction site has been active for 7 years, instead of 30 months, and is still ongoing, with completion prospected for 2023. Today the perimeter bank has yet to be built, so the Pialassa is still in physical continuity with the harbor channel and is getting

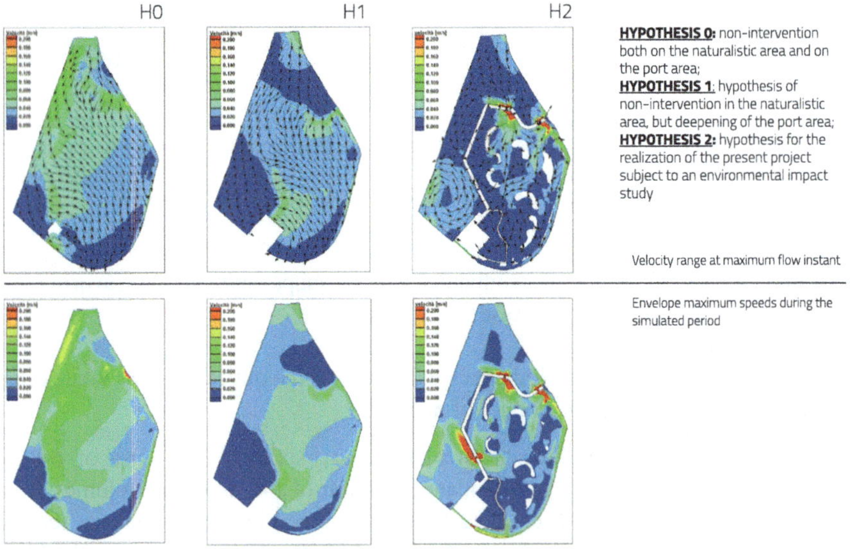

Fig. 28.8 Hydrodynamic simulation to identify and evaluate the environmental impact of the project and its alternative proposals. *Source* Pialassa's Restoration Environmental impact study General report—Definitive unified project and environmental impact study—Apr 2007—Design team: Med-Ingegneria

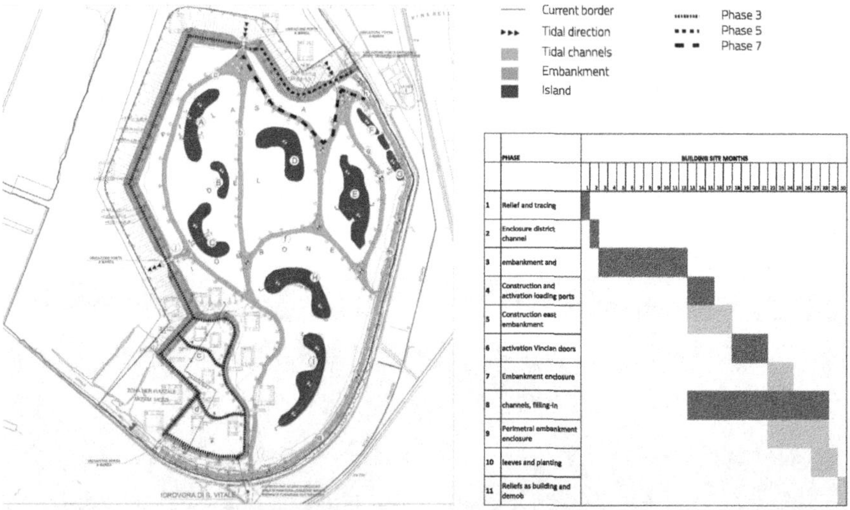

Fig. 28.9 Executive time phasing and Gannt diagram, author's reprocessessing. *Source* Pialassa's Restoration Environmental impact study General Report—Definitive unified project and environmental impact study—Apr 2007—Design team: Med-Ingegneria

covered by sand, causing the loss of the hydrodynamic and ecological balance which was the focus of the project (Fig. 28.10).

Fig. 28.10 Pialassa Piomboni today, wetland and building site. *Source* Vittoria Mencarini, Dec 2018

28.4 Conclusion

This case study shows how natural processes and peculiarities coupled with human activities can mutually contribute to building geographies. At the same time, spatially transcendent systems and regulatory apparatuses deeply affect the path to reaching project goals. Since design strategies can be conceived of as a tool for grounding metabolism (Ibañez and Katsikis 2014), their effectiveness becomes increasingly related to the control of time-based procedures and to project resilience in the face of uncertainty, risks and unforeseen new conditions (Di Giulio et al. 2018). The experiences of the Pialassa Piomboni constructed wetland also teach us that there is probably a cultural issue to overcome: by considering the protected area and harbor environment as separate entities, the regulatory framework has eliminated any possibility of making them work together as an integrated infrastructure system, as was anciently conceived. The attempt to protect the ecological value of the Pialassa by banning any exchange of material with the port channel is fostering the disappearance of the very ecosystem the law wants to preserve. Faced with this situation, a more systemic approach is needed to coordinate man-made metabolic procedures concerning inevitable urban and territorial transformation and ecosystems protection, admitting that one can be consequent and functional to the other as an integral part of a novel envisioned ecology (Hobbs et al. 2006; Belanger 2014, 2016).

From this perspective, the challenges concerning climate change and the dramatic sea-level rise facing Ravenna in the next few decades will probably shine light on the Pialassa system and the need to re-invent it as a geographic agent of adaption and defense: as a landscape infrastructure capable of grounding a new metabolism for the Ravenna territory.

References

Belanger P (2014) Ecology 5.0. New geographies—Grounding metabolism, Issue 06, pp 184–187

Belanger P (2016) Landscape as infrastructure. Routledge, London

Clement G (2013) Giardino, paesaggio e genio naturale, Italian. Quodlibet, Macerata

Clement G (2015) L'alternativa ambiente. Macerata: Quodlibet

Ellis EC (2018) Anthropocene—A very short Introduction. Oxford University Press, Oxford (UK)

Di Giulio R, Emanueli L, Lobosco G (2018) Scenarios' evaluation by design. Un Approccio per Scenari Al Tema Della Resilienza Techne 15:92–100

Hobbs RJ et al (2006) Novel Ecosystems: theoretical and management aspects of the new ecological world order. Glob Ecol Biogeogr 15:1–7

Ibañez D, Katsikis N (2014) Grounding metabolism. New geographies—grounding metabolism, Issue 06, pp 2–9

Naveh Z (2000) The total human ecosystem: integrating ecology and economics. Bioscience 50(4):357–361

Sijmons D, Hugtenburg J, Feddes F, Van Hoorn A (2014) Landscape and energy: designing transition. Nai Uitgevers Publisher, Rotterdam

Tilly N, Klijn O, Borsboom J, Looije M (2014) Urban metabolism: sustainable development of Rotterdam. Mediacenter Rotterdam, Rotterdam

Chapter 29
Mapping the Rhythm in Wetlands

Stefano Tornieri

Abstract All the processes involved in landscape transformation come from a combination of human activity and natural events, and require different periods of time to become evident.

Keywords Wetlands · Imaginary · Rhythm

29.1 Introduction

All the processes involved in landscape transformation come from a combination of human activity and natural events, and require different periods of time to become evident. For a long time, landscape was understood as a product of overlapping fixed pictures of these processes, but contemporary territorial necessities make this traditional way of thinking obsolete (McHarg 1969). In order to understand and interpret the complex physical, biological and cognitive phenomena of landscapes, it is necessary to conceptually and practically operate within a broad range of spatial and temporal scales. In coastal regions, the most delicate ecosystems of the world are concentrated and often mixed together (Martínez et al. 2007). A river delta, wetland or agricultural field are examples of time-structured systems, where different ecological processes and their "rhythms" contribute to the general complexity and dynamics.

Seasons, sea tides, river flows, natural presences, human maintenance activities, political governance decisions…they all have different repetitions or sequences, which have to be considered in an integral design approach (Dekay 2012).

This study focuses on the river mouths along the Veneto Coast and particularly on the Caorle lagoon which is taken into consideration because of its high natural ecosystem value and importance in terms of habitat diversity, infrastructure complexity and morphology dynamics. Through the representation of different "rhythms" involved in transformation processes related to their actors, this paper

S. Tornieri (✉)
IUAV University, IR.IDE (Infrastructure Research Integral Design Environment), Venice, Italy
e-mail: stornieri@iuav.it

T. Matteini et al. (eds.), *Cultivating Continuity of the European Landscape*,
Environmental History 15, https://doi.org/10.1007/978-3-031-25713-1_29

explores a potential new starting point for the landscape ecology design approach applied in the Veneto coastal region.

29.2 Rhythm and Movement

Rhythm, tempo, movement and frequency are all terms which apparently refer to the musical dimension. However, the disciplines of architectural and urban composition have also produced studies showing the perception of architectural element rhythm to be analogous to rhythm, the human body and its relations with the world as lived and perceived.

The most widely known studies read this perception of rhythm not just as the repetition of a movement in a present time, but also as a broader perceptive condition. The French sociologist Henry Lefebvre in Rhythm analysis takes elements of everyday life and thinks them through using the space-and-time-unifying concept of rhythm as a tool of analysis rather than just an object (Lefebvre 1992). Its implementation in the urban field is limited and more linked to the need to build a bond between a rhythm analysis, politics, culture and traditions of a place (Chen 2016). The applications of the study of rhythm to natural spaces are less frequent; in this case the literature is more limited and mainly concerns the question of time representation in cartography and maps, intended as a multi-dimensional approach (Rosemberg and Grafton 2012). Many transformative processes of landscape have been excluded in regional planning projects and policies because they are difficult to represent. First of all, the dynamism of landscape: its movements are sometimes visible, sometimes barely perceptible. It was the Landscape Ecology Discipline that introduced a new point of view into the study of landscape, moving the focus in urban plans from a rigidly static anthropocentric vision to the dynamic and fluid ecosystem vision (Esbjorn-Hargens and Zimmerman 2001), which is mostly applied to river systems (Mathur and Da Cunha 2001).

29.3 Case Study

The 150 km of Adriatic coastline from the modern Po Delta[1] to the Grado lagoon is considered the largest wetland in Europe (Arpav 2005).

The environmental pressures it endures, mainly subsidence and rising sea levels, seriously endanger the existence of its coastal environment, and probably only its ability to adapt can stimulate new morphological ideation (AA.VV, 2014; Bates

[1] I refer to the most recent part of the territory, formed after the exclusion of Porto Viro by the Serenissima in 1604, in order to alter the main course of the Po river and avoid lagoon cover-up. The delta as we know it today is a very young territory and its morphology has been changing fast, growing about 135 ha per year from 1600 to 1850.

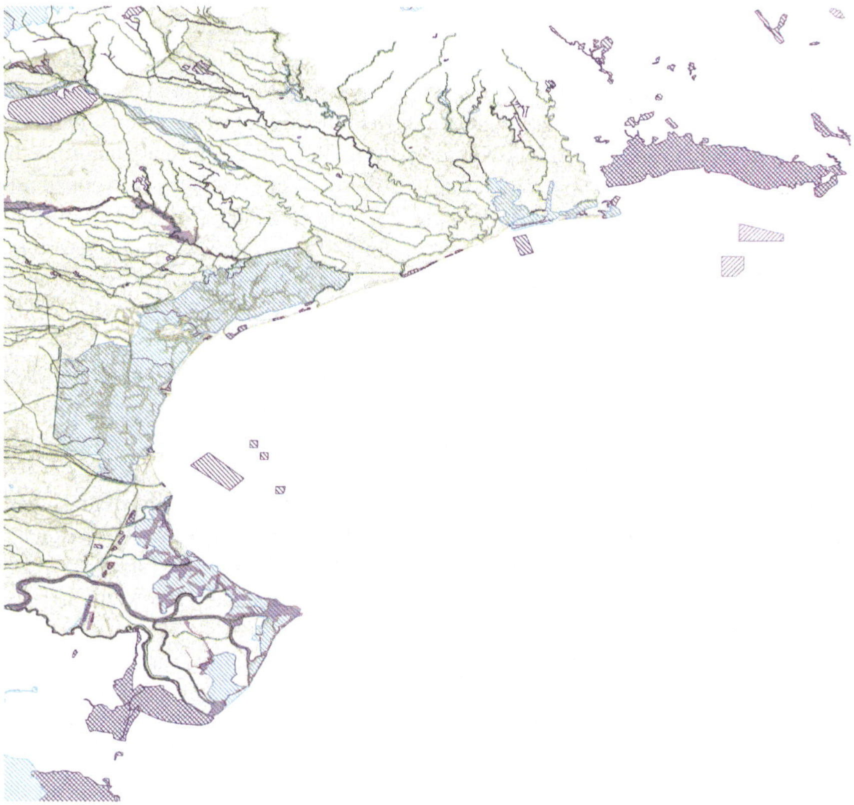

Fig. 29.1 The Veneto coast and its protected areas

et al. 2008). Along the coastline, we can identify several micro-environments and different degrees of protection: the Po Delta area (MAB Unesco area, Regional Natural Park, ZPS); the Venice lagoon (ZPS area); the Caorle lagoon (ZPS area); the Grado lagoon (SIC/ZPS area) (Fig. 29.1). However, there are some ideological forces in this territory which are not so inclined to change their point of view: the environmentalist vision, which is in favour of safeguarding and creating protected areas,[2] and the productive vision that has largely prevailed in the Veneto region and which supports reclaiming land, agriculture, farming, mass tourism and intensive urbanization. Although these ideologies appear distant, they are both conservative, seeing the environment as a good to consume, with the conservation of a static image prevailing over ecological and systemic value (Ciampi 2015).

[2] Think of the Po Delta natural park *in primis*, a territory whose existence is linked to non-stop maintenance managed by the land reclamation consortia, in contrast with its tourist vocation pursued by the Park Authority. With some differences, the Venice lagoon faces similar issues, like the Sile river regional Park. In that direction, we can place the intention to candidate the Caorle lagoon and fishermen's "casoni" as an Unesco Heritage.

A possible starting point to innovate the operational and transformative modalities of the territory is the wetland contract of the Caorle lagoon. Created in 2019, it will have to draw up a mid-term strategic plan and a short term action plan (2–3 years) aimed at safeguarding the lagoon wetland environment, protecting waters, reducing flooding risk and promoting agri-food production. The multidisciplinary nature of the contract aims to include as many "voices" as possible, using technology to constantly monitor data and coordinating the actors present on the territory. Constant monitoring and strategic maintenance orientation, key points of the contract, cannot ignore the broader time and space framework of the lagoon territory. In this context, geomorphology, hydrology and climate display long-term, sometimes repetitive, slow action hiding a secret rhythm.

The first "movement" that man opposed was that of the coastline. Historical studies on the formation of the Adriatic sea show that the coastline changed in the space of 20,000 years, a very brief period of time if compared to the telluric movements of tectonic plates (Simeoni and Corbau 2009) (Fig. 29.2).

Because the Adriatic sea is very shallow, traces of its ancient coastlines can be seen on the seabed. Other maps show us the old Caorle wetland, which was subsequently dried by intense reclamation operations; recent CNR-ENEA studies show the point to which the coastline will retreat in 50 year's time, when many areas that are now dry will be submerged.

These horizontal "movements" are useful for rebuilding some moments of territorial stability and alteration. The soil, the only continuous surface, becomes the perceptive tool of geological rhythm thanks to the water that, as it advances or retreats, reveals or hides parts of the soil itself.

By blocking this dynamism with dams, reclamations and urbanization on the coastal dunes and river deflections, human intervention altered the landscape rhythm.

Ancient maps of Caorle, the Venice lagoon (Sabbadino 1558) and the Po Delta show a much more varied landscape with intertwining wetlands, rivers and landmasses where today there is a monotonous agricultural plain.

The environmental condition reveals a more perceptible rhythm, a seasonal rhythm governing farming, temperature, humidity and soil salinity.

The rhythms of wind and tide are visible in less than 24 h: marine breezes are perceptible twice a day and tides every 6 h. The tidal rhythm is clearly visible on the whole coastline as proven by intertidal vegetation (algae and molluscs) in the case of rigid coastline, or by broad sandy or muddy intertidal plains in the case of free coastline (Fig. 29.3).

To take an example, the wetlands around the Lemene estuary and Caorle urban centre were drained by reclamation and are lower than average sea level, also due to subsidence. The reclaimed areas maintain intrinsic humidity and salinity that often need to be altered or controlled in order to enable agricultural production and make it efficient in all cultivated areas.[3]

[3] Consider the Vallevecchia case, an area reclaimed in the first half of the 1900s, today a Veneto Agricoltura experimental site, where an automated system of water-management was designed and built which allows the accumulation of high quality water (with low salinity) and the removal of high

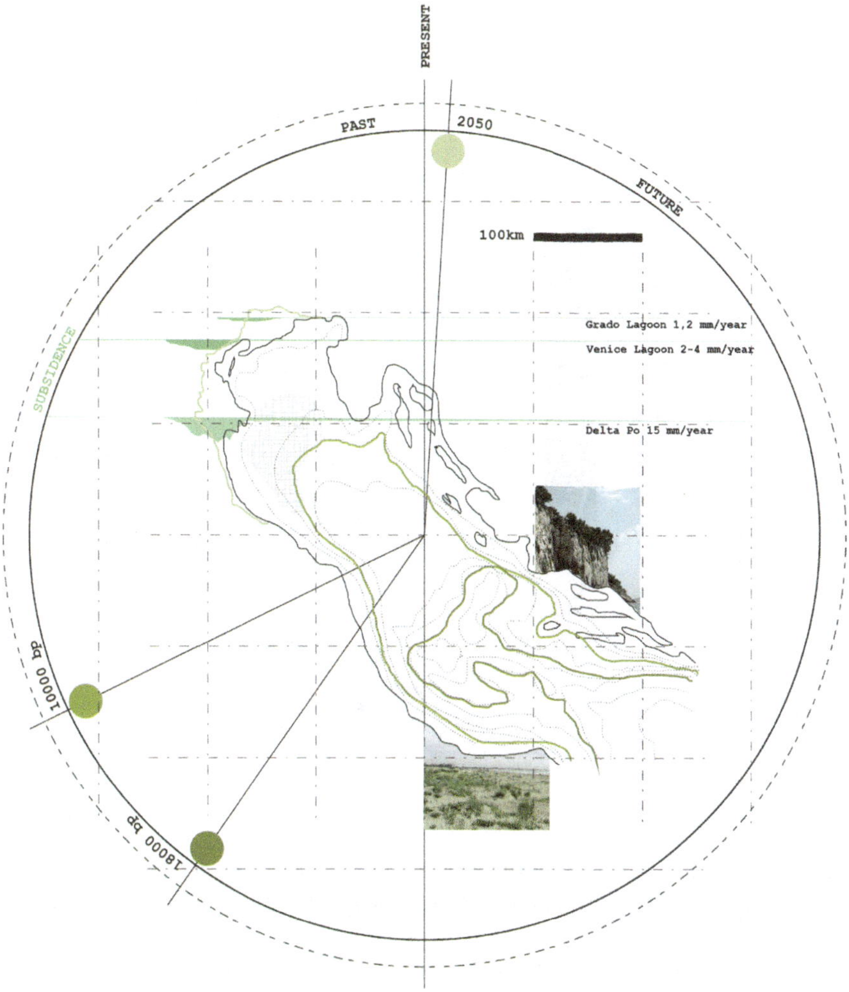

Fig. 29.2 Geological rhythm, horizontal and vertical movement

The natural rhythm could be restored by recognizing the ecosystem value of wetlands, (Costanza et al. 1997) and recreating them where necessary and useful (Bonometto et al. 2015).

Agriculture and fish farming are human activities whose rhythm is dictated by productive needs. However, it is necessary to make a distinction, which introduces the synchronization theme. Agriculture is an activity that, for economic reasons, actually tries to accelerate the natural process as much as possible. By providing nutrients

salinity water from the hydrological network based on variations in the water's chemical parameters (especially salinity).

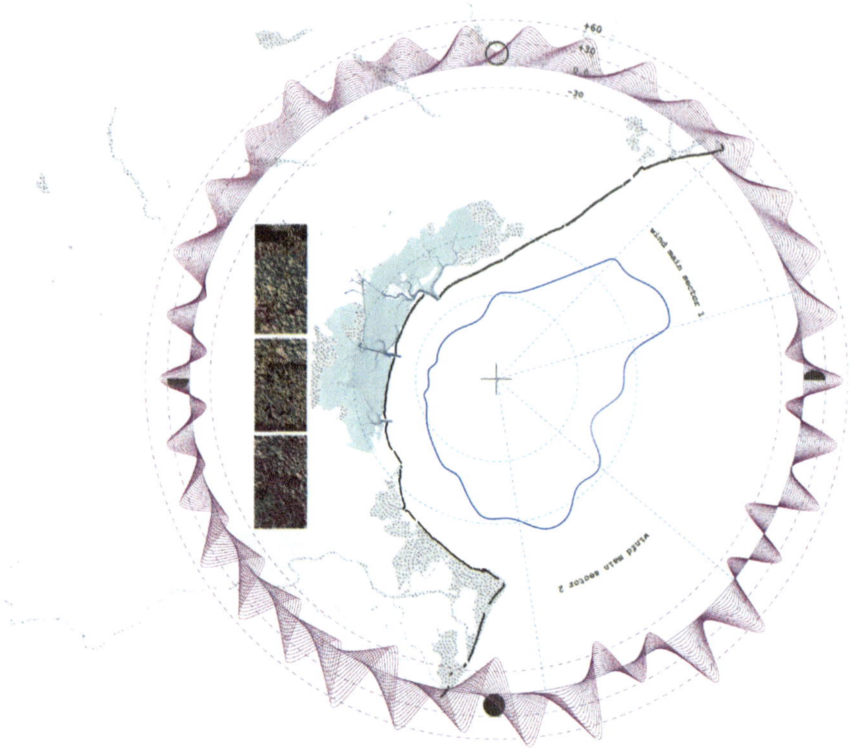

Fig. 29.3 Environmental rhythm, tides, wind, seasons

artificially, the massive use of fertilizers made crop rotation obsolete, producing chromatic monotony and an irrigation plan detached from seasonal climatic conditions (Lanzani 2003).

The non-intensive fish farming that we find along the whole Veneto coastline exploits the migratory tendencies of fish which detect the different salinity levels of the waters (Fig. 29.4).

A rhythm dictated by water exchange, salinity and fish migration which also becomes that of men's work, of those who manage the fisheries and often live in the fisheries' "casoni".

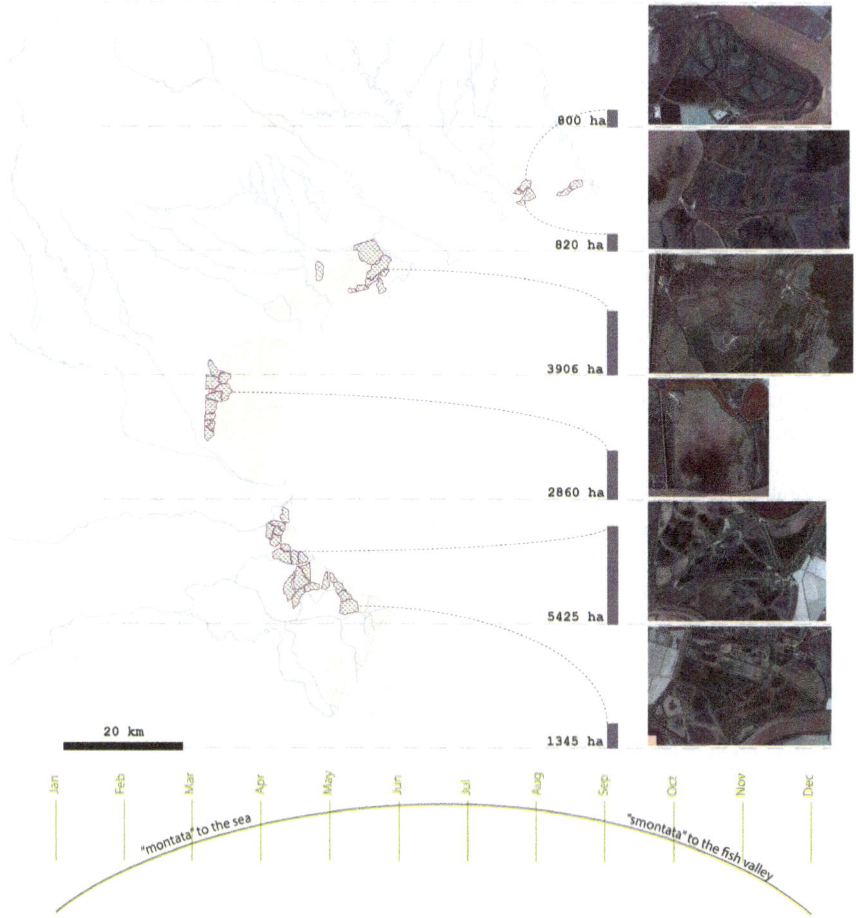

Fig. 29.4 Productive rhythm, fish follow seasons

29.4 Programming the Rhythm

From the post-war period onwards, the Italian natural environment has been undergoing a gradual dismantling process through touristic allotments and delimitation of protected areas without a scientific sieve linked to the morphologic values of the territory itself.[4]

The difficulty in coordinating institutions (state, regions, provinces, municipalities), down to the local level of those who actually live in the territory, reveals

[4] See Antonio Cederna's harsh denunciation in "La distruzione della natura in Italia" Einaudi, 1975. Still relevant today, the book examines the destruction of Italian natural resources, especially natural parks, through infrastructure, allotments, reclamation and touristic facilities in the name of progress. (Cederna, 1975).

a mismatch in the comprehension of Italian territory characteristics and evolution (Tornieri 2019). However, the environmental emergencies that have been increasingly evident in recent years are bringing attention back to environmental issues such as the reduction of land consumption, the restoration of corrupted natural environments and ecological corridors (Cunningham 2002), introducing an inversion of commission in the conception of landscape design, no longer man but landscape itself (Magnani 2016).

In this necessary paradigm change, the temporal landscape component can become a new starting point for projects.

Man acted on landscape, imposing his own rhythm by accelerating natural processes, as in the case of intensive agricultural productions, or blocking it, as in the case of land reclamation works on rivers and canals. From this asynchrony, a new balance must be found, a synchrony between natural landscape and climate changes and human activities.

Research by design and advanced imaging techniques (horizontal, vertical, temporal and composite) needs to be employed to depict a new assembling of spaces, ecologies, scales (succession/management) and economies over time (governance, actors and industries).

In conclusion, considering the specific case of the Caorle lagoon, we can say that the prevalence of agriculture in the surrounding landscape has blocked its natural rhythm, significantly reducing the shaping of ecological corridors that could issue into the lagoon and create biodiversity. Possible future scenarios should include a capacity to read landscape transformations over time in order to interpret ecological rhythms and synchronize them with man's necessities.

References

Arpav (2005) Guida agli ambienti del Veneto. Arpav, Padova

Bates B et al (2008) Climate change and water. In: Technical paper of the intergovernmental panel on climate change. Genova, IPCC Secretariat, p 210

Bonometto A, Capriolo A, Feola A, Oselladore F, Antonini C, Ponis E, Cacciatore F, Mascolo R, Ferrari C, Palumbo L, Boscolo Brusà R (2015) Servizi ecosistemici derivanti dalla realizzazione di aree umide costiere: stima ex ante dei benefici ambientali ed economici. caso studio in Sacca di Goro. Reticula, Numero Monografico. Pianificazione Integrata Della Fascia Marino-Costiera, vol 10

Cederna A (1975) La distruzione della natura in Italia. Einaudi

Ciampi G (2015) Rinaturalizzazione: tra ri-creazione e ricreazione, Limes. Available at: https://www.limesonline.com/rinaturalizzazione-tra-ri-creazione-e-ricreazione/87087. Accessed 6 Mar 2020

Costanza R, D'Arge R, De Groot R, Farber S, Grasso M, Hannon B, Limburg K, Naeem S, O'Neill R, Paruelo J, Ruskin R, Sutton P, Van Den Belt M (1997) The value of the world's ecosystem services and natural capital. Nature 387

Cunningham S (2002) In: The restoration economy: the greatest new growth frontier. Berret-Koelher Publisher

Dekay M (2012) Integral sustainable design: transformative perspectives. Routledge

Esbjorn-Hargens S, Zimmerman M (2001) Integral ecology: uniting multiple perspectives on the natural world. Integral books, Boston, London
Lanzani A (2003) I paesaggi italiani. Meltemi, Roma
Lefebvre H (1992) Rhythmanalysis space, time and everyday life. Bloomsbury Academic
Magnani C (2016) Un nuovo committente. In: Vanore M (ed) Necessità dell'oblio. Università IUAV di Venezia, Milano, Mimesis, pp 144–150
Martínez M, Intralawan A, Vázquez G, Pérez-Maqueo O, Sutton P (2007) The coasts of our world: ecological, economic and social importance. Ecol Econ 63:254–272. https://doi.org/10.1016/j.ecolecon.2006.10.022
Mathur A, Da Cunha D (2001) In: Mississipi floods, designing a shifting landscape. University of Pennsylvania
McHarg IL (1969) In: Design with nature. Doubleday & Company
Rosemberg D, Grafton A (2012) In: Cartographies of time. Princeton Architectural Press
Simeoni U, Corbau C (2009) A review of the Delta Po evolution (Italy) related to climatic changes and human impacts. Geomorphology 107(1–2):64–71. https://doi.org/10.1016/j.geomorph.2008.11.004
Tornieri S (2019) Analfabeti del paesaggio. Inl Ballarin M, Ciliberto G, Ruggeri D (eds) Italian Beauty. Anteferma ed., Conegliano

Stefano Tornieri Architect and Ph.D in architectural composition at University Iuav of Venice. He has been Visiting Fellow at FAUP University of Porto in 2014 and at Luleå University of Technology with the Lerici Foundation grant in 2022. Since 2016 he is Research Fellow at IR.IDE (Infrastructure Research Integral Design environment) at University Iuav of Venice. Within the professional activity he has been nominated at the EU Mies Van Der Rohe Award 2019, curator of Grenada Pavilion at the 17th Venice Architecture Biennale and co-curator of the Georgian Pavilion at the 18th Venice Architecture Biennale.

Chapter 30
The Creative Time of Gardens: From Microcosm to Landscape Infrastructure

Antonia Di Lauro

Keywords Project · Quality · Biodiversity

The accelerated rhythms of the Internet world greatly affect contemporary space which consequently has become more fluid and mutable. Landscape reflects the complex transformations of our time, identifying metamorphosis and dynamism as values: an "aesthetics of a creative time" that has always been the essential and intentional expressive tool of landscape projects. Landscape changes through biological and social times: today it reveals a mixture between the fast events of society and the slower evolution of places due to nature and human work. The interpretation and the arrangement of landscape transformation are a central issue in prefiguring sustainable scenarios through a shared vision of the future. Large-scale projects, marked by sprawling areas and long times, are often asynchronous with respect to the rapidity of global change; instead, a garden offers a chance to keep up with the world's "temporal and space compression".

A garden is a "microcosm" in which the short and long times of landscape—the immediacy of daily life, the slower transformation of territories, the perpetual renewal of nature—converge and amplify in a narrative of symbols and metaphors. Here, more than elsewhere, "contemporaneity" —as the temporal span of human life—expresses itself as landscape in complex past, present and future relationships.

While satisfying the needs of its own society, gardens in every time and place have represented a "laboratory of thought" anticipating the ideologies, culture and technologies through which inhabitants relate to their environment and perceive it as landscape.

A. Di Lauro (✉)
Politecnico Di Milano—Scuola Di Architettura, Urbanistica Ingegneria Delle Costruzioni, Milan, Italy
e-mail: antoniamaria.dilauro@polimi.it

T. Matteini et al. (eds.), *Cultivating Continuity of the European Landscape*, Environmental History 15, https://doi.org/10.1007/978-3-031-25713-1_30

Thus, it is a place where collective stories, urban conflicts and social aspirations meet. Here, project issues can be reconsidered and combined with a design language able to express social values.

This paper addresses how the garden as an "ethical-aesthetic image" may reveal future larger-scale scenarios of transformation. A garden project is considered a field test, i.e. a "minimum landscape" that, even if punctual and short-term, anticipates wider and longer-term strategies. It is closed and fenced by definition but open and dynamic at the same time and able to exchange information beyond its boundaries. Nowadays, it alows us to direct the creative action of time on open landscape infrastructures with an integrated vision across different scales.

The accelerated rhythms of the world reveal a rift between biological and human times. As Ricoeur reminds us about the speed that characterizes our era: «We are no longer able to produce a concept of time that is simultaneously cosmological, biological, historical and individual».[1]

According to Norbert Elias, the separation «between physical and social time, between time of nature and time as a cultural convention (evidence) the unsustainable dichotomy between nature and culture…what we call humanity and society is no less natural than atoms and molecules».[2]

The relationship between the biosphere and the "dromosphere"[3] produces implications at a global level. Certainly, the increasing speed that overwhelms every human action leads to individual crises due to the absence of certainty and projection into the future: «Timeless living also and above all concerns the radical weakening of an entire dimension of time: the future, or at least the present perception of the future, the ability to think, plan, build».[4]

Concerning this, Morin emphasizes a "thought reform"[5] and a necessary global "metamorphosis" of our "world-society"[6]—which implies the reformulation of interrelated knowledge: «We must review the absolute disjunction between the human and the natural … Homo sapiens must no longer try to dominate the Earth, but must try to care for and organize it … To do this, a knowledge reform is required. We live under the influence of disjunctive and reductionist thought … hence the knowledge paradox that produces more blindness than lucidity».[7]

"La Voie" exhorts us to change our point of view and, first of all, look at the world through communication with the Earth. Change must seek quality rather than quantity due to environmental upheaval, degradation and the finiteness of resources:

---

[1] Paul Ricoeur in Virilio (1998). Lo spazio critico, cit. p. 106.

[2] Elias (1986). Saggio sul tempo, in Pagano (2011). L'uomo senza tempo. Riflessioni sociologiche sulla temporalità nell'epoca dell'accelerazione, cit. ebook.

[3] Virilio (1998). Lo spazio critico.

[4] Pagano (2011). L'uomo senza tempo. Riflessioni sociologiche sulla temporalità nell'epoca dell'accelerazione, cit., ebook.

[5] E. Morin, La Via.Per l'avvenire dell'umanità, cit p. 71.

[6] Ivi, cit p.18. In relation to globalization, the author considers the emergence of a "world-society" in which there is "interdependence between each and all nations, communities, individuals".

[7] Belonging to an "Earth—Homeland". ibidem, cit. p.71.

«The way [to new ecological awareness] must replace the hegemony of quantity with that of quality; the obsession for more with an obsession for the best».[8]

In this overview, landscape reflects the conflict between alterations due to human acceleration and the evolution of natural places; therefore, it is characterized by a "readability crisis".[9] The contrast between natural and anthropic times is evident: two landscape production parts that move at different speeds and negatively affect the quality of places.

Landscape, a permanently changing entity, has historically expressed metamorphosis and dynamism as essential aesthetic factors in its establishment, a «tension of the original harmony of three moments: not yet (future), no longer (past), now (present)»,[10] to which culture and nature contribute to define a particular context. Nowadays, it reveals a temporal dysfunction in its scenarios as the result of the corrosive action of our species on places, as the photographer Burtynsky shows us in his "Anthropocene"[11] exhibition through the images of landscapes produced by homo sapiens in his short existence: eroded mountains, desertified lands, fluorescent waters representing the effects of humans on a habitat altered by the unbridled exploitation of resources. This process gets out of control and reveals the weaknesses of the culture that produced it.

Observing a compromised landscape, man discovers the failure of his project as a continuous desire to live and a future projection.[12] As in the social sphere, a knowledge reform is necessary. Landscape also requires a renewed design culture, identifying principles and models that replace quantity with "the best" in the search for spatial and temporal quality in human actions on places. In recent years, projects focused on landscape infrastructures[13] have reorganized sustainable and resilient territories, considering them as complex relationship systems. However, many projects are difficult to implement due to complex problems, the sprawling area and long implementation times, asynchronous with rapid place transformation. Thus, for example, ELC directives effectively remain applied to only limited contexts.[14]

A complementary and opposite point of view is the garden, a project limited in time and space but closer to the "space-time com-pression"[15] of the world.

In the garden, the daily life immediacy, the slow transformations in accordance with the nature, and the perpetual renewal of nature converge in a narrative of symbols

---

[8] Ivi, cit. p.80

[9] Ivi, cit. p.80 L. Mondada, F. Panese, O. Söderström (editors), Paysage et crise de la lisibilité, Lausanne, Presses de l'Université, 1992.

[10] Venturi Ferriolo (2009a, b). Non ancora, non più, adesso. Paesaggi luoghi del tempo.

[11] Anthropocene.mast.org.

[12] The author underlines the meaning of "project" from Latin "proiectum" as a plan towards the future. M. Venturi Ferriolo, Percepire paesaggi. La potenza dello sguar.

[13] Belanger (2017). Landscape as infrastructure: a base primer. New York: Routledge.

[14] AA.VV. (2016). Dialogo su inadempienze e ritardi nell'attuazione della Convenzione Europea, a cura di Annalisa Calcagno Maniglio, Franco Angeli: Roma.

[15] Harvey (2010). La crisi della modernità. Milano: Il Saggiatore.

and metaphors that allow us to grasp—immediately and in advance—visions of future city quality.

According to Venturi Ferriolo, a garden «is an unexplored good place metaphor overshadowed by the modern Cartesian contrast between man and nature to be recovered as a better, welcoming world with a holistic vision of nature as the whole of men, animals, plants and minerals, where responsibility ethics act against the deterioration of our unique and irreproducible planet».[16]

This "place of dreams and power"[17] keeps the best within itself. According to Clement «inside the fence you find the best: what is considered most precious, most beautiful, most useful and most balanced».[18]

Considering landscape on a temporal level—as a gap between biological evolution and human acceleration—«the garden seems the only meeting place of nature and culture where dreaming is authorized».[19] Here, outside the urban space as a place of conflicts and interests, man abandons existential troubles and synchronizes his time to that of life.

Thus premised, the garden appears as a privileged point to seek and test new contemporary project models. Clement writes: «in the garden you just have to be, getting rid of all cumbersome knowledge, little by little, as you do with useless clothes».[20] It is precisely the abandoning of political and social superstructures that has always made it possible to test technologies and aesthetics with greater freedom, audacity and creativity to create "the eternal myth of the happy place"[21]: the metaphor of a desirable future in which nature and culture coexist to mutual benefit.

In its different forms over the centuries, the garden has been characterized as a "laboratory of thought"[22] that seeks "the best" as an ideal: an "ethical-aesthetic" representation of the relationship between man and landscape.

Venturi Ferriolo writes: «From antiquity to the most recent contemporaneity, in every space and time, we can read the world in the garden, seeking and exploring its image through metaphore».[23]

Like the hortus conclusus, or the classical, baroque or picturesque garden, the contemporary garden is both a manifestation of the cultural ideals of "world-society" and a guide for its "project" of future landscapes.

In Milan, the Biblioteca degli Alberi is «a contemporary botanical garden in the heart of Porta Nuova... where you can experience culture in contact with nature».[24] It is a place where a new concept of the relationship between city and landscape is being tested: a center of urban regeneration aiming to increase biodiversity and raise

[16] Venturi Ferriolo(2019) Oltre il giardino. Filosofia di paesaggio. cit. ebook.

[17] Grimal (2000) L'arte dei giardini. Una breve storia.

[18] Clement (2013). Giardini, paesaggio e genio naturale. cit. ebook.

[19] Ibidem.

[20] Ibidem.

[21] Grimal (2000) L'arte dei giardini. Una breve storia.

[22] Zagari (1984) Architettura/arte dei giardini. Note e progetti su giardino moderno e spazio civico.

[23] Venturi Ferriolo (2019) Oltre il giardino. Filosofia di paesaggio.

[24] www.bam.milano.it.

user awareness of ecological themes through educational gardens, artistic events, outdoor activities. Designed by Petra Blaisse's Inside Outside studio, the park is a landmark and the green heart of a future city vision.

Thus, a garden—when it is not expressed as "equipped greenery" or a "green space", arid surrogates of our time[25]—lets us glimpse the ideals and values of a society that is reorganizing itself around the ecological paradigm.

This is the foundation of Clement's gardens occupying urban interstices with the "friche apprivoisée".[26] The "Jardins du foyer Laeke-nois" rises on the degraded spaces of the Cité Modèle in Brussels, a 1960s residential district. Developed by the Coloco and built in 2007, they constitute a new access to the city and contribute to landscape enhancement and neighborhood regeneration based on the "jardin en mouvement" thesis. Also the "Jardin du tiers paysage" in the former submarine base of Saint Nazaire implements the concept of biodiversity and the use of native species that is consistent with the ecological spirit of a new, more aware society. Clement points out that the garden extends its enclosure to the whole planet, while the gardener transforms into a people.[27] Emerging ecological awareness signifies that everything is interconnected in a "closed common space": the biosphere that «forces the whole neighborhood to share… Sharing obliged for obvious reasons related to finiteness».[28]

Thus, «any fence within the planetary garden … is attributable to an archaic vision of dominium» and the concept of "the best" in the project is redefined according to new paradigms: «the idea of "the best" has changed throughout history…Consequently the garden architecture which translates this idea … What once stood outside the fence—the wild, the weed—penetrates the garden today. In-deed, it may be the main subject».[29]

This is what happened in New York's High Line where Piet Oudolf expressed a new garden aesthetic and an overall feeling of reconciliation with nature. The project supports biodiversity creating an ecological corridor, a refuge for native species that coexist with citizens. The conversion of a former railway into a park has demonstrated environmental and economic benefits over the years. It arose from the will of inhabitants animated by a new ecological sensitivity, and today, the "people gardener" still collaborates in park management.

As a place of entertainment and decoration, a garden takes on a multitude of forms and functions, a metaphor for the cultural melting pot of cities. In its variety, it is always a meeting place for man and nature, but also communities.

The Superkilen park in Copenhagen summarizes the aesthetics and values of the multicultural people of Nørrebro, acting as an effective redevelopment device for

---

[25] Venturi Ferriolo (2019). Oltre il giardino. Filosofia di paesaggio.

[26] Venturi Ferriolo (2019). Oltre il giardino. Filosofia di paesaggio. Clement (2005), Manifesto del Terzo paesaggio. Macerata: Quodlibet.

[27] As Clement writes, the ecological paradigm highlights how "living beings are connected within a complex system … a system in which each element influences all the others within a finite space: the planet". Clement (2013). Giardini, paesaggio e genio naturale, cit. ebook.

[28] Ibidem.

[29] Ibidem.

a neglected area. Through community involvement in the multi-ethnic furnishings selection, the park is a synthesis of the world's gardens, an entertainment space for the 60 settled communities and, above all, a place of identity and integration. Superkilen highlights the strategic role of gardens as spaces for convergence and resolution of conflicts, where social issues can recombine starting from dialogue to integrate new aesthetics and behaviors.

This is inherent in a garden which «escapes cultural divisions ...everywhere in the world it means both enclosure and paradise»[30] allowing us to rediscover the common roots of living beings.

Shared gardens, multi-ethnic parks and biodiversity spaces support ecological system restoration and demonstrate the widespread sensitivity toward urban space care: a common practice to rediscover a sense of community and identity of place, restoring economic, environmental and social value to neglected areas.

A project's social, aesthetic and perceptive aspects—in an interdisciplinary dimension—focus on the enhancement of natural and identity features and take on a "therapeutic" role that acts on a small area but spreads over time to the city and its inhabitants, structuring urban spaces and conveying values and behaviors.

Thus, a garden is a "minimum landscape",[31] the smallest unit of intervention in which a project measures itself to overcome current problems that require a mentality change relating to nature, environment and people and a plural confrontation of actors. As Enric Batlle suggests: «The metropolis garden ... inspires a new model for territorial order ... it can give rise to a new city geography ... it wants to be simultaneously the ideal garden, a public park, a system of free spaces, a hybrid of different landscapes. It is the expression of a new spirit that tries to guarantee sustainability without renouncing its formalization starting from modernity. It wants to inspire a new model of territorial structure, based on the landscape. It wants to offer conventional urban planning new tools to face the new metropolitan situations project».[32]

The closed and fenced system of a garden is simultaneously an open and dynamic entity that exchanges information with the landscape, influencing its shape and organization. Therefore, within its microcosm, it is possible to intercept reforming principles and models of a design culture that directs "the best" through time, reconfiguring future territories in a "planetary garden".[33]

Clement argues for the "time-object": «the garden is a privileged place of the future, a mental territory of hope».[34]

---

[30] Ibidem.

[31] Zagari (1984). Architettura/arte dei giardini. Note e progetti su giardino moderno e spazio civico.

[32] Bocchi, Designing landscape in Di Lauro (2017) Sharing Landscape. Partecipazione e progetto tra spazio fisico e flussi virtuali.

[33] Clement (2013). Giardini, paesaggio e genio naturale, cit., ebook.

[34] Ibidem.

References

Belanger P (2017) Landscape as infrastructure: a base primer. Routledge, New York
Clement G (2013) Giardini, paesaggio e genio naturale. Macerata, Quodlibet—Ebook
Clement G (2005) Manifesto del Terzo paesaggio. Macerata, Quodlibet
Dardel E (1986) L'uomo e la terra. Natura della realtà geografica. C.Copeta (a cura di), Milano
Di Lauro A (2017) Sharing landscape. Partecipazione e progetto tra spazio fisico e flussi virtuali. Aracne, Roma
Elias N (1986) Saggio sul tempo, in Pagano, U. (2011). L'uomo senza tempo. Riflessioni sociologiche sulla temporalità nell'epoca dell'accelerazione, cit. ebook
Grimal P (2000) L'arte dei giardini. Una breve storia. Roma, Donzelli
Harvey D (2010) La crisi della modernità. Milano: Il Saggiatore
Morin E (2012) La Via. Per l'avvenire dell'umanità. Raffaello Cortina, Milano
Pagano U (2011) L'uomo senza tempo. Riflessioni sociologiche sulla temporalità nell'epoca dell'accelerazione. Franco Angeli—Ebook, Milano
Prigogine I, Stengers I (1999) La nuova alleanza. Metamorfosi della scienza. Piccola biblioteca Einaudi
Venturi Ferriolo M (2019) Oltre il giardino. Filosofia di paesaggio. Giulio Einaudi—Ebook, Torino
Venturi Ferriolo M (2009) Percepire paesaggi. La potenza dello sguardo, Torino, Bollati Boringhieri
Venturi Ferriolo M (2009) Non ancora, non più, adesso. Paesaggi luoghi del tempo. In: "Kronos & l'Architettura", Istant Book 3 Attilio Nesi (a cura). Reggio Calabria, Centro Stampa d'Ateneo
Virilio P (1998) Lo spazio critico. Dedalo, Bari
Zagari F (1984) Architettura/arte dei giardini. Note e progetti su giardino moderno e spazio civico. Casa del libro, Roma

Antonia Di Lauro An architect since 2010, she achieved ASN as Professor in SD 08/D1-ICAR 15 in 2018. In 2012 she had a training period at the Katholieke Universiteit of Leuven and obtained her PhD in Architecture in 2014 with the thesis "Sharing landscape. Participation and project between physical spaces and virtual flows" which investigated the relationship between landscape projects, community and ICT. From 2013 to 2016 she was a member of the Research Unit of Reggio Calabria with PRIN (2013/16) "Recycle Italy. New life cycles for city and landscape architecture and infra-structure". From 2011 to 2018 she collaborated in Landscape Architecture Courses at the Mediterranea University of Reggio Calabria and participated in projects with local authorities and administrations, as a coordinator of the laboratory "RizemuLaB" with the Municipality of Condofuri, as a delegate in the debate on the demolition of highway A3 in Bagnara—Scilla, with the province of Reggio Calabria. She collaborated with the cultural association "A di Città" in the urban regeneration workshops of Rosarno and participated at the proposal for the protected landscape of the "Costa Viola", promoted by Gal Batir. In 2015 she coordinated the Public Art workshop communities of the "Area Grecanica" with the respective Gal and the "Aniti" Association. Since 2019 she has been a member of the Italian Scientific Society for Landscape Architecture (IASLA) and Adjunct Professor at the Politecnico di Milano.

Chapter 31
Underground Heritage Valorization of Camerano's Caves in Center Italy: A Case of Transition Toward Projects Integrating the Local Community and Landscape

Alici Antonello, Andreoni Luca, Chiapparino Francesco, De Seriis Marco, Fioretti Ilaria, Galli Andrea, Malinverni Eva Savina, Marcheggiani Ernesto, Paci Francesco, and Pierdicca Roberto

Abstract The village of Camerano (Central Italy) is a local heritage and a landmark for its network of connected built underground spaces. This Underground Built Heritage represents an economic leverage for cultural tourism and an occasion to preserve local identity. The self-initiative of the local community, and the determination and far-sightedness of the local authority, allowed to differentiate the local tourism offer leading to success in terms of tourism attractiveness, with more

A. Antonello · M. E. Savina · P. Roberto
DICEA Department, Università Politecnica delle Marche (Italy), LRC/CIRP, Ancona, Italy
e-mail: a.alici@staff.univpm.it

M. E. Savina
e-mail: e.s.malinverni@staff.univpm.it

P. Roberto
e-mail: r.pierdicca@staff.univpm.it

D. S. Marco · F. Ilaria
Municipality of Camerano, Ancona Province, Italy
e-mail: ilariafioretti@comune.camerano.an.it

A. Luca · C. Francesco
DISES Department, Università Politecnica delle Marche (Italy), LRC/CIRP, Marche, Italy
e-mail: l.andreoni@staff.univpm.it

C. Francesco
e-mail: f.chiapparino@staff.univpm.it

G. Andrea (✉) · M. Ernesto · P. Francesco
D3A Department of Agricultural, Food and Environmental Sciences, Marche Polytechnic University, Italy LRC/CIRP, Marche, Italy
e-mail: a.galli@staff.univpm.it

M. Ernesto
e-mail: e.marcheggiani@staff.univpm.it

T. Matteini et al. (eds.), *Cultivating Continuity of the European Landscape*, Environmental History 15, https://doi.org/10.1007/978-3-031-25713-1_31

than 25,000 visitors per year. With the occasion of an ongoing EU project (Cost action CA18110) involving a research unit of the Marche Polytechnic University, a collaborative network is being set up, involving actors from the whole territory. Special attention is being given to the digital dimension of Cultural Heritage, an asset for the next decade. Nonetheless, the idea of a cyberization of the thematic tours of Camerano's Cave sounds attractive if considering, especially, that direct access to the cavities is not always possible given the intrinsic vulnerability of caves.

Keywords Underground built heritage · Local community · Local history · Digital cultural heritage · Transition management

31.1 Introduction

The underground complex of Camerano is a known local heritage and a landmark for this small town of Central Italy. Although the first reliable record dates 1327 AC, shreds of evidence suggest that the original complex can be traced back to the eleventh–third century BC. Along the centuries, the cavities expanded, creating a network of connected built spaces, under a strict family-owned real estate regime. In the second half of last century, the owners began to leave the caves in a state of disarray, its only thanks to determination and far-sightedness of the local authority that a recovering work started in the late 90 s (Mariano 2000). The aim was valorizing the caves as potential economic leverage for cultural tourism, and an occasion to preserve local identity. The self-initiative of the local community led to a first strategic outline to strengthen back and revive the older people's attachment to the underground patrimony. This leads to the first success in recent years, with more than 25,000 visitors per year. To progress with that strategic outline, in 2019 a multidisciplinary working group has been set up with the participation of the Università Politecnica delle Marche—Landscape Research Center (LRC/CIRP). In addition to the insights and studies on the critical issues, the partnership supports a collaborative network involving actors from the whole territory (undergrounds and archaeological sites, wineries and typical traditional food). Special attention is given to the digital dimension of Cultural Heritage, which is going to become an asset for the next decade, in particular after the COVID-19 pandemic. Besides, the recent evolution of critical European territorial policies created excellent pre-conditions for framing this landscape local planning operation at the very local scale. Finally, the ongoing Cost action CA18110 "Underground Built Heritage as a catalyzer for Community Valorisation" represents an opportunity to steer such an idea of a landscape's approach. One of the characteristics of this project, in which the local community of Camerano and the LRC/CIRP of the Università Politecnica delle Marche are involved, is its possible implementation in terms of heritage management and exploitation.

P. Francesco
e-mail: f.paci@staff.univpm.it

31.2 Origins and Use of Camerano's Caves

The presence of the caves can be connected to the historical events of settlement in the slopes of Mount Conero. Among the possible hypotheses on the origin and meaning of the toponym "Camerano" itself, it is possible to detect a link with the idea of a closed place (Morroni 1979),[1] connected with the presence of the caves. It is difficult to suggest a specific date of origin. As they appear, they date back to the remodeling of the eighteenth and nineteenth centuries, but a shred of certain evidence goes back to the medieval period (1327). This can put in relation with the existing testimonies on the presence of the caves of the nearby city of Osimo (1276) (Morroni et al. 1990, 100; Recanatini 2000, 170). Starting from the detection of the correlation between the performance of the underground network and the historical layout of the inhabited center prior to the contemporary expansion, the hypothesis of an archaic dating of the caves was formulated, to be anticipated until the Picente's era, starting from two elements: (a) the comparison with other Italian underground complexes; the context of the population of the territory before the medieval period, attested by archaeological finds (Recanatini 2000). Although the urban dimension of the Picena civilization was minimal (Naso 2000, 234–250), this hypothesis (Recanatini 2000) appears stimulating, because it finds an indirect confirmation in the case of Numana, where there are testimonies of two pits from the Picente's era, attributable to probable purposes of worship (Finocchi 2018, 256).

Some structural features of the caves are, at the same time, a sign of the close link with the overlying population, beyond the silence of the documents, and a map of the main elements of historical and artistic interest in these places. For the structure and aim of this paper, it suffices to recall two: their recursive use in different historical periods, even distant (which contributes to explaining the difficult dating, because the continuous remodeling was carried out outside the official path of management of the inhabited areas, and therefore, it was poorly documented) and the simultaneous presence of multiple uses, even in the same era. Among the latter, at least four should be mentioned: place (Figs. 31.1, 31.2 and 31.3) of worship (e.g., Ricotti cave, Mancinforte cave), possible place (Fig. 31.2) of water supply (Mancinforte cave),[2] place of refuge (for example, in July 1944, during the passage of the WWII front) (Recanatini 1997, 121; Recanatini 1999), place of storage and conservation of agricultural products, probably belonging to the owners of the buildings above.

---

[1] But see, for a different interpretation, which traces the toponym to a Roman era predial, therefore with reference to a noble agricultural property, Paci (1988).

[2] The testimonies of underground structures for water supply in the Conero area are numerous. Among the most remarkable, from the Roman era, the case of Osimo (Finocchi 2018, 301–305) and Numana (Bilò-Francioni 2015).

Fig. 31.1 Cave Mancinforte, a place of worship (Camerano, Ancona, Italy)

Fig. 31.2 Cave Mancinforte, a possible place of water supply (Camerano, Ancona, Italy)

Fig. 31.3 Cave Ricotti, a place of worship (Camerano, Ancona, Italy)

31.3 Transformations of the Surrounding Landscape

The agricultural colonization of the periadiatric Marche hill, where Camerano stands (Fig. 31.4), begins already in Roman times. However, it is the transformations between the fifteenth and sixteenth centuries that shape the typical agricultural landscape of the area. By comparing the land registers of the time, we could see the almost complete elimination of woods and of uncultivated lands, as well as the reduction of lawns and pastures, and the diffusion of mixed systems of arable and vineyards or threes; the "arativo vitato" and "arativo arborato", at the expense of the bare arable "arativo nudo", which, however, still kept occupying considerable surfaces. During the sixteenth century, the sharecropping (mezzadria classica) spread, and during the seventeenth century, the system expanded. The construction of farmhouses intensifies, overcoming the first forms of settlement in the countryside (tower-house-dovecote, inspired by urban architectural canons), assuming the typology of the "Italic house", with the rustic part on the ground floor and the living part on the first floor. The spread of the "sharecropping model" determines the formation of the characteristic agrarian landscape of Marche region, with an ever-closer mesh of "farms" materializing the symbiotic relationship between the land, the farmhouse and the sharecropper family. Already during the eighteenth century, however, the first signs of imbalance between available resources (environmental and technological) and social needs (food and economic) appeared (Paci 1987). A new and intense period of deforestation and land clearing and tillage begins, which also affected the most marginal lands. An agrarian survey of that time (Jacini 1884) highlights an increase in arable land, almost all vineyards, at the expense of woods, meadows and pastures in all the Marche during the previous forty years. After five centuries, the typical sharecropping landscape remained almost entirely unchanged over much of the provincial hilly territory in the immediate post-war period. However, sharecropping goes into crisis and is abolished by law in the 1960s. In the early 1970s, the new economic growth and technological industrialization began a profound transformation of the rural landscape. To the city of Camerano, this led to a progressive consumption of agricultural land, due to the expansion of impervious surfaces (industrial and commercial zones), especially in the "external" areas to the historic town (Figs. 31.5 and 31.6). In the same period, the change of the hilly agricultural landscape began to take place throughout the whole region. Following the progressive intensification of agricultural production systems and the specialization of viticulture and local winemaking industries, also in Camerano, the average cultivated field size increases to facilitate mechanical operations with the almost disappearance of the arable with trees "seminativo arborato".

Hilly landscape near Camerano

Fig. 31.4 Hilly landscape near Camerano (Ancona, Italy)

Fig. 31.5 New urbanization near Camerano (Ancona, Italy)

Fig. 31.6 Area of Camerano (Ancona, Italy) from a recent topographic map (scale 1:10,000)

31.4 Learning from the Local History to Designing a New Vision for the Near Future of Camerano

The transformations in the territorial setting of Camerano with its caves testify the stratification of economic, social and cultural models over the centuries. The peculiar urban and rural landscapes and new economic and social behaviors of the local community resulted from these transformations. However, the current situation is characterized by a strong identity link between the local community and the history of caves, a considerable uncertainty for the future concerns local communities, considering the recent economic downturn and cultural turmoil for the further development, especially the post-pandemic scenario.

There are two main issues: (a) a lack of experts with a broad vision of the different possibilities for enhancing the territory/landscape; (b) the need to reconnect the places overcoming the marginalization from which it suffers today. The first issue is being investigated as part of a Cost action CA18110 "Underground Built Heritage as a catalyzer for Community Valorisation" aiming at promoting balanced and sustainable approaches for the conservation of the underground heritage, unlocking its potential for regeneration policies. During the action, a network of junior experts will be trained along with three strands: 1. underground conservation and monitoring; 2. heritage-led urban and rural regeneration; 3. planning and governance tools. A tailored methodology of "living Lab" (http://underground4value.eu/living-labs/) is being implemented during the Cost action.

Although the Marche region can boast a rich heritage of underground spaces, including Camerano (Fig. 31.7), the development of thematic networks capable of attracting tourist flows is not yet complete. If the reconnection will be implemented would become a landmark reaching out to new tourism segments.

From a first analysis based on the historical functions, five potential thematic networks could be defined:

- Water supply function: aqueduct's tunnel, wells, cisterns, emissaries.
- Military-strategic function: tunnels used to defend against sieges, secret passages to exit the walls of fortified centers.
- Food storage function: storage in underground cavities of food and wine that has always been a rural tradition
- Cultural or religious function: caves that housed hermitages or small religious structures intended for the hospitality of pilgrims and wayfarers.
- Mines or quarries: hypogea used in the past as mines or quarries of stone, or other minerals.

Noteworthy is the geographic distribution of functions. While the religious use uniformly distributed along the Apennine belt from north to south, hypogea with multiple functions (defense, food storage and water supply) lay on the coast, in particular in the province of Ancona (Fig. 31.8).

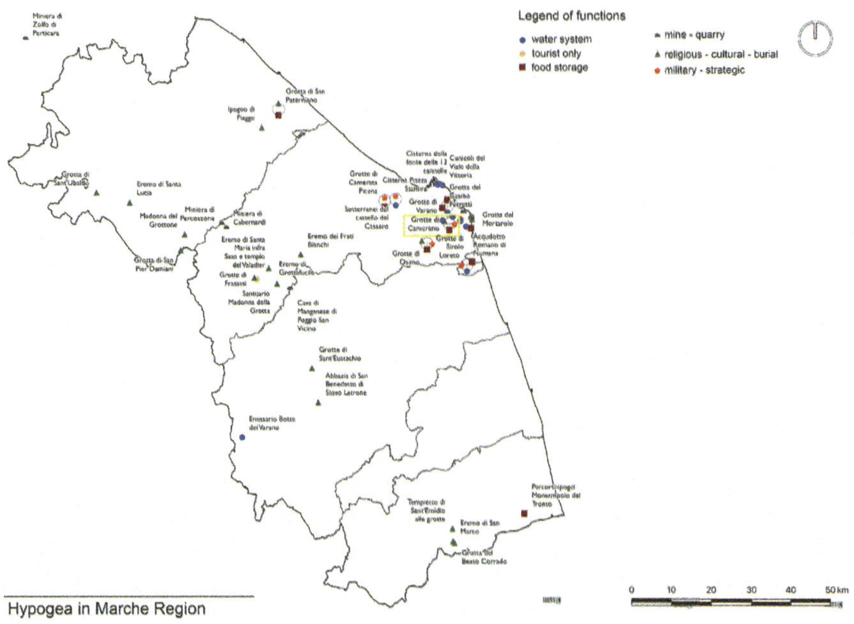

Fig. 31.7 Hypogea in Marche region

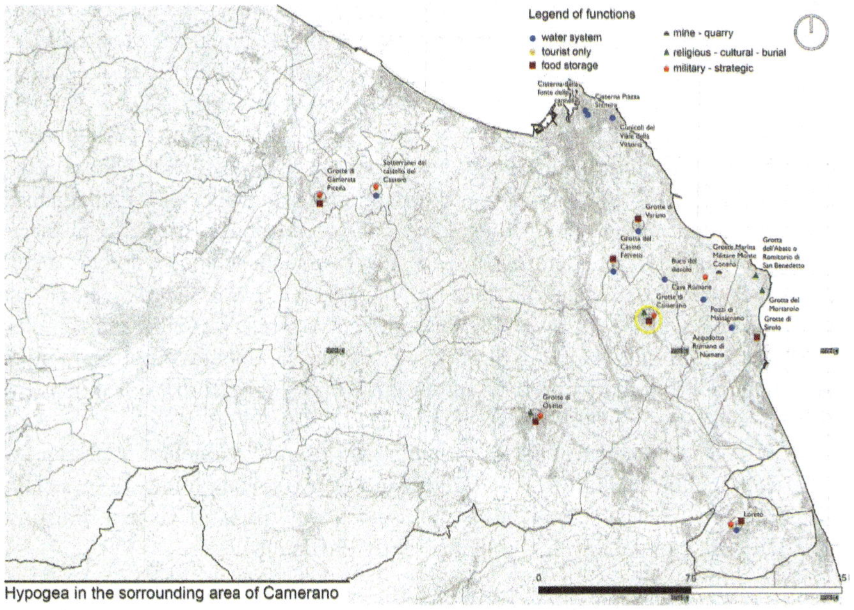

Fig. 31.8 Hypogea in the surrounding area of Camerano (Ancona, Italy)

31.5 What Opportunities from Ue Policies and ICT Tools?

The evolution of European territorial policies, the Common Agricultural Policy and the Development and Regional Cohesion Policy, has objectively created excellent conditions for framing the landscape planning at the very local scale. All current Operational Programs activated at the policy level (Leader, Rural Development Program, Inner Areas strategy, etc.) consistently show this trend by focusing on the local community and supporting a widened accessibility to financial resources. A whole set of thematic objectives and priorities foster the idea of a strengthened landscape project (care of places, enhancement of resources and sustainable production systems). In this vision, transition management, generative welfare, accessible and sustainable tourism, technological innovation are just examples of the items that could be seized but require the development of initiatives at the local and regional levels.

At the cutting edge of this debate, Camerano could represent an opportunity to steer such an approach to the landscape. One of the potentialities of this project is in terms of management and exploitation. Modern management protocols and innovative technological tools are critical issues for an effective valorization of Cultural Heritage. Sophisticated Digital Cultural Heritage solutions, such as augmented reality, 3D visualizations, virtual tours, are shaping the next generation of products in the tourism industry. Attracting visitors by new interaction among experts and visitors needs an innovative flow of data and dedicated surveys. Well-known acquisition techniques are applied in UBH surveying: Photogrammetry, Terrestrial Laser Scanner, Mobile Mapping Systems (MMSs). All allow for quick time collecting of high volume of data with reasonable accuracy (Remondino 2011). However, UBH surveying presents several issues (narrow spaces, reduced illumination, irregular and coarse surfaces), and new technologies and surveying approaches emerge providing local managers with effective supports to monitoring, management, maintenance, conservation and restoration. Among them, 3D point clouds allow for an accurate virtual rendering of underground spaces. Of course, the choice among different sensors and techniques depends on several elements (scale, costs, accuracy, etc.). Data acquisition science is a fast-evolving environment. Mobile mapping and Simultaneous Localization and Mapping (SLAM) are nowadays available solutions for quick time 3D data collection, with reduced survey post-processing efforts (Makkonen et al. 2017; Maboudi et al. 2018; Malinverni et al. 2018).

Indeed, it is not enough to make a destination successfully. Nonetheless, the idea of a cyberization of the thematic tours of Camerano's Cave sounds attractive, especially considering that direct access to the cavities is not always possible given the intrinsic vulnerability of caves, or as taught by the recent pandemic crisis.

References

Bilò M, Francioni O (2015) *L'acquedotto romano di Numana: nuove acquisizioni dallo scavo di Via Morelli.* Picus. Studi e Ricerche Sulle Marche Nell'antichità 35:89–119

Finocchi S (2018) *Numana (AN).* Picus. Studi e Ricerche Sulle Marche Nell'antichità 38:253–282

Jacini S (1884) *Atti della Giunta per l'inchiesta agraria e sulle condizioni della classe agricola,* vol XI, tomo II, Le Marche, Roma

Maboudi M, Banhid D, Gerke M (2018) Investigation of geometric performance of an indoor mobile mapping system. Int Arch Photogrammetry Remote Sens Spatial Inform Sci 42(2)

Makkonen T, Heikkilä R, Tölli P, Fedorik F (2017) Using slam-based handheld laser scanning to gain information on difficult-to-access areas for use in maintenance model. In: ISARC. Proceedings of the international symposium on automation and robotics in construction, vol 34. IAARC Publications

Malinverni ES, Pierdicca R, Bozzi CA, Bartolucci D (2018) Evaluating a slam-based mobile mapping system: a methodological comparison for 3D heritage scene real-time reconstruction. In: 2018 Metrology for archaeology and cultural heritage (MetroArchaeo), IEEE

Mariano F (2000) *Il progetto di restauro e valorizzazione delle Grotte Ipogee di Camerano.* In: Campagnoli M, Recanatini A (eds) "La memoria del sottosuolo", Atti del Convegno di studi (Camerano, 17–18 luglio 1999), Jesi

Morroni M, Recanatini A, Toccaceli F, Donzelli N, Pincini C (1990) *Terra di provincia. Tra Aspio e Gradina,* Ancona

Morroni M (1979) *Ipotesi varie sull'origine ed il significato del toponimo Camerano,* Osimo

Naso A (2000) *I piceni. Storia e archeologia delle Marche in epoca preromana,* presentazione di Giovanni Colonna, Longanesi, Milano

Paci R (1987) Nascita, sviluppo e morte della mezzadria. In: Anselmi S (ed) "La provincia di Ancona Storia di un territorio", Laterza, Bari

Paci G (1988) *Schede per l'identificazione di antichi predii in area picena.* In: Janni P, Lanzillotta E (eds) *Geographia,* Atti del secondo Convegno maceratese su geografia e cartografia antica (Macerata, 16–17 aprile 1985), G. Bretschneider, Roma

Recanatini A (1997) *Le grotte del Conero. Ricerche di speleologia archeologica nel Parco del Conero,* s.n., Jesi

Recanatini A (1999) *Il paese e la guerra. Camerano e gli avvenimenti della Seconda Guerra Mondiale: appunti, documenti, testimonianze e memorie,* 1939–1945. Biblioteca Comunale, Camerano (AN)

Recanatini A (2000) *La città segreta, correlazioni tra sviluppo urbano di superficie e rete ipogea nei centri storici alle falde del Conero. I casi di Camerano e Osimo.* In: Campagnoli M, Recanatini A (eds) "La memoria del sottosuolo", Atti del Convegno di studi (Camerano, 17–18 luglio 1999), Jesi, pp 169–188

Remondino F (2011) Heritage recording and 3d modeling with photogrammetry and 3d scanning. Remote Sensing 3(6):2011

Chapter 32
Compromised Landscapes

Barbara Angi and Alberto Soci

Abstract In the suburbs, the project for the redevelopment of public spaces conflicts with pragmatic demands for action linked to a permanent 'state of emergency'. These are places where resources are scarce; where critical issues can overshadow any positivity of this important, though uncomfortable, heritage. The chapter focuses on some examples of bottom-up urban regeneration, where the triggering of unseen human relations has become a main element for the metamorphosis of the peripheral urban landscape. The 'fuse' for developing previously unknown public spaces, constantly, in the process of becoming.

Keywords Compromised landscapes · Amnesty for the built environment · Combined redevelopment · Re-assembling urban blocks · Human relationships

In suburbs, the project—and its intangible aspects—clashes with pragmatic requests for intervention linked to a permanent "state of emergency". These are places where resources are scarce and where criticalities (hydrogeological instability, environmental contamination, structural and/or energy deficits, social unease) can overshadow any positivity of this important, although uncomfortable, heritage. The concept of heritage implies the idea of transmitting one's property to posterity, the idea of a value that goes beyond ordinary income, but which has to do with people's lives, with their loved ones. In short, with life and with the traces that in its continuous mutations, it also leaves on buildings and open spaces.

What is not always immediately clear is that in suburbs, it is possible to find an out-and-out "draw of the city" which becomes the scheme, the "pentagram" on which the elementary units—urban blocks even in their uniqueness—find a complete order and reason.

B. Angi (✉) · A. Soci
DICATAM Department, Università degli Studi di Brescia, Brescia, Italy
e-mail: barbara.angi@unibs.it
URL: https://architecture-construction.unibs.it/

A. Soci
e-mail: alberto.soci@unibs.it

T. Matteini et al. (eds.), *Cultivating Continuity of the European Landscape*, Environmental History 15, https://doi.org/10.1007/978-3-031-25713-1_32

In peripheral urban contexts, it is possible to identify principles of "re-design" that can promote a "second life" of these urban territories starting from those tiny fragments in which there are still positive characteristics of sociability. Operationally, it is time for change and adaptation with the aim of responding to shocks and stresses that are difficult to predict in current socio-economic conditions and, unfortunately in recent times, also of a health nature, which, undeniably, influence the choices of designers. For us, it is necessary to apply design tools capable of reassembling degraded contemporary urban habitats, which, metaphorically, includes 'the art of embracing damage', *Kintsugi.* The skilful Japanese technique of repairing ceramics—the deteriorated contemporary urban blocks—by welding them with precious materials—the human relationships that can be triggered through regeneration projects.

Furthermore, we think that the planning attitude necessary to operate in peripheral urban territories implies reflections on the discrepancies to be solved and the available resources, including economic ones. In order to achieve a good outcome, it seems desirable to implement urban redevelopment plans that embrace the thought presented by Aureli (2013) in his essay *Less is Enough* and are capable of «redefining what is truly necessary and what is not, outside the regime of scarcity imposed by the market». Above all, in the suburbs, it is necessary to reach a sort of "compromise" in order to determine an "agreement" between the stakeholders which, despite involving reciprocal renunciations, does not presuppose the existence of a dispute or a real conflict, but a common will to reach a shared goal by overcoming any differences.

In the last 30 years, some authors with the highest competence in the architectural field, even if not completely devoted to landscape architecture, have, in our opinion, reflected on this, with great results for the redevelopment of suburbs through completed projects that seemly incorporate sensitivity toward the context like in the Japanese art of *Kintsugi.*

Among these, the work of the French studio directed by Anne Lacaton and Jean Philippe Vassal certainly represents an operational reference for acting in the suburbs and trying to unravel the peculiarities of some urban margins and/or voids using minimalist design attitudes. In 1996 this emerging Parisian group won first place in the competition for the *Embellissement plan* of some squares in the city of Bordeaux. One of these, Place Léon Aucoc, has a triangular shape bordered by trees, with benches and a space for playing the pétanque game: a glimpse of the countryside that perfectly recalls its public dimension. On the first visit, the French architects had the feeling that the place was already "beautiful" because of its authenticity and lack of refinement. For the authors, the square possesses «the charm of what is obvious, necessary, right. Its meaning is clear. People seem at home in an atmosphere of harmony and tranquillity shaped by many years of coexistence» (Huber 2015).

During the design process, Lacaton and Vassal observed the activities taking place in it at various times of the day, interviewed the inhabitants and concluded that an "embellishment" project was not necessary (Fig. 32.1).

Fig. 32.1 Anne Lacaton e Jean-Philippe Vassal, *Place Léon Aucoc, Bordeaux, 1996* [Lacaton and Vassal]

In this place, "beauty" is made of simple architectural elements—a bench, a pétanque court—capable of wielding—again—all the potentials of human relationships. Therefore, the Parisian architect's proposal indicated simple maintenance works: replacement of gravel, a more frequent urban cleaning program, etc. The project set seems to take as reference some suggestions of the architect Berend Bakema who, during the post-War C.I.A.M. (*Congrès Internationaux d'Architecture Moderne*), proposed alternative methods of investigation and design to the functionalist approach.

In fact, for the Dutch author, «every day we discover [we rediscover, Ed.] that the only things that exist are relationships. Perhaps, we can even be said the aim of life is to become aware of the fundamental principles for an existence complete of relationships. It seems to me that this is the reason why, when we deal with architecture and urban planning projects, we so often speak of "continuity of space". For us [...] the relationships between things are more important than the things themselves» (Bakema 1954).

Even today, in our opinion, to develop a virtuous system of re-assembling urban blocks privileging human relationships, a multidisciplinary approach is necessary, but the design process can also include the way of tiny adjustments. Design practices adapted to the context that have the capacity to overcome the mentality of degraded suburbs and capable of restoring an urban landscape to the city for new relationships between people (Fig. 32.2).

The same design beginnings can be identified in the work of an increasing number of designers engaged in the afterthought of suburbs through "flimsy" tools, temporarily defined and built with basic construction techniques. Actions for the urban commons, which can be created through participatory processes. In fact, this approach is the art of *Kintsugi*: the 'junction' between the physical form of the urban regeneration project and the new social mechanisms that will be activated in public spaces. This practice can trigger the reappearance of citizens in forgotten and underused urban areas. In other cases, however, urban re-assembly interventions can be, even in this case, temporarily defined but with the aim of denouncing degraded situations that require immediate interventions by the central administration.

Fig. 32.2 Todos por la Praxis, *Jardín San Roque, Madrid, 2015* [Todos por la Praxis]

Fig. 32.3 NL. Architects, *Greening the City*, Amsterdam, 2013 [NL. Architects]

As part of the *Design as Activator* initiative organized by *Platform Openbare Ruimte* in 2013, the Dutch studio NL. Architects prepared a temporary mini-golf course called *Greening the City* in a port area of Amsterdam used nightly as a parking area for the surrounding trendy clubs in order to activate the space during the day. With a tight budget of 2500 euros, the Dutch team opted to construct nine islands of synthetic green grass scattered in an apparently causal way on the square (Fig. 32.3). With this project, NL. Architects wanted to show that it does not take great means to revitalize peripheral areas, even temporary works capable of giving meaning to these places that tend to get lost in the urban fabric.

The .NL Architect's thinking on urban landscape in this project is linked, in our opinion, to that of Also van Eyck, another author who fueled the disciplinary debate after the World War II, also through a humanistic approach. The interventions on public spaces by this Dutch architect, his famous *playgrounds* built in Amsterdam, tiny spaces for play and social aggregation in urban areas devastated by the war became, in fact, works by *Kintsugi*, in its previously stated acceptation. Architectural works that sought to 'repair the damage' of urban decay not so much with major

Fig. 32.4 Aldo Van Eyck, *Bertelmanplein*, Amsterdam, 1947 [Barbara Angi]

redevelopment projects as with minimal, almost anonymous structures. Small interventions on public space, capable of 'triggering' relationships between inhabitants and, at the same time, denouncing the need for a more active involvement on the part of the municipal administration.

Lefaivre and Alexander Tzonis, authors of an in-depth monograph entitled *Aldo van Eyck: humanist rebel* (1999), underline how the attitude of the Dutch designer was the same as those who, in different fields of culture and knowledge immediately after World War II and also today, chose to narrate actual situations of everyday life urgency with the "ordinary" language of their time.

In our opinion, it seems possible, through a careful re-reading of the Dutch architect's work regarding the construction of the *playgrounds*, to identify some key concepts for urban re-assembly through the "power" of human relationships. The first refers to the poetics of "as found" (Mumford 2000). Described and applied by Alison and Peter Smithson—other major players in post-war C.I.A.M.—the principle of "as found" aims to deal with the "as it is", the here and now, the real and the everyday, the evident and the tangible, the recent past, the present and the immediate future. In van Eyck's work, this theme is evident, but it is combined with the search for a design language for small neighbourhood spaces serving of the inhabitants. The Dutch architect's playgrounds are public places designed for shared everyday life and for inhabiting the present, but they are also evocative, a historical spaces designed to expand the imagination beyond time and space (Fig. 32.4).

The second is the "magical power" of in-between spaces (Postiglione 2015). Van Eyck underlines the strategic role of the small scale, of the voids between the built, of the margins, of all the ambivalent places, in any case on a human scale, questioning the idea of a city based on projects exclusively planned with top-down processes (Fig. 32.5).

On this topic, the in-between spaces, the marginal areas, or those without an apparent function have design relevance and represent one of fields of any credible urban redevelopment plan. In conclusion, functionless in-between spaces have design relevance and represent the focus of human-centred urban regeneration programmes

Fig. 32.5 Collectif Etc, *Place au Changement!*, Saint-Etienne, 2011 [Collectif Etc]

through the concept of public space as a "common good". It is necessary to intentionally head toward the "joint", toward the repair and, consequently, beyond the criticalities that arise from time to time. In our opinion, triggering unprecedented human relationships is one of the main elements for suburb metamorphosis, the "fuse" for building previously unknown public spaces, steadily, progressively.

References

Angi B (2018) Ricomposizioni architettoniche. Parallelismi e analogie. Lettera Ventidue Edizioni, Siracusa

Angi B, Soci A (2020) *Gioco di squadra. Tra centralismo e decentramento: l'empirismo eretico e la città.* In *EdA, Esempi di Architettura* vol 7(1). Roma, Aracne Editrice

Angi B, Botti M, Montuori M (2016) Urban Eutopia. Strategies for the combined redevelopment of social housing. In: Caverzan A, Lamperti Tornaghi M, Negro P (ed) SAFESUST workshop. A roadmap for the improvement of earthquake resistance and eco -efficiency of existing buildings and cities. Ispra Joint Research Centre, EU Publications Office, pp 193–201

Aureli PV (2013) Less is enough: on architecture and asceticism. Moscow, Strelka Press

Angi B (2014) *Manipulations anti-table rase. Stratégies architecturales adaptatives.* In Le Philothope, Réseau Scientifique Thématique Philosophie Architecture Urbain, n. 10. Clermont-Ferrand, Le Philothope

Bakema JB (1954) *Rapporti tra uomini e cose.* In: Rogers EN, Sert JL, Tyrwhitt J (eds) *Il Cuore della città.* Milano, Hoepli

Huber D (2015) Beyond relief. Artforum International, New York

Lefaivre L, Tzonis A (1999) Aldo van Eyck: humanist rebel. Rotterdam, 010 Publishers

Montuori M (2016) E pleribus unum. In: Angi B (ed) Eutopia urbana. La riqualificazione integrata dell'edilizia sociale. Eutopia Urbanscape, the combined redevelopment of social housing. Siracusa, Lettera Ventidue Edizioni

Mumford E (2000) In: The CIAM discourse on urbanism, 1928–1960. Cambridge, The MIT Press

Postiglione G (2015) *A+P Smithson. Una piccola antologia della critica.* Siracusa: Lettera Ventidue Edizioni

Soci A (2015) Architettura Esposizione Città. La storia delle grandi esposizioni nella costruzione della città. Libraccio Editore, Milano

Chapter 33
An Open Project for the Trembling Landscape of the Inner Areas: Fragile and Wounded

Angela D'Agostino and Giovangiuseppe Vannelli

Abstract The Italian landscape of the inner areas is going through a phase of change: the connoting evolutionary 'slow linear time' of these places has undergone a sudden acceleration corresponding to impending gentrification. The exponential time of abandonment threatens the integrity of this heritage because of the loss of people and values. The fragile landscape of inner areas is often a 'wounded landscape' too. In the case of Italian territory—which has always been significantly affected by seismic phenomena that have been a connotative evolutionary factor—there is an almost total overlap between the inner area's map and the seismic hazard map. Natural catastrophes mark a leap in the linear evolutionary process of these places; this temporal leap leads back to a new starting point where design actions on the landscape make sense only if in close connection with the principles and contents of the European Convention. The viewpoint inversion and the inclusive perspective at the basis of the convention lead to an approach that no longer considers only the historical-artistic values of small villages. It has become fundamental to promote interventions that consider different times and possibilities and are configured as open projects which develop a procedural logic and also look at intangible heritage: cultural, economic, social and productive. This contribution is part of PhD research investigating post-earthquake reconstruction conceived as a process linked to the different design instances related to the phases following the catastrophe. The complex condition of these fragile and wounded landscapes requires a reconstruction approach considered as the ultimate (possible) scenario of an open design process. Through case studies, the contribution investigates the role of landscape as an interpretative paradigm of the procedural nature of complex problems that must foresee multiple temporalities.

Keywords Inner areas · Reconstruction · Process

A. D'Agostino (✉) · G. Vannelli
Department of Architecture - University of Naples Federico II, Naples, Italy
e-mail: angdagos@unina.it

G. Vannelli
e-mail: giovangiuseppe.vannelli@unina.it

T. Matteini et al. (eds.), *Cultivating Continuity of the European Landscape*, Environmental History 15, https://doi.org/10.1007/978-3-031-25713-1_33

33.1 The Italian Landscape of Inner Areas: Visions for a Possible Future

The Italian landscape of inner areas is going through a phase of change: its connoting evolutionary 'slow linear time' has undergone a sudden acceleration corresponding to the impending gentrification that has exponentially sped up abandonment time. Hence, in recent decades, there has been rapid depopulation in a considerable portion of Italian territory. Abandonment threatens the future of heritage and landscape due to the loss of people and values and the interruption of caring for tangible and intangible goods. Therefore, inner area landscapes are in a fragile condition.

Architectural culture has been focusing its attention on this condition for several years, both pointing out the urgency and emphasising the values of inner area landscapes and small towns. The last two Venice Biennale of Architecture international exhibitions, in 2016, curated by Alejandro Aravena[1] and 2018, where Mario Cucinella[2] curated the Italian Pavilion exhibition, highlighted marginal territory conditions and perspectives. From both points of view, it has emerged that initiatives and projects oriented towards the care of places, cultures and communities based on an idea of belonging, involvement and sharing are proliferating against the abandonment phenomenon.

Given the evident fragility of these landscapes, one definitely looks in the direction pointed out by the European Landscape Convention: «If people are given an active role in decision-making on landscape, they are more likely to identify with the areas and towns where they spend their working and leisure time. If they have more influence on their surroundings, they will be able to reinforce local and regional identity and distinctiveness and this will bring rewards in terms of individual, social and cultural fulfilment. This in turn may help to promote the sustainable development of the area concerned, as the quality of landscape has an important bearing on the success of economic and social initiatives, whether public or private».[3] Simultaneously, we are looking at the past—to the communities that have inhabited these places—and the present—to those who have remained[4]—but also to the future—to those who may return.[5]

At the same time, policy in reference to the Italian inner landscape has put in place the Strategia Nazionale per le Aree Interne (SNAI)[6] which provides for the planning of interventions in marginal areas by identifying areas to fund to relaunch territories. The SNAI proposes a territorial model that aims to create a sort of shared self-sufficiency generated from territorial synergy between municipalities that are

[1] Cfr. Aravena (2016).

[2] Cfr. Cucinella (2018).

[3] Explanatory Report to the European Landscape Convention, Firenze, 20.10.2000. https://rm.coe.int/CoERMPublicCommonSearchServices/DisplayDCTMContent?documentId=09000016800cce47.

[4] Cfr. Teti (2017).

[5] Cfr. De Rossi (2018).

[6] Cfr. Barca et al. (2014).

called on to relinquish their local autonomy in the name of the common good. Thus, the role of communities, individuals and all other stakeholders who spontaneously or by programmatic choice are involved in this process is fundamental. In this case too, adherence to the principles of the Convention is worthy of note. «Landscape must become a mainstream political concern, since it plays an important role in the well-being of Europeans who are no longer prepared to tolerate the alteration of their surroundings by technical and economic developments in which they have had no say. Landscape is the concern of all and lends itself to democratic treatment, particularly at local and regional level».[7]

33.2 Fragile and Wounded Landscapes: A Time of Risk

The fragile landscapes of the inner areas are often wounded as well. In the Italian context of frequent earthquakes that represent a connotative evolutionary factor, there is a remarkable overlap between the maps of the inner areas and of seismic hazard (Fig. 33.1). In contrast to a slow landscape construction time, following a process of sedimentation of signs,[8] there are extraordinary events that stop time, determining opportunities for new beginnings and new processes.[9] Disasters cause an interruption in these place's evolutionary process, a temporal shift leading back to a degree 0, to a re-start.

Therefore, the inexorable long-term depopulation and the consequent abandonment of these internal and marginal fragile landscapes are often overwritten by catastrophe that brings complexity to the contextual realities from which a conscious reconstruction project must start.[10] Added to contextual vulnerability is a seismic hazard determining a considerably higher overall risk.[11]

This paper focuses on the reactive approach to disaster, understanding the event as an opportunity to rethink the damaged landscapes.[12] to rethink the damaged landscapes.

In addition to the risk outlined as the sum of seismic hazard and local vulnerability, the research identifies and focuses on a second risk, a consequence of the first, which we will call 'reconstruction risk', by which we refer to those choices that, in the interim, generate ballasts and scars that slow down or even prevent the reconstruction process.

---

[7] Explanatory Report to the European Landscape Convention, Firenze, 20.10.2000. https://rm.coe.int/CoERMPublicCommonSearchServices/DisplayDCTMContent?documentId=09000016800cce47.

[8] Cfr. Ferlenga (2014).

[9] Cfr. Bruni and D'Agostino (2004).

[10] Cfr. Caravaggi (2018).

[11] Cfr. Blaikie et al. (1994).

[12] Cfr. Charlesworth (2006).

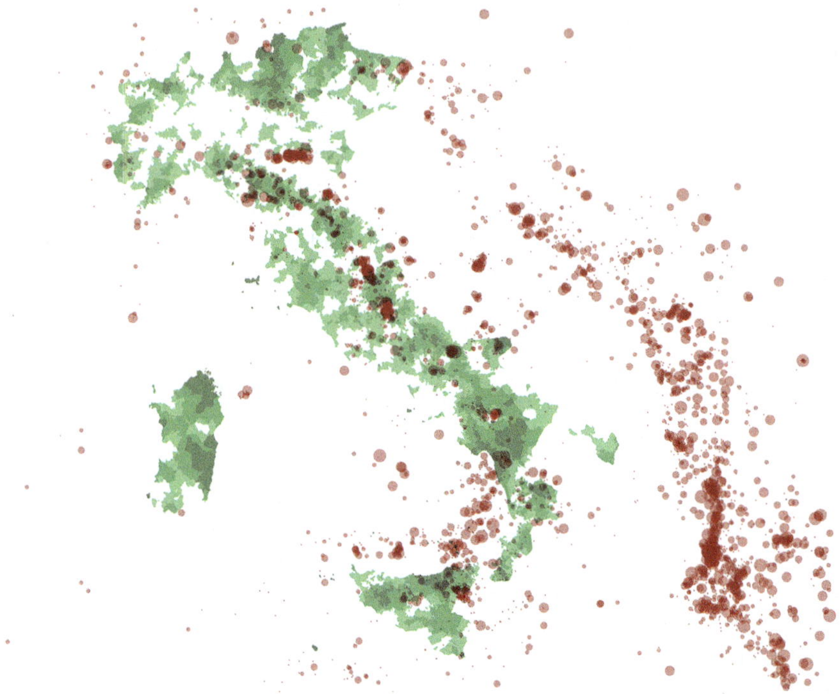

Fig. 33.1 Geolocalization of earthquakes over the last 40 years (in red) and identification of the Italian inner areas: 'intermediate', 'peripheral' and 'ultraperipheral' (in shades of green)

33.3 Time: A Design Paradigm Against Reconstruction Risk

The scientific literature generally identifies three main steps after a catastrophe: the first, relief, represented by the provision of temporary shelters; an intermediate phase, generally represented by the 'delivery' of temporary housing; and finally, permanent reconstruction.[13] Of these three phases describing—perhaps too rigidly—the post-catastrophe, the second one is the most problematic. This is because the relationship between the two fundamental parameters of space and time is merely consequential, extremely simplified. Time does not really become a design paradigm, but only a management parameter (Fig. 33.2).

While seismic hazard threatens the vulnerable architecture of minor centres, reconstruction risk threatens the landscapes of internal areas. In order to avoid the thorny issue of reconstruction risk and to take up the (though tragic) opportunity represented by a disaster, a transdisciplinary and transcalar, projective and metabolic approach to post-disaster design is needed: a four-dimensional design[14] that uses

[13] Cfr. Lizarralde et al. (2018).

[14] Bishop and Williams (2012).

Fig. 33.2 Falerone (FM), a public space of a fragile and wounded minor centre

time as an element of the project. Setting time as a paradigm for designing in regions affected by earthquakes means not precluding possible alternative scenarios that can only be identified after the disaster.

33.4 A Menaced Landscape: The Permanence of Temporary

In the inner areas, countless traces bear witness to the ambiguity of the temporariness of the phase preceding the reconstruction. The houses that have dotted Irpinia and Basilicata since the 1980 earthquake,[15] as well as those that have altered central Italy's landscapes in regions such as Marche, Umbria and Abruzzo—repeatedly affected by catastrophic seismic events in recent decades—should have been temporary.[16]

Looking at the damaged landscapes altered by temporary settlements in the Italian scenario, the need to rethink interventions following the first aid phase seems evident.

After a disaster, there is inescapably a need to provide a solution based on the «notions of speed, performance and numbers»[17] where the time and cost-saving advantage of using prefabrication systems appears to be undeniable.[18] However,

[15] Cfr. Mazzoleni and Sepe (2005).

[16] Cfr. Anzalone (2008).

[17] Boano and Hunter (2012).

[18] Cfr. Bologna and Terpolilli (2005).

Fig. 33.3 Pizzoli (AQ), the box effect at the scale of the landscape

architecture in post-disaster practice cannot continue to be «a mere deliverable service-oriented producing autonomous subject».[19]

The objectual conception of post-emergency architecture has to change to a vision interpreting architecture in relation to the landscape, especially in the specific context of the Italian inner areas. «Writing off the spatial element as something separate and unrelated severely limits the transformative potential of reconstruction and rehabilitation in the post-disaster context. Architecture [must] distance itself from the obsession of the object and recognise its dependency on outside forces and influences».[20] Due to the impossibility of precisely foreseeing emergency settlement outcomes and destinies—with respect to complex social, economic and urban dynamics—an architectural project cannot rely on closed design solutions. The technical solutions used in this transition phase must guarantee reversibility and disassembly in order not to leave any indelible traces. The project must become a process; it must be strategic, adaptive and co-evolutive: «shelter must be considered as a process, not as an object» (Fig. 33.3).[21]

[19] Boano C., Hunter W., *op.cit.*, p. 11.

[20] *Ivi*, p. 10.

[21] Davis (1978).

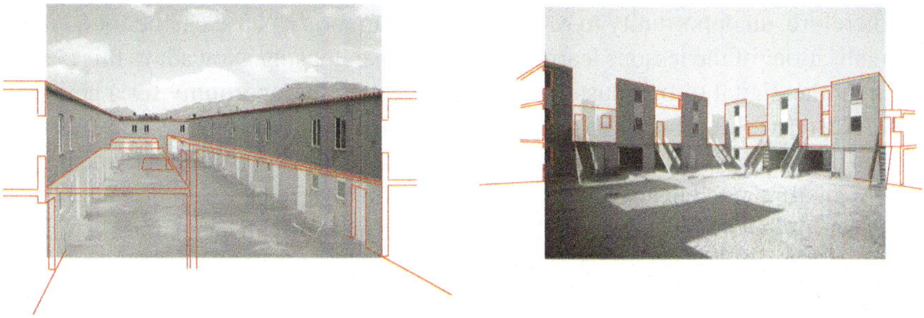

Fig. 33.4 Facatativà (left) and Quinta Monroy (right), incremental strategies

33.5 Learning from the Informal: Open Project and Multiple Temporalities

In the light of the above, a viewpoint shift, a paradigm shift, is needed. The proposed approach aims to develop an open project that accepts the unexpected, modification and change. Once the need for a process-oriented project nature has been identified, there is a useful lesson to be learned from informal housing, an opportunity that should not be intended as a transposition of logic based on a deregulated approach to architecture but rather as a reinterpretation of values and advantages as highlighted in research investigating the topic.

Through an investigation comparing informal housing and formal project case studies, Lizarralde clearly shows patterns that demonstrate possible elements to be learnt from informal architecture, such as the flexible use of enclosed and open spaces, the possibility of integrating the original core with later additions or modifications, great variety between housing units, the intensive use of recycled materials and components, the variety of functions and uses, the hierarchy of streets and paths and the variety of open spaces, plot sizes and forms.[22]

Two South American projects are proposed as examples because through different design actions, they 'embody' the same incremental strategy and make the abovementioned patterns explicit in different ways. In 2005 a subsidised housing project of 1263 residential units was carried out in Facatativà (Colombia) in an area at risk. The residential units, built with a core unit, are arranged around an elongated central courtyard which can be used for potential increases built by the users.[23] Another example is the well-known Aravena project, Elemental. Here, the incremental strategy is proposed in four buildings placed around a collective central courtyard by alternating a core unit with a void module that can be used for extensions (Fig. 33.4).[24]

[22] Cfr. Lizarralde (2010).

[23] Cfr. Lizarralde (2011).

[24] Cfr. Aravena (2007).

Therefore, an opportunity to re-evaluate the temporary phase can be found in the 'formalisation' of the lessons learned from informal housing. Nowadays this is facilitated by the digital turn[25]: customization[26] becomes an opportunity for a new form of participation. Moreover, the components and construction systems selected to be used in temporary housing projects from the wide range of the latest technical innovations are fundamental as they should permit further incremental interventions and/or dismantling of the temporary housing units and also provide plausible alternatives for reuse, recycling, upgrades and upcycles.

The proposed approach is an opportunity to avoid the alienating homologation of temporary settlements caused by the «box effect»[27] and favours a transition from the concept of house to that of home «to unite the physical components with the social, cultural and emotive ones».[28]

This approach, where the architect is a mediator,[29] proposes providing the tools and not the objects for projects englobing multiple temporalities in which reconstruction is the ultimate (possible) scenario.

References

Bishop P, Williams L (2012) The temporary city, Routledge, Abington, pp 182

Boano C (2009) Housing anxiety, paradoxical spaces and multiple geographies of post tsunami housing intervention in Sri Lanka. Disasters 34(3)

Boano C, Hunter W (2012) Architecture at risk (?) The ambivalent nature of post-disaster practice. Architectoni.ca 1:1

Boano C, Zetter R (2009) Gendering space for forcibly displaced women and children: concepts, policies and guidelines. In: Martin SF, Tirman J (eds) Women, migration, and conflict. Springer, Londra, p 210

Anzalone M (2008) L'urbanistica dell'emergenza. Progettare la flessibilità degli spazi urbani, Alinea Editrice, Firenze

Aravena A (2007) Alejandro Aravena. Progettare e costruire. Electa, Milano

Aravena A (2016) Reporting from the front: 15th international architecture exhibition, Marsilio, Padova

Barca F, Casavola P, Lucatelli S (2014) Strategia Nazionale per le Aree Interne: definizione, obiettivi, strumenti e governance. Collana Materiali Uval, Roma

Blaikie P, Cannon T, Davis I, Wisner B (1994) At risk: natura! hazards, peop!e's vu!nerability, and disasters. Routledge, London and New York

Bologna R, Terpolilli C (2005) (eds) Emergenza del progetto. Progetto dell'emergenza. Architetture con-temporaneità, Federico Motta, Milano

Bruni F, D'Agostino A (2004) Le occasioni della città. Al di fuori del piano, dentro una politica di piano. Eventi straordinari e trasformazioni urbane. Edizioni Scientifiche Italiane, Napoli

Caravaggi L (2018) Ricostruzione di territory. In: Abitare la terra, Terremoti 48

Carpo M (2012) In: The digital turn in architecture 1992–2012. Wiley, New York

[25] Cfr. Carpo (2012).

[26] Cfr. Naboni and Paoletti (2014).

[27] Lizarralde G. (2010), *op. cit.*, p. 31.

[28] Boano and Zetter (2009).

[29] Boano (2009).

Charlesworth E (2006) In: Architects without frontiers. War, reconstruction and design responsibility. Architectural Press, Oxford

Cucinella M (2018) Arcipelago Italia. Progetti per il futuro dei territori interni del paese. Catalogo della mostra Padiglione Italia alla Biennale di Architettura 2018, Quodlibet, Macerata

Davis I (1978) Shelter after disaster. Oxford Polytechnic Press, Headington, p 33

De Rossi A (2018) (ed) Riabitare l'Italia. Le aree interne tra abbandoni e riconquiste, Donzelli Editore, Roma

Ferlenga A (2014) 'Segni'. In: Capuano A (ed) Paesaggi di rovine. Paesaggi rovinati, Quodlibet, Macerata

Lizarralde G (2010) Post-disaster low-cost housing solutions. Learning from the poor. In: Lizarralde G, Johnson C, Davidson C (eds) Rebuilding after disasters. From emergency to sustainability, Spon Press, London and New York

Lizarralde G (2011) Stakeholder participation and incremental housing in subsidized housing projects in Colombia and South Africa. Habitat Int 35

Lizarralde G, Johnson C, Davidson C (2018) (eds) Rebuilding after disasters. From emergency to sustainability, Spon Press, London and New York

Mazzoleni D, Sepe M (2005) (eds) Rischio sismico, paesaggio, architettura: l'Irpinia, contributi per un progetto, Centro Regionale di Competenza Analisi e Monitoraggio del Rischio Ambientale, Napoli

Naboni R, Paoletti I (2014) Advanced customization in architectural design and construction, Springer, Londra

Teti V (2017) Quel che resta. L'Italia dei paesi, tra abbandoni e ritorni. Donzelli Editore, Roma

Chapter 34
Abandoned Army Barracks in Friuli Venezia Giulia (Italy) as New Time-Shaped Community Landscape Potential

Luca Maria Francesco Fabris and Federico Camerin

Abstract The dismantling of Italian Army Barracks in northeast Italy in the last 20 years has left a series of large abandoned areas where nature has freely operated as a designer, creating new transitional landscapes in places devoted during the Cold War to protecting the Italian border. We could call them 'auto-regenerative' landscapes. This paper aims to briefly illustrate the making of this kind of unconventional landscape in the Friuli Venezia Giulia Region over the second half of the XX century through the different scales of intervention that have activated multiple spatial relations over time. After the fall of the Berlin Wall in 1989, this exceptional infrastructure-based landscape, adapted to the needs of the Italian Army, has been reconfigured and gradually abandoned, becoming almost completely unused from the year 2000. The authors propose a reading of these derelict spaces and formulate possible new scenarios of sustainable regeneration and inclusive reintegration that aspire to promote a kind of open project to return them as social pay-back to Friuli Venezia Giulia communities after more than 100 years of militarization.

Keywords Transition · Post-military · Open project

L. M. F. Fabris (✉)
DASTU, Politecnico di Milano, Milano, Italy
e-mail: lucamariafrancesco.fabris@polimi.it

Beijing University of Civil Engineering and Architecture, Beijing, China

F. Camerin
Departamento de Urbanística y Ordenación del Territorio - Grupo de Investigación en Arquitectura, Urbanismo y Sostenibilidad (GIAU+S), Universidad de Valladolid - Universidad Politécnica de Madrid, Madrid, Spain
e-mail: federico.camerin@uva.es

T. Matteini et al. (eds.), *Cultivating Continuity of the European Landscape*, Environmental History 15, https://doi.org/10.1007/978-3-031-25713-1_34

34.1 Cold War Military Landscape Expansion in Friuli Venezia Giulia

Militarisation has historically shaped the Italian Friuli Venezia Giulia Region territory. During the first half of the last century, this was boosted throughout WWI (Pascoli and Vazzaz 2005; Bernasconi and Muran 2009) and reached a peak after WWII, when the Region was considered as a 'buffer' zone by NATO to deal with a possible conflict with the Soviet bloc (Baccichet 2015). This led to the creation of a myriad of various military infrastructures where, for decades, a large part of the Italian Army was concentrated, inevitably creating a specific kind of military-landscaped territory, entirely off-limits to civil society. The so-called 'Cold War Military Landscape' was created by various infrastructures, such as airports, barracks, armouries, deposits, and training fields, plus a quantity of military easements. This army presence, which increased in numbers for more than four decades, together with regional economic development based more on the agricultural sector than industry acted as a sort of 'environmental restriction' preserving large portions from the massive post-WWII urbanisation that occurred over most of the Italian territory. In 2006, the Friuli Venezia Giulia Region had 119.205 km^2 of military surface, representing 1.67% of its total territorial surface area (Pinotti 2006).

After the fall of the Berlin Wall in 1989, this exceptional infrastructure-based landscape, adapted to Italian Army needs, had reconfigured and been gradually abandoned, becoming almost completely unused since the year 2000. The amount of area released by the military in Friuli Venezia Giulia is up to 9,811,245 m^2, of which 3,716,371 m^2 are barracks (38%) and so the biggest part of this derelict military territory. Nevertheless, installations still belonging to the Ministry of Defence were not taken into account even though they are now underused, partially abandoned, or even entirely abandoned and in ruins. Among these, the ruins of the partially abandoned 23.5 hectare 'Trieste' barracks in Casarsa della Delizia (Pordenone) stand out as one of the biggest barracks ever built in Italy (Fabris and Camerin 2017) (Fig. 34.1).

34.2 'Deactivated' Army Barracks: How to Shift Them into 'Active' Infrastructures?

The 'military landscape' issue has been tackled from various points of view in the frame of the military–civil society relationship (well analysed by Woodward 2014). Here, the authors focus on the fact that the dismantling of Italian Army Barracks and other infrastructures in the Northeast of Italy in the last 20 years has left a series of large abandoned areas where nature has freely operated as a designer, creating new transitional landscapes in places formerly devoted to protecting the Italian border during the Cold War. We could call them 'auto-regenerative' landscapes as they are still intact and are characterised by a strong historical/anthropic

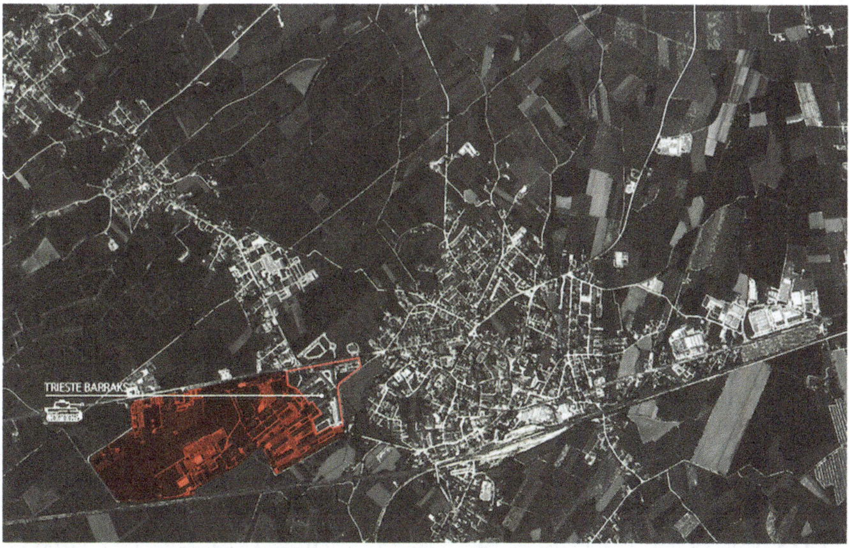

Fig. 34.1 Trieste Barracks area in Casarsa della Delizia. The town's historic urban centre has the same surface area as the former military compound. Table elaborated during the 2015 workshop

identity as well as an important process of botanical and faunal restoration. Moreover, their intrinsic military characteristics offer us a reading of these derelict spaces to formulate possible new scenarios of sustainable regeneration and inclusive reintegration that aspire to promote a kind of open project to return them to the Friuli Venezia Giulia communities as social pay-back after more than 100 years of militarisation (Fig. 34.2).

The characteristics and possibilities for barrack reuse are the following: large-scale surface, usually around 100,000 m^2 or more, provided with both built and unbuilt spaces organised on a military scheme-basis; many buildings with valuable architectural style; strategic localisation at the edge of existing urban landscape; the status of public property; considerable public interest for its individual and collective memory aspects; aesthetic value as natural areas. All these features could be the foundations for defining a particular kind of unique military memory landscape under the European Landscape Convention. However, no landscape policy has taken these 'auto-regenerative' landscapes into consideration in the last 20 years, not even as a community value. Therefore, we propose an original approach to Army barracks and their regeneration on the grounds of more than 5 years of experience in developing new proposals for the abandoned area of the Trieste barracks in Casarsa della Delizia.

Fig. 34.2 View of the unused spaces in the former Trieste Barracks (photo by Luca MF Fabris 2015)

34.3 The Trieste Barracks in Casarsa Della Delizia as a New Time-Shaped Community Landscape Sample

Most of the barracks located in Friuli Venezia Giulia's municipalities housed hundreds of soldiers, usually followed by their families, and were thus an important element in local communities. In the 1970s, the Trieste barracks housed over 4000 soldiers plus their families and so remarkably influenced the social, economic, and urban aspects of the small municipality of Casarsa. However, their gradual abandonment resulted in the loss of vitality and local administrations have produced no concrete ideas or policies to re-integrate these Cold War military landscapes into the urban fabric.

Given the challenge of long-standing economic crises which have meant significant budgetary deficits, scarcity of public resources, and difficulties in the real estate market in the last decade, the authors propose a new approach based on landscape architecture-project solutions. The 2015 international design workshop 'Restarting from Bacchus and Dionysus, with the help of Oyamatsumi' (Fabris and Camerin 2017), plus further elaborations in the ambit of the Master of Science in Architecture at the AUIC School of the Politecnico di Milano (Italy), promoted and proposed project ideas to create a new infrastructural landscape based on three key projects missing in the Friuli Region. They are a Cold War museum, a wine museum, and a pilot vineyard which would recreate the typical Friulian country landscape and promote organic farming and local social cooperative winery history. The goal has been to meld citizen's and other stakeholder's desires to integrate sustainability into a public park as a new active infrastructure. Key elements were maintaining places

Fig. 34.3 Adaptive reuse of the open spaces present in the former Trieste Barracks according to citizen's requests. Table elaborated during the 2015 workshop

and their historic value as memory; wine tasting to attract tourism; new cultural events based on the preeminent figure of Casarsa's poet, Pier Paolo Pasolini. Such project-based principles can be also exported to other former military barracks in Friuli to feed coordinated actions for local community transformation (Fig. 34.3).

34.4 The Post-Covid-19 Landscape Twenty Years After the Landscape European Convention. Inertia and Opportunities for the Army Barracks

The Covid-19 pandemic will have wide-ranging effects on the practice of public landscape. Whilst most of the state-provided proposals for the redevelopment of abandoned Army barracks have relied on real estate solutions to extract the maximum financial benefit (Gastaldi and Camerin 2019), nowadays public-owned assets should guarantee the 'right to the city' (Lefebvre 1968).

Our 5-year experience of the Trieste barracks has shown us that the creation of an infrastructural landscape would guarantee the care of abandoned places, the configuration of new sustainable realities, and the enhancement of local resources and production systems in a progressive adaptation to continuously changing conditions. The most recent example of such an (unexpected) changing situation is the challenging and long-lasting repercussions across the society due to the Covid-19 pandemic, many of which will surely influence the way we approach the design of landscape. In this framework, the Army barracks, as public properties belonging to the community, should endorse the creation of an articulated system capable of

Fig. 34.4 Productive landscape as connection between Casarsa's urban area and the abandoned former military area. Table elaborated during the 2015 workshop

being inclusive. On the one hand, people are likely to become more involved in their communities and the decision-making process of future interventions. On the other hand, the post-Covid-19 landscapes should guarantee open spaces for physical and mental health and environmental benefits to respond to citizen's needs.

The new Trieste barracks scenario as a public open space would create an attractive memorial landscape upgrading resident's quality of life whilst promoting slow tourism along with local products (wine). The proposals define a specific way of making landscape, still valid for building a post-pandemic landscape, based on constant evaluation of its value for the community. This approach implies a complex understanding of places, resources, spaces, natural conditions, built construction conditions, and physical and cultural relationships in order to harmonise multidisciplinary skills to build a shared vision of the future. Architecture, engineering, and urban planning, as well as other disciplines, contributed to the definition of this specific open auto-regenerative landscape (Fig. 34.4).

34.5 Conclusions

Working on the redevelopment of 'auto-regenerative' landscapes such as the Army barracks implies recognition of the values of a specific community. The size and historical characteristics of former military sites could trigger public environmental regeneration and landscape design integrated with local sustainable development. The experiment of elaborating scenarios for the Trieste barracks provided original landscape-based solutions not related to a mere real estate approach (Camerin and

Fabris 2020). Results show that the regeneration of former Army barracks combines different scales: on a local scale by enhancing socio-economic features; on a territorial scale as it can be replicated by following the same patterns to find new solutions for the well-being of other communities. Also, the transition from military to civil society could solve the new post-Covid-19 pandemic urban and landscape design issues as 'common' public-owned property should promote. Open projects for open spaces with the involvement of people would definitely foster the application of the Landscape European Convention (Fig. 34.5).

Fig. 34.5 Landscape Masterplan table from the Master in Architecture Thesis of Edoardo Del Conte and Erica Nonis, supervisor Luca Maria Francesco Fabris, Politecnico di Milano, AUIC School, 2018

Note

This paper was set up and carried out in collaboration with the two authors. In this context, the paragraphs 3 and 4 are attributed to Luca Maria Francesco Fabris, and the paragraphs 1 and 2 to Federico Camerin. The conclusions are common.

Federico Camerin has participated as co-author within the research project "La Regeneración Urbana como una nueva versión de los Programas de Renovación Urbana. Logros y fracasos". This project is co-funded by the Spanish Ministry of Universities in the framework of the Recovery, Transformation and Resilience Plan, by the European Union – NextGenerationEU and by the Universidad de Valladolid. This work was carried out at the Grupo de Investigación en Arquitectura, Urbanismo y Sostenibilidad (GIAU+S).

References

Baccichet M (ed) (2015) Fortezza FGV. Dalla guerra fredda alle aree militari dismesse. Monfalcone (Go), Edicom Edizioni

Bernasconi A, Muran G (2009) Il testimone di cemento. Le fortificazioni del «Vallo Alpino Littorio» in Cadore Carnia e Tarvisio, Udine, La Nuova Base

Camerin F, Fabris LMF (2020) Conflict landscape: memory and future. The Casarsa della Delizia Case. In: Dalzero S et al. (ed) Boundary landscapes, Roma, Tab Edizioni, pp 343–351

Fabris LMF, Camerin F (eds) (2017) Ripartire da Bacco e Dioniso con l'aiuto di Oyamatsumi. Proposte sostenibili per la riqualificazione dell'ex caserma Trieste di Casarsa/Restarting from Bacchus and Dionysus, with the help of Oyamatsumi. Sustainable proposals for the renewal of former Trieste Barracks in Casarsa. Fossalta di Portogruaro, Lions Club Sesto al Reghena in Sylvis

Gastaldi F, Camerin F (2019) Aree militari dismesse e rigenerazione urbana. Innovazioni urbane, potenzialità di valorizzazione del territorio, inerzialità legislative e di processo. Siracusa, Lettera Ventidue

Lefebvre H (1968) Le droit à la ville. Éditions Anthropos, Paris

Pascoli M, Vazzaz A (2005) I forti e il sistema difensivo del Friuli: itinerari sconosciuti nel più grande campo di battaglia italiano della Grande Guerra. Udine, Gaspari

Pinotti R (2006) Indagine conoscitiva. Rome, IV Commissione Difesa. http://www.camera.it/_dati/leg15/lavori/stencomm/04/indag/militari/2006/1025/s000r.htm

Woodward R (2014) Military landscapes: agendas and approaches for future research. Prog Hum Geogr 38(1):40–61

Luca Maria Francesco Fabris journalist and architect, obtained a PhD in Architectural and Environmental Technology, and a Master in Urban Planning and Environment, is 'Haiju' Expert at the Beijing University of Civil Engineering and Architecture (China) and professor in Architectural Technology at the Department of Architecture and Urban Studies of the Politecnico di Milano (Italy) where he's teaching since 1997 at the AUIC School's MS in Architecture and at the MS in Landscape Architecture—Land Landscape Heritage. He's writing for the international architectural review The Plan and ABITARE, and is Scientific Director of the Italian technical architectural magazine 'YouBuild' and of the 'Environscapes' book series published by Maggioli Editore (since 2008) and has been Past-Coordinator of the design section of the Italian architectural magazine Costruire (Abitare-Segesta/RCS Media Group; 2004-2011). Visiting professor at European, American and Asian Universities, he focuses on researches related to contemporary built environment, sustainability and landscape. Editor and scientific referee of various university publications, he has written several books and essays about the above-mentioned subjects and contributes to some specialized reviews.

Federico Camerin city planner, obtained in 2014 an Inter-University-Programme Graduate Degree in "City and Environment: planning and policies" which included a European postgraduate degree in "Planning and Policies for Cities, Environment and Landscape" at Università IUAV di Venezia (Italy). He has been awarded in 2014-15 and 2016-17 with two "Urban Planning" fellow research grants at the Department of Design and Planning in Complex Environments of the Università IUAV di Venezia. In June 2020, he has been awarded with the degree of Doctor and Philosophiae doctor in the frame of European Programme "UrbanHist" (2017-2020) between the Instituto Universitario de Urbanística of the University UVA of Valladolid (Spain) and the Fakultät Architektur und Urbanistik of the Bauhaus-Universität Weimar (Germany). Camerin's research interest focuses on the understanding of the European city-making process starting from the late XIX century onwards and the impacts on landscape of dismissal-regeneration of military assets, great events and urban transformation, and the role of star-architects in the built environment management. He lectured in European countries (Germany, Italy, Slovak Republic, Czech Republic, and Spain) and attended more than 20 national and international conferences. He also published several papers among international and national journals, conference proceedings, and reviews of essays in Italian, English, Spanish, and Catalan.

Chapter 35
Rhythm and Continuity in Interventions to Reactivate Vacant Urban Spaces. Recognizing the Stratification of Places to Foster an Autonomous Process of Integration Between the Parts

Chiara Caravello and Elisa Baldin

Abstract The combination of architectural intervention for the recovery of disused post-industrial complexes—necessary to respond to urgent and necessarily impatient social demands—with work on the landscape—accustomed to slowness and natural rhythms—leads us to consider the aleatory, the sense of uncertainty, which demands respect for places and materials as well as for time, at the basis of a project in constant evolution. Landscape, as an 'upstream condition' of social and ecological actions, requires recognition and intervention strategies with very high standards of flexibility, dimensioned with relatively low financial resources and capable of meeting an increasingly wide range of needs. The authors examine two post-industrial landscapes as logical links of places and signifiers, ongoing processes of stratification in continuous evolution, where the recovery project infiltrates itself in a space of experimentation in the form of versatile strategies for a gradual development that can be sustainable in the long term.

Keywords Landscape stratification · Infiltration strategy · Open project

---

C. Caravello (✉) · E. Baldin
Université de Liège – Lab VTP, Liège, Belgium
e-mail: chiara.caravello@uliege.be

E. Baldin
e-mail: elisa.baldin@uliege.be

C. Caravello
Politecnico di Milano, DAStU, Milano, Italy

T. Matteini et al. (eds.), *Cultivating Continuity of the European Landscape*,
Environmental History 15, https://doi.org/10.1007/978-3-031-25713-1_35

35.1 Introduction

The term landscape designates "an area, as perceived by people, whose character is the result of the action and interaction of natural and/or human factors" (ELC 2000). The materials and dispositions of urban voids communicate instability, absence of continuity. Answers must be provided to a society that still has the need—and the right—to love its places. How to cure the landscape of a shipwrecked generation is looking for itself in wastage, with the unfair task of rising again from the rubble of an obsolete system? "How to (re)insert these forbidden and contaminated islands into the space and time of urban life?" Occhiuto (2012) The physical and cultural limits that confine these lands need to be overcome, opening up new opportunities to involve and integrate complementary parts. The phase between decommissioning and reuse proves to be a precious moment for observing the specific characteristics of a site, the local mutations, the perception of different actors, and for experimenting with temporary uses as re-qualification vectors. The site as a relational construct (Burns and Kahn 2005) acquires meaning through interaction with its context and the continuous relationship between reality, its perception, and its representation. Thus, observing ongoing processes emerges as an important basis of design approaches, where interventions are accompanied over time to enhance what exists: the "déjà-là".

35.2 Discussion Notes

The conservative landscape approach

In a context where man's unconscious actions have been a (pro)motor of environmental catastrophe, leading to reduced flora and fauna and precarious conditions of survival for the human species itself, the need to bring abandoned urban spaces back to a state of utility according to a contemporary model responds once again to an anthropocentric system. In general, the adoption of re-functionalization programs imposed from above excludes any consideration of natural processes underway and the local character determined layer by layer over time, which, instead, should be recognized as essential components of landscape which is formed "not only on the line of topographical or political decisions but through the organization of people on-site and the development of spaces serving the community" (Jackson 1984). In this sense, the priorities assigned to the built and natural environment as implemented by the conservative approach might be nothing more than a deceptive classification of a system whose attitude to observation and ability to acknowledge the other is limited.

If industrial activity represents a reversal of the pre-existing agricultural landscape, abandonment is not immutable but, on the contrary, constitutes a second revolution. Regeneration, the property of living beings that allows them to continuously self-repair, occurs in abandoned sites through spontaneous re-appropriation processes. From ruderal vegetation to wooded *milieux*, the time of abandonment expresses new mutations of the site through the living system. This spontaneous transformative process can suggest new value in brownfields: not only as a refuge for rare species activating ecosystems but also as a place of rediscovery and informal appropriation by inhabitants, local associations, and passionate explorers. Just as pioneer species enliven the bare earth, these new experiences of brownfields let other qualities emerge, beyond mere land value.

Supporting these local processes through punctual interventions during the long time of a redevelopment project gives voice to local resources, defines a site-specific strategy, and maintains flexibility between the transformation phases according to a logic of "intermediate nature" (Desvigne 2011).

The landscape approach encourages the acknowledgment of value before engaging in change, freeing the object from judgment that condemns it while refusing to newly frame it in models that reduce it to a parody of itself, a Quasimodo crowned at the "Fête des Fous".

35.3 Research Notes

The cases of the Haut Fourneau B à Ougrée in Wallonia-Belgium and the P.D. Rasspe Industry in Solingen, North Rhine-Westphalia, Germany.

The Ougrée blast furnace (HFB) site is a case study of a post-industrial landscape in the territory of Liège (VTP lab, and doctoral thesis 2017–2020, co-direction of the Faculty of Architecture, ULiège and LACTH, ENSAPL). Active until 2011, the site is the result of long-term stratification of industrial activities and is interesting today as the last vestige of hot steel production in the province of Liège. In the absence of a specific project, the study carried out observation activities and interaction with local actors who experience the site in everyday life to identify specific potential re-signification value.

First of all, Urbex explorations allowed us to get to know the site as a complex set of singular elements and environments, perceived as a place of interest by a community of explorer–photographers at both a local and cross-border level.

In addition, collaboration with the ATD Quarte Monde association allowed us to experience the reality of Ougrée bas—a neighborhood that physically and symbolically lives in the shadow of the disused blast furnace. In this degraded context, spatial appropriation forms express the need for dialogue between different social groups and for founding new meeting places based on existing elements such as the wall of the ancient blast furnace.

In this case, the lack of a project means new site values need to be traced starting from the site itself. The informal uses and relationships of the inhabitants suggest characteristics to be preserved and integrated into an open redevelopment process. Starting with the identification of these cornerstones (selection) and defining the *genius loci* as a trajectory, the project can be developed over time by defining and implementing intervention priorities (program) while supporting temporary activities according to a logic of continuous implementation (implementation) to keep attention focused on the site and develop its increasingly identifiable reuse potential.

The P. D. Rasspe industry was decommissioned in 2009. In 2014, two buildings belonging to the plant were placed under protection: the deposit (1902) and the administrative building (1922). This case study (Master Thesis April 2019, Politecnico di Milano, co-direction Lab VTP ULiège, RWTH Aachen) combines two themes: industrial heritage protection and the reactivation of landscape as an autonomous system, each developed independently. In this sense, preserving architectural assets offers an opportunity to re-see the landscape, a pretext to reconstruct the meaning of the place to reactivate its evolving relationship with the context.

Adopting a conservative approach, the architectural heritage was recognized as a priority. Firstly, a remediation and consolidation operation was carried out inserting flexible use to guarantee usage, and therefore constant maintenance, of the protected asset. In the case of the P.D. Rasspe warehouse and administrative building which fronts a public road, a social service function is foreseen to launch the reopening of the complex.

Later, the study of the landscape became a tool for in-depth research by putting the built material in relation to the physical and cultural characteristics of the place. By investigating the meaning and the transformation processes of the existing geography, elements marking the environmental characteristics of the site were enhanced to underline the architectural composition, relating interior uses to exterior features and landscape settings. Transitional green solutions are envisaged for the interstitial spaces of the former industrial plant to promote usage of the site's open spaces by making them available while engaging a dialogue for different uses over time.

The proposed design strategy thus questions the strictly conservative approach that was previously adopted by providing a gradual sequence of reorientation design actions in a dynamic design protocol that considers landscape and the environment fundamental to fostering change over time toward new forms of balance. The strategy has been developed in stages with increasing economic capacity thanks to targeted investments distributed over a sequence of specific iterative actions, which allows for continuous monitoring and adaptation. Programs and intentions that had initially excluded reactivating living materials have now been brought back in new qualitative hypotheses informed by landscape needs.

Improving the initial objectives, which were limited to the conservation and functional conversion of the architectural heritage, disclosed a new way of thinking of and designing the space. Everything previously considered marginal compared to the built area is now seen as a bearer of landscape characteristics and plays a crucial role as an intersectional space between the different milieus coexisting at the plant and context levels. Thus, the recovery proposition focuses on implementing a gradual global process of rehabilitating the post-industrial landscape by grafting a pilot project articulated in a series of non-characterizing and non-limiting inputs on to the conservative architectural design action to integrate and re-direct it. Taking into account the irregularity and indefinite duration of a recovery and context reintegration project for an entire industrial complex, action is taken in a time of experimentation, based on the suggestion of conservative though temporary hypotheses are offering recovery stimuli that allow for settling time. Accompanying the conservative project with actions that involve grafting explicitly provisional landscape immediately betters site quality by recovering natural living materials, integrating the knowledge of landscape dynamics as re-forming principles for the architectural project.

35.4 Conclusion

The search for a strategy to reactivate abandoned industrial complexes where nature is already at work unambiguously shows patience and attention to be lost values that have progressively vanished before the full-speed monetization approach. Reacting to the ineffable manifestation of our decaying economic system, our taking responsibility puts us before not only architectural and environmental patrimony but also cultural heritage needing prompt recovery because, even more than goods, human values are about to succumb, overcome by those of silent nature, "whose forces, however, are—and always remain—more powerful than all human intelligence" (Riegl 1903). Only by establishing a finally mutual relationship based on respect and consideration of the other will it be possible to restore our haven on Earth, where we will be able to defend our values by having questioned them and have found our own identity again.

References

Burns CJ, Kahn A (2005) Site matters: design concepts, histories and strategies. Routledge, New York

Council of Europe (2000) European Landscape Convention. ETS No. 176, Florence

Desvigne M (2011) The landscape as precondition. Lotus Int 150:20–27

Jackson JB (1984) Discovering vernacular landscape. Yale University Press, New Haven and London

Occhiuto R (2012) La frammentarietà come risorsa del sistema d'open space in Wallonie (BE). Faculté Architecture Université de Liège, Belgium

Riegl A (1903) Der moderne Denkmalkultus. Sein Wesen und seine Entstehung. W. Braumüller, Wien

Chapter 36
What Future for Italian Historical Urban Landscape—A Veneto Case Study

Daniela Ruggeri

Italy has a unique historical urban landscape, tangible testimony of the many cultures that have built the country.

New economic and demographic dynamics are changing historical Italian urban landscape, such as buildings transforming to commercial use only, depopulation, and abandonment. However, even historical centres can offer a sustainable and qualitative standard of life: the beauty of the historic landscape, the organization of spaces "on a human scale" which allows travel without a car, and the proximity of facilities and houses.

The foothills landscape of Veneto, the Pedemontana area, is studded with a series of historical centres, "polar" settlements within the prevalent urban sprawl pattern. Part of this territory, the Prosecco hills, recently achieved a landscape identity recognized by UNESCO.

This paper aims to identify specific critical issues as a basis for new strategies to enhance the landscape of one of these centres, Vittorio Veneto. This town is placed between the hills and Cadore mountains, at the edge of Prosecco area, but despite its important artistic and urban heritage, it is not included in the UNESCO core zone.

The growth of Vittorio Veneto, formed of two ancient urban entities, Serravalle and Cenda (united in 1866), is closely intertwined with the Meschio river water system, which allowed the first industrial settlements; the new town was an example of nineteenth-century urban planning, with a tree-lined avenue (now Viale della Vittoria) and the symmetry of the Town Hall Square in axis with the Public Gardens. However, this intervention did not erase the individuality of the two ancient centres. In addition, it is possible to find a certain duality of land use within the territory, both industrial and agricultural. Over time, territorial strategies shaped the landscape favouring the industrial sector. The Meschio river became infrastructure and at the same time the structural axis of the city. Nowadays, the city cores are quite well

D. Ruggeri (✉)
Università Iuav di Venezia, Venice, Italy
e-mail: daniela.ruggeri@iuav.it

T. Matteini et al. (eds.), *Cultivating Continuity of the European Landscape*, Environmental History 15, https://doi.org/10.1007/978-3-031-25713-1_36

conserved, but there are empty buildings and factories. Vittorio Veneto is an example of "shrinking Italy".[1] From 2009 to 2019 the population dropped from 29,234 to 28,148 inhabitants.[2] These data do not directly concern the process of depopulation of the historical centres of Vittorio Veneto, since they refer to the entire population of the territory. A further study should verify whether the phenomenon is widespread or limited to the ancient centres of the Municipality. However, it is possible to find a substantial amount of empty or underused buildings and infrastructure heritage.

This is evident in the centre of Serravalle, which other research has highlighted[3] but which needs further investigation. In fact, at the moment there is no study on the historical centres of Vittorio Veneto that has identified and quantified empty buildings and spaces, and this represents one of the main goals of this research.[4]

The origin of the empty or abandoned building problem probably dates back to post-World War II; oversized projections can already be found in the first Plan of 1953, known as the Torres Plan.[5] Certainly, the recent industrial sector crisis has had an impact on this dynamic, so Vittorio Veneto is now a town where centrality and marginality coexist and no longer correspond to the duality of centre and periphery.

This research investigates how to enhance what may appear marginal in the ancient urban landscape through projects capable of re-introducing empty buildings and spaces into a new life cycle by rethinking them not only on the basis of a functional change but as a resource available even in times of crisis.

To complete the reflections on the economic and cultural dynamics of this centre, Vittorio Veneto and the surrounding area are part of the UNESCO buffer zone and seem to benefit from this recent classification, which is trying to appeal to wine-gastronomic tourism.[6] On the other hand, the UNESCO classification has changed the landscape of the hills, where a single type of intensive grape cultivation is causing pollution. This phenomenon has recently lead to abandonment by the population living in settlements spread across the hilly territory. Therefore, paradoxically, the ancient urban landscape of Vittorio Veneto could become an attractive place to live as it is far away from Prosecco pollution. Of course, the hilly landscape of Prosecco should be protected from environmental pollution and the dichotomy between the

[1] For further information on the meaning of the term referring to the Italian territory, see Arturo Lanzani, Francesco Curci, *Le Italie in contrazione,* in Antonio De Rossi Ed., *Riabitare l'Italia. Le aree interne tra abbandoni e riconquiste*, Donzelli editore, Roma 2018, pp. 72–94.

[2] ISTAT data, reworked by the author in collaboration with Matteo Basso of Lab. Vault IR.IDE.

[3] See Stefano Munarin, Elisa Beordo, *Le energie della valle del Piave,* in Lorenzo Fabian, Stefano Munarin Eds., *Re-Cycle Veneto*, Aracne, Roma 2015, pp. 57–63.

[4] This article is the result of ongoing research being carried out by the author at the IR.IDE Research Centre, Pride laboratory, Iuav University of Venice.

[5] Moreno Baccichet, in Anna Marson, Moreno Baccichet, Matteo Basso, *Vittorio Veneto, Piani e progetti per la città,* Kellermann, Vottorio Veneto 2019, pp. 123.

[6] There is a general tendency to think of the positive effects of designation as a UNESCO site, which are actually having benefits for the site and its buffer zones, but a direct relationship between UNESCO designation and an increase in visitors has not yet been proven. Gaetano M. Golinelli, Fabiola Sfodera, *World Heritage brand,* in Gaetano M. Golinelli Ed., *Patrimonio culturale e creazione di valore. La componente naturalistica*, Kluwer CEDAM, Padova 2015, p. 176–177.

countryside and the city should not exist as an opposing relationship, but be restored to the original complementary relationship. In order to achieve such a change, this research proposes to deal with the issues concerning the reactivation of the historical urban landscape of Vittorio Veneto with an inter-scalar and integral approach, considering the complexity of the urban organism within the territorial framework. In particular, multi-layer inter-scalar landscape monitoring is presented: from field observation, using photographic surveys, sketches, videos, and interviews, to the interpretative elaboration of existing information.

While the existing planning tools (PTCP, PRG, PP) are necessary in some respects to protect urban and territorial heritage, they are limited in that they are not able show the different components of the landscape all together in an integrated multi-temporal inter-scalar vision. To test how to overcome this limit, this research seeks cooperation with the Municipal Administration of Vittorio Veneto—which has already expressed interest in this project[7]—and participation by other territorial stakeholders.

This research suggests some design practices at different scales of intervention:

A map of the "emptiness" of the ancient urban cores is the first step to monitoring the "health" of the ancient urban landscape, which must precede any operation of "care", keeping in mind that whether "mapping or looking directly at "landscape", meditation, or analysis in view of an intervention, the relationship with the object–subject will always remain partial and intermittent, which means open".[8]

This mapping process implies an analytical dimension, part of the design project itself, as it recomposes the available urban heritage into an overall view in order to carry out a critical classification of the buildings and urban spaces which need to be recovered or restored, define possible demolitions and new interventions, and increase the green areas within the ancient city.[9] Experimenting with contemporary multimedia technologies, this study aims to develop an active, digital, and updatable map to represent the ongoing landscape transformations, in other words, a map intended as "a tireless and relentless monitoring of redesign work, [that] can consider all the works of landscape transformation in the context of relationships that will to act".[10]

Through collaboration with the local Municipality, the research proposes a territorial strategy through actions aimed at activating integrated dynamics between natural and artificial factors: Launching rental and sales offers at fixed and affordable prices for residential buildings located in the historic centres of Vittorio Veneto, favouring those who leave their houses on the hillsides due to pollution.

---

[7] The procedure for an agreement between Iuav University of Venice and the Municipality of Vittorio Veneto is underway.

[8] André Corboz, *Il territorio come Palinsesto,* «Casabella» 1985, n. 516, p. 27.

[9] Currently, the Municipality has carried out a classification of public buildings and abandoned areas, but the data are not represented on a map that can give an overview.

[10] Renato Bocchi, *The Space of Landscape - Time and movement factors,* point 11 of the *Manifesto of Canary Islands for the European Landscape Project,* European Landscape Convention (CIVILSCAP, UNISCAPE, RECEP-ENELC), Universidad de Las Palmas de Gran Canaria, March 23–24, 2011 promoted by Observatorio del Paisaje de Canarias. See www.observatoriodelpaisajedecanarias.es, and www.sites.google.com/a/edumail.iuav.it/rbweb (last access September 2020).

Starting reclamation of the Prosecco hills from pollution by reconverting to organic cultivations and by recovering and protecting traditional cultivations.

This project sets the basis for a vision in line with the "2030: Regional Strategy for Sustainable Development",[11] which needs to be enriched by interaction with the inhabitants and other administration and cultural institutions.

In conclusion, by widening the view to a territorial scale, this project also intends to enhance the existing network of relations between small historical centres, natural areas, and the agricultural landscape in a comprehensive sustainable strategy, a future perspective which will be possible only if we consider the landscape "as infrastructure, an instrumental strategy across a range of jurisdictions, interests and stakeholders, inviting new inventors, users, and agents. [...] No longer we are just talking about roads, sewers or power plants anymore. We are referring to a systematic field of biophysical resources, socio-technological services, and exchange spaces, held together by a mesh of hardware and software that calibrates and condition urban economies".[12] Hence, it might be possible to reactivate not only the ancient urban tissue of Vittorio Veneto but also take care of the entire cultural landscape of the Prosecco hills, where the Meschio river landscape can rediscover a new meaning of infrastructure, the "backbone" of a slow fruition of the ancient city and its surroundings from which several paths can branch off, to the south towards Conegliano and to the west towards Valdobbiadene and the Prosecco hills.

Such actions must deal on one hand with the landscape's feature of change and on the other with certain forms of territorial resistance; their achievement requires a necessary spending of time, in a territory where settlement sprawl is part of a long history of infrastructure that, however, made it inhabitable.[13]

If it is possible to hypothesize an intervention divided into temporal phases, in which the elaboration of a "map of emptiness" corresponds to a first step, at the same time it is necessary to recognize that such a transformation can not only proceed in a linear way but also in a multidirectional and, therefore, multi-temporal way. So, it must be intended as having no fixed form[14] in which each direction takes care of the territory participating in a long-term adaptive recovery process.

[11] In particular see point 5: Reducing soil consumption, increasing green areas in urban and peri-urban areas, protecting and enhancing the system of protected natural areas and biodiversity.

[12] Pierre Belanger, *Landscape as Infrastructure: A Base Primer*, Routledge, New York 2017.

[13] Cfr. Stefano Munarin, Maria Chiara Tosi, *Tracce di Città. Esplorazioni di un territorio abitato: l'area veneta*, Franco Angeli, Milano 2001, p. 10.

[14] See, Renato Bocchi, *op. cit.*

Chapter 37
Green Infrastructure for the Landscape Reconfiguration of the VII Municipio of Roma: An Open Project, Beyond Urban Regeneration

Giorgio Caprari, Matej Gulic, and Elio Trusiani

Abstract The paper returns the results of research work, both applied and theoretical, concerning the reconfiguration of the city planning and landscaping of the VII Municipio of Rome. This is achieved through the use of green and blue infrastructure. The proposition gives answers to the need to give value to the "landscape system" as an urban structure for cities more sustainable, liveable and resilient to the impact of climate change. The study proposes scenarios through the reading of the physical/natural elements and the anthropic, historic and contemporary, that at one time marked and today define the structure of one of the most populous neighbourhood of the city. The methodology founded on a multiple systems approach, multi-scalars, an in depth investigation of the historic archives and direct and detailed knowledge of the Municipio took into examination three principle macrocategories: a) the park system and of the areas next to aqueducts and railways; b) main connection axis; c) three linear identity and historical testimony elements. The critical issues which emerged are reinterpreted in propositions of short/medium/long timescales that integrate the programmes of the institutions, the requirements of the citizenship and in depth botanical study. The theoretical investigation has been the basis for a proposal of scenarios that give back the "landscape system" as a scheme which is composed of interconnected infrastructure that constitutes itself as an open project for the entire territory. The reconfiguration of the landscape joins together and implements the green infrastructure of the Municipio by activating processes of social inclusion and new public spaces. The scenarios answer to both the implicit and explicit questions of the territory prospecting forms and uses, even temporary-experimental, in areas

G. Caprari (✉) · E. Trusiani
School of Architecture and Design, University of Camerino, Camerino, Italy
e-mail: giorgio.caprari@unicam.it

E. Trusiani
e-mail: elio.trusiani@unicam.it

M. Gulic
Environmental Protection Department - Green Areas Design Office, Roma Capitale, Italy
e-mail: matej.gulic@comune.roma.it

T. Matteini et al. (eds.), *Cultivating Continuity of the European Landscape*, Environmental History 15, https://doi.org/10.1007/978-3-031-25713-1_37

and property where the citizenship become co-designers of the process and of the regeneration of their daily life space.

Keywords Landscape making · Forms of time · Open project

37.1 The VII Municipality of Rome: Introduction

Research was carried out on the seventh Municipality of Rome. Its longitudinal north–south orientation, developed along two main directional roads: Via Appia Nuova and Via Tuscolana, is marked by a series of transversal axes that divide the territory into heterogeneous neighbourhoods and various landscape areas. The early 1900s neighbourhoods (Appio Latino, San Giovanni) relating to the Aurelian Wall historical–archaeological system and the planned expansions of the 1909 plan (Rossi 1991) and the other neighbourhoods (Appio Tuscolano, Tuscolano, Cinecitta, Quadraro, Tor Fiscale) differ from each other by formation age as well as settlement, architectural and social characteristics.

Created by the volcanic activity of the nearby Colli Albani, the territory is crossed by Rome's third river, the Almone, and its tributaries and features Roman aqueducts. The area of study has a wealth of environmental and landscape resources and is peppered with undervalued historic-archaeological heritage, as illustrated in Montanari (2017). Numerous writers from travellers on their Grand Tour to P. P. Pasolini have narrated its rich history of settlement while modern architecture such as the most important Project INA Casa of Rome, A. Libera's social housing and the metaphysical suggestions of Piazza Don Bosco (Rossi 1991) recount its post-WWII urban history (Fig. 37.1).

This system of expressed and unexpressed resources is linked to issues relating to society, ecology, infrastructure and public space liveability that are in turn traceable to important issues regarding urban and landscape planning: the fragmentation of formerly unified systems, inaccessibility, impermeable surfaces and land consumption, improper usage and environmental pollution, social isolation and the lack of alternative mobility.

37.2 Materials and Methods: From Knowledge to the Question of the Project

The knowledge-gathering phase was based on a re-examination of the physical-naturalistic and anthropic elements that once characterised the entire area. Direct knowledge of the territory permitted the elaboration of different multisystemic and multiscalar analyses which from the beginning were directed at five areas:

- The parks system and areas bordering and between aqueducts and railways.

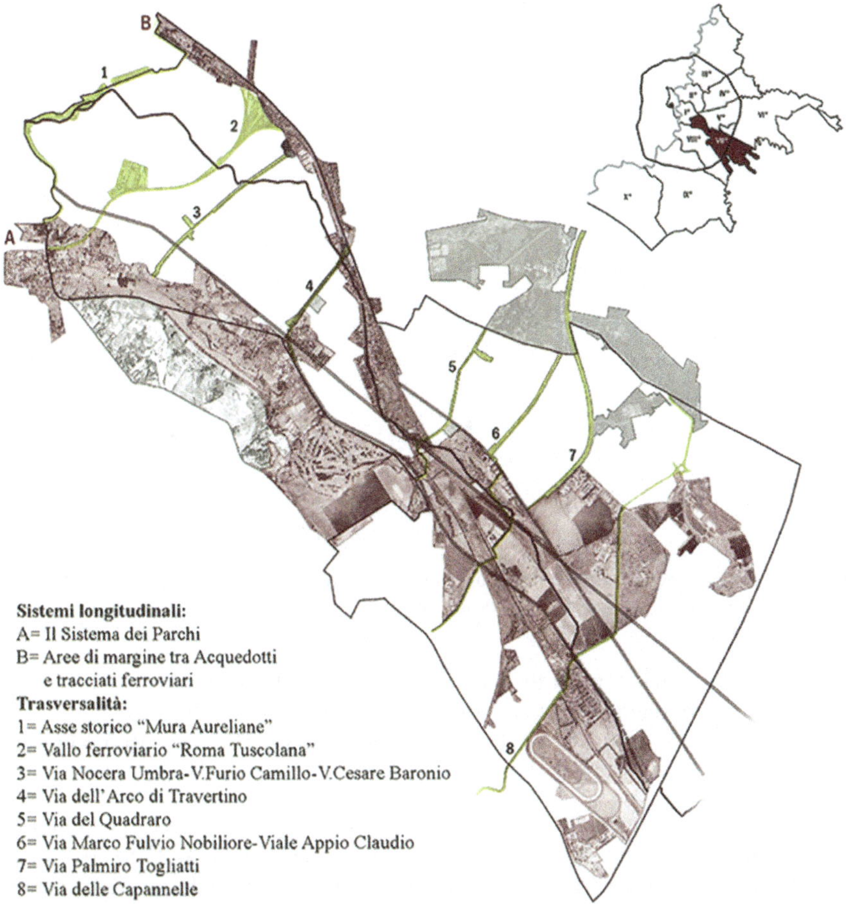

Fig. 37.1 VII Municipality of Rome and its main areas of interest

- A census of unused public heritage.
- Reconnoitring the historic-archaeological patrimony.
- The axis of transversal connections.
- The elements of linear identity and historic remains.

This phase of thorough historic-archival investigation resulted in the following: (a) re-construction of the historic evolution of the territory with attention to extant patrimony, even that of lesser historic, identity and testimony value (Comune di Roma 1988); (b) a detailed census of over ninety abandoned and unused publicly owned properties. This historic investigation based on cadastral cartographic and topographic sources[1] was interested in reconstructing three historic tracks:

[1] Mainly the "Cessato Catasto Rustico", 1870 and IGM Cartography regarding the years: 1877, 1907–1924 and 1925.

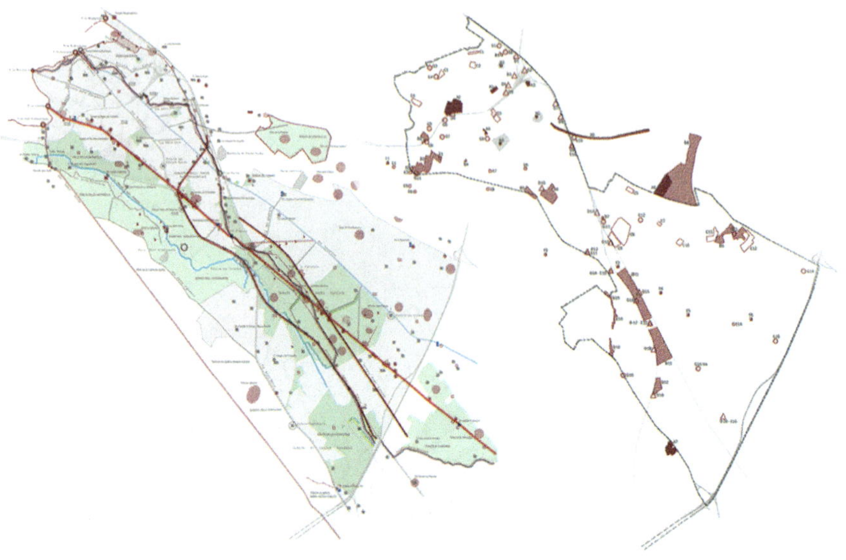

Fig. 37.2 Historic-archaeological patrimony and linear identity elements (left side of image); decommissioned public heritage (right side of image)

- The ancient roman road of the Via Latina and related patrimony.
- The Acqua Mariana channel, an artificial waterway realised in 1122.
- The abandoned railways of Roma-Albano (1888–1939) and Roma-Frascati (1856–1905) (Figs. 37.2 and 37.3).

A settlement system emerged, characterised by small historic communities (Mandrione, Tor Fiscale, Latino), intensive urban areas, humble workers' buildings, identity architecture; environmentally, the territory contains wide swaths of natural areas deriving from the estates of the Ancient Agro Romano as described in Iazzetti (2015). Inhabited since the sixth century B.C., these areas are disseminated with cultural and landscape patrimony and are mostly within the Appia Antica Regional Park.

The study revealed a very low value of naturalness[2] due to the widespread presence of incompatible activities and the association of foreign and parasitic trees and shrubs. The methodological process carefully considered phenomena related to temporary usage, classifying them by type and use and closely examining the properties and dynamics involved linked to social discomfort or to bottom-up regeneration processes.[3]

---

[2] The areas inside the Public Parks System ("Caffarella", "Tombe Latine", "Tor Fiscale", and "Acquedotti") were examined using the Index of Landscape Conservation (ILC), (Pizzolotto and Brandmayr 1996).

[3] The relation between the characteristics of a certain property and the use to which it is subject should be highlighted, such as the covered Acqua Mariana channel, previously a shared garden and spontaneous settlement and then a landscape track (Fig. 37.4).

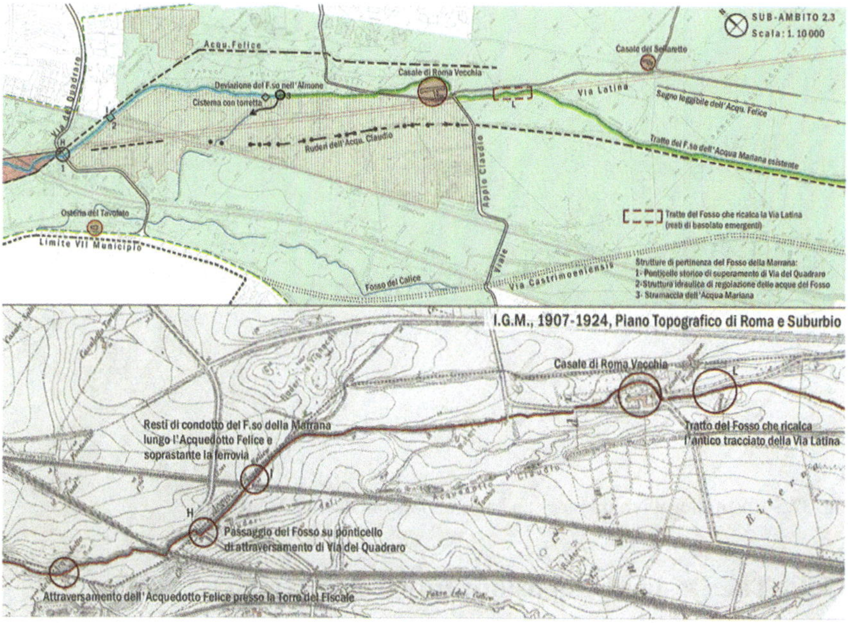

Fig. 37.3 Historical linear elements: the Acqua Mariana channel, Via Latina and related patrimony

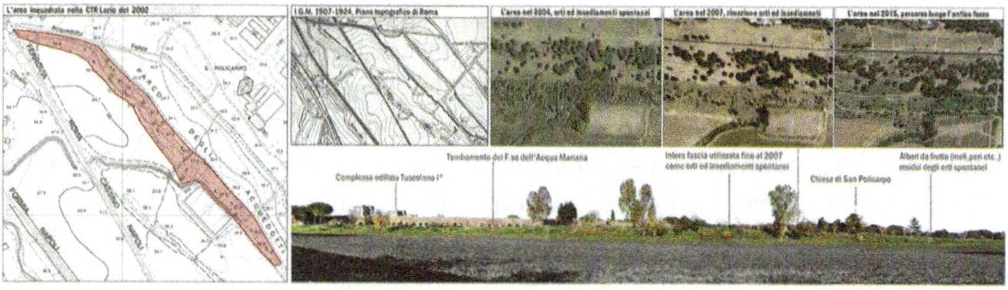

Fig. 37.4 An example of temporary use: the Acqua Mariana channel

To answer the "question of design" of the territory, the main citizens' requests which emerged from municipal urban planning conferences and local association assemblies were integrated into the study. Parallel to this, the principle tools of governance of the territory were collected in order to obtain a proper overview of needs and the prescriptions for use were analysed so as to act in accordance with or, where deemed necessary, to propose alternatives (Fig. 37.5).

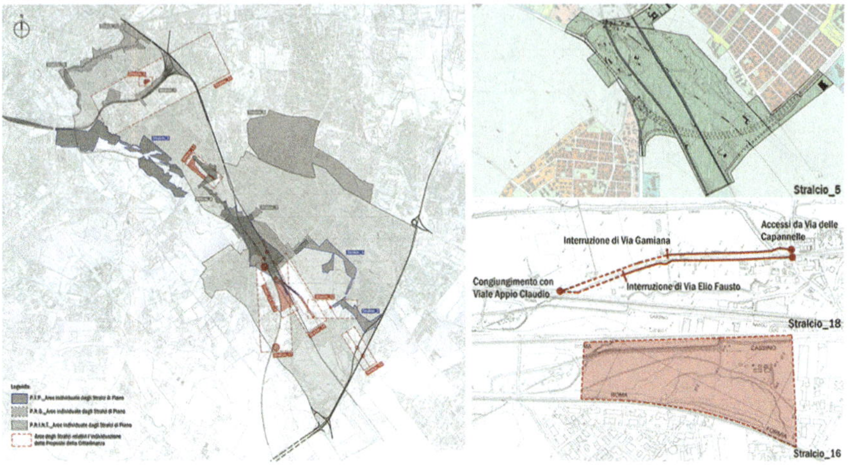

Fig. 37.5 Institutional programmes and an example of citizens' requests

37.3 The Design Dimension: Integrated Green Infrastructure

The process of critical synthesis gave rise to general strategies and specific objectives for the General Concept and the Masterplan consisting of: (a) two longitudinal macro-areas resulting from the ecological and physical re-stitching of large parks and from the requalification of the areas between the aqueducts and railways; (b) eight transversal axes, repurposed as pedestrian–bicycle paths, new tramlines, and rows of trees of non-allergenic species suitable for "stocking" fine particulates and pollution. The new territorial framework has been configured as an integrated "green and blue infrastructure" that valorizes the landscape resources present and responds to the needs of inhabitants with "light" interventions in the spirit of minimal interventions rather than a "grand project". The green infrastructure is the result of both systemic and distinct projects like the re-naturalisation of water bodies, the requalification of degraded areas through "nature based solutions", the creation of buffer zones or urban forestation interventions, etc.; there is even an articulated naturalistic system that bolsters the urban fabric against the effects of climate change by countering urban runoff and the summer heat island effect.

The slow mobility network project of urban greenways and didactic–naturalistic parkways connects neighbourhoods, becoming the centres and nodes of intermodal exchange (Fig. 37.6).

The restauration of forgotten historic-archaeological heritage, the refurbishing of the three historical linear elements with vegetation, and the re-use of unutilised public patrimony for new social–cultural functions and accommodation complete the whole vision.

Fig. 37.6 Masterplan of the VII Municipality of Rome (excerpt): the central area of the intersecting systems

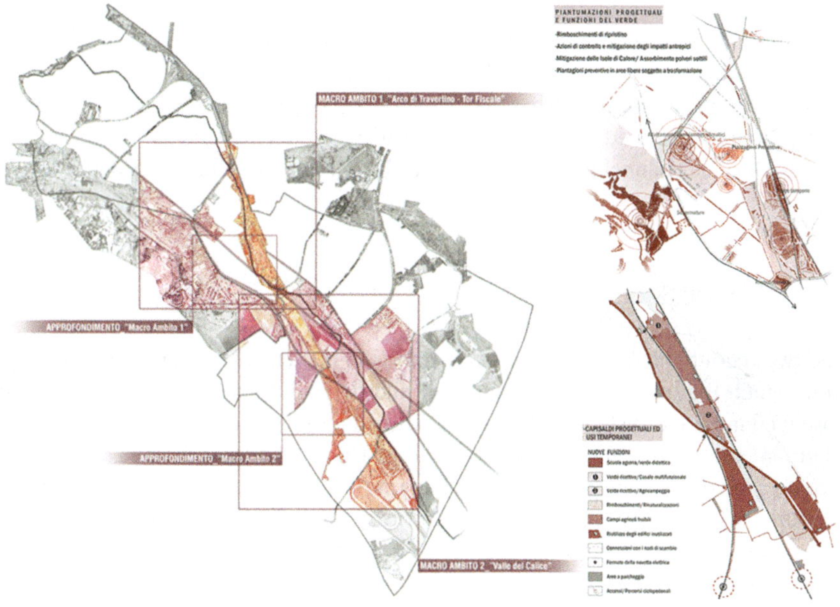

Fig. 37.7 Areas 1 and 2: identification and a project scheme example

The central area at the intersection of the longitudinal systems that is strategic for the regeneration of the landscape complex has been subdivided into two macro-areas in relation to the programme of the proposed interventions:

- 1. "Arco di Travertino – Tor Fiscale" (medium–long timescale).
- 2. "Valle del Calice" (short timescale).

In the first area, the project scenarios provide an alternative vision to the Programma Integrato (PRINT[4]) "Torre del Fiscale" (Comune di Roma 2012) that requires substantial urban reorganisation and building renovations that have yet to be realised; the proposed solutions restore the fracture between parks, creating new polarity and accesses: to the north, a tram reutilises the available capacity of the metropolitan line (known as "del Duce"); to the south, the historic area of "Via Latina-Campo Barbarico" has been redefined so that it is returned to citizens. Other proposals, for example, regard the expropriation and relocation of non-compatible activities, the realisation of a multifunctional park in a degraded public property area, a productive agricultural centre, a linear park along the aqueducts, and the reconfiguration of part of Via del Quadraro as a parkway to spatially connect the landscape between the two areas (Fig. 37.7).

In the second area, the more impactful project solutions regard the system of areas between the aqueducts and the railways. The interventions aim to integrate these areas of great landscape value into the existing park system through new functions, a new system of paths, the differentiation of the green areas, and botanical solutions. New accommodation and didactic functions have been planned for public areas and buildings in disuse; new paths are characterised by "light" interventions that recuperate degraded and inaccessible avenues, accesses and underpasses; the new usable areas are differentiated based on their uses and have been placed in the network of the three recuperated historic linear systems: the ancient track of Via Latina, the Acqua Mariana channel and the railways of Roma-Albano (Figs. 37.8 and 37.9).

37.4 Conclusions

The two reconfigured macro-areas constitute the nucleus of a "Landscape System" from which the other projects branch off and depend upon. The scenarios recompose the landscape system of the Municipality and tie it into a territorial unit with a single-integrated Urban Green Infrastructure. This "open project" landscape vision proposes scenarios functional to the changing socio-economic and environmental conditions of the city, "caring for" the territory on both short and long timescales. The new territorial framework is configured as an ecological infrastructure capable of accompanying urban planning by providing ecosystem goods and services which translate into direct and indirect benefits for human health (microclimate mitigation,

[4] PRINT, art.53 N.T.A., P.R.G. di Roma.

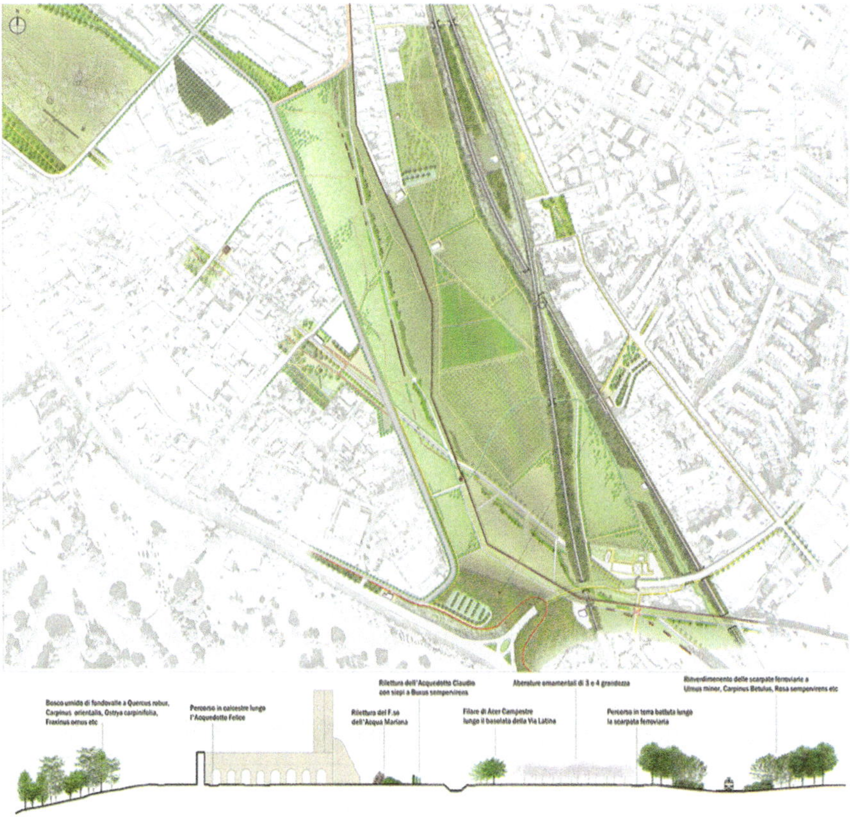

Fig. 37.8 Design scenario of the relevant part of area 1: plan and section

water management, agricultural products, sequestering of polluting elements, well-being of body and mind, etc.) and constitute an input for the start of a circular process that the transition phase imposes on us (EC 2019). This new, high-performance multi-functional landscape strengthens individual and collective identity by enriching the urban habitat with meaning and symbols perceived by inhabitants and visitors alike, thereby providing not only ecological but also social connectivity by experimenting with new production processes for "proximity consumption" through the use of shared urban gardens feeding new synergies and by recognising the added value of the growing leadership of local communities for a true "open project".

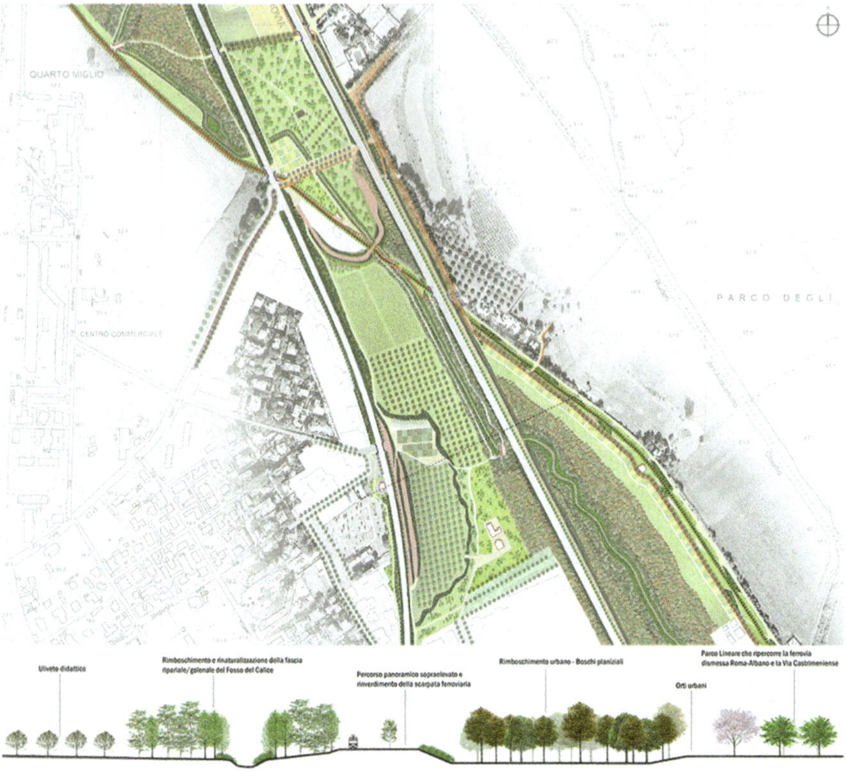

Fig. 37.9 Design scenario of the relevant part of area 2: plan and section

Fig. 37.10 New green and blue infrastructure

References

Comune di Roma—X Ripartizione Antichità e Belle Arti (1988) Carta storica archeologica monumentale e paesistica del suburbio e dell'agro romano, F. 15s-24n-16s-25n-25s, scala 1:10.000, Roma

Comune di Roma—ex Municipio IX—Roma Capitale (2012) Programma Integrato "Torre del Fiscale" Ambito 2/2a e ambito "Via Latina - Campo Barbarico". https://webmain.comune.roma.it/PCR/resources/cms/documents/print_tor_fiscale_1.pdf. Accessed on 01 June 2020

EC (2019) Commission staff working document—guidance on a strategic framework for further supporting the deployment of EU-level green and blue infrastructure, Brussels

Iazzetti S (2015) I quartieri Don Bosco e Appio Claudio a Roma. Storie e memorie tra Cecafumo e Cinecittà, p 7–31. Youcanprint, Roma

Montanari P (2017) Appio Latino Tuscolano. Alla luce delle più eclatanti scoperte, pp 174–344. Europa Edizioni, Roma

Pizzolotto R, Brandmayr P (1996) An index to evaluate landscape conservation state based on land-use pattern analysis and geographic information system techniques. Coenoses 11(1):37–44. http://www.jstor.org/stable/43461163

Rossi PO (1991) Guida all'architettura moderna 1909–1991, pp 12–15, 177–180. Laterza, Roma-Bari, Roma

Chapter 38
The De-structuring of Infrastructure Landscapes

Josep Lluís Miralles-Garcia

Abstract In very few cases, landscape has been an object of concern for the infrastructure designer, especially for transport infrastructure. Paradoxically, transport infrastructure is a large presence in the territories of all countries, especially the most developed ones, and an essential element of unstructured and disintegrated landscapes. Transport infrastructure connects extremes but, at the same time, creates barriers in its path that separate both natural ecosystems and human activities. In this way, it fragments the territory, establishing lines that are alien to the logic of the natural, rural or even urban landscape of the territories it traverses. The logic of minimizing economic cost or maximizing the transport function replaces and cancels the logic of previous landscapes. This paper is focused in the case of Spain and analyses the aspects of the current transport infrastructure design processes that explain this situation. Environmental and landscape assessment becomes an obstacle to overcome and environmental, and landscape quality is not targeted as a global objective. All these aspects can be synthesized as a one-dimensional functional vision of infrastructure that ignores its territorial transversality and the ecosystem functions that the territory produces and shows in the way of a landscape. This reductionist vision, where the territory is only a physical base to support the resistant elements of infrastructure, has reduced it to a functional engineering design for traffic. Finally, the article contributes to reflection on the criteria or methodological changes necessaries in the project process to integrate transversal values and produce another type of landscape of the infrastructure in a territory.

Keywords Transport infrastructure landscape · Territory fragmented · Environmental assessment · Territorial holistic vision · Transport infrastructure design

J. L. Miralles-Garcia (✉)
Departament d'Urbanisme, Universitat Politècnica de València, Camí de Vera, s/n., 46022 Valencia, Spain
e-mail: jlmirall@urb.upv.es

T. Matteini et al. (eds.), *Cultivating Continuity of the European Landscape*, Environmental History 15, https://doi.org/10.1007/978-3-031-25713-1_38

Fig. 38.1 Transport infrastructure forms broke traditional forms of rural propriety generating processes of abandon and degradation on borders. Picture about high speed train infrastructure in Valencia (Spain)

38.1 Introduction

Infrastructure in general, and transport infrastructure especially, usually produces strong landscape impacts on territories. This is a well-known fact. Transport infrastructure needs very specific geometric forms designed as a function of vehicle speed: geometric conditions of curves and slopes. These forms are generally badly adapted to the rural landscape, or even to an urban landscape, and therefore break up their structures. That is, in these cases, the infrastructure de-structures the landscape. These kinds of landscapes are the result of overlaying uses disconnected from each other as you can see in Fig. 38.1.

Nevertheless, it is not only an issue of form. Also transport infrastructure breaks up natural ecosystems or rural landscapes, producing fragmentation and generating a barrier between separate parts of the territory. Often roads are full of dead animals hit by vehicles, which is a way to remind society of its outstanding debt to nature, ecosystems or natural resources in the territory where it lives.

This paper is a reflection on this problem, possible trends and possible solutions.

38.2 The Landscape of Transport Infrastructure. The Case of Valencia

Many authors have dealt with this topic since Le Corbusier and the Modern Architecture Movement proposed big transport infrastructure as a way of planning regions and cities (Le Corbusier 1946). Later, many other authors studied how to integrate transport infrastructure into territories, such as Strang (1996) or Shannon and Smets (2011).

The problem is not easy because, in general, vehicle speed (that is, the main function of transport infrastructure) is not compatible with other territorial functions carried out at low speeds, where the two types of territorial functions meet is always a point of conflict. For this reason, only pedestrian traffic generally permits the design of urban landscape, i.e. public spaces, which are compatible with surrounding activities.

We need to considere the main objective, that is, how to design new infrastructure in a territory; in other words, how to design new transport infrastructure in a rural landscape. If we want to maintain rural landscapes that exist where a new transport infrastructure is projected, only this option is possible crossing both uses at different levels (by bridges or tunnels) to maintain the functions of both networks (transport and ecosystem services) and avoid conflicts. Of course, these kind of solutions are more expensive and, in any case, it is necessary to pay special attention to the design of borders, i.e. the design of contact points between different kinds of territorial usage.

However, the current situation of urban and metropolitan areas in developed countries is even more complicated. Today, European cities are generally in an equilibrium growth situation.

Cities have been growing for a long period which has produced urbanized urban areas but also the urban sprawl phenomenon in peri-urban zones (Gielen et al. 2019) and a list of infrastructure networks connecting with energy sources and natural resources and that connect urban areas for commuter travelling or other kind of trips.

In contrast, despite the low growth of urban and metropolitan areas, their extension is growing (Vheeler 2017a, b) (Feria and Martínez 2016; García-Ayllón 2013). That is, progressively more and more municipalities become part of urban and metropolitan areas. For example, in the case of Valencia, its new infrastructure projects aim to improve the main services but not to integrate the infrastructure into the territory as you can see in Figs. 38.2 and 38.3. Despite new laws in Spain and the Valencian Community on landscape integration in planning and project design, the traditional normal process of designing infrastructure continues.

38.3 Transport Infrastructure Versus Green Infrastructure in the Valencian Community

38.3.1 New Trends

Paradoxically, the European Union has developed instruments for environmental intervention on the territory or environmental controls. The 'Environmental Impact Directive' for projects, including civil works projects of infrastructure networks supplying natural resources to cities, dates from 1986, that is, 31 years ago. In addition, the 'Directive on environmental assessment of plans and programs', or strategic environmental assessment, was approved in 2001, that is, 16 years ago.

In Valencia, the first law that introduced landscape studies in planning and projects was passed in 2006 and was an important new vision.

However the results are not satisfactory. If we look at Figs. 38.1, 38.2 and 38.3 corresponding to the high-speed train infrastructure built around 2010, landscape integration is notably absent. These cases occur systematically and this example is not an exception. It is still the usual way of designing public works in which the

Fig. 38.2 The 'new' landscapes of high speed rail infrastructure on metropolitan rural areas of Valencia (Spain)

Fig. 38.3 Indefinite places between high speed rail infrastructures in metropolitan area of Valencia (Spain)

territory only plays the role of a physical base to support transport infrastructure. On the other hand, regional laws cannot be applied to state/central administration projects.

Environmental impact studies, strategic environmental assessments or landscape integration studies for civil works plans or projects are often contracted with the objective of environmentally justifying a previously decided solution or improving embankments or cut mountains with vegetation.

38.3.2 The Landscape of the Borders

As a result of the above-mentioned dynamics, large metropolitan areas such as Valencia are largely occupied by degraded rural areas, especially the urban borders. These areas have no specific vocation; they are undefined. In theory, they are rural areas but in fact they are areas of mixed use that break up the rural landscape and the economic continuity of the territory's rural uses.

Conventional planning is not intended to address and intervene in these areas primarily because of the scale of work. Intervention must be carried out practically at project level in order to define detail.

On the other hand, if the scale of intervention on these edge landscapes has to be on a project scale, the appropriate moment to define the intervention is the moment of a project's definition that produces these undefined spaces. This implies consideration of three aspects.

First of all, the scope of the project definition. For example, in the case of a railway infrastructure, the project cannot be strictly reduced to the land that is necessary to functional conditions of the infrastructure but must be extended to the residual spaces it generates, and which will potentially be degraded, with the objective of acting upon them. That is, it is necessary to rearrange uses and activities in these marginal spaces to assign new uses like urban gardens or others, even buildings for activities.

Secondly, guaranteeing intervention in these adjoining spaces implies expropriation in practice, which has an economic cost that can be significant. The application of excessive expropriation prices for land which practically no have any usefulness in a free market situation, in compliance with the current legislation in Spain, highlights the need to review the theory of land ownership.

Thirdly, for this type of landscape intervention to be effective, the intervention methodology and the delimitation of areas potentially at risk of degradation should be established as a rule. Otherwise, as is the current situation, the tendency is to minimize costs and, consequently, any type of intervention that involves additional cost. And all this, despite the fact that the cause of these undefined landscapes is clearly the civil work or intervention being implemented in the territory that breaks up the pre-existing rural landscape structure. Regulation of civil project contents (not additional studies) is possible to mandate a green infrastructure design.

38.3.3 Stakeholders

In Valencia, there are some big projects to expand on newly built infrastructures: the expansion of highway V-21 north now under construction; the expansion of Valencia's sea port now under discussion, the expansion of a highway bypass now being finished and new rail infrastructures for the Mediterranean Corridor now in the process of design. All these cases present projects that destroy landscapes or produce hard impacts even despite alternative solutions. There is an important social movement of university teachers against new infrastructures that continue to destroy natural resources and landscape. For example, the new sea port expansion will produce environmental impacts and change the skyline of the Mediterranean Sea in Valencia beach with the new port cranes for handling containers (in streaming you can see a very complete discussion with all stakeholders on the SCITO web in Catalan/Spanish). New railways can destroy the landscape of the historical peri-urban land agriculture of Valencia like the expansion of V-21 is already doing. All these projects are promoted by the Central Administration's Ministry of Public Works that still designs projects according to traditional methods previous to new landscape trends.

38.4 Conclusions

In the case of Valencia region, although the population of the metropolitan area is stabilizing, the occupation of space has increased considerably. This dynamic has typically produced degraded peri-urban spaces adjoining private and public facilities and infrastructure networks that cross the open spaces or green infrastructure to provide natural resources or ecosystem services to cities.

Among the infrastructures that generate this type of spaces are transport networks such as highways, roads and railroad, and the electric energy, water supply and waste water networks.

Studies and procedures for environmental impact or landscape evaluation that exist today have failed to generate landscape integration or landscape improvement of the areas degraded by the infrastructure.

Civil society demands another way to design transport infrastructures so as to avoid environmental impacts and create quality landscapes in peri-urban areas.

In this situation, the challenge regarding the transport infrastructure network is to redesign already-built infrastructures with the objective to rewild marginal areas next to infrastructure inside the urban and metropolitan areas and, in addition, to restrict new urban development to urban renewal operations whenever possible. On the contrary, if new urban developments are carried out by transforming more rural lands although the urban area population has stabilized, some built urban areas will be progressively abandoned and, in consequence, the urban landscape will progressively degrade.

It is possible to regulate the methodology of designing transport infrastructure to include the creation of quality landscapes, especially at crossing points between transport and green infrastructure, and in border spaces, as an alternative to the current situation where impact and landscape studies are carried out to complement infrastructure projects.

References

Boira JV (2011) Urbanismo expansivo: de la utopía a la realidad'. In: Proceedings of XXII Congreso de Geógrafos Españoles, pp 79–90

Burriel de Orueta EL (2009) La planificación territorial en la Comunidad Valenciana (1986–2009). Scripta Nova XIII(306). http://www.ub.es/geocrit/sn/sn-306.htm. Accessed 15 Jan 2017

European Comission (2017) Communication from the Commission to the European Parliament, the Council, the EESC and the CR. Green Infrastructure (GI)—Enhancing Europe's Natural Capital. http://eur-lex.europa.eu/legal-content/EN/TXT/?uri=CELEX:52013DC0249. Accessed 15 Jan 2017

Feria Toribio JM, Martínez Bernabeu L (2016) La definición y delimitación del sistema metropolitano español: permanencias y cambios entre 2001 y 2011. Revista Ciudad y Territorio. Estudios Territoriales 187:9–24

Feria Toribio JM, Martínez Bernabeu L (2016b) La definición y delimitación del sistema metropolitano español: permanencias y cambios entre 2001 y 2011. Revista Ciudad y Territorio. Estudios Territoriales. 187:9–24

Fundació SCITO (2020) https://debatport.scito.org/. Accessed 27 July 2020

García-Ayllón S (2013) Retrospective analysis of urban development in the Spanish mediterranean coast. WIT Trans Ecol Environ 179:291–302

Generalitat Valenciana (GVA) (2017) Conselleria d'Habitatge, Obres Públiques i Vertebració del Territori. Documento de Inicio del proceso de Evaluación Ambiental y Territorial Estratégica del Plan de Acción Territorial Metropolitano de Valencia. Web site, Valencia, http://www.habitatge.gva.es/web/planificacion-territorial-e-infraestructura-verde/plan-de-accion-territorial-metropolitano-de-valencia-pateval-. Accessed on: 20 Feb 2017

Generalitat Valenciana (GVA) (2017) Conselleria d'Habitatge, Obres Públiques i Vertebració del Territori. Estrategia Territorial de la Comunitat Valenciana. Web site, Valencia, http://www.habitatge.gva.es/web/planificacion-territorial-e-infraestructura-verde/estrategia-territorial-de-la-comunitat-valenciana-77496. Accessed on 20 Feb 2017

Gielen E (2016) Costes del urban sprawl para la administración local. Publicacions de la Universitat de València, El caso valenciano

Gielen E, Riutort-Mayol G, Palencia-Jiménez JS, Cantarino I (2017) An urban sprawl index based on multivariate and Bayesian factor analysis with application at the municipality level in Valencia. Environ Plann B: Analytics City Sci. https://doi.org/10.1177/2399808317690148

Gielen E et al (2019) Cost assessment of urban sprawl on municipal services using hierarchical regression. Environ Plann B: Urban Analytics City Sci

Herce Vallejo M (2008) La forma de las infraestructuras en la construcción del territorio y de su paisaje. Infraestructura y Territorio 84:38–47

Le Corbusier (1946) Manière de penser l'urbanisme. Éditions L'Architecture d'Aujourd'hui. Boulongne-sur-Seine

Miralles i Garcia JL (2014) El darrer cicle immobiliari al País Valencià. O el progrés de la misèria. Fundació Nexe, València

Miralles i Garcia JL (2015a) Revising the basis for planning a new kind of progress: the case of Valencia's city. WIT Trans Built Environ 148:3–14

Miralles i Garcia JL (2015b) Environmental management of Peri-urban Natural Resources: L'Horta de València case study. WIT Trans Ecol Environ 192:99–110
Shannon K, Smets M (2011) Towards integrating infrastructure and landscape. Topos 74:64–71
Strang GL (1996) Infrastructure as landscape. Places 10(3):10–15. (UC Berkeley)
Vheeler SM (2017) The evolution of built landscapes in Metropolitan Regions. J Plann Educ Res 27:400–416. http://www.jpe.sagepub.com/cgi/content/abstract/27/4/400. Accessed on 20 June 2020
Vheeler SM (2017) The evolution of built landscapes in Metropolitan regions. J Plann Educ Res 27:400–416. http://www.jpe.sagepub.com/cgi/content/abstract/27/4/400. Accessed on 20 Feb 2017

Chapter 39
Venice Marathon: From Landscape Brand to Healthy Infrastructure

Massimo Triches

Abstract The image of Venice is inevitably and indissolubly linked to water landscapes. However, the strong role—administrative, economic, cultural and social—that the historic centre has always played in relation to the surrounding territory has often reduced this image to a single water landscape, the lagoon. Moreover, in pursuit of ever-growing profits, the city has for a long time supported this narration by limiting itself to a reduced description, thus only partially showing itself.

Keywords Care · Landscape making · Open project

39.1 Introduction

The image of Venice is inevitably and indissolubly linked to water landscapes. However, the strong role—administrative, economic, cultural and social—that the historic centre has always played in relation to the surrounding territory has often reduced this image to a single water landscape, the lagoon. Moreover, in pursuit of ever-growing profits, the city has for a long time supported this narration by limiting itself to a reduced description, thus only partially showing itself (Fig. 39.1).

Actually, the Venetian territory is defined and characterized by other water landscapes, as well as by other urban models and other cultural and social identities. These, together with the most renowned one, contribute to providing a global image of Venice capable of representing its faceted identity.[1]

Since 1986 and with increasing popularity, a different narration is spreading, one that intends to describe and promote a Venice that is not just the well-known Tiziano Scarpa's fish.[2] This is the Venice city landscape crossed by the *Venicemarathon*

[1] Lane (1991) and Mancuso (2009).

[2] Scarpa (2000).

M. Triches (✉)
Dipartimento di Culture del Progetto, Infrastruttura di Ricerca IR.IDE, Laboratorio di Progettazione PRIDE, Università Iuav di Venezia, Venice, Italy
e-mail: massimotriches@iuav.it

T. Matteini et al. (eds.), *Cultivating Continuity of the European Landscape*, Environmental History 15, https://doi.org/10.1007/978-3-031-25713-1_39

Fig. 39.1 Venice city landscape: the agricultural system and the infrastructure of the water channels. Graphic design: M. Triches

which starts from Villa Pisani in Stra and reaches the Arsenale following the water landscapes of the Riviera del Brenta, the canals of Marghera and Mestre and the Venetian lagoon. It is a journey that tells the story of the territory in its complexity and whose narrative describes the history, culture, architecture and nature of these places. This is not just a random image. From the beginning, Piero Rosa Salva, the creator of the event, wanted to present Venice as he understood it: in its metropolitan articulation as a combination of different but strongly interconnected places and communities.[3]

Alongside this strong role as a landscape brand, the *Venicemarathon* can be read as a possible infrastructure for physical, urban and environmental health that crosses the city landscape of Venice. It defines itself as an element able to unify and systematize the active places of social sharing and individual well-being in the territory, together with those that can potentially take on the same role thanks to an integrated design vision, thus configuring a global and articulated system (Fig. 39.2).

[3] Casarin (2002).

Fig. 39.2 Venice city landscape: urban centers and infrastructure network. Graphic design: M. Triches

39.2 *Venicemarathon* from Narration to Territorial Brand

> Par son extension, le Tour est incorporé aux profondeurs de la France; en lui, chaque Français revit ses maisons et ses monuments, son présent provincial et son passé antique.
>
> On dit le Français peu géographe: sa géographie n'est pas celle des livres, c'est celle du Tour; chaque année, par le Tour, il connaît la longueur de ses côtes et le hauteur de ses montagnes. Chaque annee, il refait l'unite materielle de son pays, il recense ses frontières et ses produits.[4]

In the text for the film *Le sport et les hommes* (1961), Roland Barthes exalts the narrative power of the *Tour de France* and its ability to describe and represent the environmental, geographical, historical, cultural and social complexity of the landscapes that articulate French territory. The whole of France becomes a theatre of battle, the stage for the epic deeds of athletes/heroes and teams/armies. Their efforts fix in the collective memory images and names of mountains, rivers, villages and roads in the same way as the military conflicts of the past did (Figs. 39.3 and 39.4).

[4] Barthes (2004).

Fig. 39.3 S. Orsato view of the territory 1678

The narrative power and imaginative ability that these events exercise on the territory have proved to be incredible tools of territorial branding over time. The tourism promotion and the economic benefits guaranteed by these events trigger a "battle" between local authorities to sponsor particular stages, or to be on the path crossed by the athletes. For example, the starting point of a *Giro d'Italia* stage costs between 70 and 100 thousand euros, while the investment needed for hosting the arrival line rises up to 200 thousand euros.[5]

The *Venicemarathon*, although on a smaller scale compared to the big bicycle races, can be perfectly placed within these dynamics of territorial branding.[6] Furthermore, paraphrasing Barthes, the Venetians rediscover the variety of landscapes, the rich network of routes and water canals, the articulation of the different towns, the vastness of the artistic and cultural heritage, the modern history of the industrial hub, the contemporary transformations of the territory and even the length of the bridge

[5] *Cfr*. D. Ceccarelli, *Il giro d'affari del Giro d'Italia: costi e business della corsa più amata* in "Sole24ore" online on May 10th 2019. Source: https://www.ilsole24ore.com/art/il-giro-d-affari-giro-d-italia-costi-e-business-corsa-piu-amata-AC1Rpv?refresh_ce=1.

[6] Some representative numbers of the 2019 event: 14,000 participants at the Venicemarathon, 7500 (of which 24% foreigners) at the 42 K and 6500 (15% foreigners) at the 10 K; 18.000 participants at the Family Run Tour (Chioggia, San Donà di Piave, Dolo and Mestre); 35,000 visitors at the Expo-sport Venicemarathon Village with 70 exhibiting stands; 100,000 spectators along the route; live television broadcast in more than 90 countries. Source: Venicemarathon Dreaming Run in Venice.

Fig. 39.4 Venice city landscape: the Venicemarathon route. Graphic design: M. Triches

that separates/joins the historical city from/to the mainland, through the 42 km and 195 m of the running race (Figs. 39.5, 39.6, 39.7 and 39.8).

Venice is in fact a "landscape-city" shaped by a constellation of places which can no longer be read or described starting from the traditional narrative structure that identifies the historical city as a central hub imposing a strong network of social, economic and functional relations on the territory. On the contrary, this territorial complex is characterized by a number of articulated urban, environmental, social and cultural dynamics defining a composite but unitary spatial identity.[7] The system is similar to an archipelago: dynamic, variable and impermeable to hierarchies: different elements coexist within its area, triggering mutual coordination strategies.[8]

The *Venicemarathon* narrates this condition starting from an east-west spatial directrix, which follows the Brenta river blue and green infrastructure until the Venetian lagoon. This directrix follows the Padua/Venice vertical axis pictured by Sertorio Orsato (1678) around which the artist represents the constellation of mainland centres and the complexity of the infrastructural water system. Since its first edition (1986),

[7] Casarin (2002), pp. 36–37.

[8] Cacciari (1997).

Fig. 39.5 Venicemarathon starting point, Villa Pisani, Strà

Fig. 39.6 Venicemarathon at Ponte della Libertà

Fig. 39.7 Venicemaraton at Riva degli Schiavoni

Fig. 39.8 Venicemarathon at Zattere

the *Venicemarathon* route has aimed to present Venice in its metropolitan articulation, promoting the idea of a global city from its physical, social, identity-making, sporting, economic and cultural points of view. The route itself, which seems broadly unchanged, has been able to record and adapt to the territorial transformations occurring over time, while proposing an increasingly richer itinerary, both on a technical/sporting level and from a narrative/experiential side. Some examples are: the change of the starting point, originally located in the main square of Strà and which is now at Villa Pisani (which opens to the public for the occasion); the crossing of the M9 museum block in Mestre; the moving of the headquarters and exhibition spaces to the San Giuliano park. Therefore, *Venicemarathon* represents a dynamic territorial brand aimed at defining and promoting the Venetian tourist district and the Riviera del Brenta (which at the time of the first editions was mainly known as the "Mala del Brenta", i.e. an organized crime area). The event acts as an element unifying the energies and needs of the various local authorities, associations and enterprises for the touristic, social and cultural development of the territory. It is not just a sporting event but a set of side events, such as minor competitions for amateurs and families, exhibitions and entertainment shows, able to attract a diverse audience in terms of interests and age groups.[9]

With this in mind, the Organizing Committee has promoted an annual data collection and specific multi-sector analyses since the very beginning to assess the economic effects generated by the event and the other sporting events, both on the territory and on the communities involved (in recent years about 10 million euros). These studies, which are still scarcely used at a national level for similar events, represent a useful tool for cost/benefit evaluation, for choosing ideal promotion and communication strategies, for checking the achievement of set objectives and for identifying possible improvements to promote technical/sports and social/economic growth of the event and the territory.[10]

39.3 *Venicemarathon* as Health Infrastructure

Alongside this narrative and economic value, the *Venicemarathon* represents an important infrastructure for individual and collective health. Over time, the event has triggered many actions aimed at improving the welfare of local communities, raising public awareness about environmental issues, educating children and families about physical activity, promoting and supporting solidarity initiatives and social participation, etc. There are, e.g. direct territorial transformation operations such as the installation of walkways greatly improving the accessibility of many parts of the historic centre. This initiative, which has been in place since the first edition of the event for reasons linked to international sporting regulations, allows athletes to cross

[9] Interview with Lorenzo Cortesi (Secretary General of Venicemarathon), June 17th 2020.

[10] Cfr. The report: L'impatto economico sul territorio veneziano di manifestazioni sportive. Analisi dei dati relativi agli eventi prodotti dal Venicemarathon Club gradients. Source: http://www.fidal.it/.

14 bridges without having to climb any steps.[11] During the following years, in collaboration with the EBA (Accessibility Improvements) office of Venice City Council and some local associations,[12] a process of maintenance and adaptation of these structures was put in place, acting on their slopes and their relationship with the context in order to make them a permanent, though easily removable, asset. This has made the regularity of the sports event possible and simultaneously significantly improved the living conditions of the residents, making many areas along the Zattere and Riva degli Schiavoni more accessible.[13] Moreover, in October 2019 *Venicemarathon,* in collaboration with the Veneto Region, completed an accessibility project funded by the Interreg Italy-Croatia Programme for Villa Pisani that created a permanent ramp to improve access to the museum.[14]

Fostering accessibility is not only a technical and spatial issue, it also means involving local communities and avoiding social, cultural, gender discrimination or distinctions of age, physical abilities, interests, etc. For this reason, over the years, a 10 km race was introduced which shares the finish line with the marathon but starts from the San Giuliano park, alongside many other minor events in the area, such as the Family Run of Chioggia, San Donà di Piave, Dolo and Mestre. These are non-competitive races, organized in collaboration with schools mainly to interest children and families with the aim of bringing people closer to sport and physical activity, as well as promoting solidarity initiatives and raising awareness of ecological issues.[15] Moreover, *Venicemarathon* is the pilot event of the environmental and territorial protection project "*Zero Waste Blue sport events for territorial development*" promoted by the Veneto Region. The first experimental phase involving the 2018 and 2019 editions proposed a reduction in CO_2 impact, the adoption of sustainable transport strategies, the implementation of separate waste collection, the reduction of plastic materials and the use of paper.[16]

In future, and in continuity with the described initiatives, *Venicemarathon* will finally become an opportunity to trigger new processes of territorial transformation aimed at environmental well-being and citizens' health. In addition to enhancing local landmarks, the route of the event crosses several areas in need of redevelopment and revitalization, such as the numerous abandoned areas that occupy the edges of the Marghera district. These places represent potential shared spaces capable of building a seamless system of public services and urban attractors. In continuity with other initiatives promoted by local authorities and associations, such as the "*Passeggiate di quartiere*", the "*Sulla via del riuso*" walk, the "*Marghera sentieri nel tempo*"

[11] The international technical regulations do not allow the passage of the route on stairs and also define the maximum permitted.

[12] Peggy Guggenheim, Fondazione Vedova, Punta della Dogana. Source: https://www.comune.venezia.it/it/content/venice-art-all-2013.

[13] Tatano (2018), pp. 38–39.

[14] https://regione.veneto.it/web/turismo/progetto-zero-waste-blu.

[15] Interview with Lorenzo Cortesi (Secretary General of Venicemarathon), June 17th 2020.

[16] *Venice Marathon sempre più green: l'obiettivo è l'impatto zero* on "La Gazzetta dello Sport" on-line February 2th 2020. Source: https://running.gazzetta.it/news/21-02-2020/venicemarathon-sempre-piu-green-lobiettivo-e-limpatto-zero-62891.

itineraries and the cycling event "*Tour de Marghera*",[17] the *Venicemarathon* can raise awareness among inhabitants, administrators and investors, thus promoting a series of territorial transformations that take the form of a network, or infrastructure, of shared places for the well-being of territory and community.

References

Barthes R (2004) Le sport et les hommes. Texte du film Le sport et les hommes d'Hubert Aquin. Les Presses de l'Universite de Montreal
Cacciari M (1997) L'Arcipelago. Adelphi, Milano
Casarin M (2002) Veneza Mestre Mestre Venezia. Luoghi, parole e percorsi di un'identità. Ediciclo Editore, Portogruaro (VE)
Lane FC (1991) Storia di Venezia. Giulio Einaudi Editore, Torino
Mancuso F (2009) Venezia è una città. Come è stata costruita e come vive. Corte del Fontego, Venezia
Rubini C (2010) Marghera sentieri nel tempo. Tre itinerari in un'altra Venezia. Enel in collaborazione con Trekking Italia, Venezia
Scarpa T (2000) Venezia è un pesce. Una guida. Feltrinelli, Milano
Tatano V (2018) Atlante dell'accessibilità urbana a Venezia. Anteferma Edizioni, Conegliano

Massimo Triches An architect, he graduated with a degree in Architecture (2010) and obtained his Ph.D. in Architectural Composition (2016) at Iuav University of Venice. In 2011, he was visiting fellow at the Manchester School of Architecture (MSA) and in 2014 at ETSAB in Barcelona, during his doctoral studies.

He is co-founder of the firm Babau Bureau in Venice with which he has participated in numerous design competitions, exhibitions and international workshops, including the Venice Architecture Biennale (2016), the Kosovo Architecture Festival and the Tiblisi Architecture Biennial (2018). He has received several awards and mentions, including a nomination for the European Award for Architectural Heritage Intervention AADIPA and the EU Mies Van der Rohe Award in 2019.

He was a research fellow in PRIN 2015 "La città come cura e la cura della città".

An architect, he graduated with a degree in Architecture (2010) and obtained his PhD in Architectural Composition (2016) at Iuav University of Venice. In 2011, he was visiting fellow at the Manchester School of Architecture (MSA) and in 2014 at ETSAB in Barcelona, during his doctoral studies. He is co-founder of the firm Babau Bureau in Venice with which he has participated in numerous design competitions, exhibitions and international workshops, including the Venice Architecture Biennale (2016), the Kosovo Architecture Festival and the Tiblisi Architecture Biennial (2018). He has received several awards and mentions, including a nomination for the European Award for Architectural Heritage Intervention AADIPA and the EU Mies Van der

[17] These are events, on foot or by bicycle, organized to rediscover the territory of Marghera and stimulate the participation of the inhabitants: the *Passeggiata di quartiere*" organized by the Centro Sociale Rivolta on Saturday, 14th April 2018; "*Sulla* via *del riuso. Un percorso narrativo per condividere il progetto di un nuovo tracciato che metta in relazione i luoghi dismessi a Marghera*"" organized by the G124 group on Saturday, 8th October 2016; "*Tour de Marghera. Per le strade di Marghera*"organized by the committee Marghera Libera e Pensante Saturday, 20th April 2013. C. Rubini, *Marghera sentieri nel tempo. Tre itinerari in un'altra Venezia*, Enel in collaborazione con Trekking Italia, Venezia, 2010.

Rohe Award in 2019. He was a research fellow in PRIN 2015 "La città come cura e la cura della città". He is currently research fellow at the PRIDE pro research in integral design environment of the research infrastructure IR.IDE (Dipartimento di Culture del Progetto Università Iuav di Venezia). He is also a teaching tutor at Iuav architectural design studios and workshops and at the Iuav international doctorate. He edited (with M. Vanore) the book "Del prendersi cura. Abitare la città-paesaggio" (Quodlibet 2019).

Chapter 40
KM0. Merging Public Space and Nature

Enric Batlle i Durany and Javier Rivera Linares

Abstract The current city is full of limits formed by large private vehicles and transport infrastructures, separating the urban interior from nature and the countryside. However, can there be another way to connect these seemingly distant places? The km0 proposes another way of flowing through the metropolis, where biodiversity and productivity will be key to offering another way of merging public space and nature.

Keywords Landscape making · Transition · Open project

Nowadays, cities are full of impenetrable boundaries, barriers, and borders. Large private vehicle infrastructure has become the modern walls of the ancient city, the limit separating two presumed worlds: the dense, equipped, but increasingly polluted urban interior and the nature outside, the open countryside, pure and healthy. People live inside, but they would like to live in nature, in the nearby paradise they imagine is on the other side of the freeway.

Faced with this situation, the km0 theory proposes to create a network of interconnected urban spaces relying on three concepts to merge the public spaces of cities with the nature that still surrounds them: FLOW through the metropolis, saving the walls to allow new sustainable mobility and ecological connectivity, BIODIVERSITY networks renaturing cities and connecting them with open spaces, from the individual pot on the balcony to large natural spaces, and PRODUCTIVE cities, where any urban space can be cultivated, creating a mesh of edible landscapes that generate recognizable, beautiful, ecological and productive environments. Km0 is a set of productive and equipped spaces that promote biodiversity, sustainable energy production, water control, proximity food cultivation and the leisure of citizens.

E. Batlle i Durany
Batlle I Roig Associates, Barcelona, Spain
e-mail: enric.batlle@batlleiroig.com

J. Rivera Linares (✉)
Universidad Cardenal Herrera-CEU, CEU Universities, València, Spain
e-mail: javier.riveralinares@uchceu.es

T. Matteini et al. (eds.), *Cultivating Continuity of the European Landscape*, Environmental History 15, https://doi.org/10.1007/978-3-031-25713-1_40

40.1 Flow Through the Metropolis

Most cities and large metropolises were initially associated in some way with mobility infrastructure: around a small fishing port, along a meandering river, at important crossroads or crowding around a railway station. More recently, they have continued to flourish due to mobility but rather like airport satellites, adjacent to highway exits that lead to larger nearby cities, or linked to high-speed rail stops in the middle of nowhere. All founded on flow from the mobility of goods and people according to their time.[1]

The new approach to flow through metropolises seeks a re-foundation of those beginnings associated with rapid mobility infrastructure, with a change in perspective. A revision is needed so that public spaces and nature merge because of the new infrastructure and this time, with slow mobility. The new flows connecting km0 spaces will be sustainable mobility flows (Fig. 1).

Fig. 1 With a paradigm shift from km0 to door-to-door, the interruptions of large vehicular or railway infrastructure must be resolved by providing safer, healthier and more convenient connections. This is precisely what the new bicycle and pedestrian route between Barcelona and Esplugues de Llobregat achieves, getting past the motorway. Batlle i Roig arquitectes, 2017. *Photo* Jordi Surroca

A transmutation between connectivity and speed is proposed, with a change in model from the inherited Fordism, where the speed of moving flows of goods, people, and, today, information drifts towards the new post-industrial one of km0 and where

[1] Shannon and Smets (2010).

connectivity is of the essence, allowing slow flows. This new way of connecting calls for a new humanism of public spaces. Let information go very fast, but citizens go at a slower pace. It will be what David Mangin has called "la ville passante", a city that flows, one that would manage to go from fast speeds to slower ones.[2]

While the industrially functional society divided cities into areas according to function—residential areas, work zones, leisure and transport zones—km0 must mend and reconnect these functions in a new zone that will be continuous, the public space zone. This connectivity would soften those rough and impenetrable edges that Kevin Lynch described in his Image of the city by managing to generate an infrastructure that enables flow and movement in all directions and merges the parts, allowing slow mobility.[3]

In this sense, km0 would join those international movements and policies that claim to be able to move around the city, flow, and enjoy public spaces. The Flow Movement or the International Charter for Walking promoted by Walk21 is two magnificent examples of this type (Fig. 2).[4]

Fig. 2 The Green Ring of Igualada seeks to generate a set of itineraries for pedestrians and bicycles in the form of a green belt on the perimeter of the city, generating a sequence of rest areas. The road network also fosters landscape and biodiversity recovery through the use of water. Recovery of the old gypsum of Igualada. Batlle i Roig arquitectes, 2018. *Photo* Jordi Surroca

[2] Mangin (2006).

[3] Lynch (1964).

[4] WALK21. International Charter for Walking. Online resource: https://www.walk21.com/charter.

40.2 Biodiversity Network

In recent times, the growth of ecological awareness has led to the question of whether this wide set of public spaces with such different uses and regulations could have a second common objective: to build a metropolitan environmental network that infiltrates all corners of the city. Public spaces could be the backbone of metropolitan territories. The streets, squares, and parks of cities can be re-naturalized and connected to still preserved agricultural and natural spaces as a network of urban spaces with every possible open connectivity as well as ecological, urban, and metropolitan continuity, making productivity compatible with public use.[5]

Thus, km0 acquires a new property: the function of ecological connections, of flows of plants and other living species, of water, of cold air, of biodiversity. After all, Km0 maintains the city as an ecosystem, as a shared habitat with other species, as a hybrid place between nature and city. The search for the recovery and propagation of biodiversity demonstrates the necessary re-naturalization of the metropolis, the sustainability of the system through new readings of km0, and the metropolitan ecological matrix connecting them and giving them continuity (Fig. 3).

[5] Batlle E. Flowing through the City: Continuity, Ecology and Production. Digital edition: https://www.publicspace.org/web/guest/multimedia/-/post/flowing-through-the-city-continuity-ecology-and-production.

Fig. 3 The revitalization of existing geographies serves as a link between the biodiversity of the natural environment and that of public spaces. The recovery of the river space of the Llobregat river in Sant Boi del Llobregat manages to understand the river as a natural space with its own dynamics and functioning and as a green and peri-urban leisure space for the enjoyment of citizens. Batlle i Roig arquitectes, 2015. *Photo* Jordi Surroca

A heterogeneous biodiversity system will be more effective if it is seen as a large, almost unlimited matrix that unites the different existing infrastructures where they connect and overlap as a network of networks to create a global ecosystem, like the planetary garden imagined by Gilles Clément. A system that would unite and intertwine agroforestry spaces, natural protected areas, fluvial and coastal areas, the more urban open space system (parks, gardens, streets, squares, roofs, etc.), and thus with the different elements that make up the territory. In other words, a large ecosystem that would encompass and merge the environmental matrices of ecology, the park systems of landscaping and the civic spaces of urban planning (Fig. 4).[6]

[6] CLÉMENT, G. Le Jardin Planetaire. Online resource: http://www.gillesclement.com/cat-jardin planetaire-tit-Le-Jardin-Planetaire.

Fig. 4 The Santander Atlantic Park incorporates an area of high ecological value on the outskirts of the city into the network of public spaces by incorporating vegetation and water into the urban core, thus exponentially increasing biodiversity in the most built-up part of the territory and achieving a more resilient landscape. Batlle i Roig arquitectes, 2008. *Photo* Jordi Surroca

40.3 Productive Urbes

Productive cities are a third fundamental leg of the km0 theory, almost a continuation of the discourse on networked biodiversity, a requirement to recover and reinvent ties that should never have been broken between urban space and agricultural or productive space. However, today cities expel their productive soil and only entrust their supply system to the global market. The new km0 network means re-appropriating the importance of the strategic element that guarantees being able to provide city inhabitants with a local supply of food and other necessary products.

This km0 network would incorporate what André Viljoen and other authors have defined as a "continuous productive urban landscape", the city of CPUL. This landscape links the urban scale agriculture of inner cities with that of suburbs and periphery areas on a metropolitan scale, and these with their counterparts of other cities and towns on a national and international scale, to generate a continuous cultivated landscape intermingled with the built part of cities (Fig. 5).[7]

[7] Viljoen (2005).

Fig. 5 The transformation and landscape restoration of the Vall d'en Joan landfill has been achieved by means of ditching, generating a series of productive plots that, in addition to providing changing and adaptable processes, originate an identity landscape of agricultural valorization. Batlle i Roig arquitectes/Teresa Galí-Izard, 2003. *Photo* Jordi Surroca

Km0 can add an idea to how to address the agriculture in cities that is spreading little by little from different countries and from different states. From small parcels or farming communities with more or less isolated urban gardens to government strategies for large cities (such as Cultivating the Capital developed in London for that entire British city), this could create an urban productive network.[8]

The km0 theory wishes to join other efforts in this direction, such as the Milan Urban Food Policy agreement, an urban food policy pact begun in 2015 to promote policies on how cities should be fed. Today, this pact has been signed by 210 cities around the world. It guarantees healthy diets through social and economic food labelling and by ensuring quality food production as well as the distribution of products on a local scale so as to be competitive with that on a global scale and reduce environmental impact.[9]

All this unites to reclaim the factor of small scale proximity, promoting the creation of contiguous production units to generate an "immediate designation of origin" within urban settings that also orients the commercial model towards prioritizing km0 products. According to the FAO, this urban agriculture system, together with flow and biodiversity, has extraordinary potential for the transformation of cities

[8] London Assembly. Planning and Housing Committee. Cultivating the Capital. Food growing and the planning system. Greater London Authority, 2010. Digital edition: https://www.london.gov.uk/sites/default/files/gla_migrate_files_destination/archives/archive-assembly-reports-plansd-growing-food.pdf.

[9] Milan Urban Food Policy Pact. Online source: http://www.milanurbanfoodpolicypact.org.

since it provides fresh food, generates employment, recycles urban waste, creates green belts, and strengthens the resilience of cities to climate change (Fig. 6).[10]

Fig. 6 In the Finestrelles sector, in Esplugues de Llobregat, a green corridor has been created with a central strip of continuous meadow connecting the Collserola park with the city of Barcelona. Planted with alfalfa, this area is used as pasture for the flocks of sheep that still subsist on the mountain. Batlle i Roig arquitectes, 2003. *Photo* Alejandra Liébana

With km0, limits become doors which open in both directions, nearby spaces unite and link public space with the agrarian, rural or natural environment that it previously only saw, or sensed, in the distance above a grey infrastructure wall. Km0 materializes as an antidote to current non-places, making margins intelligible and endowing them with identity and character so that people, their inhabitants, not only live in them, but also identify with them and beyond. These initiatory marginal points that end up being part of all the components of the system guarantee connectivity and new mobility flows and are ecological, biodiverse, and productive. Thus km0 functions, on all scales and with art as a vehicle, as a creator of idyllic places, transforming living in the city into a paradise of living in nature at the same time as part of a network of complex, heterogeneous, diverse, connected multifunctional spaces that manage to merge public space and nature in a flexible, dynamic, and changing way.

References

Lynch K (1964) The image of the city. MIT press, Massachusetts

Mangin D (2006) Pour une ville passante et métisse. Le Journal de l'École de Paris, Vies collectives, n° 58

[10] FAO. Framework for the Urban Food Agenda. Digital edition: http://www.fao.org/3/ca3151en/CA3151EN.pdf.

Shannon K, Smets M (2010) The landscape of contemporary infrastructure. Nai Publishers, Rotterdam

Viljoen A (2005) Continuous productive urban landscape: designing urban agriculture for sustainable cities. The Architectural Press, Oxford

Chapter 41
Project Perspectives for Making Landscape

Rita Occhiuto and Margherita Vanore

Abstract *Being in project* that promotes *landscape making* is an extended and integrated knowledge process which involves different areas of investigation and skills. It requires the ability to intertwine analyses on the quality of places with interpretations of the morphological, environmental and historical stratification characters of hybrid and complex territorial schedules. Movement is the condition that gives landscape the profound meaning of a *process of change in progress*. This is not only a way to characterize and appropriate territory, but it is a way to understand what has happened in order to intervene in ongoing transformation and to continue its evolution and the writing of the future. If we consider space as a set of environments in progress or in fieri, that is, contexts whose composition can remain open or not completed, *making landscape* becomes above all the regaining of an ability to continue existing territorial writings. Making landscape re-emerges the ability to immerse oneself in places to evoke physical and mental relationships that, above all, make it possible to restore life to the imagination that distinguishes local cultures. To achieve more ambitious objectives of critical conscience, it is necessary to re-elaborate the concept of transition to enhance its spatial and temporal values. The concept of open project expresses a need to adapt to changing conditions, but also a responsibility to transmit resources and potential to the future of territories

Keywords Open project · Landscape making · Transition · Being in project · Forms of time

The *Landscape Design and Time* session opens a reflection on the territorial continuum of European landscapes, already under the aegis of the European Convention for 20 years.

R. Occhiuto (✉)
Department of Architecture, University of Liège, Liege, Belgium
e-mail: r.occhiuto@uliege.be

M. Vanore
Università Iuav Di Venezia, Venice, Italy
e-mail: vanorem@iuav.it

T. Matteini et al. (eds.), *Cultivating Continuity of the European Landscape*, Environmental History 15, https://doi.org/10.1007/978-3-031-25713-1_41

Through a critical re-reading of the notions of *design and time*, the debate has allowed us to re-examine landscape practices which have lost both dynamic potential and being in action or in motion. The critical thread of the session was introduced by two strong design approaches: the first, based on the study and critical interpretation of traces, read as signs and drawings of the earth, narrates the relational values of landscape, investigating the depth of the material; the second, attentive to the multiple territorial components, responds to landscape complexity with actions of rewriting and explication of systems which accompany the reconstruction of the relationships between places and inhabitants over time.

Both show the appeal of recovering the potential of time as a necessary condition to act and create landscape in continuity with existing materials and characters and in a decidedly future-oriented perspective.

This objective leads to a recurring and implicit concept to be rediscovered: **Making**. This, in turn, opens up reflection on *being in progress*, or *being in action*.

"Making" for Tim Ingold is a cognitive action that involves "being inside", "doing to learn" or "learning by doing". He states that: "To know things you have to grow into them, and let them grow in you, so that they become a part of who you are.

It is, in short, "by watching, listening and feeling—by paying attention to what the world has to tell us—that we learn." This implies understanding "what to look for, how to look at things, and that knowing is a process of active following, of going along". This attentive approach to the existing also allows us to draw out the reasons for the changes and transformations of what is around us, movement and territory in constant evolution.

Intention, acting and doing are presented above all as cognitive modalities that imply an awareness that *knowing is movement*. And if landscape needs to consider movement as a vital dimension for its recognition and development, it can be deduced that action situated or acting on or in the earth must necessarily be conceived in motion, in continuous becoming and change: an indispensable condition for carrying out a project; not an action that fixes, limits or freezes the state of the earth, but an aptitude to insert oneself into a growing or life-changing process.

In this way "*The world itself becomes a place to study [...] by people everywhere. [...] Learning to learn means shaking off, instead off applying, the preconceptions that might otherwise give shape to their observations*".[1]

Starting from a desire to question the way of relating to a project, to be considered more as a tendency towards a condition of being in progress or of engaging or getting involved in the dynamics at work in places of landscape transformation, our reflection concerns the development of critical thinking, built through the following keywords: landscape making, care, transition, forms of time, open project.

Being in project that promotes *landscape making* is an extended and integrated knowledge process which involves different areas of investigation and skills. It requires the ability to intertwine analyses on the quality of places with interpretations of the morphological, environmental and historical stratification characters of hybrid and complex territorial schedules. The shapes of the earth, subjected to

[1] Ingold (2013).

the incessant intervention of natural elements, make man one of the many components of a global system in continuous change. Movement is the condition that gives landscape the profound meaning of a *process of change in progress*. This is not only a way to characterize and appropriate territory, but it is a way to understand what has happened in order to intervene in ongoing transformation and to continue its evolution and the writing of the future. If we consider space as a set of environments in progress or *in fieri*, that is, contexts whose composition can remain open or not completed, *making landscape* becomes above all the regaining of an ability to continue existing territorial writings. Making landscape re-emerges the ability to immerse oneself in places to evoke physical and mental relationships that, above all, make it possible to restore life to the imagination that distinguishes local cultures.

These give character and shape to the taking care of territories, building narrative plots and leaving open various possibilities of interaction between different materials, places, stories and communities. Care must inevitably be entrusted to those who live in places and feel them as their own, aware of the responsibility of preserving their values through sustainable lifestyles, uses and activities.

The methods of treatment require an ability to identify degradation, detachment and fragility factors. But they also make it possible to reveal the values to be protected, the resources to be preserved and the potential for qualification.

Instability, accumulation or excess in the use of the earth's resources cause bewilderment and require an ability to intervene in the process of environmental degradation. It is therefore necessary to *know how to take care* of a territory in order to re-acquire forgotten knowledge, temporalities and relationships: to reconstitute the most appropriate connections in the current equilibriums, but also to recover lost knowledge to re-launch more balanced dynamics. The intention to "take care, nurture, grow, and guarantee balanced relationships and their continuity is a cyclical movement"[2] and depends on the "ability to feed dialogue"[3] according to the interdependence that exists between man and his environment.

Care, therefore, goes beyond protection and in landscape, it affects heritage as much as ordinary places. But it enters a project as an essential part so that the redevelopment of degraded places can generate or integrate new landscapes.

It opens up a cycle of renewed interest in the "forms of time", which a landscape project interprets and returns to the territory as different types: those of historical stratification and cultural heritage, environmental variation, seasons, relationships and the articulation of the plant and mineral components, construction and its development, cultivation, exceptional events and calamities, maintenance and care. Moreover, we could continue to point out many areas of formal expression of time. Here, it is particularly interesting to underline how the time of becoming is the material of a project that produces landscape. This is, in reality, a collective creation, and constitutes, today more than ever, a terrain for the re-composition of social and political legitimacy. It is a public space shaped by plants, slopes and topography but also by

---

[2] McHarg (1969).

[3] Whiston Spirn (1984).

walls and supporting infrastructure. Although it is possible to guess the interrelationships between the passage of time and the variations in the configurations of the earth, the visible shapes of soils transformed by the continuous actions of modelling and accumulation of materials often arouse amazement, disbelief and even fear or disorientation. In fact, these temporal configurations speak of human intentions to use land to exhaustion or to orient its becoming until it is transformed into new artefacts of city or countryside corresponding to a "Third Nature" which has even become unrecognizable.

But the most interesting question for us is to understand whether time plays an active role in the processes of reconstruction or regeneration of new landscape and how this fits into the design process.

Unfortunately, a static interpretation of time is very often recognized in the integration of the evidence of history and stratigraphy, reduced to documents to be preserved for an unspecified time. In other cases, time is interpreted to construct a sort of time schedule which ends in the realization of works or prefigures variations of the plant component alone.

Time as a conformation factor of a project for landscape is little investigated by research but constitutes a fundamental sphere of interaction and integration with territorial dimensions. How the time of landscape, in its multiplicity, modifies the quality of a project is to be found in its ability to manage transition and conform to places as an open project.

It is therefore necessary to rediscover and rewrite Time in research processes to observe, understand, and transform experience into knowledge. This will be interpreted to relaunch new possibilities with far-sightedness (forward-looking ELC) in order to change and re-establish new relationships and visions *in perspective* as agents increasingly aware of the importance of their territory.

To achieve more ambitious objectives of critical conscience, it is necessary to re-elaborate the concept of *transition* to enhance its spatial and temporal values. In fact, welcoming and enhancing transition leads a project to assign a mediating role to some systems.

The characters of intermediate systems are capable of providing ecosystem services and sustaining variations. The search for permeability and articulation can accommodate different conditions and influence the approach to the design of places and infrastructural systems. The role of margins as media and transition areas therefore brings into play an attention that involves physical factors but above all a different attention of the project towards the built environment. In transition areas, different configuration modes are concentrated at the edges.

The places of transition inevitably highlight temporality, passage and process, dynamism, the interactions between elements and environmental factors, but also the need to be able to welcome transformation.

Transitions of various kinds involve techniques and arts, times and environmental factors, and generate possibilities and new landscapes. In transition territories, particular production processes are formed which in the wise composition of artifice and nature govern the rhythms and modalities of variation.

A project that produces landscape works with attention on transition areas to understand the rules and to insert them in different levels of permeability, resistance and interaction between recurring and exceptional factors. In places that recognize themselves as part of transition territories, the adaptation of forms to environmental as well as urban and socio-economic dynamics experience an opening to new scenarios which make it possible to overturn the classic concept of design, to open new ways to more flexible conception methods.

Among these, the concept of *open project* expresses a need to adapt to changing conditions, but also a responsibility to transmit resources and potential to the future of territories. Operating in the complexity of a system of relationships that are consolidated, modified, integrated or interrupted over time always responds to various conditions, uses, sensitivities and cultures.

The condition of *open project* is typically what prefigures and activates processes that generate living places and landscapes, capable of changing without its primary character and values being altered.

The inclusiveness of a project is a difficult feature to ensure over a long time and requires awareness of cultural, social and environmental evolution, but also a sense of responsibility for what constitutes the physical and relational structure of the modified context.

A project can be understood as an *open work*[4] when it is conceived and articulated as a process or an iterative development of intentions based on different phases of reading and writing. In fact, to give rise to a project, it is necessary to read the existing one and interpret it. From these first traces of design thinking, ideas or mental images arise as an embryonic form of design. This, in its establishment, requires more types of writings, drawings and sketches from which design choices gradually emerge to be shared and subjected to other phases of dialogue and critical review. Subsequently, the writings on paper migrate towards the ground giving the sign or design intention the sense of an engraving of the earth. This progressively takes shape until it reaches different stages of stabilization. This way of affecting or building a landscape artefact over time passes through different stages, which in the case of a landscape composition include the time of growth, maturity and decline. Therefore, a project can no longer be understood as a time-limited action that generates immutable objects or cities and landscapes similar to design objects to be consumed. A project, as defined by A. Berque, is the dynamic expression of a sign or trace (*jet*), which is given a perspective value, that is, a dimension of propensity towards the future (*pro*). So a project is the action of reading the existing that gives continuity to a system of materials or forces that in nature can never be considered closed or concluded. Accepting and recognizing these dynamics therefore firstly requires an ability to *resume writing in progress* and a desire to *leave the discussion open*. This is because a landscape project is not a formal, fixed and immutable solution.

On the contrary, a landscape project requires an ability to recognize and welcome ongoing movement: movement that animates and gives life to natural systems in

[4] Eco (1967).

action. This recognition gives rise to a commitment to maintain an attitude of wakefulness or care in which doing a project becomes knowing how to move, accompanying mutation in its making, through small changes in space over time like a *guardian*[5] of landscape.

References

Corajoud M (2000) Le projet de paysage: lettre aux étudiants, coll., direct. Brisson J-L Le jardinier, l'artiste et l'ingénieur. Les Editions de l'Imprimeur
Eco U (1967) Opera aperta, Milano Bompiani (1a edizione 1962)
Ingold T (2013) Making. Anthropology, archaeology, art and architecture. Routledge, New York
McHarg I (1969) Design with nature. National History Press
Whiston Spirn A (1984) The granite garden. Urban nature and human design. Perseus Books Group, New York

[5] Corajoud (2000).

Part IV
Observing Landscape

Chapter 42
Observing Landscape: General Reflections and Linkages with the UNISCAPE2020 Conference

Juanjo Galan and Juan Manuel Palerm

Abstract In this text, we intend to look again at the Earth, or rather we are obliged to understand globality (globalization), to look again at our planet, which is the same as always, although deeply modified and wounded. And that is why it appears diverse and different. The landscape of our environment, real and imaginary, requires this time of observation and with it, through its perception and understanding, project it again.Uniscape, as a network of European universities on the landscape and a framework for reflection of the landscape observatories in Europe, requires and proposes new instruments for its management and offers, in this context and publication, a space for debate.

Keywords Landscape · Research · Observatories · Perception · Diagnosis · Monitoring · Transversality · Multidisciplinarity

The dynamic, ubiquitous and participatory qualities of landscape are at the core of the European Landscape Convention and have triggered many of the reflections that have emerged around the Convention since the year 2000. On the 20th anniversary of the ELC, the Cultivation of the Continuity of the European Landscape was proposed as the central motto of the UNISCAPE2020 Conference in order to emphasize the evolving and collective quality of landscape and evoke, at the same time, new conceptual possibilities for the landscape concept. Cultivation requires intentionality, interpretation of external and internal signs and a conscious understanding of time, processes and cycles. Cultivation implies care, attachment, commitment and attentive observation.

J. Galan (✉)
Department of Architecture, Aalto University, Espoo, Finland
e-mail: juagavi@urb.upv.es

Department of Urban Planning, Universitat Politecnica de Valencia, Valencia, Spain

J. M. Palerm
University of Las Palmas de Gran Canaria, Las Palmas, Spain
e-mail: jpalerm@paltab.com

T. Matteini et al. (eds.), *Cultivating Continuity of the European Landscape*, Environmental History 15, https://doi.org/10.1007/978-3-031-25713-1_42

In this chapter, we would like to reflect on the observation of landscape as a way to interact with it and as a necessary tool to facilitate its cultivation (understood in this text as an umbrella concept comprising protection, management and planning). The etymology of the Latin verb "observare" reveals that it was a synonym of "watch over, note, heed, look to, attend to, guard, regard and comply with". Nowadays, the act of observing has kept many of these meanings, can be conducted in multiple ways and can serve different purposes, both in practice and education. Thus, a phenomenological approach to the observation of landscape would emphasize its experiential qualities, both as a lens to study the landscape and as a way to understand how specific individuals or groups experience particular landscapes (Nogue 1993). Conversely, the observation of landscape can be aimed at the study or description of physical landscape, quite often from a positivist perspective in which knowledge is derived from the rational and logical interpretation of sensory experience. Differently again, a cultural approach would place more emphasis on the observation of the socio-cultural processes underlying the evolution of landscape and the ways in which those processes frame the interaction between individuals, communities and the environment. Obviously, all these types of observations are interconnected through the multifaceted character of landscape and its historical evolution, but their specific weight can vary depending on the purpose or intention of the study.

From a planning or design perspective, the observation of landscape provides essential information for the definition of policies, strategies, actions and projects aimed at its protection, management or planning. On the other hand, the monitorization of landscape has gained increasing relevance as a continuous, diachronic and systematic way of observing landscape in order to provide solid evidence of its evolution (Kienast 2019). In this regard, by stating the need to identify, characterize and assess landscapes with the participation of the interested parties, the ELC highlights the importance of promoting rigorous and co-participated landscape observation. Similarly, by explicitly affirming the importance of "monitoring landscape changes and the effectiveness of (landscape) operations", Recommendation CM/Rec(2008)3 of the Council of Europe Committee of Ministers defines a clear purpose for monitoring processes and introduces Landscape Observatories/Institutes/Centres as potential key actors for this endeavour and for the exchange of landscape information (ELC 2000; CM/REC 2008).

However, the observation of landscape by researchers and scholars exceeds utilitarian or goal-oriented approaches and oVen dives into the theoretical, intellectual, pedagogical and artistic investigation of landscape phenomena in which both the observer (subject) and the observed (landscape) can establish different types of dialogue and can influence each other in multiple ways. In this wider context, both holistic and specialized studies coexist and retrofit each other, enriching at the same time the possibilities of the landscape concept to foster reflections on the interactions between humans and the environment (Antrop and Van Eetvelde 2017). Moreover, in this interdisciplinary context, the many "dimensions of place" can facilitate the development of a "socio-ecological", "place conscious education" (Gruenewald 2003: 619) and subsequently foster multiple ways of looking at the landscape.

All these different ways of observing landscape and all these different purposes for conducting a systematic observation of it can be found in contemporary landscape research and were represented at the UNISCAPE2020 Conference. The importance of the "eye" of the observer, the experience of landscape by people with different sensibilities or the multiple relationships that observers can establish with a particular site are amongst the phenomenological aspects that have been investigated and attested by the Fondazione Benetton (Fondazione Benetton 2024). In the same line of thought, Andersen (2024) argues that the experience of physical landscape together with personal backgrounds and socially constructed understandings define a lens to critically observe landscape and to collectively envision its future evolution (Andersen 2024).

On the other hand, the act of observing has been used as a vehicle to create different types of landscape observatories committed to the systematic observation of landscape as a means to support the implementation of the European Landscape Convention at a national, regional or local level. Some landscape observatories have been able to integrate multifaceted observation of landscape and to effectively link their activity with spatial planning, landscape governance or community building. The Landscape Observatory of Catalonia stands out as one of these observatories and its structure, composition and activities demonstrate how the contents of the European Landscape Convention, together with a clear legal and political framework, can define an effective road map for landscape protection, management and planning (Sala 2024). Similarly, the Landscape Observatory of Finland is an example of a national observatory co-participated by governmental, civil, economic and academic organizations whose activity was initiated with a collective diagnosis of the implementation of the ELC in that Nordic country and the subsequent definition of strategic actions and projects (Galan 2024).

Landscape Observatories can also emerge from bottom-up processes. In these cases, the characteristics of their constituent members and the conditions of the physical and social context in which they operate can strongly influence their activity. For instance, a critical modification of landscape can mobilize different types of actors and awaken social interest in observing and monitoring the transformation of landscape. Thus, the formation of the Landscape Observatory of Alqueva was triggered by the construction of the Alqueva Dam in Portugal, and this observatory is now a clear example of a multisectoral platform supporting the development of policies and strategies for landscape and territorial planning at a regional scale (Freire et al. 2024). Similarly, the Landscape Observatory of Charneca appeared in central Portugal to support the continuity of a unique type of agricultural landscape. Due to the complex and multimodal composition of the Charneca observatory, a study was conducted to specifically "observe" its internal functioning as a social network. The results of this study were quite revealing and provided key information to recognize the strengths and weaknesses of the network and to improve its future performance and social impact (Saraiva et al. 2024).

According to the spirit of the ELC, the identification of landscapes, their characterization and assessment, the formulation of landscape quality objectives and the definition of landscape policies or actions should count on the active participation of the interested parties.

Consequently, it has been argued that landscape observatories should be shaped following the shapes of civil society and that their work should be based on a deep understanding of individuals' expectations and issues and of the citizens' actions that give places meaning (Lorenzzeto 2024).

Moreover, the multiple meanings of the act of observing have fostered the creation of different types of landscape observatories and have prompted their connection to other organizations or their involvement in different kinds of studies. Thus, the observation and monitoring of landscape have been often related to the protection and management of exceptional sites. In these cases, landscape observatories can become co-participated stewards of the landscape and their activity can be linked to that of associations, foundations or museums (Żołnierczuk 2024). Similarly, the observation of landscape and the activity of landscape observatories can focus on specific types of landscape elements, landscape processes or landscape types. In this regard, Larsson (2024) suggests that Landscape Observatories can play a key role in the coordination of large-scale infrastructures by providing place-centred logic and by offering a regional perspective (Larsson 2024). Alternatively, Dam, 2021 proposes incorporating new narratives and topics into the activity of regional landscape observatories by working on the study and management of drystone walls as vernacular manifestations of the rich and site-based material culture of European landscapes (Dam 2024). This proposal highly resonates with the thematic study on terraced landscapes conducted by the Landscape Observatory of the Autonomous Province of Trento. This extensive and detailed work proves how knowledge of a landscape can consistently inform its protection, management and planning, and which culminated in the publication of a special Atlas (Tecilla and Cosner 2024).

If there is no landscape without an observer, this short text has tried to show how the multiple meanings of observation can foster different ways of studying the interactions between landscape and people. Furthermore, these different approaches have generated new possibilities for the observation of landscape by linking it with its protection, management and planning and by making multifaceted, rich observation of landscape an opportunity for dialogue and collaboration between people with different types of perceptions.

References

Andersen NM (2024) Aesthetically performed landscape criticism. In: Agnoletti M, Dobricic S, Matteini T, Palerm JM (eds) Cultivating continuity of the European landscape: new challenges, innovative perspectives. Springer, Berlin

Antrop M, Van Eetvelde V (2017) Landscape perspectives: the holistic nature of landscape. Springer, Berlin/Heidelberg

CM/REC (2008) Recommendation CM/Rec(2008)3 of the Committee of Ministers to member states on the guidelines for the implementation of the European Landscape Convention; Council of Europe

Dam T (2024) Cultivating drystone walls—regional observatories for European landscapes. In: Agnoletti M, Dobricic S, Matteini T, Palerm JM (eds) Cultivating continuity of the European landscape: new challenges, innovative perspectives. Springer, Berlin

ELC (2000) European landscape convention, Council of Europe

Fondazione Benetton (2024) Landscape. https://www.fbsr.it/en/landscape. Accessed 30 Apr 2021

Freire M, Ramos IJ, Rego MC (2024) Landscape observatory in Alqueva, Alentejo region (Portugal). Why it matters? In: Agnoletti M, Dobricic S, Matteini T, Palerm JM (eds) Cultivating continuity of the European landscape: new challenges, innovative perspectives. Springer, Berlin

Galan J (2024) Landscape observatories: an analytical framework and a case study at the landscape observatory of Finland. In: Agnoletti M, Dobricic S, Matteini T, Palerm JM (eds) Cultivating continuity of the European landscape: new challenges, innovative perspectives. Springer, Berlin

DA Gruenewald 2003 Foundations of place: a multidisciplinary framework for place-conscious education Am Educ Res J 40 3 619 654

Kienast F, Wartmann F, Zaugg A, Hunziker M (2019) A review of integrated approaches for landscape monitoring: report prepared in the framework of the work program of the Council of Europe for the implementation of the European landscape convention. Sine Nominee, in CEP-CDCPP (2019) 7E

Larsson A (2024) Regional coordination of large scale infrastructure projects—a new role for landscape observatories? In: Agnoletti M, Dobricic S, Matteini T, Palerm JM (eds) Cultivating continuity of the European landscape: new challenges, innovative perspectives. Springer, Berlin

Lorenzzeto E (2024) The observation of informal landscape public participation. In: Agnoletti M, Dobricic S, Matteini T, Palerm JM (eds) Cultivating continuity of the European landscape: new challenges, innovative perspectives. Springer, Berlin

Nogue i Font J (1993) Toward a phenomenology of landscape and landscape experience: an example from Catalonia. In: Seamon D (ed) Dwelling, seeing and designing: toward a phenomenological ecology. SUNY Press, Albany, New York

Sala P (2024) The landscape observatory of Catalonia. A journey of fifteen years. In: Agnoletti M, Dobricic S, Matteini T, Palerm JM (eds) Cultivating continuity of the European landscape: new challenges, innovative perspectives. Springer, Berlin

Saraiva MG, Loupa-Ramos I, Van Eetvelde V (2024) Landscape observatory in Alqueva, Alentejo region (Portugal). Why it matters? In: Agnoletti M, Dobricic S, Matteini T, Palerm JM (eds) Cultivating continuity of the European landscape: new challenges, innovative perspectives. Springer, Berlin

Tecilla G, Corner A (2024) The Atlas of terraced landscapes of Trentino, Italy. In: Agnoletti M, Dobricic S, Matteini T, Palerm JM (eds) Cultivating continuity of the European landscape: new Challenges, innovative perspectives. Springer, Berlin

Żołnierczuk M (2024) The museum as an active observatory: protecting and shaping the landscape. In: Agnoletti M, Dobricic S, Matteini T, Palerm JM (eds) Cultivating continuity of the European landscape: new Challenges, innovative perspectives. Springer, Berlin

Chapter 43
The Landscape Observatory of Catalonia. A Journey of Fifteen Years

Pere Sala i Martí

Keywords Landscape Observatory of Catalonia · Landscape values · Landscape perception · Landscape planning · Landscape knowledge · Landscape participation

In December 2000, the Parliament of the autonomous region of Catalonia (Spain) signed the European Landscape Convention (Council of Europe 2000), promoted by the Council of Europe and adopted on 20 October 2000. Without the institutional (and moral) impact of this Convention, many of Europe's subsequent political, legislative, academic and professional initiatives on landscape would have been unthinkable, and Catalonia is no exception.

Five years later, in 2005, Parliament passed the Landscape Protection, Management and Planning Act 8/2005 (Generalitat de Catalunya 2006) as the basis for landscape policy in Catalonia. With the passing of this Act, landscape ceased to have an exclusively aesthetic connotation and became an active element in regional planning and management and a driving force in development. A few months before the Act was drawn up, the Landscape Observatory of Catalonia was set up (and eventually recognized in the Act) as the Generalitat of Catalonia's assessment body and as a way to create public awareness on landscape matters. The observatory has been operating since 1 March 2005.

The book "Quinze anys" (Observatori del Paisatge de Catalunya 2020) was published to mark the fifteenth anniversary of the Landscape Observatory. It contains fifteen concepts and reflections that resonate and connect with the entity's objectives, tasks and *raison d'être*, and the raw material with which it works: the landscapes of

P. S. i Martí (✉)
Landscape Observatory of Catalonia, Catalonia, Spain
e-mail: pere.sala@catpaisatge.net

T. Matteini et al. (eds.), *Cultivating Continuity of the European Landscape*, Environmental History 15, https://doi.org/10.1007/978-3-031-25713-1_43

Catalonia and their values. The fifteen concepts are: journey, place, laboratory, knowledge, action, mirror, dialogue, community, bridges, atmospheres, values, emotions, gazes, colours and edges.

43.1 Fifteen Concepts and Reflections for Fifteen Years

Since the Landscape Observatory started this amazing **journey** fifteen years ago, it has been joined by many travelling companions from Catalonia, Spain and around the world (institutions, organizations and networks, from local to international levels). With their help, it has generated an enormous body of knowledge about landscape that has reached institutions and society in general and guided public policy. After fifteen years, level zero has been covered; in other words, the Observatory has laid the foundations that are indispensable to take a leap forward and go even further afield.

The Observatory is the **place** where theoretical reflection and practical application come together to serve landscape policy, and it has been located from the beginning in the city of Olot. It is a shared, open, receptive meeting point, midway between civil society and government and academic and professional sectors, in all areas related to landscape culture. The Observatory is completely open to new and emerging approaches, to what is happening today, and not just what happened yesterday, and has come down to us through that extraordinary palimpsest of our landscape.

The Observatory is constantly evolving and redefining itself; by applying the principles of pluralism and diversity, it has become a **laboratory**, an incubator for innovative projects. It is open to creative, imaginative initiatives, and it explores and creates new tools and instruments that often go beyond the limitations that inevitably constrain government activity (Nogué and Sala 2018).

One of the main functions of the Landscape Observatory is **knowledge** generation. Knowledge is fundamental for taking decisions both locally and nationally regarding landscape; knowledge that, by definition, must be holistic and cross-cutting. The Observatory generates, shares and transfers knowledge and methodologies to all parts of society, inviting participation from all its stakeholders. And it does so in the conviction that expert knowledge requires the interchange of many and diverse inputs, from the knowledge of specialists to the perceptions and opinions of civil society. To this end, landscape catalogues have become the main instruments for generating knowledge about landscape in Catalonia and for introducing landscape quality objectives into urban and regional planning as well as into sectorial policies, such as agriculture, heritage, tourism and energy, with the cooperation and participation of all stakeholders. The catalogues mark a turning point in landscape knowledge and management (Sala 2010) due to both their ability to document and guide public policies and their potential to educate and raise public awareness (Sala 2010; 2012).

Knowledge must always precede **action**. The Observatory is a space for reflection and action on landscape issues; its purpose is not merely to provide support for public policies but also to educate, imbuing society as a whole with knowledge and increased awareness. In addition to landscape charters and plans, projects, guidelines, documents, symposia, seminars, courses and publications, the landscape catalogues continue to be one of the tools we use to directly impact society by fostering more specific planning, management and protection measures. Despite the fact that Catalan Landscape Policy has been in existence for fifteen years, landscape is only beginning to penetrate public and political debate, so the task ahead of raising awareness is immense.

We could also say that the Observatory acts as a **mirror** of landscape. Through its action, the Observatory reflects landscape's true state of health, detects emerging dynamics and provides input for future agendas.

The next three concepts (dialogue, community and bridges) are absolutely crucial. The daily **dialogue** that contemporary societies have with their everyday landscapes demands continuous interdisciplinary methodology exchange and spaces for sharing ideas, cooperation and consensus between government and civil society and between the public and private spheres, leading to new forms of participative democracy in territorial management. Such experiences as the UNESCO Candidature of Priorat or the Cross-border Plan of La Cerdanya (Observatori del Paisatge de Catalunya 2013) show that the Landscape Observatory can promote new forms of participatory democracy in all aspects of territorial governance and management. Closely related to this issue, landscape is generating increasing local interest as local stakeholders perceive landscape as a catalyst for development and a way to increase self-esteem, identity and quality of life. The current COVID-19 crisis is reinforcing this situation.

There is no landscape without **community**. We live in a community and it is with this community that we share our perceptions and experiences of landscape. This is why landscape provides such an excellent tool for addressing the complexity of territorial management and for promoting community organization by constantly eliciting new forms of engagement and governance. Collectively, we cannot be indifferent to those landscape values that may help us transition towards a new model of society inspired by cohesion, solidarity, diversity, cooperation, the common good, integration, peaceful coexistence, tradition, innovation or beauty.

Dialogue and community are inseparable**.** Integrated visions of landscape have the capacity to build (political, cultural, economic…) **bridges** between broad sectors of society and find a middle ground and areas of consensus between often opposing visions. The aim is to promote landscape quality that has positive effects on people's well-being, generates economic opportunities and invigorates territories both socially and culturally.

Another function of the Observatory is the creation of **atmospheres,** spaces for sharing diverse, contrasting experiences and sociability that, in turn, generate new atmospheres… in an infinitely repeating cycle. Building landscapes implies creating synergies, complicities and interdependencies from a sense of proximity and everyday experience.

One of our main efforts during these 15 years has been the identification of landscape **values**. That is right; landscapes treasure an extraordinary diversity of values: natural, aesthetic, historic, social, symbolic and productive. These values are acknowledged both by the scientific community and by the general public. And some of them are increasingly recognized as crucial in addressing the challenges and uncertainties of today's world, such as climate change, health crises or forced migrations.

The concept of *values* is closely linked to another: **emotion**. Landscape can generate individual **emotions** and, at the same time, be a depositary of scientifically recognized and socially shared values. Landscape is full of meaning; it summons memories and awakens emotions that have a direct impact on our well-being and health. This is why the methodologies and work of the Observatory have always considered the emotional and affective dimension resulting from the relationship between people and their landscape to be inseparable from the physical and material dimensions.

Values and *emotions* lead us directly to the idea of **gazes**. The landscape is the result of the confluence between different gazes, perceptions, interpretations and sensibilities. Every one of these gazes is needed to understand a landscape's logic and idiosyncrasy and to ensure its adequate management, as the European Landscape Convention establishes.

And of the final two reflections related to the last fifteen years, the first is **colours** because blending a few primary colours in different proportions gives an extraordinary range of complementary colours. The Observatory is a primary **colour**, a space in which extremely diverse sensibilities converge, interacting to generate new projects and new relational frameworks. The 134 landscape units identified by the Landscape Observatory (which are understood as areas that share the same landscape character) define the first map of Catalonia's landscapes (Observatori del Paisatge de Catalunya & ICGC 2019) and provide a metaphorical expression of this chromatic diversity. Beyond being the basic territorial units for the application of specific landscape policies, or even new political-administrative entities rethinking landscape governance, they are spaces for life, meeting points, repositories of collective memories.

The last idea is **edge**. The Observatory is at the interface between science and management. Similarly, landscapes are thresholds between different territorial situations and—sometimes—mental attitudes. To act on these fringes, we need to find transitional spaces, the edges between their time and space scales, and understand their social and symbolic references.

43.2 Looking Ahead

Now, fifteen years after the approval of the Catalan Landscape Act, is a good time to take stock. The landscape policies of Catalonia were drawn up and took their first steps in a context of an economic boom which drove great, although not entirely well accepted, dynamism that substantially impoverished the landscape in many parts of the territory. Now that the 134 landscapes of Catalonia have been identified and defined, that the main instruments of the Landscape Act have been rolled-out and that the Landscape Observatory is a consolidated body supporting public administrations and raising public awareness; now that new bridges have been built between the landscape and land planning policies; and now that the importance of the landscape and the work of the Observatory has gradually been internalized by the administrations, organizations and economic and social agents of Catalonia, it is time to make a leap forward in landscape policy to make it more transversal, pragmatic and designed for action in order to reinforce public–private partnerships and cooperation and ensure that we are better oriented towards our current challenges and today's areas of great landscape debate.

Some of these challenges and areas are landscape contribution to well-being, equality, health, social cohesion and quality of life; the growing role of landscape as an indicator of, and now as a response to, climate change in a situation of global climate emergency; major dialectics: Local–global, rural–urban, authenticity *versus* artificialization or transformation *versus* destruction; the question of identities in a globalized and hyper-mobile context; the creation of new benchmark landscapes; ordinary landscapes and their *patrimonialization*; the relationship between landscape and new tourism; the common good; city limits; participatory democracy; virtual landscapes.

Finally, these challenges show us that landscape policies must be cross-cutting, interdepartmental and participatory, which is never very easy to achieve. At a time of global challenges and social uncertainty, there is an increasing need for in-depth discussion on the role of landscape in contemporary societies.

References

Council of Europe (2000) European landscape convention, Strasbourg, Council of Europe.

Generalitat de Catalunya (2006) Llei 8/2005 i Reglament de protecció, gestió i ordenació del paisatge, (Quaderns de Legislació; 65). Generalitat de Catalunya, Barcelona

Nogué J, Sala P (2018) Landscape, local knowledge and democracy: the work of the land-scape observatory of Catalonia. In: Fairclough A, Graham S-H, Ingrid i Swanwick, Carys (eds) Landscape character assessment: current approaches to characterisation and assessment. Routledge Handbook, Oxon

Observatori del Paisatge de Catalunya (ed) (2013) El paisatge transfronterer de la Cerdanya. Un projecte de futur en comú = Le paysage transfrontalier de la Cerdagne. Un projet d'avenir commune

Observatori del Paisatge de Catalunya 2016. [en línia]. https://www.cat-paisatge.net/cat/catalegs.php. consult: 8 Jan 2021

Observatori del Paisatge de Catalunya, Institut Cartogràfic i Geològic de Catalunya (ICGC) (2019) [Mapa]. 1:250.000. Barcelona: Institut Cartogràfic i Geològic de Catalunya

Observatori del Paisatge de Catalunya (2020) Quinze anys. Olot: Observatori del Paisatge

Sala P (2010) Els catàlegs del paisatge. In: Oriol Nel·lo (dir) La política de paisatge a Catalunya, Barcelona, Departament de Política Territorial i Obres Públiques de la Generalitat de Catalunya. (Guia d'Integració Paisatgística; 4).

Sala P (2012) Regional and local participation and co-operation in implementing the European landscape convention. The experience of the landscape observatory of Catalonia. In: Raasakka N, Sivonen S (ed) Northern landscapes. Implementation of the European landscape convention in the north CaloPe municipalities. Conference in Inari 7-9.9.2011, Inari, Centre for Economic Development, Transport and the Environment for Lapland (Reports 48), pp. 55–66

Chapter 44
Aesthetically Performed Landscape Criticism

Thematic Session C: Observing Landscape. Topic 2: Landscape Observatories: Post 2020 Agenda and Future Roles

Nina Marie Andersen

Abstract The role of landscape analysis in a professional planning situation is to convey a common, imagined, future landscape. To represent the complexity of qualities of an area acknowledged as landscape, it is necessary to perform an aesthetic landscape criticism. Within this understanding, an aesthetic act of thinking and writing is required to capture landscape qualities adequately. In meeting the European Landscape Convention's aim to raise landscape awareness, a hermeneutic approach is therefore more appropriate than a positivistic one. However, the aesthetical approach is barely recognized in the Norwegian planning system today. Landscape Observatories introduce apparently neutral mapping systems made to categorize and document areas. Unfortunately, they fail to capture the distinctive qualities that create landscape character and may thus reduce the complexity of a landscape. By offering a positivistic alternative intended to map landscape qualities and raise landscape awareness, these mapping systems will weaken the position of the aesthetical approach even more. In the future, we need to focus on how to convey information provided by mapping systems by performing landscape criticisms aesthetically in order to raise landscape awareness.

Keywords Landscape analysis · Aesthetics · Imagination · Landscape writing · Landscape observatories

---

N. M. Andersen (✉)
Associate Professor, Norwegian University of Life Sciences, Ås, Norway
e-mail: nina.marie.andersen@nmbu.no

T. Matteini et al. (eds.), *Cultivating Continuity of the European Landscape*, Environmental History 15, https://doi.org/10.1007/978-3-031-25713-1_44

44.1 Introduction: Landscape Constructed by Words

Landscape is constituted of the way we think about it, talk about it and write about it. It emerges through the experience of its physical layout, but our personal background, memories and imagination affect the perception of it. Thus, experiences are individual but also rooted in a common and culturally shared understanding.

Managing landscapes today—as projects—involves a planning process where public management and administration, professional planners and investigators as well as political decision-makers play important roles. The professionals prepare subject-specific reports that form the basis for the choice of solutions or development of specific landscapes. This process even includes an impact assessment of the landscape's visual properties. To avoid a purely subjective analysis, it is important to implement this within a professional framework rooted in landscape theory, history, and norms of understanding.

The point of departure of this paper is landscape analysis, understood as a critique which works as a tool to define a *shared landscape* in a planning context. The role of such a critique is to convey an imagined, future landscape in a text, i.e. a recognized instrument representing both its value and potential. Such shared landscape requires and implies a collective act of imagination. Nevertheless, it is based on professional landscape theory. Thus, as Fiskevold and Geelmuyden argue, the expert leads the performance of this landscape in a planning process. "Raising landscape awareness means making visible a piece of land under a current circumstance, thus making it available for shared recognition and comprehension following the visual rationality of an analytical narrative" (Fiskevold and Geelmuyden 2019: 3).

44.2 Scope

Hermeneutics forms the basis for the acquisition of insight and understanding. Paul de Man explains hermeneutics as "a process directed towards the determination of meaning" (de Man 1982: ix). Thus, it acts as a driving force towards understanding and is crucial for professional landscape assessment. Such an approach also serves as the premise for this paper.

I will argue that a hermeneutic and aesthetical approach is more fruitful and appropriate than a positivistic one in meeting the European Landscape Convention's aim of raising landscape awareness. Landscapes are complex, and it is essential that the qualities and complexity of a landscape are adequately captured in a landscape analysis to provide an adequate basis for the imagination.

I will also argue that if fixed indicators are applied for assessing the effectiveness of landscape policies within Landscape Observatories, the hermeneutic approach may be threatened and, consequently, the prospect of raising landscape awareness will be weakened.

44.3 Aesthetics to Convey Complexity

The role of landscape analysis in a professional planning situation is to convey an imagined, future landscape. To represent its complexity in that respect, it is necessary to perform an aesthetical landscape criticism (Rendell 2010). In order to capture landscape qualities adequately, an aesthetic act of thinking and writing is required. Aesthetics, in this regard, is performed at two levels: firstly, through the expert's *engagement* with the object of analysis—a landscape—and secondly, through the expert's *act of representing* the landscape.

There are several means of highlighting the characteristics of landscape. Not least, visual art has a strong tradition of representing landscapes (Andrews 1999). However, this paper's focus is on written text, due to the availability of anyone to perform landscape assessments and, more importantly, the opportunity the text medium provides to represent the dimensions of time and space, both essential for landscape.

In meeting the European Landscape Convention's aim to raise landscape awareness, a hermeneutic act is more appropriate than a positivistic one (Fiskevold and Geelmuyden 2019). And both the described levels (engagement and representation) are part of a hermeneutic process which supports qualitative assessments. They are not straightforward tasks, but need *thought, understanding* and *time* to be performed.

An expert's engagement with the object of analysis relates to the experience of landscape. To make this action productive, he or she must be open when approaching an analysis and mentally present in experiencing the place. This means being aware of and paying close attention to the actual situation and daring to challenge the conventions of how landscapes are perceived. In this way, the expert can recognize and identify essential landscape qualities. The recognized landscape then needs to be represented in text, accompanied by photographs and other graphics. Representing the appreciated landscape by describing defined landscape properties, rather than by normative concepts such as defining the landscape value as "good" or "bad", requires that the critique be performed within a professional framework and a clarified context. Hence, the analysis becomes academically precise.

The text should give the reader access to a future landscape which is made available through the reader's imagination. Also, the act of writing is an aesthetic performance which is carried out hermeneutically. The approach is expert-based, but aims to appeal to readers without a professional landscape background. This requires language which can equip the reader with images of a future situation. To achieve this, there is no obstacle to using literary means.

This aesthetic and hermeneutic process implies that landscape qualities be defined professionally but also acknowledged by non-professionals because of the aesthetic performance of the writer. Thus, this approach fits the objectives of the European Landscape Convention on addressing the aspirations of the general public.

Landscape mappings may tell us about the condition of the land surface but do not represent *landscape*. Unfortunately, they fail to capture the distinctive qualities that create a landscape's character and may therefore reduce the complexity of a landscape. Thus, if Landscape Observatories introduce apparently neutral mapping systems as a tool to categorize and document areas, the result may be an unfortunate simplification of the object described. Consequently, the basis for creating landscape awareness in managing landscapes will be deficient.

44.4 The Threat of Fixed Indicators

However, the aesthetical approach is barely recognized in the Norwegian planning system today. Rather, seemingly neutral mapping is held to be more reliable and relevant than qualitative assessments based on landscape architecture theories and practice. Norway does not have a Landscape Observatory. But in 2019, *Nature in Norway—landscape (NiN—landskap)*, a system that produces machine-generated landscape maps, was launched with the aim, among others, of registering changes over time (Erikstad et al 2019).

A post in a seminar on this methodology explained how through this system Nordland county municipality gained a knowledge base they could refer to in planning processes. The example cited was the islands of Helligvær off Bodø in northern Norway (see Fig. 44.1). Because the area was proven to be "a rare landscape type" by the *NiN-system*, the mapping could serve as an argument against the suggested development envisioned for the islands (Davidsen 2019). However, anyone with a touch of landscape understanding would be able to comprehend that the proposed change would impair the strong landscape character of the islands. But why then do the authorities need such a system to rely on? It may seem that the possibility of referring to the allegedly neutral mapping is necessary to give force to the argument.

Based on my own experiences from working with landscape assessments, including visual analyses, within the planning system, I know it can be challenging to argue qualitatively. For example, in an impact assessment of a new railway through the Norwegian city of Hamar, a limited area defined in the Register of Directorate for Cultural Heritage (2019) was included as an affected area. One of the railway lines in question was to be located on an embankment in a bay which constitutes a strong landscape feature in the medium-sized city. The location was outside the mapped polygon of the registered heritage environment (see Fig. 44.2). However, the *visual impact* on that urban structure was substantial and the visual character of the urban environment would have been significantly weakened if the large-scale railway facility was constructed in the described context (Andersen et al. 2019). Yet, as the proposed transformation was outside the mapped delimitation of the heritage values, it was difficult to argue that the railway facility would weaken the landscape value of the whole area because of its visual impact. The qualitatively rooted argument of spatial and visual connections in the urban area did not seem valid for a section of the public administration involved. It seemed like the registration system itself had become an authority that required nothing more than a referral, its meaning appeared to be irrelevant.

Fig. 44.1 Helligvær. (Foto: Ernst Furuhatt)

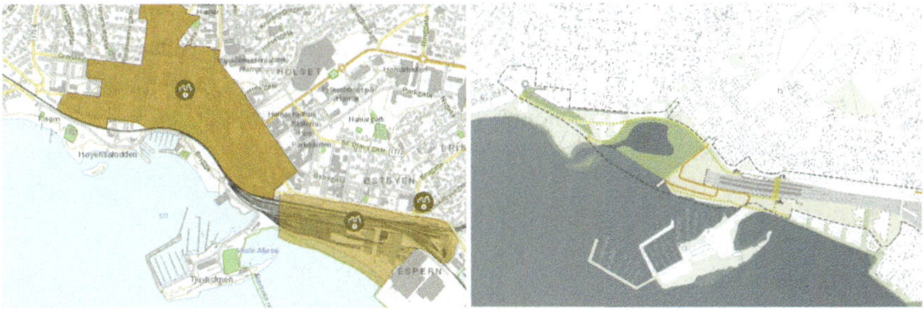

Fig. 44.2 Section from the National Heritage Register map (left). (https://riksantikvaren.maps.arcgis.com/). Section from the plan of the planned situation by Bane NOR (right)

These two examples illustrate the problems that can emerge from the rigid use of inadequate indicators or reductionistic methods. By offering a positivistic alternative for mapping landscape qualities and raising landscape awareness, these mapping systems will weaken the position of the aesthetical approach even more. When the object of description contains more complexity than the system has the ability to cope with, the system ends up reducing the complexity (Fiskevold 2019).

44.5 Discussion

I have argued that a hermeneutic or aesthetical approach is more appropriate than a positivistic one in meeting the European Landscape Convention's aim of raising landscape awareness. This is because of the approach's potential to promote an imagined, future, and complex landscape. In addition, the approach also presents the present and the past in a more adequate way for this purpose. To fulfil potential and capture the blend of landscape qualities, an aesthetic act of thinking and writing is required. Otherwise, adequate landscape awareness is difficult to achieve. Nonetheless, by simply referring to concrete, measurable conditions or fixed indicators, we may reduce landscape complexity because of the system's inability to capture "landscape as perceived by people".

Defining fixed indicators for assessing the effectiveness of landscape policies within Landscape Observatories means offering a competing alternative that may threaten a hermeneutic approach and, consequently, threaten the raising of landscape awareness. This threat arises principally because when such systems exist as an alternative, decision-makers and public management seem to favour measurable and quantifiable information bases rather than qualitative assessments. There is also the risk that implementing such systems in the planning process may transfer the focus from the actual qualities of a landscape to the possibility of referring to the system.

So, what do we do? Mapping systems offer important information about the land surface, but we need aesthetical analysis to raise landscape awareness. In the future, it will be necessary to promote the conveyance of information provided by mapping systems by performing landscape criticisms aesthetically. It is crucial to maintain the *will to understand* as a driving force to produce landscapes in an aesthetically adequate way, which is a hermeneutic process and aesthetic act, in order to raise landscape awareness.

References

Andersen et al (2019) InterCity-prosjektet Dovrebanen, Konsekvensutredning ikke-prissatte konsekvenser (Impact assement) for Bane NOR

Andrews M (1999) Landscape and western art. Oxford University Press, Oxford

Davidsen G (2019) Presentation at a conference under the auspices of the Ministry of Local Government and Modernisation, Oslo, 2 Apr 2019

de Man P (1982) Introduction in: Hans Robert Jauss, Towards an aesthetics of reception. University of Minnesota Press, Minnesota, pp. i–xxv

Directorate for Cultural Heritage/Riksantikvarens NB!-register. NB!registeret - Kulturmiljø av nasjonal interesse i byer og tettsteder - Riksantikvaren Acceced 8 May 2019

Erikstad L et al (2019) Natur i Norge (NiN) versjon 2.2. Inndelingen i landskapstyper. Artsdatabanken, Trondheim. 2019. Available via Artsdatabanken: Natur i Norge (artsdatabanken.no)

Fiskevold M (2019) Landskapskartlegging, systemtenkning og faglig dømmekraV. Plan 51:48–51. Available via idunn: https://www.idunn.no/plan/2019/02/landskapskartlegging_systemtenkning_og_faglig_doemmekraV

Fiskevold M, Geelmuyden AK (2019) Arcadia updated. Raising landscape awareness through analytical narratives. Routledge, New York and London
Rendell J (2010) Site-writing: the architecture of art criticism. I.B. Tauris, London

Chapter 45
Cultivating Drystone Walls—Regional Observatories for European Landscapes

Torben Dam

Abstract Drystone walls are a tangible and eye-catching structural element in the landscape. They bear witness to natural and human action and interaction and integrate the prefiguration of territory with aspects of space, life and agricultural production. How should the landscapes that they create be conserved and maintained in the future—and by whom? How can a territory accommodate considerations of space, nature, geology and material culture? This chapter aims to analyse and identify how landscape observatories can detect novel and relevant data and knowledge, and how the involvement of landscape observatories can be more effective. While there is a growing body of data and mapping, there has been little attention on particular site-specific perspectives, and how to visualise and communicate with the public. To some degree, the feeling of loss obstructs a nuanced and realistic strategy for future magic and attractive drystone landscapes. Based on cases from Småland, Sweden, Britain and Mallorca, and on three heritage cases from Jelling (Denmark), Kalkriese (Germany) and Olot (Spain), I critically analyse and identify a potential model for the future role and involvement of landscape observatories in the elaboration and management of drystone landscapes and suggest selecting particular sites in association with drywalls.

Keywords Planning · Management · Governance · Association · Site · Abstraction

T. Dam (✉)
Landscape Architect MDL, Landscape Architecture and Planning, KU-Science, University of Copenhagen, Copenhagen, Denmark
e-mail: toda@ign.ku.dk

T. Matteini et al. (eds.), *Cultivating Continuity of the European Landscape*, Environmental History 15, https://doi.org/10.1007/978-3-031-25713-1_45

45.1 Introduction

Heritage landscapes engage with the genius of site. Since 1980, ICOMOS has supported the shift from restoration to association with the site.[1] The European Landscape Convention (ELC) has acted as a catalyst for interest in regional landscapes. In short, the genius of site has become a central issue in the intersection of landscape architecture and landscape preservation.[2]

Today, drystone walls represent an attractive landscape component, revealing how space, life and agricultural production are framed by and can co-exist with crumbling walls (see Fig. 45.1). They have evolved and diversified in response to changing societal challenges. While there is a growing body of data and mapping,[3] there has been little attention on particular site-specific perspectives, and how to visualise and communicate with the public.

How can landscape observatories (LO) detect emergent dynamics and how can they define new models for involvement? By addressing these questions, this chapter aims to contribute to a critical debate that takes in ELC, LO and critical heritage cases.[4]

Guided by interesting aspects of research on drystone walls in Sweden, Britain and Mallorca, and three critical heritage cases from Jelling (Denmark), Kalkriese (Germany) and Olot (Spain), I discuss how landscapes could enhance their current values and historical legacy without resorting to exhaustive reconstruction and maintenance, to the benefit of LO. My aims here are twofold. First, to reflect on how LO

Fig. 45.1 Drystone walls combine architectural aesthetics with a landscape's historical, cultural, geological and biological aspects[8]

can contribute to change the approach to preservation. Second, to find models for drystone landscapes that support the involvement of LO in ways that allow people to associate with the site.

45.2 The Site

Drystone walls are a manifestation of agricultural thinking. Pursuing that line of enquiry, it becomes possible to reconstruct the living conditions of people at the time the walls were built. A farmer, ploughman or poorly-paid hired hand removed stones from the land and built the walls. Their history also encompasses skilled workers—the 'margers' of Mallorca[5] and local quarriers in Britain.[6]

The sources describe the working conditions and outcomes of these invisible and partly forgotten human factors.[7] Farmhouses are the epicentre of the agricultural set-up—those travelling from outside arrive at the periphery, while household and agricultural logics are understood from the centre.

45.3 Regional Landscapes

It takes a regional perspective to understand the massive amounts of stone, the extent of the labour force,[9] wall dating and the epistemological legacy of the most significant periods of drystone construction.[10] In the English Midlands in 1700, 30% of the land was walled off. In Småland,[11] clearing fields and raising drystone walls peaked in the 1800s.[12] In Mallorca, the regional landscape is characterised by the various personal styles of the 'margers'.[13]

45.4 Geology

The raw material for drystone walls is either quarried on-site or cleared from land adjacent to the wall, as lifting and transporting stones is labour-intensive.[14] Drystone walls therefore express proximity to the origin of the stone material and expose their extent in a regional perspective.[15] In a European perspective, the stone provides the clearest delineation between the regions of Europe. Differences in colour, texture and form between indidivual stones become especially evident when they are stacked in a drystone wall.

The stones used in the walls embody both geology and the site. Their shape is, in fact, a prerequisite for the stacking of dry stones. In Mallorca (see Fig. 45.2), the 'margers' named each of the various shapes.[16] In Småland, stones were defined by their sides. Dictated by mineralogy, they become[17] tangible at site level. See Fig. 45.3 for the British case.

Fig. 45.2 Drystone wall, Mallorca. The yellow pedra viva (live stone) adds variety to the regional landscapes of Europe[18]

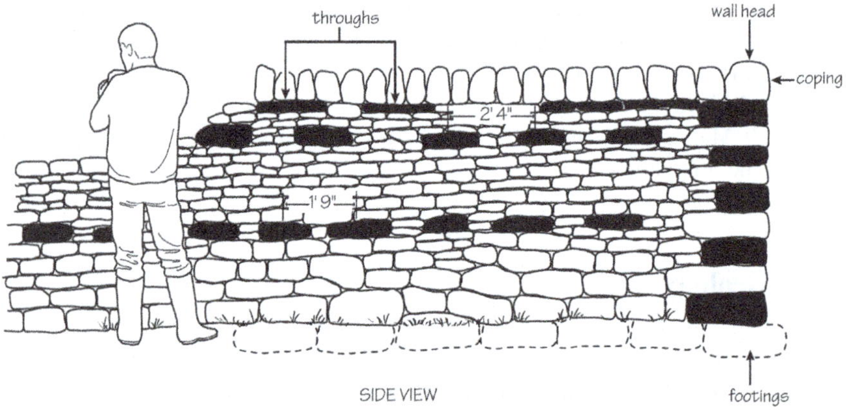

Fig. 45.3 Many authors have sought to recreate the walls' construction principles, seemingly to address deterioration. In Britain, the construction principle is based on the number of 'throughs'[19]

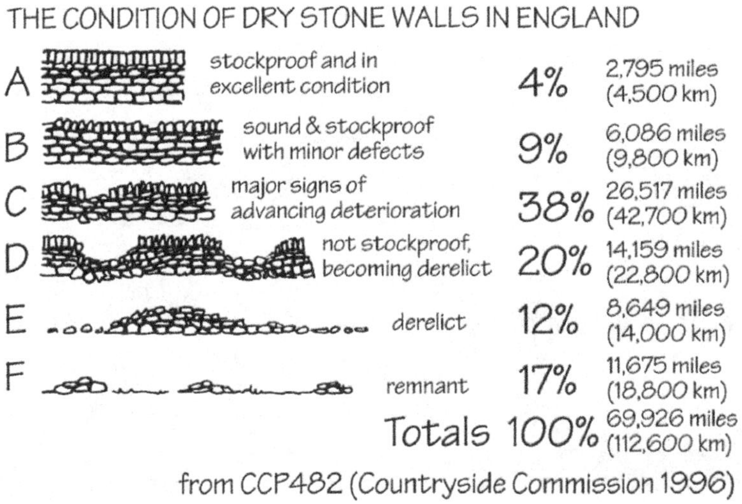

Fig. 45.4 In 1996, 58% of drystone walls showed signs of advancing deterioration or were no longer stockproof, 29% were derelict or with only remnants. Twenty-five years later, the situation is even worse[21]

45.5 Construction and Decay of Drystone Walls

Drystone walls deteriorate, partly due to their sheer number and the absence of practical reasons to maintain them, and partly due to conflicting interests in terms of land use. Many authors have sought to recreate the construction principles, seemingly to address deterioration,[20] see Fig. 45.4. From a preservation perspective, the situation is depressingly out of control. Although scientists, local history groups and volunteers aim to remedy this deterioration—and minor reconstruction projects have taken place—a more comprehensive and multifarious approach to preservation and reconstruction is required.

45.6 Contemporary Approaches to Landscape and Heritage

While the above-mentioned sources—similar to the contributions of LO[22]—bear witness to deterioration, the association with the site, as ICOMOS recommends for historical gardens, remains open. The critical heritage cases of Jelling, Varusschlact and Pedra Tosca exemplify association with the site—a model, where story-telling and public attention could involve LO[23] more effectively.

Fig. 45.5 Monument area in Jelling. The site's history is presented in a new and ambitious way that preserves and visualises the landscape on a 1:1 scale[27]

45.6.1 The Monument Area in Jelling

Until 2013, the monument area in Jelling[24] consisted of two burial mounds, two runestones and a stone ship beside a church and cemetery. Despite being small in size and lacking in contextual information, the area has long been recognised as a national monument. Recent archaeological surveys revealed a much larger historic site under the nearby streets and residential buildings. The monument area now preserves and visualises the whole landscape on a 1:1 scale.[25] The strategy[26] gives the visitor an overview of the whole setting and highlights hidden archaeological finds while walking in areas with streets and residential buildings. Although the monument area does not focus on drystone walls, the case acts as a model for creating a sense of entity and acceptance of the dynamics of landscape that could also be used on a farm with drystone walls scattered across younger cultural landscapes (see Figs. 45.5 and 45.6).

45.6.2 Varusschlacht in Kalkriese

The Varusschlacht Museum at Kalkriese, near Osnabrück, focuses on Roman-era Germany. It reveals how landscape conditions and topography influenced the battle between Roman soldiers and local Germanic tribes (see Figs. 45.7 and 45.8). After archaeologists uncovered evidence of palisades, the architects Zulauf and Schweingrüber[29] proposed clearing a pathway through the forest to illustrate a narrow spatial corridor. Corten steel poles[30] mark the palisades, while steel plates indicate

Fig. 45.6 Monument area in Jelling. Recent archaeological surveys revealed a much larger historic site.[28] An abstract visible structure now creates an association with the site

the orderly formation of the Roman forces, and a more disordered arrangement of steel plates signals the defeat, dissolution and disorder of the Romans following the ambush.[31] The steel plates capture a narrative, and spatial changes visualise the historic landscape at the time of the battle. The spatial changes and the strong narrative of Romans trickling through the landscape convey a human factor that is also relevant in drystone landscapes. Removed or deteriorated drystone walls could similarly be made visible without interfering with the landscape as it is today.

45.6.3 Parc de Pedra Tosca in Olot

Parc de Pedra Tosca[34] is crossed by drystone walls, but the magic of the area comes from its multiple layers, with excavated basalt, enclosed low-lying sunken gardens and fields, and winding systems of trails that cross the beautiful landscape. The landscape architects RCR have created an entrance[35] that encourages visitors to explore Pedra Tosca for themselves and create their own individual experience of the site. A previously unknown area thus becomes ‘a park’.[36] In carefully screened regional drystone landscapes, similar entrances could also invite the public to freely experience and individually engage with all the aspects of drystone walls, without necessarily focusing on loss and deterioration (see Figs. 45.9 and 45.10).

Fig. 45.7 Varusschlacht, Kalkriese, scale model of the terrain. Between the Wiehen Hills and the marshland is a narrow corridor. The heights are exaggerated on the model[32]

Fig. 45.8 Varusschlacht, Kalkriese. The narrow corridor is made spatially visible by felling trees, with Corten steel poles marking the palisades[33]

Fig. 45.9 Parc de Pedra Tosca, Olot. A previously unknown area becomes 'a park', which is revealed through marked entrances, exits and signposts[37]

Fig. 45.10 Parc de Pedra Tosca, Olot. The magic of the area comes from its multiple layers, with excavated basalt, enclosed low-lying sunken gardens and fields and winding systems of trails[38]

45.7 Cultivating Drystone Landscapes and Supporting Landscape Observatories

The UNISCAPE 2020 conference asks how we can cultivate continuity. For drystone walls, a discussion of continuity starts by acknowledging what the ELC, LO and other institutions have already achieved[39]—primarily, registering drystone walls and documenting the incredible numbers of stacked local stones. LO and other institutions have thus provided local knowledge. However, shifts in agriculture and deterioration in the walls engender a feeling of loss, resulting in parallel and uncoordinated data and information.

It is questionable whether it will be possible to comprehensively preserve drystone walls. In the future, LO could focus on particular regional perspectives and suggest particular sites in which a comprehensive context or a particular aspect of drystone landscapes may potentially be experienced.

A strategy is required to allay the feeling of loss, and instead focus broadly on the values of drystone landscapes. The examples from Jelling, Kalkriese and Olot comprise a trend that can attract public attention and secure funding for future studies and research. Likewise the trend can utilise drystone walls' associations with a site. Without that, we cannot understand the effort and logic behind drystone walls, or enjoy natural life and geology in landscapes with deteriorating walls. Knowledge is simply not enough—we also need imagination.

The presented critical heritage cases supplement and visualise studies, indicators, data and compiled historic documents[40] and can serve as models for a more effective involvement of LO. The public engage with history in ways that create enthusiasm and demand further knowledge. Landscape policy, studies and research depend on public interest. Similarly, the future involvement of LO also depends on general public interest.

In the words of senior archaeologist Steen Hvass, 'new ways to do things rarely develop within the individual profession'. He continues: 'in the interfaces between professions, the most astonishing things can happen'.[41] Human interaction at these sites takes place via 'landscapes defined as dynamic areas that participate in a cultural exchange with society'.[42] This is vital for accepting the landscape as a 'result of the action and interaction of natural and/or human factors'.

45.8 Conclusion

I propose that LO incorporate into their agendas and participate in the development of a shared European material culture of drystone walls, and that they screen and suggest particular sites of interest for a visible and tangible elaboration of the landscape dimension—the association with the site. The spatial and narrative ideas in the presented critical heritage cases convey the potential to support the involvement of

LO in the evolution of the landscape dimension in general and of drystone landscapes in particular.

Notes

1. https://www.icomos.org/en/179-articles-en-francais/ressources/charters-and-standards/158-the-florence-charter.
2. Riesto and Tietjen (2019:248).
3. The history is based mainly on material from Småland, Sweden, from the Conservation Volunteers (TCV) in Britain and from the Mediterranean project MEDSTONE's report from Mallorca. Additional literature from Iceland, Catalonia: 2009, Actes de la V Trobada d'Estudi per a la Preservació del Patrimoni de Pedra en Sec als Països Catalans, Denmark: Holgersen & Dam, 2002 Befæstelser; Fischer, 2012 Herregårdshaver and Switzerland: Tufnell: Trockenmaurer.
4. Winter (2013).
5. Di Rosa (ed) (2002:22).
6. Brooks (1977:56).
7. Andersen (2011:85; Brooks (1977).
8. Photo: Torben Dam.
9. The total volume of stone in Småland is calculated at 200 million m^3. In one particular county—Kronobergs län—the drystone walls in the forest are of an extent equivalent to 1/3 of the circumference of the globe (app. 14,000 km) (Andersen 2011:13). In England, the Countryside Commission in 1996 listed 112,600 km of drystone walls, approximately 2.5 times the circumference of the globe (CCP 482, Countryside Commission 1996 in TCV, 2017).
10. https://www.conservationhandbooks.com/dry-stone-walling. (Accessed 5 May 2020).
11. Brooks (1977:24).
12. Andersen (2011:46).
13. Di Rosa (ed) (2002:22).
14. Andersen, 2011:98.
15. Brooks, 1977:57.
16. Di Rosa (ed.) (2002:32) and Drystone work (2017).
17. Each stone has a building side (byggsida), a laying side (liggsida) and a façade side (visesida). The building side determines the continuous stabling on top, the laying side how stable the stone will lay, and the façade side how the stone will fit into the façade of the drystone wall. A good stone should have the shape of a 'horse head' (Andersen 2011:107).
18. Di Rosa (ed.) (2002:34).
19. Brooks (1977:63) and TCV https://www.conservationhandbooks.com/dry-stone-walling/rebuilding-a-free-standing-wall/general-features/. Accessed 29 June 2020.
20. Volunteers from Britain and Switzerland and researchers from Mallorca clearly explain the principles and tools (TCV, 2020; Tufnell et al. 1996; MEDSTONE 19xx).
21. CCP482 (Countryside Commission, 1996) in TCV, 2020.
22. https://rm.coe.int/16802f80c9. Accessed 6 May 2020.
23. https://www.uniscape.eu/landscape-observatories. Accessed 6 May 2020.
24. Jelling has about 3500 inhabitants and is located in the middle of the peninsula of Jutland, on the north side of the Vejle River Valley, in close proximity to 'Hærvejen' (the Ox Road). In Jelling, two important rune-stones indicate the birth of Denmark as a Christian kingdom circa 950–970 AD.
25. Steen Hvass in Jensen et al. (2015:59).
26. In cooperation with the sculptor and artist Ingvar Kronhammer (Jensen et al. 2015).
27. Aerial photo: https://natmus.dk/museer-og-slotte/kongernes-jelling/oplevelser-ude/.
28. Photo: Torben Dam.

29. The Swiss architects Zulauf and Schweingrüber won a competition in 1998 (Baumeister 2007:154).
30. Where archaeological studies had been successful, the distance between the poles was narrow. The distance was larger when the marking of the palisades was a qualified guess. New afforestation behind the palisades emphasised the historic spatial conditions and the narrow corridor through which the Roman soldiers had to pass (Baumeister 2007:155).
31. Defeat, dissolution and disorder are represented by steel plates that are no longer parallel with the stipulated line. Some plates are off to the side, in the grass.
32. Photo: Torben Dam.
33. Photo: Torben Dam.
34. Parc de Pedra Tosca is an old mining area with fertile horticultural areas. The site lies by a closed railway line between Girona and Olot.
35. The entrance possesses a tectonic logic that conveys the physical properties of the materials. Corten steel and volcanic rock define the entrance and exitways, and express both the cultural and natural history of the region (Herrington 2017:125). 500 mm-wide slabs of steel stand approximately two metres tall, arranged in palisade formation, with 75 mm gaps between them, through which the rocks behind are visible (Raxworthy in Herrington 2017: 126) (TOPOS 55 2006 91–94).
36. Now part of a larger regional system of tourist paths.
37. Photo: Torben Dam.
38. Photo: Torben Dam.
39. https://areeweb.polito.it/LOD/project.htm. Accessed 6 May 2020.
40. https://rm.coe.int/16802f80c9. Accessed 6 May 2020.
41. Steen Hvass in Jensen et al. (2015:59).
42. ELC, article 1, https://rm.coe.int/16802f80c9. Accessed 6 May 2020.

References

Andersen I (2011) Stenminnen, Grodan Förlag

Baumeister N (2007) New landscape architecture. Braun

Brooks A (1977) Dry stone walling. British Trust for Conservation Volunteers Ltd.

Di Rosa M (ed) (2002) 12 drystone work—materials, tools and techniques in the Mediterranean Islands. MEDSTONE

Drystone Walling (2020) https://www.conservationhandbooks.com/dry-stone-walling/. Accessed 20 May 2020

Drystone Work (2017) https://caminsdepedra.conselldemallorca.cat/en/dry-stone-work. Accessed 2 Jul 2020

European Landscape Conventions (2000) https://rm.coe.int/1680080621. Accessed 6 May 2020

Herrington S (2017) Landscape theory in design. Routledge

Icomos (1982) https://www.icomos.org/en/179-articles-en-francais/ressources/charters-and-standards/158-the-florence-charter. Accessed 6 May 2020

Jensen K, Cronhammer I, Krath L, Vesterbæk T (eds) (2015) The monument area in Jelling, Arkitekt Kristine Jensens Tegnestue

Landscape Observatory Documentation (2014). https://areeweb.polito.it/LOD/project.htm. Accessed 6 May 2020

Riesto S, Tietjen A (2019) Planning with heritage. In: Steiner, Braae (eds) Routledge Research Companion to Landscape Architecture. Routledge

Winter T (2013) Clarifying the critical in critical heritage studies. Int J Herit Stud 19(6):532–545

Chapter 46
Landscape Observatory in Alqueva, Alentejo Region (Portugal). Why It Matters?

Thematic Session C. Observing Landscape Topic 2 | Landscape Observatories: Post-2020 Agenda and Future Roles

Maria Freire, Isabel Joaquina Ramos, and Maria da Conceição Rego

Abstract Landscape Observatories are important for the continuous and systematic observation of landscape, so as to provide information and design guidelines for landscape protection, management and planning. In Portugal, various institutions have recently created landscape observatories. The Alqueva Dam, located in Portugal on the Guadiana River in the Alentejo Region, was built for multiple purposes in 2002. The dam changed the landscape, land use and the way of working and managing the land in a way that must be monitored and evaluated, which underlies the importance of studying the region and assessing its economic, ecological, environmental, socio-cultural and aesthetic domains. A landscape observatory for the Alqueva Region would constitute a fundamental platform for understanding its landscape dynamics in space and time and support public policies and strategies for regional landscape and territorial planning. It must involve central and regional institutions, public and private organizations, the University of Évora and other research centres such as the Alqueva Development Company (EDIA).

M. Freire (✉)
Department of Landscape, Environment and Planning, Centre for Art History and Artistic Research–CHAIA-UÉ, School of Sciences and Technology, University of Évora (UÉ), Évora, Portugal
e-mail: mcmf@uevora.pt

I. J. Ramos
Department of Landscape, Environment and Planning, Interdiscliplinary Centre of Social Sciences–CICS.NOVA.UÉvora, School of Sciences and Technology, University of Évora (UÉ), Évora, Portugal
e-mail: iar@uevora.pt

M. da Conceição Rego
Department of Economics, Centre for Advanced Studies in Management and Economics–CEFAGE–UÉ, School of Social Sciences, University of Évora (UÉ), Evora, Portugal
e-mail: mcpr@uevora.pt

T. Matteini et al. (eds.), *Cultivating Continuity of the European Landscape*, Environmental History 15, https://doi.org/10.1007/978-3-031-25713-1_46

Keywords Alqueva · Alentejo · Portugal · Landscape transformation · Landscape observatory

46.1 Introduction

Landscape Observatories are recommended as one of the instruments for the implementation of the European Landscape Convention (ELC) (Council of Europe 2000, 2008). They are important tools to study landscapes and to monitor their transformation, linking scientific research with the expectations of citizens and landscape policies. They provide significant information to design guidelines for landscape protection, management and planning and, simultaneously, they promote landscape awareness in society.

In the last two decades, several different types of landscape observation centres have emerged throughout Europe with an important role in diverse landscape areas, as encouraged by the Council of Europe (2008). They operate at local, regional or national levels, either independently or as part of a broader observation system, some focused on increasing landscape knowledge (identification, characterization and monitoring), others more determined to promote cooperation between the public and private sectors, and still others more involved in serving education (e.g. disseminating knowledge and awareness of landscape values) (AA.VV. 2015, 2017). Some have been established by governments, others by institutions or by civil society groups. (Larcher and Cassatella 2015; Saraiva et al. 2017).

Some networks and projects also emerged: RECEP-ENELC (European network of local and regional authorities for implementation of ELC, now extinct), UNISCAPE (European network of universities and scientific production)[1] and CIVILSCAPE (International association involving civil society with landscape)[2] (Souza 2019). The LOD project (Landscape Observatory Documentation), started under the auspice of such networks, promotes scientific cooperation among existing European and international Landscape Observatories to reduce local initiative fragmentation and coordinate potentials.[3]

Landscape use and transformation are inevitable. However, new trends and development must take into account the importance of ensuring the preservation of the landscape values (ecological, aesthetic, economic and socio-cultural) underlying European landscape identity continuity. This issue is particularly important in Mediterranean landscapes where changes have been very fast, and at times radical, disregarding these values in the present without any respect for the past when envisioning the future.

The aim of this article is to reflect on the potential interest in setting up an Alqueva Regional Landscape Observatory to reflect on the problems, potentials and challenges

[1] https://www.uniscape.eu.

[2] https://civilscape.eu.

[3] https://areeweb.polito.it/LOD/index.htm.

associated with dams and lakes in contemporary Mediterranean landscapes. The article also attempts to draw attention to the present requirement for observatories to be tools supporting more efficient, integrative and operative landscape planning and policies.

46.2 Landscape Observatories in Portugal

In 2005, the ELC was ratified in Portugal. Between 1999 and 2004, Portugal nationally delimited landscape units for the continental territory according to ELC recommendations (Fig. 46.1) in the study 'Identification and characterization of the landscapes of mainland Portugal' (Abreu et al. 2004).

Since 2015, different types of landscape observatories have emerged (Guimarães Landscape 2015, Landscape of University of Porto 2017, Charneca Landscape 2018, Madeira Island Landscape 2018) or have been explored (e.g. Landscape Tagus River or Landscape of Serras do Socorro e Archeira) centred on several objectives and supported by diverse activities. In 2019, the Portuguese government decided to set up a national Spatial Planning and Urbanism Observatory to monitor and evaluate territorial dynamics and management systems and to support strategic decision-making.

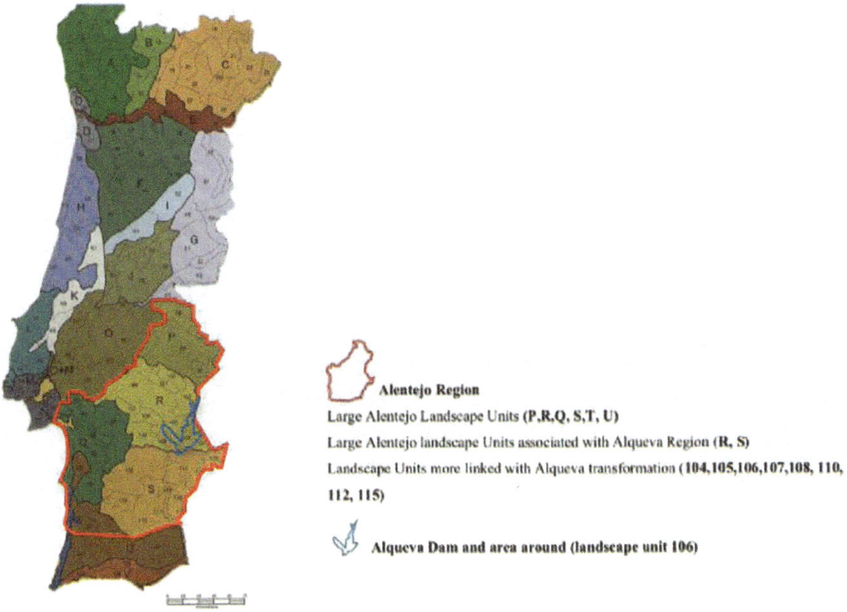

Fig. 46.1 Portugal landscape units and Alentejo region. *Source* Adapted from Abreu et al. 2004

46.3 Alqueva Landscape

The Alentejo landscape is mediterranean in nature: undulating relief with high temperatures, low rainfall and intense light, depopulated, with a concentrated, aged population and extensive agricultural exploitation (Ribeiro 1993; Abreu et al. 2004). Large spaces prevail, with open fields of cereal, expanses of *montados*, woods and forests and even vineyards and olive groves (Abreu et al. 2004; Jorge 2007). The severe lack of water has been one of the main constraints to development and has limited the possibilities for modernizing agriculture and for achieving a sustainable public water supply.

The Alqueva Dam was built in 2002 to facilitate public water supply, electricity production and irrigation. It is the largest in the country and the main strategic water reserve in Europe (Fig. 46.1). The reservoir is 83 km long and covers an area of 250 km^2. Its total storage capacity is 4 150 million m^3. The dam introduced significant changes to the landscape especially in the biophysical, cultural and aesthetic domains, but also in agriculture and agro-industry. Water is the new transformation agent for this landscape and the new development paradigm (Fig. 46.2). Today, the landscape is different; the plains are unrecognizable (Fig. 46.3). The transformation has involved new agricultural uses, tourist and leisure activities and significant environmental impacts, demanding enormous organizational and management efforts (Abreu et al. 2004) as well as effective public policies aimed at adequate landscape management. From a rain-fed landscape where cereals and fallow land were dominant, an irrigated landscape has emerged with intensive crops such as olives, corn, vegetables, oilseeds, forage and almonds.

Fig. 46.2 Alqueva Region landscape around the lake. *Source* © Maria Freire (2020)

Fig. 46.3 Landscape transformation in Alentejo Alqueva Region. *Source* EDIA (https://www.edia.pt/pt/o-que-e-o-alqueva/o-territorio/)

46.4 A Landscape Observatory for the Alqueva Region (LOA)

Considering the major forces affecting the Alqueva Region, we believe that the creation of a landscape observatory for Alqueva (LOA) can provide a useful and efficient tool for landscape protection, management and planning and also help us harmonize the interests of man and nature. Increasing change, transformation and destruction over the last century, particularly in the last two decades, determine the following objectives for LOA:

(1) To document and monitor landscape transformation: intensification of crops (agro-industry), desertification, depopulation and simplification of ecological infrastructure;
(2) To generate and articulate knowledge, strategies and policies: a comprehensive understanding of recent landscape dynamics;
(3) To become a point connecting all stakeholders involved in the government and management of the landscape at local and regional levels (local and regional administrations and institutions, schools and universities, professional groups and especially farmers);
(4) To achieve a sustainable landscape that is ecologically, economically and culturally balanced and aesthetically beautiful;
(5) To strive to establish a new regional planning initiative capable of influencing new planning tools and policies.

To address these aims, some aspects related to its functions, structure and composition should be underlined as is usual in other similar recent structures (AA.VV. 2015, 2017). LOA should be provided with a documentation centre and promote research activities as part of a global strategy driven by the overarching themes of biological biodiversity and ecological stability. LOA is conceived as an instrument for characterizing and evaluating transformations and enhancing evolutive visions for the future, a tool to support public policies and integrate landscape management and evaluation into public decisions. The observatory should foster integrated collaboration between scientists, professionals, technicians, private sectors and NGO organizations, national and regional public entities and academic institutions and strongly encourage social involvement, particularly by farmers. The conditions undoubtedly exist for the University of Évora (Alentejo) to start this process with the Alqueva Development Company (EDIA).

46.5 Final Considerations

Changes in land use and exploitation derive from inadequate planning options, mainly grounded in a productivist view, leading to destruction and/or contamination of natural elements and flows and the simplification of landscape mosaics with no consideration for the ecological structure.

Any landscape land use or transformation activity (*e.g.* agriculture, agro-industry, tourism, forestry and industry) should be based on the concepts of landscape restoration and quality and the conservation of natural resources, ecological structure, multifunctionality and adaptation to climate change.

Portugal has a very extensive legal framework concerning land use planning, environment and nature conservation. Nevertheless, landscape has been taken into account in a sectoral way instead as a comprehensive entity identified and characterized by its unique character that should determine the trends, potentials and problems to be addressed in land use planning instruments as a way to preserve the continuity of this mediterraenan landscape. The proposed observatory can contribute to envisioning Alqueva Landscape as a whole in order to overcome the massive gap affecting decision-making processes related to public landscape policies.

Acknowledgements This work was financed with National Funds through FCT—Portuguese Foundation for Science and Technology, within the scope of the project: (UI&D 112)/2020—[Ref.ª UIDB/00112/2020] of CHAIA/UÉ.

This work is financed by national funds through FCT - Foundation for Science and Technology, I.P., within the scope of the project «UIDB / 04647/2020» of CICS.NOVA - Centro Interdisciplinar de Ciências Sociais da Universidade Nova de Lisboa.

References

AA.VV. (2015) Landscape Observatories in Europe II. In: Cassatella C, Larcher F (eds) Proceedings of the UNISCAPE En-Route International Seminar, UNISCAPE En-Route, a.I, n.I, Torino, Italy

AA.VV. (2017) Landscape Observatories in Europe II. In: Zanaica T (coord) Proceedings of UNISCAPE En-Route International Seminar, UNISCAPE En-Route, vol 2, no 5, Amersfoort, Netherlands

Abreu A, Correia T, Oliveira, M (2004) Contributos para a Identificação e Caracterização das Paisagens de Portugal Continental. DGOTDU, Lisboa

Council of Europe (2000) European Landscape Convention
Council of Europe (2008) CM/Rec (2008)3 of the Commijee of Ministers to member states on the guidelines for the implementation of the ELC.
Jorge F (2007) Portugal visto do Céu. Argumentum EdiØes, Lisboa
Larcher F, Cassatella C (2015) Landscape Observatories in Europe: the Action Plan. UNISCAPE En-Route, a.I, n.I, p 118–119
Ribeiro O (1993) Portugal, o mediterrâneo e o atlântico. EdiØes João Sá da Costa, Lisboa
Saraiva M, Loupa-Ramos I, Bernardo F (2017) Avaliar paisagens e sua identidade–Observatório(s) como resposta metodológica? In: Fidalgo P (coord.) Estudos de Paisagem, vol III. Instituto História Contemporânea da FCSH/Universidade de Lisboa, Lisboa, p 276–293
Souza R (2019) Os Observatórios de Paisagem como uma plataforma de envolvimento da sociedade civil a longo prazo. Universidade do Porto, Dissertação Mestrado em Arquitetura Paisagista

Webgraphy [Consulted 15–06–2020]

EDIA https://www.edia.pt/pt/o-que-e-o-alqueva/o-territorio
UNISCAPE https://www.uniscape.eu
CIVILSCAPE https://civilscape.eu
LANDSCAPE OBSERVATORY DOCUMENTATION. https://areeweb.polito.it/LOD/index.htm

Chapter 47
Landscape Observatories: An Analytical Framework and a Case Study at the Landscape Observatory of Finland

UNISCAPE2020 Conference; Thematic Session C: Observing Landscape Session/Topic 1: Landscape Observatories: Concept, Models and Main Contributions

Juanjo Galan

Abstract Landscape observatories are one of the instruments foreseen by the Council of Europe for the implementation of the European Landscape Convention. They can be defined at different scales and can play a strategic role in the intermediation between administrations, civil society, researchers, and the economic sector. However, a systematic study of existing landscape observatories suggests a need to define a more coherent framework to understand their specific characteristics, evaluate their performance, facilitate comparative studies, promote the generation of inter-scalar networks, and support future changes. Accordingly, this paper proposes that such a framework should be based on different institutional, organizational, and operational variables that can be flexibly adjusted and combined to respond more effectively to the specific contexts and aims of each landscape observatory as well as to the new methodological and topical agendas emerging from landscape studies. Secondly, the paper briefly introduces the landscape observatory of Finland as a potential case to evaluate the utility of the proposed analytical framework. The preliminary results recommend advancing in the proposed study and complementing the delineated framework with an additional tool to more objectively evaluate the performance and impact of landscape observatories.

Keywords European landscape convention · Landscape planning · Landscape policies · Landscape governance · Landscape observatory of Finland

---

J. Galan (✉)
Department of Architecture, Aalto University, Espoo, Finland
e-mail: juagavi@urb.upv.es

Department of Urban Planning, Universidad Politécnica de Valencia, Valencia, Spain

T. Matteini et al. (eds.), *Cultivating Continuity of the European Landscape*,
Environmental History 15, https://doi.org/10.1007/978-3-031-25713-1_47

47.1 Introduction

The role and functions of landscape observatories and their contribution to the implementation of the European Landscape Convention (ELC hereafter) has been a topic of heated debate since the creation of the pioneering observatories after the year 2000, and more intensively, after the publication of Recommendation CM/REC(2008)3 of the Committee of Ministers of the Council of Europe (CM/REC hereafter). This recommendation includes a set of flexible and general suggestions concerning the territorial scope, functions, methods, and composition of landscape observatories, and simultaneously, positions the observatories as one of the types of instruments supporting the implementation of landscape policies together with landscape plans, landscape studies within spatial or sectoral plans, landscape charters, landscape impact assessments, protected sites, allocation of resources, and landscape awards (ELC 2000, CM/REC 2008).

Therefore, landscape observatories are one of the many pieces of the legal, institutional, and societal frameworks promoting landscape protection, management, and planning through the integrative, participative, and dynamic approach defined in the ELC. Furthermore, the potential of a specific landscape observatory can only be fully understood after realizing its role within these frameworks. That is why there is no general recipe and each landscape observatory needs to respond to its specific context.

Due to their novel character and nodal position between different types of stakeholders, landscape observatories have provided a new and multifaceted lens for the study and use of the landscape concept. Thus, from an operational point of view, the activity of the observatories has been increasingly associated with the development and implementation of new methods of observation, new models of governance, the generation of democratic processes, or the perception of the landscape as the broadest and most integrative type of heritage (Nogue and Sala 2018). Additionally, most of the landscape observatories have promoted the holistic, proactive, and ubiquitous approach to landscape proposed by the ELC, as well as its capacity to foster connections between people and places on the basis of the manifold and interrelated intrinsic, inherent, or instrumental values of a landscape (Antrop and Van Eelvelde 2017).

47.2 Functions, Agendas, and Challenges

According to the CM/REC, landscape observatories should facilitate the implementation of the ELC through the continuous and rigorous observation of landscape using appropriate protocols and indicators and through the exchange of information on policies and experiences "*concerning protection, management and planning, public participation and implementation at different levels*". Landscape observatories are site-specific and, therefore, these functions should be implemented in the overall spatial scope covered by each observatory (local, regional, national, or international), keeping in mind that any area within that scope has an associated landscape that deserves study and consideration. A comprehensive study of European landscape observatories reveals that most of them include among their functions the generation, exchange, and dissemination of landscape knowledge, the involvement in the definition and follow-up of landscape quality objectives and policies, and the implementation of landscape actions as defined in the CM/REC. In addition, landscape observatories are often described as meeting points or platforms vertically and horizontally connecting the many groups with agency or influence on the landscape, facilitating participatory processes, and supporting the political, cultural and planning changes promoted by the ELC (Gambino et al. 2015; Nigro 2015). However, the actual performance of landscape observatories can be highly influenced by many factors such as their connection (or disconnection) from other instruments or organizations involved in the implementation of the ELC, by their composition and structure, or by their level of autonomy. Despite the fact that these factors can evolve over time, they strongly determine the history and activities of each landscape observatory and affect their level of influence, their holistic or specialized focus, their geographical scope, or even the balanced or unbalanced relationship between different types of stakeholders (Staniscia et al. 2015; Salevid 2015; Oliveira 2015; Nogue and Sala 2018).

In the year 2000, the ELC justified the re-conceptualization of landscape on the grounds of different challenges that, 20 years later, still remain active and have even gained new dimensions (e.g., sustainable development, acceleration of landscape transformations, loss of bio-cultural diversity, climate change, threats to democracy, humanitarian crises, etc.). These global and European challenges, together with specific problems and processes affecting the spatial scope of each landscape observatory, play a key role in the definition of their working agendas. However, the existing literature suggests that landscape observatories are also facing methodological and conceptual challenges that affect their activities and the adequate fulfillment of their functions.

Thus, in a comprehensive study of landscape characterization methods, Fairclough et al. (2018a, 2018b) consider the effect that the generalized weakening of planning systems might have had on landscape planning or how effective the ELC has actually been in overcoming some of the prejudices that people have traditionally held against the "landscape approach" (e.g., a barrier for development or a refuge for nativists/nimbyists).

Similarly, the critical analysis of the activity of landscape observatories can be also a key source of information to understand landscape challenges. According to Nogue and Sala (2018), methodological inertias, corporate interests, and professional lobbying are some of the barriers hampering the development of more innovative and interdisciplinary approaches to the landscape. Nevertheless, the activity of the Landscape Observatory of Catalonia provides an inspiring example of how these barriers can be overcome or how landscape characterization can produce functional units informing spatial and sectoral planning (Nogue and Sala 2018). Moreover, this observatory shows how a supportive legal and political framework, together with the official recognition of the functions and tasks of landscape observatories, can transform landscape into a collective and social instrument, foster genuine and effective collaboration between a wide range of stakeholders, and refute the sterile dichotomy between bottom-up and top-down approaches to landscape.

47.3 Typological Studies on Landscape Observatories

The systematic and comparative study of landscape observatories reveals some of the key factors affecting their functioning and facilitates their typological classification. According to Cassatella (2017), these factors would include the existence of a legal foundational act or law (if any); the functions and activities assigned to the observatory; its territorial scope; its juridical nature (public body, consortium, association, NGO, etc.); its financial and human resources; and the disciplines or competences involved in its activity. Similarly, Kienast et al. (2019) propose a classification of landscape observatories based on how they monitor the landscape. In addition, further studies on landscape observatories uncover their potential to be integrated into multi-scalar structures such as the local, regional, and national observatories in Italy (Devecchi 2015) or the local and regional observatories in some Italian regions like Veneto (Castiglioni 2015).

Based on existing typological studies and the critical reflection on key functions and challenges presented in this paper, Table 47.1 proposes a framework to study the character and performance of landscape observatories. The framework includes three types of factors: institutional, organizational, and operational, and defines different options for each factor. Although these options usually represent two extreme situations, there might be many intermediate positions between the extremes presented. The proposed framework is expected to facilitate the systematic study, diagnosis, and improvement of existing observatories, as well as the more consistent creation of new ones, with the firm belief that landscape observatories are dynamic organizations that are site-specific and have the capacity to evolve over time.

47.4 A Case Study: The Landscape Observatory of Finland

Following the initiative of Aalto University, the Landscape Observatory of Finland (LOF henceforth) was officially created in 2016 as a consortium including different governmental, academic, professional, and economic organizations. The LOF was the first cross-party collaboration in the field of landscape planning and management in Finland and its main aim is to support the implementation of the ELC in that country by promoting research, discussion, participation, and actions, and by assuming that any Finnish landscape, no matter its character, quality, or scale, can become a positive asset for sustainable development and well-being (Galan, 2017a, 2017b).

Table 47.1 Key factors for the analysis of landscape observatories

Type of Factor	Factor	Main options for the factor
Institutional	Legal status, roles and responsibilities	– Legal recognition of the observatory, its roles and responsibilities – Voluntary organization with self-defined roles and responsibilities
	Legal personality	– Independent legal and contractual personality – No independent legal and contractual personality
	Institutional integration	– Clearly integrated into a broader structure working on the implementation of the ELC in the addressed landscapes. Existence of complementary functions and mutual interdependences within that structure – Highly independent, isolated, or unclearly integrated, overlapping with other organizations
	Links with other landscape observatories	– Strong and systemic collaboration (e.g., methodologies, joint projects) – Weak and sporadic collaboration
Organizational and Managerial	Composition	– Diverse and balanced composition covering different dimensions of the landscape and types of stakeholders (civil groups, authorities, private sector, academia, etc.) – Specialized composition focused on some dimensions or approaches to the landscape
	Structure and management	– Self-governing management and structure. Independent decision-making processes and self-defined agendas responding to the interests of different groups – Subordinate structure and management
	Autonomy, Financial and Human Resources	– High level of administrative autonomy. Availability of financial and human resources to cover the basic functioning of the observatory – Absence of financial and human resources and dependence on volunteering work or support from partners
Operational	Scale or geographical scope	– Local – Regional – National – International – Independent landscape unit or area
	Spatial coverage	– Integral Coverage: All the landscapes within the geographical scope – Partial Coverage: Focus on a particular type of landscape within the geographical scope

(continued)

Table 47.1 (continued)

Type of Factor	Factor	Main options for the factor
	Agendas and thematic scope	– Integral Agendas (as per ELC): Covering the various dimensions of the landscape, usually through highly integrative tasks: characterization, assessment, integral monitoring, awareness raising, training and education, participation, spatial and inter-sectoral planning – Specialized Agendas: Focus on specific dimensions or elements of the landscape
	Research and methods	– Development and validation of new methodologies – Implementation of existing and predefined methodologies

The main task included in the first annual plan of the LOF was the diagnosis of the level of implementation in Finland of the ELC (see Table 47.2). This diagnosis permitted the identification of potential synergies with national agendas on Sustainable Development and Cultural Environments, as well as the opportunity to more clearly integrate the landscape concept into the national Planning and Building Act in order to provide landscape with the legal status required by the ELC (Cultural Environments Strategy 2014–2020). Interestingly, it was possible to detect similar issues in other Nordic countries, quite probably as a consequence of the direct and joint involvement of the Nordic Council of Ministers in the implementation of the ELC (Sarlöv Herlin et al. 2018).

In 2017, the high-priority topics displayed in Table 47.2 were selected to define the structural agenda of the Observatory for the coming years. Due to the volunteer nature of the LOF and the lack of internal resources, different "ad-hoc" groups were constituted to work on each of them and to explore potential sources of funding. The academic partners of the LOF usually took a leading role within each group under the supervision of the Steering Group of the Observatory. In particular, the project for a "*New Landscape Characterization of Finland*" was unanimously considered the most relevant task due to its integrative nature and its capacity to incorporate most of the other topics. The key aim was to develop a new type of characterization adapted to current societal needs, the principles of the ELC, and the new methodological and technological possibilities available now for landscape studies. The "*Adaptation of Finnish Landscapes to Climate Change*" led to the activation of the AELCLIC-pathfinder, a European project funded by EIT-Climate KIC that was implemented during the years 2018 and 2019 (AELCLIC project 2020). In addition, the development of specific actions on "*Landscape Sustainability and Sustainable Communities*" was expected to be framed as part of a collective and conceptual discussion. Moreover, as the activities of the LOF progressed, it was agreed to promote a wider discussion on the "landscape" concept and other associated terms in order to align

Table 47.2 Identification of key topics for the activity of the landscape observatory of Finland

Priority	Topic
High	New landscape characterization of Finland
	Participatory governance in landscape conservation
	Climate change and its effect on Finnish landscapes
	Landscape sustainability and sustainable communities
Medium	Landscape indicators and monitoring
	Finnish urban landscapes
	Landscape, People, Governance and Politics
	An international project with other landscape observatories
	A National/regional/local landscape policy and strategy
	Landscape services
Low	Landscape quality objectives
	A Finnish landscape laboratory
	Landscape and nature-based solutions
	Rural Finnish landscapes
	Didactic materials about Finnish landscapes

Result of an internal survey and workshop in which the partners of the observatory proposed and ranked the topics according to their level of importance and urgency, 2017

the ELC with other scientific or sociocultural agendas. (Landscape Observatory of Finland 2020).

A preliminary analysis of the institutional, organizational, and operational characteristics of the LOF, according to the factors listed in Table 47.1, reveals its singularity and partially explains its performance during the last four years. Thus, its internal diversity and the collective definition of the thematic agenda guiding the activity of the Observatory constitute two of its main strengths, whereas the lack of juridical personality and the absence of internal resources might have severely affected its capacity to initiate and manage projects. Nevertheless, a more detailed diagnosis of the LOF through Table 47.1 would provide new ideas for its transformation into a more operative and impactful organization, as well as for its future role in the promotion of regional and local landscape observatories in Finland.

47.5 Conclusions

The research presented introduces a framework for studying the structure and functioning of landscape observatories in order to facilitate a clearer diagnosis of their characteristics and their more effective contribution to the implementation of the ELC. The proposed framework was designed to allow for the rich variety of contexts, goals, and conditions in which landscape observatories operate and is expected to support public authorities, civil organizations, academic institutions, and other landscape stakeholders in the generation and management of landscape observatories. In fact, the studied case (Landscape Observatory of Finland) reveals how a systematic understanding of its characteristics, strengths, and weaknesses can facilitate its future improvement and inform future actions. Moreover, 20 years after the official presentation of the ELC, Europe faces new landscape challenges and opportunities that permeate all the territorial scales, from the global to the local, and that bring new agendas and possibilities to landscape observatories. The proposed framework is therefore perceived as a tool to work on the institutional, organizational, and operational characteristics of the observatories in order to prepare them for current and future landscape challenges.

References

AELCLIC Project (2020) Adaptation of European landscapes to climate change. Available from: https://aelclicpathfinder.com. Accessed 30 June 2020

Antrop M, Van Eetvelde V (2017) Landscape perspectives: the holistic nature of landscape. Springer, Berlin/Heidelberg

Cassatella C (2017) Overview of landscape observatories. In: International conference on landscape observatories, 9–10 February 2017, Amersfoort, Netherlands. Landschaps Observatorium

Castiglioni B, Fontanari E, Reho M, Varotto M (2015) Three years of landscape observatories in the Veneto region: an evaluation. In: Proceedings of the UNISCAPE en-route international seminar landscape observatories in Europe II, 22–23 September 2014, Torino, Italy. UNISCAPE, pp 65–70

CM/REC (2008) Recommendation CM/Rec(2008)3 of the Committee of Ministers to member states on the guidelines for the implementation of the European Landscape Convention; Council of Europe

Cultural Environments Strategy 2014–2020. Available from: https://www.ym.fi/en-us/land_use_and_building/Programmes_and_strategies/Cultural_environment_strategy. Accessed 31 Aug 2020

Devecchi E (2015) Landscape observatories in Italy: the state of the art. In: Proceedings of the UNISCAPE en-route international seminar landscape observatories in Europe II, 22–23 September 2014, Torino, Italy. UNISCAPE, pp 10–13

ELC (2000) European Landscape Convention, Council of Europe

Fairclough G, Sarlöv Herlin I, Swanwick C (2018a) Landscape character approaches in global, disciplinary and policy context: an introduction. In: Fairclough G, Sarlöv Herlin I, Swanwick C (eds) Routledge handbook of landscape character assessment. Routledge, London, pp 3–20

Fairclough G, Sarlöv Herlin I, Swanwick C (2018b) Conclusion: seeing obstacles and finding ways ahead. In: Fairclough G, Sarlöv Herlin I, Swanwick C (eds) Routledge handbook of landscape character assessment. Routledge, London, pp 279–284

Galan J (2017a) Towards a landscape observatory of Finland? International conference on landscape observatories, 9–10 February 2017. Amersfoort, The Netherlands. Available from: www.landschapsobservatorium.nl/Uploaded_files/juanjo-galan.6c89b6.pdf. Accessed 31 Aug 2020

Galan J (2017b) Finland aims to establish a Finnish landscape observatory. In: Proceedings of the International conference on landscape observatories, 9–10 February 2017. UNISCAPE, Amersfoort, The Netherlands, pp 11–13

Gambino R, Cassatella C, Larcher F (2015) Mission and role of the landscape observatories in Europe: an introduction. In: Proceedings of the UNISCAPE en-route international seminar landscape observatories in Europe II, 22–23 September 2014, Torino, Italy. UNISCAPE, pp 6–9
Kienast F, Wartmann F, Zaugg A, Hunziker M (2019) A review to integrated approaches for landscape monitoring. In: 10th Council of Europe conference on the European landscape convention, Report number: CEP-CDCPP (2019) 7E, 6–7 May 2019, Strasbourg, France. Council of Europe
Landscape Observatory of Finland (2020) Available from: http://www.landscapeobservatoryfinland.com/ongoing-activities/. Accessed 30 June 2020
Nigro C (2015) Planning beyond the institutions: landscape observatories as key- medium- bodies between communities and institutions. In: Proceedings of the UNISCAPE en-route international seminar landscape observatories in Europe II, 22–23 September 2014, Torino, Italy. UNISCAPE, pp 95–102
Nogue J, Sala P (2018) The landscape observatory of Catalonia: increasing knowledge and promoting cooperation in landscape. In: Fairclough G, Sarlöv Herlin I, Swanwick C (eds) Routledge handbook of landscape character assessment. Routledge, London, pp 265–278
Oliveira R (2015) The Tagus river cultural landscape. Building up a landscape observatory. In: Proceedings of the UNISCAPE en-route international seminar landscape observatories in Europe II, 22–23 September 2014, Torino, Italy. UNISCAPE, pp 14–20
Salevid E (2015) When landscape is a cause: implementing the ELC effectively II. In: Proceedings of the UNISCAPE en-route international seminar landscape observatories in Europe II, 22–23 September 2014, Torino, Italy. UNISCAPE, pp 45–50
Sarlöv Herlin I, Nord J, Qviström M (2018) Landscape characterisation in Sweden: landscape in the planning system. In: Fairclough G, Sarlöv Herlin I, Swanwick C (eds) Routledge Handbook of Landscape Character Assessment. Routledge, London, pp 279–284
Staniscia S, Altieri G., Tecilla G (2015) The activities of the landscape observatory of Trentino: a five-year report on the state of the landscape. In: Proceedings of the UNISCAPE en-route international seminar landscape observatories in Europe II, 22–23 September 2014, Torino, Italy. UNISCAPE, pp 71–77

Chapter 48
Regional Coordination of Large-Scale Infrastructure Projects—A New Role for Landscape Observatories?

Anders Larsson

Abstract Regional coordination should be of the utmost importance when working with large-scale infrastructure projects such as roads, railroads, wind farms, etc. However, when studying infrastructure planning in Sweden, it has become evident that large-scale infrastructure projects are ruled by project-centred logic rather than place-centred logic (regional coordination). Each project has its specific budget and project management, leading to, for example, duplication of landscape related analyses and poor consideration regarding possible synergy effects and resulting in a scattered and fragmented infrastructural landscape rather than a landscape where the historical continuity and regional qualities have been safeguarded. Present regional authorities act mainly upon project initiatives and are limited by their formal, jurisdictional framework. Thus, there is a role to fill when it comes to regional and landscape-centred coordination of landscape interventions. This is where landscape observatories, as suggested by the European Landscape Convention, could provide an alternative solution. Landscape observatories could also be incorporated into formal planning processes, but with the very specific task of arguing from a landscape point of view, rather than from a perspective of economic growth, exploitation or individual project handling.

Keywords Regional planning · Infrastructure planning · Landscape observatory · ELC (European Landscape Convention)

48.1 Background

Regional planning has its specific kind of limitations, depending, e.g. on the specific country and planning context. However, problems are quite often related to the fact that all parts of a region are not valued equally (e.g. Björling and Fredriksson 2018). There are regional "centres" and rural "hinterlands" with infrastructure planning

---

A. Larsson (✉)
Department of Landscape Architecture, Planning and Management, SLU, Alnarp, Sweden
e-mail: anders.larsson@slu.se

T. Matteini et al. (eds.), *Cultivating Continuity of the European Landscape*, Environmental History 15, https://doi.org/10.1007/978-3-031-25713-1_48

based primarily on nodal principles connecting core areas and promoting commuting between the countryside and urban districts. Infrastructure development is often justified in terms of efficiency regarding transportation of goods and people and with the argument that better infrastructure could make distant rural areas more "livable" in terms of commuting. Thus, there seems to be an "urban norm" in both transport planning and other parts of society affecting planning responsibility, the organization of planning practice and the concrete outcome.

The ESPON report, urban–rural relations in Europe, (ESPON 2006) focuses on how the urban and rural relate to each other with examples from several countries in the European Union. One interesting aspect of the report is when it differentiates between "structural" and "functional" properties. Structural properties comprise, e.g. land-use patterns, settlement structures and population distribution, i.e. characteristics that are comparatively stable over time, while functional properties comprise, e.g. production, consumption and communication, i.e. activities taking place within the structures (ESPON 2006; Nordregio 2006). The prevailing planning system, biased towards a dichotomy between the urban and the rural, might be the result of a structural rather than a functional division of responsibilities. In such a structural perspective, especially in the context of an urban norm, the countryside is not evaluated on its own merits. Instead, rural areas are seen as places from where resources can be extracted and to where the urban population can go for recreation. Thus, when it comes to "regional" development, when compared with "countryside" or "urban" development, planning also needs to start from functional properties. Rather than speaking of the urban and/or the rural, we should talk about regions and regional sub-divisions with varied functional possibilities (Nordregio 2006).

The need for territorial coordination in the perspective of large-scale investments in infrastructure has also been recognised for many years (ESPON 2006). However, recent studies of infrastructure planning in Sweden (road and railroad construction) has illustrated that nothing much has happened in this respect in recent decades (Persson et al. 2015; Persson and Larsson 2014). There is no regional coordination involved when, for example, performing landscape analysis, environmental impact assessments (EIAs) and when suggesting compensation measures. Large-scale infrastructure projects are carried out using project-centred logic rather than place-centred logic. Each project has its specific budget and project management with no, or very few, considerations regarding possible synergy effects, even though several parallel projects could be affecting the same habitat (ibid.). One case study from northern Sweden outside the city of Umeå revealed that there were so many ongoing road and railroad projects affecting, for example, the very same bird protection area that it was difficult get a proper overview of the processes and outcome. Another case study close to Stockholm revealed that there was no coordination at all between two ongoing sub-projects affecting the same area, except occasionally and unintentionally. Both cases would have gained in both efficiency and transparency from regional coordination, where not only project specific landscape issues could have been discussed (ibid.). Other studies have illustrated how efficient policies for regional development need to be both "place-aware" and "people-based" (Barca et al. 2012). A "people-based" and a more participative approach to regional development further indicates an increased need for "landscape literacy" among both the general public and planning professionals (Whiston Spirn 2005), but also within any possible new organisations undertaking the complex task of sustainable regional planning.

48.2 Problems Related to Landscape Assessment in a Project-Centred Planning Perspective

Some authors have mentioned that planners are much more skilled at evaluating ecological rather than cultural heritage values (e.g. Eliasson et al. 2018; Stenseke 2016). Swensen and Jerpåsen (2008) have gone even further and stated that cultural heritage and the protection and caretaking of heritage values is "more to be seen as a derivate interest", when compared with the protection and caretaking of "green interests". Closely related to this is the division between tangible and intangible values. Ecological values are often more tangible than values related to cultural heritage, but there is also a difference within each category. One sub-region could, for example, have a long historical continuity of farming as its main quality, while another sub-region might have less historical continuity, but at the same time more protected "objects" within its delimitation (churches, estate buildings, ancient remains, etc.). The latter could then get more "points" within a traditional evaluation scheme (project-centred perspective), even though the former landscape might be much more valuable in a holistic landscape perspective (see more in, for example, Germundsson 2005; Eliasson et al. 2018; Swensen et al. 2013; Larsson 2004). Also, according to Swensen et al. (2013), intangible values (narratives) matter most to laymen, while experts tend to focus more on tangible values. Case studies from three different Norwegian towns indicated that non-visible elements such as sound, smell and memories of people and activities were important for the local population (ibid.). This is why an increased focus on (continuous) public participation is essential for landscape assessment.

Other problems relate more to the process in itself. Systems constructed for handling values and value assessment within a traditional EIA process should, for example, be possible to use in a coherent way, notwithstanding the specific company or consultant that has the responsibility of carrying out the assessment. Comparability and simplicity are of the utmost importance in project-centred logic where there is a clear demarcation between client and contractor, while a more place-centred process would focus more on transparency, continuity and ongoing discussion within a group of specialists and between experts and the general public, dealing foremost with specific (non-comparable) landscape values and always be ready to re-evaluate along the way (Larsson 2020).

When discussing road and railroad planning in Sweden, concepts related to compensation measures such as "in-site", "off-site", "in-kind" and "out-of-kind" have been the focus. When compensating for damage done, mitigation and compensation measures are normally only undertaken within the formal road and railroad areas (approx. equal to the "in-site" safety zone) and by compensating only the very specific value which has been negatively affected ("in-kind"), often to a much lesser degree than 1:1 (size, quality, etc.) (Larsson 2020; Persson et al. 2015; Persson and Larsson 2014). The present project-centred planning process does not allow for wider, place-specific and out-of-site and out-of-kind solutions, which could be more contextually relevant on a (regional) landscape scale and over a longer period of time. Thus, there is most certainly a need for a planning system where cumulative effects and synergy effects between different, ongoing and overlapping infrastructure projects could be better understood and evaluated (Jones 2016; Persson et al. 2015; Persson and Larsson 2014).

48.3 Observing Landscapes from a Place-Centred Planning Perspective

In consideration of the foregoing, a regional landscape administration office should be able to coordinate many different projects in a specific region, involving, for example, landscape character analyses, impact assessments, public participation and compensation measures, focusing on both functional and structural properties in order to move away from project-centred logic towards more place-centred logic. Landscape observatories have been discussed within the context of the European Landscape Convention (ELC) (see for example Appendix 1, paragraph 10 in the guidelines for implementation of the ELC (Council of Europe 2008)). The ELC came about as a reaction to the manifold, often negative, driving forces affecting the landscapes of today, together with an insight about the need for a trans-sectorial landscape policy considering citizens' rights to information and participation in landscape-related decisions. The protection, management and planning of landscapes in accordance with the statutes of the ELC require continuous observation and suitable forums for the exchange of information, such as landscape observatories, centres or institutes for landscape. A fundamental value is progress towards a holistic perspective by handling the entire landscape and not only specific objects one by one (Sarlöv Herlin et al. 2019; Larsson 2020).

Possible tasks for a landscape observatory could include gathering information about the status of present landscapes, developing indicators for observing landscape changes or implementing such observations, collecting information about landscape policy and experiences on protection, management and planning through the collection and use of historical documentation on the evolution of landscape over time, or providing data to increase the understanding of current trends (Council of Europe 2008; Sarlöv Herlin et al. 2019; Larsson 2020). Some of the existing landscape observatories have a local focus or handle very subject-specific tasks, while others are of much higher complexity, covering a larger geographical area and with professional governance supported by, for example, regional political bodies (Sarlöv Herlin et al. 2019; Larsson 2020). The landscape observatory in Catalonia is thus far probably the most advanced in Europe and could provide inspiration for many others (http://www.catpaisatge.net/eng/index.php). This observatory, as well as the general intentions specified by the ELC, certainly promotes place-centred logic over project-centred logic, while handling both structural and functional properties without favouring one over the other. Thus, the Catalonian landscape observatory could act as a role model when developing organisational framework, content and specific methodologies for the creation of other national and regional landscape observatories, but it could also inspire the European Union to recognise the strategic value of continuing the work already done by the Council of Europe and to integrate the ambitions of the ELC within EU policies.

48.4 Some Final Words

To sum up, there is a strong need for regional landscape "hubs" from which landscape projects could be dealt with in a more holistic way than within the present project-centred system. In fact, Sweden has already ratified the ELC, where landscape observatories are mentioned as a method towards reaching a more effective, place-centred, democratic and constructive discourse on landscape change and preferred development, all in line with the ambition to cultivate the continuity of the European landscape and improve landscape literacy among both experts and the public in general. Why not start that with an acknowledgement of the need for regional coordination of large-scale infrastructure projects, where there is such an obvious drawback regarding the present project-centred planning system? This could be the seed from which landscape observatories grow, with the ambition of looking even further down the road towards other current large-scale landscape interventions, where consideration of past and future landscape qualities is urgently needed, and where incremental sector planning is not sufficient, such as urban–rural relations regarding energy and food production versus energy and food consumption, growth, de-growth, sprawl and much more.

References

Barca F, McCann P, Rodríguez-Pose A (2012) The case for regional development intervention: place-based versus place-neutral approaches. J Regional Sci 52(1):134–152. https://doi.org/10.1111/j.1467-9787.2011.00756.x

Björling N, Fredriksson J (2018) Relationer mellan stad och land i det regionala stadslandskapet. Mistra Urban Futures Report 2018:1. Göteborg

Council of Europe (2008) Recommendation CM/Rec (2008)3 of the Committee of Ministers to member states on the guidelines for the implementation of the European Landscape Convention, Council of Europe, Strasbourg

Eliasson I, Knez, I, Fredholm S (2018) Heritage planning in practice and the role of cultural ecosystem services. Heritage Soc 11(1): 44–69. https://doi.org/10.1080/2159032X.2019.1576428

ESPON (2006) Urban-rural relations in Europe. In: Bengs C, Schmidt-Thomé K (eds) ESPON 1.1.2 final report. Centre for Urban and Regional studies, Helsinki University of Technology

Germundsson T (2005) Regional cultural heritage versus national heritage in Scania's disputed national landscape. Int J Heritage Studies 11(1):21–37. https://doi.org/10.1080/13527250500036791

Jones FC (2016) Cumulative effects assessment: theoretical underpinnings and big problems. Env Review 24:187–204. https://doi.org/10.1139/er-2015-0073

Larsson A (2020) Place logic rather than project logic: landscape observatories as regional coordinators of large scale projects and compensation measures. In: Rönn M, Grahn Danielsson B (eds) Cultural heritage compensation: approaches to transformation of sites with cultural values and architectural qualities. Kulturlandskapet, Chalmers University of Technology, Göteborg

Larsson A (2004) Landskapsplanering genom jordbrukspolitik—en kritisk granskning av EU:s agrara jordbrukspolitik ur ett planeringsperspektiv. Dissertation. SLU Alnarp

Nordregio (2006) Den storstadsnära landsbygden—regionala funktioner och intressenter. En förstudie om landsbygden i Stockholms, Uppsala och Södermanlands län ur ett regionalt perspektiv. PM 3:2006. Stockholm

Persson J, Larsson A, Villarroya A (2015) Compensation in Swedish infrastructure projects and suggestions on policy improvements. Nat Conserv 11:113–127. https://doi.org/10.3897/natureconservation.11.4367

Persson J, Larsson A (2014) En nationell kartläggning över miljökompensationsåtgärder i väg- och järnvägsprojekt In: Persson J (ed) Miljökompensation vid väg- och järnvägsprojekt: identifiering av status, problem och möjligheter, Rapport 2014:24. SLU Alnarp

Sarlöv Herlin I, Klintborg Ahlklo Å, Larsson A, Lindahl A, Peterson A (2019) Landskapsobservatorier i Europa: Organisationsformer och relevans för kulturlandskapet. Sveriges lantbruksuniversitet, Fakulteten för landskapsarkitektur, trädgårds- och växtproduktionsvetenskap, Rapport 2019:6. SLU Alnarp

Stenseke M (2016) Integrated landscape management and the complicating issue of temporality. Landsc Res 41(2):199–211. https://doi.org/10.1080/01426397.2015.1135316

Swensen G, Jerpåsen GB, Saeter O, Sundli Tveit M (2013) Capturing the intangible and tangible aspects of heritage: personal versus official perspectives in cultural heritage management. Landsc Res 38(2):202–221. https://doi.org/10.1080/01426397.2011.642346

Swensen G, Jerpåsen GB (2008) Cultural heritage in suburban landscape planning: a case study in Southern Norway. Landsc Urban Planning 87(4):289–300. https://doi.org/10.1016/j.landurbplan.2008.07.001

Whiston Spirn A (2005) Restoring Mill Creek: landscape literacy, environmental justice and city planning and design. Landsc Res 30(3):395–413. https://doi.org/10.1080/01426390500171193

Chapter 49
The Observation of Informal Landscape Public Participation

Elena Lorenzetto

Abstract The discussion about Landscape Observatories might be advantaged by an analysis about what civil society currently does and how it relates to the landscape. Only if we understand the actual practices of citizens towards the places they live in will we be able to more effectively develop a relevant platform of public participation such as Landscape Observatories. In this perspective, the paper will illustrate research on citizens' actions which give meaning to, make connection with and develop empathy towards places. Though many surveys have concluded that individualism, urban sprawl and consumerism are the paradigms of contemporary society, this study showed an increase in citizen awareness and common activity. The research was conducted in 2019 and focused on case studies located in the Venetian area of Italy. The paper describes the specific languages and features of these cases; thus, it illustrates how they involve and connect people and generate symbolic and physical transformations. The thesis is that these actions can be assessed as innovative landscape practices inspiring methods and solutions as they allow institutional platforms to understand more about the social landscape. The paper shows how Landscape Observatories can support and foster these informal movements and groups.

49.1 Introduction

How can Landscape Observatories be shaped if not by following the shape of civil society landscape? How can they improve the relationship between people and places if not by understanding the expectations and issues of individuals?

This paper claims that only by understanding actual citizen actions towards the places they live in will we be able to help them fulfil their aspirations through open participation platforms such as Landscape Observatories.

In this perspective, this paper will illustrate research on citizen actions which give places meaning and develop empathy and a sense of belonging towards them.

E. Lorenzetto (✉)
Fondazione Benetton Studi Ricerche, Treviso, Italy
e-mail: lorenzetto.elena@gmail.com

T. Matteini et al. (eds.), *Cultivating Continuity of the European Landscape*, Environmental History 15, https://doi.org/10.1007/978-3-031-25713-1_49

Fig. 49.1 Vaghe Stelle group trekking between Bassano del Grappa and Valstagna. *Photo credit* PierangeloMiola

Fig. 49.2 Sentieri Indivisibili event in Cima Monte Tomba (TV). *Photo credit* Elena Lorenzetto

The research was conducted in 2019[1] on case studies located in the Venetian area of Italy and was related to two specific macro-attitudes: walking and narrating. The cases were heterogeneous and included: groups that organise research-action walks; festivals that regenerate abandoned villages; cultural and social activities such as writing workshops in nature; committees aiming to defend ancient or natural places at risk of development; groups raising awareness towards landscape by proposing film exhibitions or hikes (see Figs. 49.1 and 49.2).

The thesis is that despite their diversity, these actions can be assessed as *innovative landscape practices* allowing institutional platforms to understand more about perceived and social landscape and thus develop methods and solutions. The research aimed to analyse the semiotic dynamics and languages implicit in these actions.

[1] The research was supported by Fondazione Benetton Studi Ricerche by an annual scholarship. It is inspired by the project "Luoghi di valore", conducted by FBSR from 2007 to 2012 (Zanon 2016).

Fig. 49.3 Vaghe Stelle in 2018 in Bassano del Grappa. *Photo credit* Pierangelo Miola

49.2 Analysis

49.2.1 The Languages

The first important aspect is that inhabitants take part in "making landscape" in their own specific ways and languages, which are often different from the ones used by urban planning and policy. As the architect Anna Lambertini states, people are "changing change … that invite[s] us to introduce alternative times, procedures and tools of working into the planning, designing and management of habitable space" (Lambertini 2017, 108). Our body plays a key role in this approach. First of all, our own steps are a way of re-appropriating the places we are walking in; secondly, a body on stage in shows and performances reveals unexpected spatial uses and discloses planning potentialities and feasible transformations; moreover, all the ways of using words, poetry and the arts contribute to transform a space into a cultural landscape (see Figs. 49.3, 49.4, 49.5 and 49.6).

Walking is a favourite action that satisfies a growing need to reconnect to the physical world, a compulsion working against the trend of the automobile city, of *movism* (Girot 2004), of accelerated contemporary digital society. From an overview, we can assess different types of walking: walking alone; marching and parading; collective walking as an instrument for the re-appropriation of public space by citizens and wayfaring as a process of acquiring knowledge of the current actual territory (Careri 2006).

49.2.2 The Paradigm

Contemporary western society is often described as selfish and not inclined to collaboration (Sennett 2012), or narrow, cynical and rancorous (Bonomi 2008). Contrary to that vision, this constellation of empathetic and positive actions redefines territorial

Fig. 49.4 Autobiographic writing workshop "Mettiti in Poesia" on a creek in Cison di Valmarino (TV). *Photo credit* Elena Lorenzetto

Fig. 49.5 Performance "Il genio dei luoghi" by Farmacia ZoÈ, during Sentieri Indivisibili. *Photo credit* Rec Photo

balance, proposing a cultural paradigm based on caring for places instead of their consumption and expressing desire for change and reconnection to places. These actions express positive attitudes and feelings such as courage, freedom and dedication, in contrast to the aforementioned unhappy and negative feelings (Benasayag and Schmit 2003).

Another fact is the mutuality of self and place in the practice of caring and valuing. On one hand, actions which feed interpersonal relationships and create a sense of community lead to taking care of places and dealing with them as relational platforms. On the other hand, activities that manage, preserve, improve and regenerate spaces lead to dealing with ourselves as individuals, groups and humankind because taking care of places implies listening to our own soul, dealing with our own identity and thinking about our attitude towards the world.

Fig. 49.6 Lago Film Fest in Revine Lago, a depopulated little village, now regenerated by the cultural festival. *Photo credit* Elena Lorenzetto

49.2.3 The Subjects

The case studies analysed confirm how the concept of community is difficult and complex to understand and use in an appropriate manner (Aime 2019, 7). In fact, when we observe bottom-up processes we see that they might not involve the whole local community. The actions are designed, managed and carried out by minorities that might share one or several traits, such as culture, age, social class, professional environment or ideology. Even inside a group there is often a core of five or six people dedicating a huge and admirable amount of time and talent. This is a normal condition as it is voluntary work that not everyone can afford to do. However, this causes vulnerability and instability in the activities. Indeed, groups are frequently not able to communicate with and involve the larger population and risk addressing only a niche of inhabitants.

Moreover, getting involved and creating a single "we" means sharing values and goals and, above all, pushing beyond ourselves, which is not an easy process. Thus, these actions might also have limitations, as many experiences of tactical urbanism have demonstrated, since they might be ephemeral and less vivid if not connected to a landscape governance system.

49.3 Conclusions

49.3.1 The Need for Landscape Observatories

The groups I interviewed and observed demonstrate what is hardest for them: moving from the stage of design and cooperation to a transformation stage actually dedicated to planning, conservation and the transformation of spaces. In particular, the people in charge of projects told me in interviews that they are sometimes frustrated because as group they are unable to achieve the goals they set themselves, accomplish the projects they draft, produce effective communication to citizens and institutions or report their experiences as storytelling. These failures are related to the voluntary nature of their work whose financial resources and time are almost always insufficient, rather than to actual group inefficiency. However, this situation underlines a more general problem: the lack of institutions actively involved in the design, financial support and transformation phases. Rather, the expectations of associations are often unfulfilled as institutions sometimes lack a strong awareness of landscape, the technical expertise to manage community projects, or even capital. In other situations, political actors might exploit local group efforts to their own advantage or shirk responsibility for issues they should be dealing with and managing (Gangemi 2019).

All these problems point to the absence of a public, institutional actor handling the management of places. Civil society is expressing the need for an institution such as a Landscape Observatory with its network of local stakeholders. The results of this research show that the process approaches I studied indicate that for citizen groups; Landscape Observatories would be perfect platforms of reference for hosting and supporting local activities regarding places and landscape as they could provide means, methods and instructions for drafting and dealing with complex projects and mediate between local communities and municipalities, provinces and regional institutions.

As the "guidelines for the implementation of the European Landscape Convention" state, Landscape Observatories can "exchange information on policies and experience concerning protection, management and planning public participation and implementation at different levels" and should allow for collaboration between the different subjects involved in landscape planning: "scientists, professionals and technicians from the public authorities and the public"(Council of Europe 2008, 21). Even though the various administrative levels should be in charge of landscape planning, we believe that Landscape Observatories should have a more operative role than that of observation and data collection. They can also coordinate and lead participation actions and community activities regarding landscapes.[2]

Indeed, a third subject such as a Landscape Observatory could assess the actual results of civil activities on the whole population. A Landscape Observatory could

[2] A case study that tested this model of Observatory Landscape is the local Observatory of Canale del Brenta in the province of Vicenza (Italy), lead by the Region of Veneto and the Geography Department of the University of Padova (Castiglioni 2015; Castiglioni and Varotto 2013).

catalogue and supervise these bottom-up processes without hindering their civic passion for places.

49.3.2 How This Translates into Landscape Policies

Therefore, how can we organise all that energy and desire for transformation into a landscape policy framework? If we want to address these issues, we should ask: what is the actual—and non-rhetorical—sense of civil participation in landscape policies? And then, how can we actually involve these groups in shared landscape planning processes, approaching landscape as a common good? An analysis of these participative actions leads to a better definition of roles and levels of participation.

Experimentation by the Landscape Observatory of Catalonia (Nogué et al. 2016) shows us an interesting approach that fits with the results of this research. The public could be involved in the starting phase, which would include active listening and analytical knowledge of places. These results would become useful material for the planning process by experts. The experiences I analysed produced more awareness and understanding of the territory and, in some cases, they were actual instruments of action-research (see Fig. 49.7). Indeed, they show different ways and means of understanding and analysing the places we live on or cross over. Moreover, they foster more creative actions involving the body and senses that connect at a slow rhythm, preferring words rather than numbers, a feet-on-the-ground rather than bird's eye view, dancing rather than calculating. This *geopoetical* approach could be adopted by a Landscape Observatory to involve local communities in landscape issues and receive information from inhabitants. As Joan Nogué said referring to the Landscape Catalogues of Catalonia, participation in the starting phase could be an effective strategy for landscape policies. First, the participation could help discover intangible values and cultural elements that are precious for planning and management; secondly, if the population is engaged, it becomes aware of place matters and develops trust towards politicians and experts. Then, the citizenship could be involved at a subsequent stage when the project would be explained and tested, so that inhabitants have the chance to play an active role in transformations, but not in the position of planners. This is exactly what the groups I interviewed asked for. From the experiences of these minorities, we can learn how a qualitative sensorial method could be the basis for collectively working on landscape. As we observed, the more we take care of places and construct personally passionate connections, the more community feelings and social relationships develop. This could also be valid in the context of landscape practices: the more we propose an empathetic understanding of places to citizens and politicians, the more this approach is translated into a positive, dialectic relationship between these subjects.

Therefore, not only could a Landscape Observatory be an institutional platform where these experiences are collected but also a place to test how this geopoetical and tactical approach towards landscape can become the first step in its planning and protection.

Fig. 49.7 Workshop "Che paesaggio sono" during Sentieri Indivisibili, at Collina Nervo (TV). *Photo credit* Rec Studio

References

Aime M (2019) Comunità. Il Mulino, Bologna

Benasayag M, Schmit G (2003) Le passion tristes, souffranse psychique et crise sociale. Èditions La Découverte, Paris

Bonomi A (2008) Il rancore: alle radici del malessere del Nord. Feltrinelli, Milano

Careri F (2006) Walkscapes. Camminare come pratica estetica. Einaudi, Torino

Castiglioni B (2015) Three years of landscape observatories in the Veneto region: an evaluation. In: UniscapeEn-Route, y. 1 n. 1

Castiglioni B, Varotto M (2013) Paesaggio e osservatori locali. L'esperienza del Canale di Brenta. Franco Angeli, Milano

Council of Europe (2008) Guidelines for the implementation of the European Landscape Convention, 6 Feb 2008

Gangemi S (2019) Common landscape. Processi di educazione, partecipazione ed empowerment in paesaggi ordinari. Quodlibet, Macerata

Girot C (2004) Movism: prologue to a new visual theory. In: Koll-Schretzenmayr M, Nussbaumer G, Keiner M (eds) The real and virtual worlds of spatial planning. Springer, Berlin

Lambertini A (2017) The gift as a paradigm of the ordinary landscapes project. In: Boschiero P, Latini L, Zanon S (eds) Caring for the land. Places, practices, experiences. Fondazione Benetton Studi Ricerche - Antiga Edizioni, Treviso

Nogué J, Sala P, Grau J (eds) (2016) The landscape catalogues of Catalonia. In: Methodology. Observatori del paisat, Olot

Sennett R (2012) Together. The rituals, pleasures and politics of cooperation. Penguin Books, London

Zanon S (ed) (2016) Outstanding places. An experiment in the province of Treviso, in the wake of the European Landscape Convention. Fondazione Benetton Studi Ricerche-Antiga, Treviso

Chapter 50
Governance Challenges of a Local Landscape Observatory Initiative in a Remote Landscape: Ways Ahead

Maria Graça Saraiva, Isabel Loupa-Ramos, and Veerle Van Eetvelde

Abstract Following the recommendations of the European Landscape Convention for the establishment of landscape observatories (LO), there has been a plethora of governance models put in place. This article reports on the governance challenges of the local LO of Charneca (Portugal) arising from its bottom-up approach in a peripheral and low-density territory. The objective of this paper is to analyze the stakeholders in the Charneca landscape and its landscape observatory. Methodologically, a social network analysis is used to identify the position of actors in the network and the density of relations. The results disclose the position of the LO within the network, notably its close relationship to the estate where it is located and to the local administration, leading to the conclusion that including other actors and local governments to go beyond the borders of the Chamusca municipality in which the observatory is located is pivotal to enabling action at a landscape scale.

Keywords Landscape governance · Landscape observatory · Actors · Montado · Portugal

M. G. Saraiva (✉)
CIAUD, Research Center for Architecture, Urbanism and Design, Lisbon School of Architecture, Universidade de Lisboa, Lisbon, Portugal
e-mail: gsaraiva@sapo.pt

I. Loupa-Ramos
IST, CiTUA, University of Lisbon, Lisbon, Portugal
e-mail: isabel.ramos@tecnico.ulisboa.pt

V. Van Eetvelde
Department of Geography, Ghent University, Ghent, Belgium
e-mail: Veerle.VanEetvelde@UGent.be

T. Matteini et al. (eds.), *Cultivating Continuity of the European Landscape*, Environmental History 15, https://doi.org/10.1007/978-3-031-25713-1_50

50.1 Introduction

Landscape is the arena in which people live and on which they act in multiple ways. These actions take place at a local level and, despite the importance of policies as drivers, the role of actor interplay on landscape change locally cannot be neglected (Nagendra and Ostrom 2012). This calls for a shift in view point from a 'government' to a 'governance' perspective. Governance has been defined in multiple ways, depending on the disciplinary backdrop, context and objectives. Meuleman (2008) defines governance as the "totality of interactions in which government, other public bodies, private sector, and civil society participate, aiming at solving societal problems or creating societal opportunities". When applying the concept to landscape, "governance deals with the interconnections between socially constructed spaces and 'natural' conditions of places" (Görg 2007). Using the governance concept, landscape is both a territory and also scenery and a construct (Antrop and Van Eetvelde 2019). So landscape can be considered as a mediator between people and place, bridging social and ecological relations. Landscape governance can be defined "as the set of rules (policies and cultural norms) and the decision-making processes of public, private and civic sector and actors with stakes in the landscape that affect actions in the landscape" (de Graaf et al. 2017). Landscape governance is used both as an analytical framework and as a normative approach based on the idea of 'good governance', aiming to achieve sustainability and involving local communities in negotiations and decision-making (e.g. Reed et al. 2015; Oosten et al. 2018). As landscape is a meeting-point for policies, activities and people, landscape governance is multi-level, multi-sector and multi-actor (Kozar et al. 2014). The argument arises that by not having their own jurisdiction and transcending administrative and sectoral boundaries, landscapes provide a sort of functional space (van Oosten 2018) which is more fit to serve as the locus of policy, sector and actor integration. In these networks, the actors, their power relations and involvement are shaped by institutional structures that influence decision-making and by the institutional and cultural context which can be critical (Ostrom 2007).

Landscape observatories can be significant in making the landscape approach operational on a local level, engaging different actors and drivers of landscape change. The ELC recommendation to develop landscape observatories does not prescribe any specific model of LO: "The choice of the composition of observatories is a matter for the administrative bodies concerned". Landscape observatories can emerge as new actors in a variety of ways (Cassatella and Larcher 2015). The top-down and bottom-up approaches represent two different worldviews, the first one seeing laws as back bones, formalizing institutions with authority to bind others, and the second seeing "institutions emerging spontaneously from the social norms, customs, traditions, beliefs, and values of individuals within a society" (Easterly 2008). Landscape observatories with a top-down worldview are typically led by a public administration institution, working with a previously allocated mainly public budget or private foundation funds. This kind of observatory tends to rely on expert insight for decision-making and landscape change approaches and methods. Bottom-up initiatives are

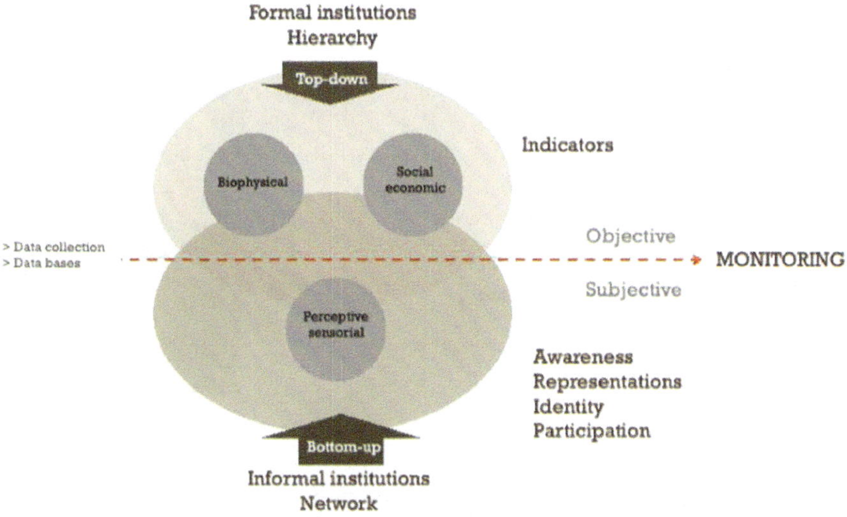

Fig. 50.1 Ways of approaching landscape based on the type of initiative creating the LO

place-based and grounded on a participatory model of interaction between actors. They tend to rely on voluntary work and there might be no fixed allocated budget (Cassatella and Larcher 2015).

The type of initiative that creates the LO tends to resonate in the way it operates. Top-down approaches tend to emphasize the objective components of the landscape based on structured data collection. Due to lack of stable resources, bottom-up approaches tend to be more project-based, conduct less systematic data collection and rely on voluntary contributions (Fig. 50.1). Both approaches can act as observation and knowledge sharing platforms as recommended by the ELC, currently merging towards a participatory model.

50.2 Research Objective and Methodological Approach

To fully grasp the governance model of a bottom-up landscape observatory, it is crucial to define the different stakeholders and understand their relationships. The objective of this paper is to analyze the stakeholders in the Charneca landscape and its landscape observatory. The results of the social network analysis feed into the identification of ways to strengthen the position of the LO and its governance model in the future. First, all actors related to the LO were identified. The strength of the linkages between the LO and the actors was weighted on a 0–4 scale. Subsequently, the actors were grouped. Data were compiled into a matrix identifying network nodes and edges and computed using Gephi software to plot a final graph making the strength and density linkages and node ranking visual.

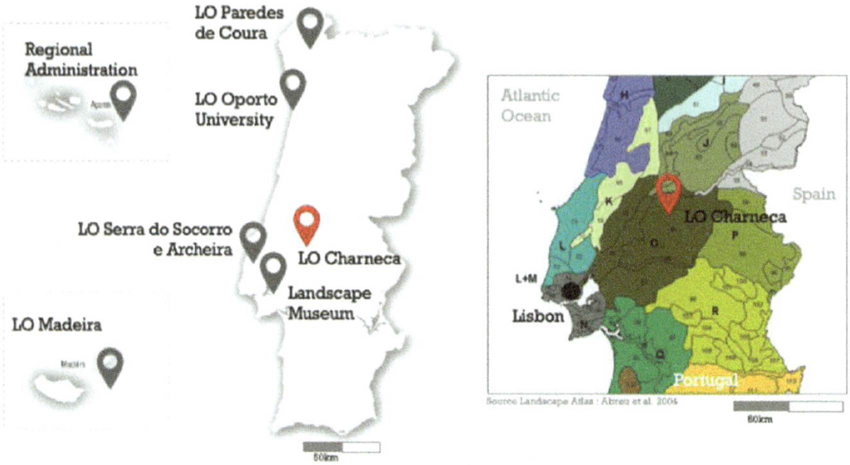

Fig. 50.2 **a** Geographical location of the Portuguese LOs; **b** Location of CLO in the Charneca landscape unit (86), included in landscape unit Group 'O'—Ribatejo

50.3 The Case of the Charneca Landscape Observatory (CLO)

50.3.1 Placing the CLO in the LOs of Portugal

Portugal has seven LOs situated on the mainland and archipelagos (Fig. 50.2). One type of LO has the dual objective of community mobilization and scientific activities, as a kind of knowledge broker using a bottom-up approach (e.g. the Charneca LO and the *Museu da Paisagem*). The others essentially follow a top-down approach led by local or regional public administrations whose main objective is to support landscape planning and management. Madeira and *Paredes de Coura* follow a mixed approach based on partnerships. Oporto University LO is focused on research and technical advice in multiple landscapes. The landscape observatories involving or led by public administration bodies are focused on their area of jurisdiction. The *Museu da Paisagem* and the Charneca LO are focused on specific landscapes.

50.3.2 The Context, Origin and Aims of the CLO

The Charneca landscape observatory is located on a silvo-pastoral private farm in the lower Tagus region (Chamusca, Portugal). The observed landscape is called 'Charneca', a type of matorral dominated by cork oak woodlands used in the past as the king's hunting ground. Since the beginning of the twentieth century, it has been developed for wheat production, forest and pasture management. This landscape is a distinct unit in the Landscape Atlas of Portugal developed following the ELC signature by Portugal (Abreu et al. 2004). The Charneca landscape is a vast area of circa 5000 km^2 on the interface between the Ribatejo and Alentejo region. It is part of the Mediterranean climate region, shaped by sedimentary processes related to the Tagus river evolution, with poor, acid soils and a vegetation cover dominated by cork oak woodlands (*Quercus suber*) (Saraiva et al. 2017). It is a high natural-value farming system, rich in biodiversity. The population is concentrated in small settlements such as villages and farmsteads, known locally as *casais* (Fig. 50.3). Agriculture and forestry are the main activities, carried out in large estates. The area has a low population density with a high percentage of elderly people. Younger people are leaving the countryside in search of better opportunities in larger cities. This is a paradigmatic remote landscape, marked by poor accessibility to the main highway network, despite the relatively short distance to Lisbon (150 km). However, many natural and cultural qualities still exist in the area. Local communities strongly identify with and are attached to this landscape, its history and the traditions related to its vernacular management, heritage and values. The landscape is characterized by a visible diversity of contrasting land forms, such as valleys and hills, and various land covers, such as woodlands and open fields, with their contrasting textures. The landscape is therefore intriguing and shows interesting seasonal variety. It is considered a hidden treasure whose quietness, naturalness and character show great potential for future tourism activities.

The Charneca LO builds on the concept of a landscape observatory following the ELC recommendations and uses a bottom-up approach involving landscape researchers and practitioners and local communities. It was formally created as an association in 2018, although many activities had been developed since 2015. The

Fig. 50.3 Image of the landscape of Charneca and a traditional farmstead, 'Casal'

aims of LO are (i) to promote and disseminate knowledge of the Charneca landscape; (ii) to foster communication among people and organizations interested in landscape-related topics and activities, so as to create actions and experiences based on the landscape, cultural heritage and territorial identities; (iii) to collaborate nationally and internationally with public and private organizations on common goals through the establishment of partnerships and (iv) to contribute to an increased awareness in civil society of the value of the Charneca landscape and its transformations (see: https://opc-paisagem.pt/en/).

The CLO is located in the 'Casal do Gavião', a traditional farmstead and a rural enterprise (Fig. 50.4) in the municipality of Chamusca. This 500 ha estate is based on traditional *montado* management using sustainable practices. Montado is a silvo-pastoral system with high natural and cultural value based on evergreen oak woodlands, adapted to the southern European mediterranean climate (Pinto-Correia et al. 2013). Its managers were involved in landscape research and practice. Their intention was to share this particular setting with different actors in order to produce and disseminate this specific knowledge. The landscape of the *casal* acts as an experimental pilot area for different activities. Some public and demonstration activities were developed in an informal way, beginning with visual arts workshops focusing on the *montado* and the landscape and hosting academic visits for fieldwork. This led to the development and creation of a formal landscape observatory, which has attracted more participants and partnerships and creative tourism activities. An old barn was rehabilitated with a contemporary architectural design and sustainability concept. The hall is now hosting LO activities such as workshops, field trips, landscape painting sessions and tourism activities (Fig. 50.5). In 2018 and 2019, almost 500 people visited the LO.

A strategy was developed using the activities and projects in the framework of a flexible 'constellation' model gaining momentum according to arising opportunities,

Fig. 50.4 **a** Farmstead 'Casal do Gavião', the 'Charneca' LO headquarters; **b** Hall of the observatory (credit of photos FG + SG)

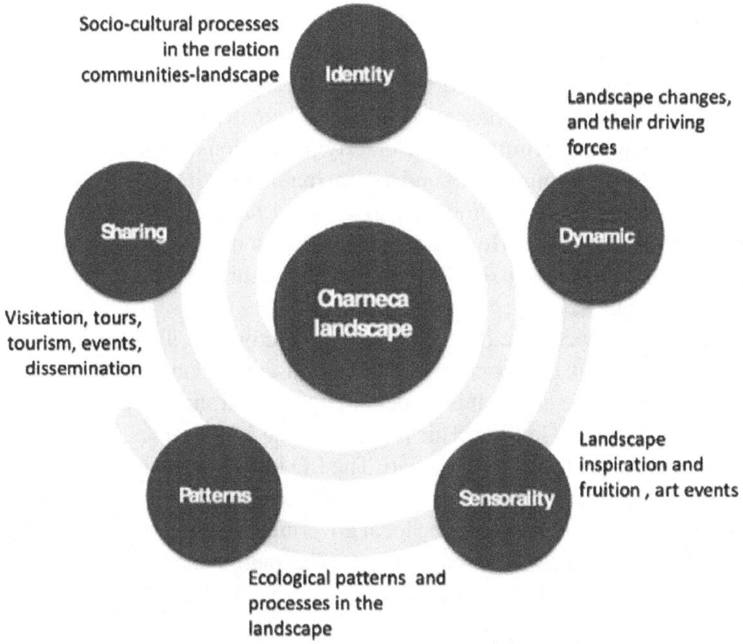

Fig. 50.5 Articulation model of projects and networks developed at the Charneca LO

actors and available resources. This model includes five domains of interactions (Fig. 50.5).

50.4 Actors and Initiatives of the Charneca LO

The Charneca LO brings together different actors involved in a variety of projects and activities. In total, 29 actors have been identified and clustered into twelve main actor groups shown in Fig. 50.6. Three actors are particularly essential to the network: the **Charneca LO** itself, the **Casal do Gavião**, as the 'place of action' for activities, and the **local government**, represented by the Chamusca municipality and its administrative parishes. The latter is crucial for processes such as permits, culture, relations with local entities and access to information. Two groups of actors are related to education and research: the **local school** organizes field trips and joint projects in its teaching curricula; **universities** are important partners that enrich the experience and knowledge of the LO and include international academic institutes, such as the University of Ghent which has conducted an annual excursion for several years in collaboration with the University of Lisbon (IST). **NGOs** are significant actors: the **nature-culture** field includes the collection and dissemination of biodiversity information as well as geocaching activities; the field of **farming and forestry** plays a role by providing technical support and insights from their members. Two groups are related to economic aspects: **local enterprises** are involved in the business of buying and selling products, such as timber or crops or provide services and the **regional tourism agencies** promote nature and creative tourism activities in the region. **Social support groups** aid the elderly population and develop activities for leisure and health, promoting visits and tours to natural environments. **Artists** realize visual arts workshops, getting together to work *en plein air* in the landscape. The **Observatory of Cork and Montado**, promoted by a neighbouring municipality, is similar to the LO and aims to develop the region's natural resources and values with a collaborative mindset.

The relations between the different groups of actors are illustrated in Fig. 50.6. It presents the network of actors active in the landscape relating with Charneca LO. From an analysis of the graph, it is evident that there is a shared centrality of the LO and its estate (Casal do Gavião). This means that the LO accesses some actors, such as local NGOs, through Casal do Gavião. The LO invests in educational activities in both local schools and universities and uses the activities to connect with the local government. The graph shows that the local government is a centrality in its own right which interacts with many landscape actors. The shared centrality of the Charneca LO and Casal do Gavião is synergetic in the sense that the former established links to outside institutions and other centralities, while the latter focuses mainly on creating links within the community.

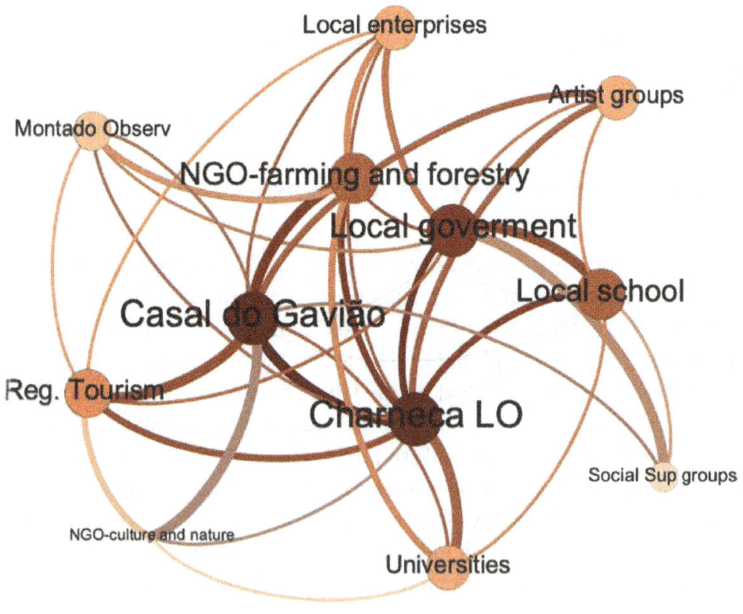

Fig. 50.6 Network of landscape actors in the sphere of the Charneca LO

50.5 Discussion and Conclusions

Landscape observatories in Portugal, and elsewhere, show a range of models according to Cassatella and Larcher (2017) typology, depending on their governance arrangements and strategies to achieve specific goals. The CLO occupies a role of mediator between local and expert knowledge. Moreover, it also increases the critical mass of a low-density landscape by creating opportunities for outsiders to reflect on the landscape. The LO's links to multiple universities (national and international) are pivotal in ensuring the dynamics of co-production of knowledge. Under the umbrella of the CLO, several projects are under development, e.g. an educational project with local schools called 'School in the forest', or educational walking trails. This shows the great potential of the CLO.

The Charneca LO has a bottom-up approach driven by a group of individuals. The strength of this adaptative governance model is its flexibility and independence from administrative power structures. However, this flexible structure has two main downsides: no stable income and funding require constant investment in network nurturing to keep ongoing activities in progress. A shortage of resources and the difficulties of ensuring long-term commitments with the actors in the network are also considered a major threat to the sustainable long-term future of the LO. The idea of landscape as a functional space has become tangible in the Charneca LO. Multiple actors need to be gathered to engage in co-producing knowledge that generates co-benefits. As the administrative system does not formally operate at the landscape level, governance needs to find arrangements that fit into the specific dynamics of landscapes. Since the landscape is changing, the arrangements also need to be dynamic, in line with the changing actors and issues at stake in the landscape. There might be no "good" landscape governance but instead, the most appropriate one based on existing values and available resources. Landscape observatories show a potential to play a pivotal role in facilitating the process of finding a suitable governance model that fosters more sustainable landscapes.

The way ahead?... There is an urgent need to consolidate the existing CLO network by formalizing partnerships that ease fundraising activities to develop projects and to go beyond the borders of the municipality of Chamusca by including other actors and other local governments at the landscape scale.

References

Abreu AC, Pinto Correia T, Oliveira R (2004) Contributos para a IdenDficação e Caracterização das Paisagens de Portugal ConDnental [ContribuDons for IdenDficaDon and CharacterizaDon of Landscapes in mainland Portugal]. Universidade de Évora, Direção-Geral de Ordenamento do Território e Desenvolvimento Urbano

Antrop M, Van Eetvelde V (2019) Territory and/or scenery: concepts and prospects of western landscape research. In: Muller, Eulenstein (eds) Current trends in landscape research. Springer Nature, pp 3–39. https://link.springer.com/chapter/. https://doi.org/10.1007/978-3-030-30069-2_1

Cassatella C, Larcher F (2015) Landscape observatories in Europe: the action plan. Uniscape En-Route, a. I-n. I, pp 118–119. ISSN 2281-3195

Cassatella C, Larcher F (2017) Overview of landscape observatories in Europe. In: Presentation at the international conference of landscape observatories, Amersfoort. www.landschapsobservatorium.nl/Uploaded_files/casatella.b7eb5f.pdf

de Graaf M, Buck L, Seth Shames S, Roderick Zagt R (2017) Assessing landscape governance: a participatory approach. Tropenbos International and EcoAgriculture Partners. ISBN: 9789051131383. https://www.tropenbos.org/resources/publications/

Easterly W (2008) Institutions: top down or bottom up? Am Econ Rev Pap Proc 98(2):95–99 http://www.aeaweb.org/articles.php?. https://doi.org/10.1257/aer.98.2.95

George C, Homsy ZL, Mildred EW (2019) Multilevel governance: framing the integration of top-down and bottom-up policymaking. Int J Public Adm 42:572–582. https://doi.org/10.1080/01900692.2018.1491597

Görg C (2007) Landscape governance: the 'politics of scale' and the 'natural' conditions of places. Geoforum 38:954–966

Kozar RLE, Buck EG, Barrow TCH, Sunderland DE, Catacutan C, Planicka AK, Willemen L (2014) Toward viable landscape governance systems: what works? EcoAgriculture Partners, on behalf of the Landscapes for People, Food, and Nature Initiative, Washington, D.C.

Meuleman L (2008) Public management and the Metagovernance of hierarchies, networks and markets: the feasibility of designing and management governance styles. Springer, Heidelberg

Nagendra H, Ostrom E (2012) Polycentric governance of multifunctional forested landscapes. Int J Common 6:104–133

Ostrom E (2007) Beyond markets and states: polycentric governance of complex economic systems. Am Econ Rev 100:641–672

Pinto-Correia T, Ribeiro N, Oliveira I (2013) The MONTADO/DEHESA as high nature value farming systems: implications for classification and for policy support. In: ICAAM international conference 2013. Acknowledging the MONTADOS and DEHESAS as high nature value farming systems. 6–8 Feb 2013, Mitra Campus, University of Évora, Évora Portugal

Reed J, Deakin L, Sunderland T (2015) What are 'Integrated landscape approaches' and how effectively have they been implemented in the tropics: a systematic map protocol. Environ Evid 4:2. 2014. http://www.environmentalevidencejournal.org/

Saraiva MG, Loupa-Ramos I, Van Eetvelde V (2017) Towards a local landscape observatory in a Montado landscape. Which dimensions to explore? Uniscape en-Route 2(5):37–42. ISSN 2281-3195

Sayer J, Sunderland T, Ghazoul J, Pfund J-L, Sheil D, Meijaard E, Venter M, Boedhihartono AK, Day M, Garcia C, van Oosten C, Buck LE (2013) Ten principles for a landscape approach to reconciling agriculture, conservation, and other competing land uses. Proc Natl Acad Sci 110(21):8349–8356

van Oosten C, Uzamukunda A, Runhaar H (2018) Strategies for achieving environmental policy integration at the landscape level. A framework illustrated with an analysis of landscape governance in Rwanda. Environ Sci Policy 83:63–70

Chapter 51
The Atlas of Terraced Landscapes of Trentino, Italy

Giorgio Tecilla and Alberto Cosner

Abstract In recent years, the Autonomous Province of Trento (Trentino, Italy) has carried out a series of studies and projects centred on the relationships between landscape and agricultural activities. In 2014, the Landscape Observatory Forum approved the programme "Ten actions for the rural landscapes of Trentino" in which the terraced agricultural areas which characterise many valleys of this Alpine province received special attention. The programme soon took the form of an Atlas of terraced landscapes. An operational methodology for the identification of terraced areas, often hidden by recent forest expansion, was developed. It is a predictive model based on LiDAR data, which allowed the identification and characterisation of rural terraced landscapes, and which was followed by an on-site inspection that produced more than 7000 geo-referenced pictures. The Atlas comprises 16 volumes with several thematic maps covering all of the provincial territory that has been terraced. The research identified around 10,500 ha of terraced areas is 1.7% of the total provincial surface. Some 5800 ha are still in use, while the rest are abandoned and covered by forests. The retaining structures of the terraces are composed mainly of walls and slopes and extend for some 4500 km.

Keywords Terraced landscapes · LiDAR data analysis · Mountain agriculture · Agricultural areas abandonment · Trentino

G. Tecilla
Landscape Observatory, Autonomous Province of Trento, Trento, Italy
e-mail: giorgio.tecilla@provincia.tn.it

A. Cosner (✉)
President Research Group TeSto, Trento, Italy
e-mail: alb.cosner@gmail.com

T. Matteini et al. (eds.), *Cultivating Continuity of the European Landscape*, Environmental History 15, https://doi.org/10.1007/978-3-031-25713-1_51

51.1 Mountain Agriculture and Terraced Landscapes in Trentino

Terraced landscapes, which characterise many areas of the Alpine region, deserve special attention because they represent an extraordinary human artefact which is in danger due to its intrinsic fragility and the abandonment of marginal agricultural areas.

Over the past five years, the Autonomous Province of Trento (Trentino) in Northern Italy has carried out a series of studies and technical analyses and has developed several projects focusing on the complex relationships between landscape and agricultural activities. In November 2014, the Landscape Observatory Forum, the main participatory body on landscape issues, approved a programme named "Ten actions for rural landscapes of Trentino" (Osservatorio del paesaggio, 2014). This document devoted special attention to terraced agricultural areas, particularly emphasising the need for their mapping and for the preservation of their productive role, countering the ongoing abandonment process.

Given the technical complexity of these actions, in 2015 the Observatory developed preliminary research directed by Giorgio Tecilla and supervised by Alberto Cosner, which produced a "methodology for the identification and classification of the terraced landscapes in Trentino" (Osservatorio del paesaggio 2015).

In October 2016, the Province of Trento consolidated its interest in this issue by participating in the World Congress of the International Terraced Landscapes Alliance (ITLA). A dedicated session was held in Trentino, during which the role of local authorities in monitoring the dynamics of terraced landscapes and in supporting their rehabilitation and reuse was discussed in depth (Osservatorio del paesaggio 2016).

Extensive application of the method developed by the Observatory together with on-site visits and a photographic campaign made it possible to produce the "Atlas of Terraced Landscapes of Trentino", recently completed for the entire provincial territory and published in 16 volumes (Osservatorio del paesaggio 2020). The volumes, available both in paper format and as a digital document downloadable from the Observatory's website, are supported by a photographic archive of around 5000 georeferenced images (Fig. 51.1). In 2021, another volume is expected, providing a summary of the documentation work carried out.

51.2 The LiDAR Method of Terrace Identification

The methodology for mapping the terraced landscapes was developed taking into account the fact that many terraced areas are not completely visible because of their abandonment and the extension of the forest. Therefore, a semi-automatic predictive model was developed to identify and analyse active and ruderal terraced landscapes

Fig. 51.1 Terraced landscapes in the southern part of Trentino

by processing LiDAR data (LiDAR stands for *Light DetecEon and Ranging*, a tri-dimensional laser scanning technology). The method benefitted from discussion on the use of such data in archaeological research, while special attention was paid to the numerical consistency. The methodology has been designed to be fully reproducible by any user of LiDAR data with a minimum resolution of 1 m using open source[1] software along a simple and consistent workflow.

The team of Mauro VaroNo (University of Padova) greatly contributed to the development of the project by analysing the first pilot results (Ferrarese et al. 2019). The accuracy of the method was successfully tested in a few areas and then extended to the entire province, providing the basis for the Atlas.

The analysis process was structured into the following steps:

- delimitation of the study area (Geomorphological Survey Unit);
- extraction of raw data from the DTM LiDAR (Digital Terrain Model based on LiDAR elevation data);
- cleaning up of DTM LiDAR data distortions (removal of spurious or non-relevant signals);
- creation of curvature profiles concerning the bases and the edges of containment structures (walls, ramps, slopes);
- first vectorisation of the terrace boundaries;
- identification of retaining walls and terraces;

[1] GRASSGIS (version 6.4.2 and 7.0 https://grass.osgeo.org) and QuantumGIS (version 2.14.10 Essen www.qgis.org).

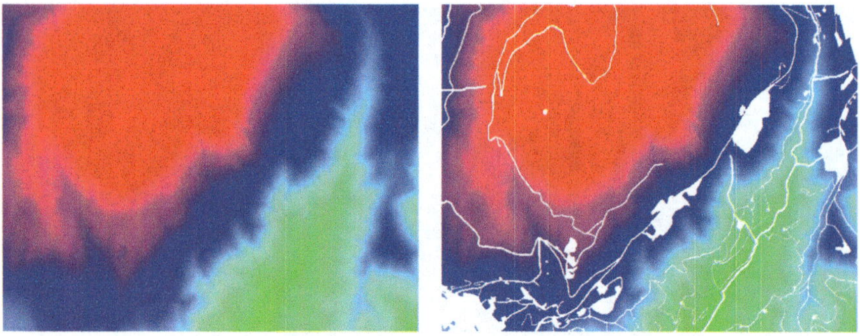

Fig. 51.2 Cleaned DTM map. On the left a standard DTM map, on the right all parts of the territory which cannot host terraces have been eliminated: streams, riverbeds, lakes, bare rocks, roads, quarries, urbanised and built-up areas, etc.

- characterisation of the terraced landscape within the municipal territory.

The study areas (geomorphological survey units) were defined following the limits of the Valley Communities (16 municipal consortia) whose sizes vary, but do not influence the effectiveness of the operations. A DTM was produced for each unit and the most important morphological features were processed, namely: *slope*, *aspect*, and *shaded relief*. In order to minimise the survey area and reduce data complexity, all parts of a territory which could not host terraces were eliminated: streams, riverbeds, lakes, bare rocks, roads, quarries, urbanised and built-up areas, etc.

The DTM, therefore, was cleared of the information corresponding to the cells where terraced areas are not possible or the retaining structures are not relevant for the investigation (Fig. 51.2).

The main feature of a terrace is the presence of a containment structure, generally vertical or very steep and of variable height, supporting an upper surface, nearly horizontal, the size of which depends on the steepness of the valley side. From a morphological point of view, the containment structure (wall, slope, ramp) is the most evident feature, which can be detected on the basis of a sudden change in altitude. This difference in height can be measured using a DTM by considering the curvature profile, a characteristic that describes the quantity and the nature of the steepness in a specific point.

In the model, the bases of the terrace walls are characterised by highly positive values of the curvature profile (marked concave angles), while the upper edges of the vertical containment structures are characterised by highly negative values (pronounced convex angles), see Fig. 51.3. Peaks of both positive and negative values in the curvature profiles are indicators of the probable presence of a containment structure.

The first step consists of filtering the curvature profile data to highlight positive and negative peaks which can be potential bases and edges of the containment structures. In this way, all areas with a low value of the curvature profile, oscillating around zero, can be safely excluded, as they are too low to be related to any vertical containment

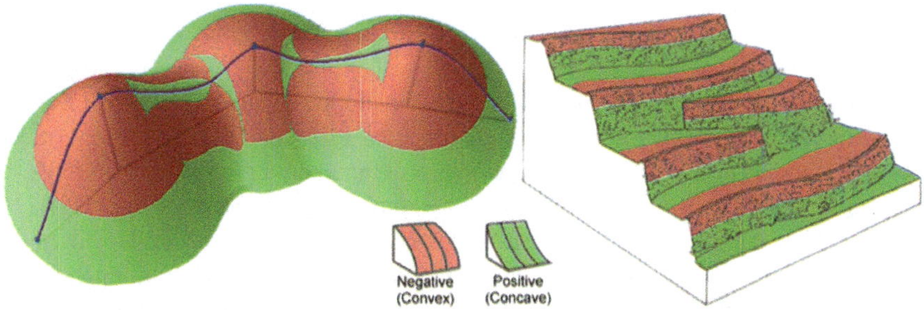

Fig. 51.3 Curvature profiles. The negative profiles (convex angles) are in red; the positive ones (concave angles) are in green

structure. A second cleaning is then carried out excluding all the small-sized surfaces, produced by faults in the data or reflecting minor elements, such as small rocks or natural steps in the terrain. The results, after applying the two filters, are definitely realistic and suitable for investigating the territory at a provincial scale.

The initial vectorisation of the terrace boundaries was performed through video recognition—the only non-automatic process used in the whole methodology—by superimposing the sum of negative and positive curvature profiles on the shaded relief map. This process was done on the cadastral scale of 1:2.800. Then, the areas identified were cleaned according to the density of the containment structures, which is an important indicator (Scaramellini and VaroNo 2008), removing those that were too small or too isolated, as well as terraces with only one or two walls, generating low-density values. In this way, the method favours densely terraced areas, with high ratios between vertical elements and horizontal surfaces, rather than isolated walls or edges.

The resulting map represents the borders of the terraced areas, which are characterised by pairs of highly negative and positive values of the curvature profile with a surface of at least 20 m^2 and which are sufficiently dense. This vectorial map is not the final one, as it must be subsequently improved through quality control checks.

Once the terraced areas were identified, it was possible to focus on the retaining walls. The surface representing the potential vertical containment structure is the line/area between a line of positive peak values of curvature profile and a line of negative peaks. In the space between these two lines, there are cells with a high steepness value. The distance between the lines of the curvature profiles can significantly influence the recognition of "real" walls: the closer these lines are, the larger the change in the slope and the greater the probability of the presence of a vertical man-made element rather than rocks or boulders.

The result is a map of the vertical structures. This analysis supports the last cleaning step of the map of the terraced area boundaries: only contexts containing sufficiently vertical surfaces are taken into consideration, while those without appreciable vertical structures and linear elements are excluded. The outcome is the final map of the terraced area borders.

51.3 The Atlas

The documentation and study programme was developed during the years 2017–2020 and allowed the identification and characterisation of around 10,440 ha of terraced territory, accounting for 1.68% of the whole provincial area. Some 5750 ha (55% of the total) are still in use, while 4690 ha (45%) have been abandoned (Fig. 51.4).

The Atlas comprises in-depth surveys on a municipal scale collected in 16 volumes, each corresponding to the aforementioned "Valley Communities" into which the provincial territory is divided. For each of the 146 municipalities where the presence of terraces was detected, the Atlas provides seven thematic maps, accompanied by graphics, tables, and explanatory texts (Fig. 51.5).

The thematic maps are the following:

- *ruderal/active terraced landscape*: it reports the location and extension of the terraced areas and highlights their main characteristics, with particular emphasis on the abandoned but still cultivable areas;

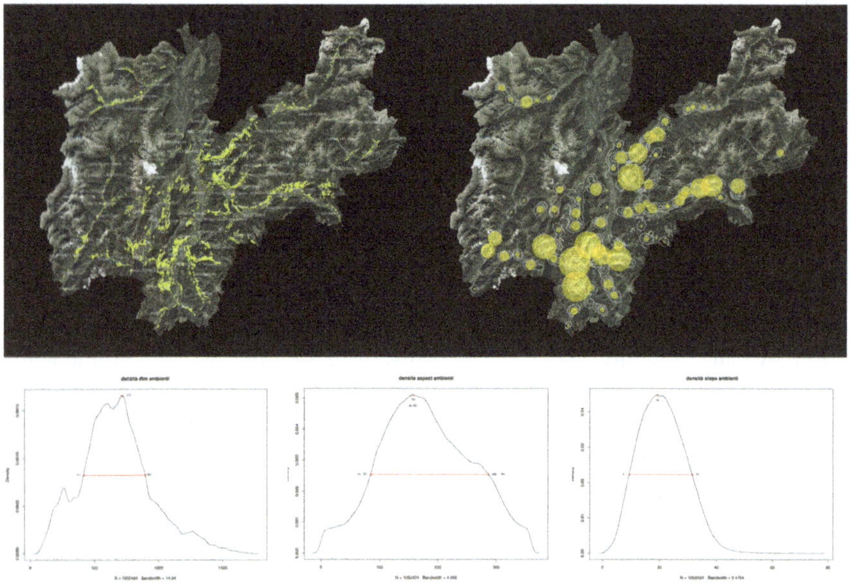

Fig. 51.4 The image on the left shows the geographical distribution of terraced environments in the autonomous province of Trento. On the right, the geographical density of these environments reconstructed through the nearest neighbour analysis. The three graphs below the images show the density curves and the full width at half maximum (FWHM) of the altimetry, slope, and aspect values of the terraced environments. For altimetry, the highest density is at 717 m a.s.l., the FWHM is between 411 and 897 m a.s.l. For aspect, the highest density is at 157° (S-SE exposure), the FWHM is between 84° (E) and 286° (W). For slope, the highest density is at 19°, the FWHM is between 9° and 31°

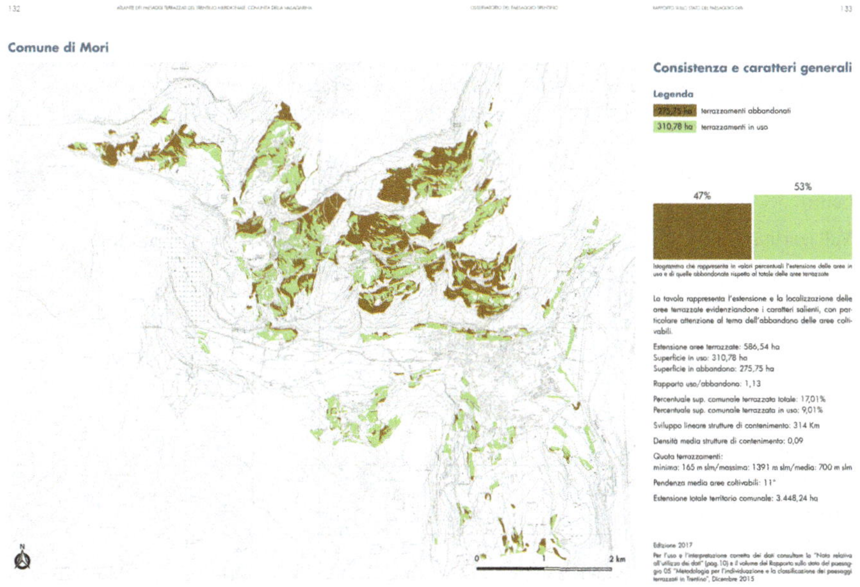

Fig. 51.5 Thematic map. Atlas of terraced landscapes of southern Trentino

- *density of containment structures*: it documents the terraced landscape by highlighting the different levels of density of containment structures;
- *altimetry*: it points out the elevation above sea level;
- *land use*: it classifies land use in: "annual crops and grassland" (arable land, horticultural areas, and permanent grass), "permanent crops" (fruit trees and vineyards), "forests" (deciduous and coniferous forests, mixed forests, and neo-forests), "other" (bushes, shrubs, and uncultivated land);
- *aspect*: it highlights the orientation of terraced areas;
- *slope of terraces*: it describes the slope of the cultivated spaces and of the containment structures;
- *accessibility*: it reports different levels of accessibility to the terraces by road.

Table 51.1 briefly summarises some data that emerged during the preparation/development of the Atlas. The most important data concern the abandonment phenomenon. While the average of abandoned terraced areas at the provincial level is 45%, in some valleys there are much higher values, which can reach 60% at the level of the Valley Community, with peaks of 80% at the municipal level.

The Atlas considers "abandoned terraces" to be those which were originally cultivated and are currently covered by forest. The numerous cases of underutilisation or transitioning of many terraced areas are not considered abandoned.

Considering the commitment towards the protection and regeneration of terraced landscapes, the data concerning the linear extent of the containment structures, which are mostly made of dry-stone walls and thus require continuous maintenance, are of the utmost importance (Table 51.2).

Table 51.1 Percentage of terraced areas and abandoned areas

Valley community	Terraced area (ha)	Percentage of terraced areas over the total	Abandoned terraced areas (ha)	Cultivated terraced areas (ha)	Percentage of abandoned areas (%)	Percentage of used areas (%)
Alto Garda e Ledro	952	2.7	278	674	29	71
Vallagarina	3139	5	1338	1801	43	57
Altopiano Folgaria Lavarone Luserna	227	2.1	141	86	62	38
Primiero	118	0.3	67	51	57	43
Alta Valsugana e Bersntol	1324	3.7	771	553	58	42
Valsugana e Tesino	1192	2.1	632	560	53	47
Cembra	852	6.3	373	479	44	56
Valle di Fiemme	219	0.5	57	162	26	74
Ladino di Fassa	71	0.2	5	66	7	93
Val di Non	34	0.1	7	27	21	79
Rotaliana	125	1.3	12	113	9	91
Val d'Adige	416	2.2	163	253	39	61
Valle dei Laghi	418	3	196	222	47	53
Paganella	58	0.6	21	37	37	63
Val di Sole	492	0.8	215	277	44	56
Giudicarie	803	0.7	414	389	52	48
Total	10,440	1.68	4690	5750	45	55

The extension of these structures is almost 4.469 km in total. This figure provides a first indication of the investment in terms of financial and labour resources required to preserve the function and vitality of these particular rural landscapes.

51.4 Conclusions

The main aim of the Atlas is to address the issue of agricultural area abandonment by quantifying and localising the phenomenon. The analyses developed and the documentation provided by the Atlas represent an important contribution in terms

Table 51.2 Main characteristics of the terraced areas

Valley community	Average elevation (m.a.s.l.)	Average steepness of terraces	Average density of containment structures	Length of containment structures (km)	Percentage of permanent cultures (%)	Percentage of non- permanent cultures (%)	Percentage of terraced areas with difficult accessibility (%)
Alto Garda e Ledro	472	13°	0.07	404	38	32	57
Vallagarina	606	13°	0.09	1599	25	30	70
Altopiano Folgaria Lavarone Luserna	987	16°	0.06	88	1	36	73
Primiero	833	22°	0.06	46	0	43	83
Alta Valsugana e Bersntol	721	17°	0.06	526	15	26	71
Valsugana e Tesino	682	15°	0.05	271	8	38	83
Cembra	650	18°	0.1	610	48	8	77
Valle di Fiemme	1036	14°	0.05	55	0	71	82
Ladino di Fassa	1385	16°	0.03	12	0	93	93
Val di Non	845	16°	0.03	8	37	42	81
Rotaliana	351	14°	0.05	47	87	1	79
Val d'Adige	545	15°	0.06	171	34	25	76
Valle dei Laghi	500	11°	0.05	137	27	24	69
Paganella	780	16°	0.03	10	17	44	85
Val di Sole	1195	22°	0.08	244	2	54	83
Giudicarie	705	17°	0.06	241	3	44	76
Total	673	15°	0.07	4469	22	32	73

Fig. 51.6 Terraced landscapes in the eastern part of Trentino

of knowledge, raising awareness of the crisis of terraced landscapes and outlining a scenario of the abandonment process. The Atlas has laid the foundations for a recovery action targeting these extraordinary rural landscapes which have characterised the alpine and pre-alpine landscape for centuries (Fig. 51.6). In particular, the data concerning the altitude above sea level and the steepness and accessibility of terraced areas support policies and actions aimed at the recovery of such agricultural areas taking into account management difficulties and production potential.

References

Ferrarese F, Pappalardo SE, Cosner A, Brugnaro S, Alum K, Dal Pozzo A, De Marchi M (2019) Mapping agricultural terraces in Italy. Methodologies applied in the MAPTER project. In: Varotto M, Bonardi L, Tarolli P (eds) World terraced landscapes: history, environment, quality of life. Springer, Berlin, pp 179–194

Osservatorio del paesaggio del Trentino (2014) Dieci azioni per il paesaggio rurale del Trentino, a cura del Gruppo di lavoro Agricoltura e paesaggio, Documenti 02, Trento. http://www.paesaggiotrentino.it/documenti/Documentazione/Documenti%20dell%20Osservatorio/D02Dieci%20azioni%20paesaggio%20rurale.pdf

Osservatorio del paesaggio del Trentino (2015) Metodologia per l'individuazione e la classificazione dei paesaggi terrazzati in Trentino, Rapporto sullo stato del paesaggio 05, Trento.

https://www.paesaggiotrentino.it/documenti/Documentazione/Rapporto%20sullo%20stato%20del%20paesaggio/R05_Metodologia%20individuazione%20paesaggi%20terrazzati%20in%20Trentino.pdf
Osservatorio del paesaggio del Trentino (2016) Terraced landscapes choosing the future. Documenti 05, Trento. https://www.paesaggiotrentino.it/documenti/Documentazione/Documenti%20dell%20Osservatorio/D05_Terraced%20landscapes%20choosing%20the%20future.pdf
Osservatorio del paesaggio del Trentino (2020) Atlante dei paesaggi terrazzati del Trentino. http://www.paesaggiotrentino.it/it/rapporto-stato-del-paesaggio/atlante-dei-paesaggi-terrazzati-del-trentino/
Scaramellini G, Varotto M (2008) Paesaggi terrazzati dell'arco alpino, Marsilio, Venezia

Chapter 52
The Museum as an Active Observatory: Protecting and Shaping the Landscape

Maciej Żołnierczuk

Abstract The Museum of King Jan III's Palace at Wilanów is located in Warsaw. Wilanów district has been experiencing growing urbanization pressure for over 20 years. As a result, an historical old big-scale Polish farm was changed into townscape due to intensive housing development. Despite applying measures of nature and monument protection, these are now in a degraded condition and stand as proof of the ineffectiveness of Polish law in the field of landscape protection. The following article briefly discusses the main landscape threats identified by the Museum. It also presents a model of activities, as well as examples of tools for a Museum institution to serve as an observatory. The conducted analyses have proven that museums and landscape observatories largely share the same tasks. The SWOT analyses compiled for the Wilanów landscape and a prospective landscape observatory justify the need to introduce a local landscape observatory which would coexist with the Museum. The presented diagram clearly shows the interconnectedness of the opportunities and the growing threats to, i.e., loss of, spatial, cultural and natural values arising in the absence of such coexisting units. Unfortunately, Poland has yet to regulate the establishment of landscape observatories. Perhaps, the Museum's initiative and this article will change that.

Keywords Museums · Landscape heritage · *Genius loci*

52.1 Introduction

In a rapidly changing world, obtaining reliable and useful information about heritage poses an enormous challenge. Heritage-related data influences landscape perception, evaluation and management. Museums should constitute a major source of knowledge about these values as their role is not limited to typical exhibit storage but also includes serving as an observatory which works toward preserving history and tradition.

M. Żołnierczuk (✉)
Development Department, Museum of King Jan III's Palace at Wilanów, Warsaw, Poland
e-mail: mzolnierczuk@muzeum-wilanow.pl

T. Matteini et al. (eds.), *Cultivating Continuity of the European Landscape*, Environmental History 15, https://doi.org/10.1007/978-3-031-25713-1_52

Although Poland ratified the European Landscape Convention of the Council of Europe (ELC) in 2004, it took a long time for any effects of this commitment to become visible. It was not until April 24, 2015, that Poland adopted its Landscape Act, a law which amended certain acts by strengthening landscape protection instruments. Four years later, the Regulation of the Council of Ministers regarding the Preparation of Landscape Audits was approved. However, the above steps are only two drops in a vast ocean of needs. There is still an overall impression that Poland is not ensuring proper protection of its landscapes. Moreover, Poland lacks the legal tools for effective landscape protection and law enforcement.

The Museum of King Jan III's Palace at Wilanów is one of Poland's most important museums. It used to be a royal summer residence belonging to Jan III Sobieski. The surrounding landscape is part of a large-scale farm (modeled after the *villeggiatura* in the 17c. and the *ferme ornée* in the 18c.). In the course of time, the Wilanów land estates grew through the purchase of fragmented noble property. By 1894, their surface area had already reached 6623.5 ha (Szpanowski 2015). Currently, the Museum grounds house a palace with accompanying farm buildings, as well as 92 ha of gardens and parks, including a 53.5 ha nature reserve. It should be emphasized that despite the passage of time and changes, the research area is still of unique historical and artistic value (Sikora 2017). The need to protect not only the residences, but also their surroundings, results from the fact that local physiography determines how historic complexes are formed (Kaczyńska 2011).

The purpose of this article is to present the potential of Museum of King Jan III's Palace at Wilanów as a local landscape observatory. The author formulates and addresses the research question of whether it is possible for a Museum and a landscape observatory to coexist under the umbrella of a single institution.

52.2 Study Areas and Methodology

The scope of research is limited to the Wilanów district, which is a legacy of the former rural use of this area. Scientific work fell into four stages: (1) identifying the main landscape problems; (2) comparing the objectives of the Museum and those of a landscape observatory; (3) presenting examples of Museum efforts aimed at implementing the entrusted statutory objectives; and (4) compiling two SWOT analyses—one of the Wilanów landscape observatory, and one focused on the Wilanów landscape itself. Attention was paid to the interrelationships between these two analyses.

52.3 Results

52.3.1 Identification of Landscape Hazards

One of the most serious identified threats is urban pressure. The year 1996 saw the announcement of a competition for a West Wilanów development concept. Implementation of the design began in 2002. Currently, the new buildings are coming dangerously close to the palace and garden arrangement (Figs. 52.1 and 52.2). Construction activity has resulted in dying trees, destruction of ecological and ventilation corridors, and reduction of the biologically active area.

What is important to note is the lack of holistic and coherent spatial planning, which results in: fragmentation of historic compositional axes, insufficient green areas and an undeveloped network of public educational centers (most are located along one street). The absence of alternative routes and connections with neighboring districts leads to traffic congestion in other parts of Wilanów. An alternative would be to provide adequate public transport solutions; however, the local spatial development plans do not foresee any space for a bus terminus. Heightened air and noise pollution levels are associated with the transportation issue.

Water pollution is another equally serious problem (Fig. 52.3). The waters of the Służewiecki Stream flow into Wilanowskie Lake, which forms part of the palace and garden arrangement. The stream basin covers 10% of Warsaw's area. Water transports

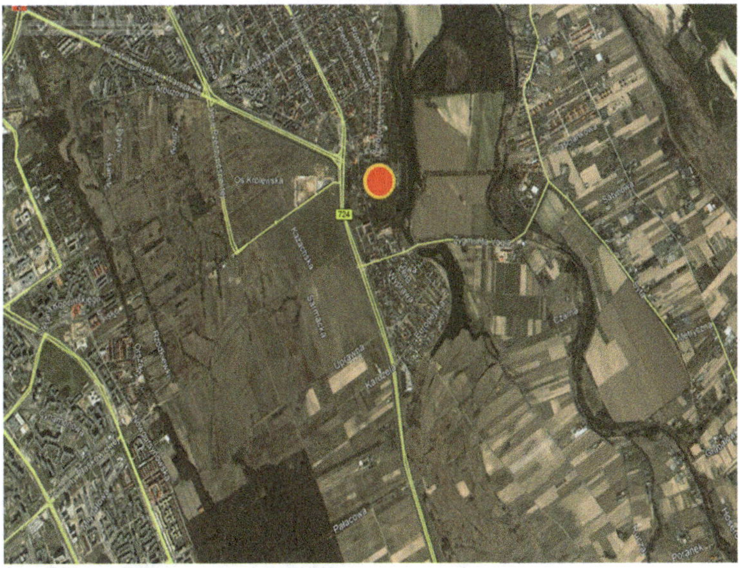

Fig. 52.1 Wilanów Landscape in 2002 (left) (*source* Google Earth, access: 05/04/2020). The red point marks the location of the Museum of King Jan III's Palace at Wilanów

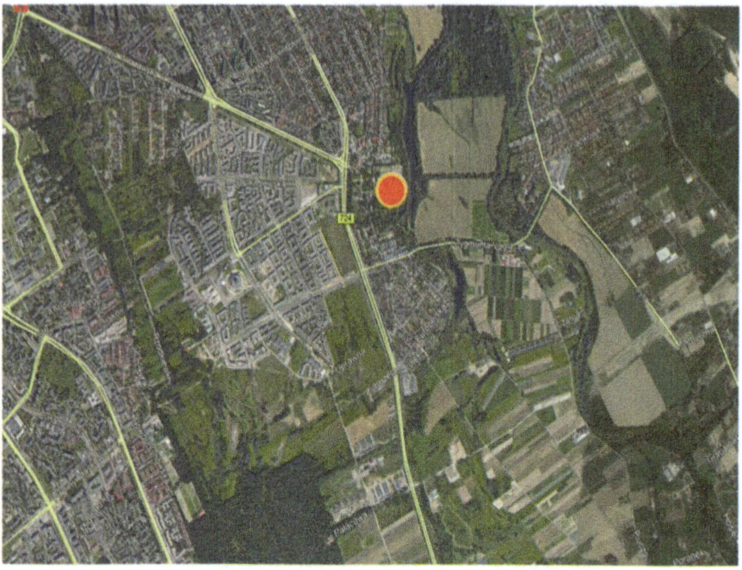

Fig. 52.2 Wilanów Landscape in 2019 (right) (*source* Google Earth, access: 05/04/2020). The red point marks the location of the Museum of King Jan III's Palace at Wilanów. Owing to continuing urban pressure, any delay in planning activities only works to the detriment of the landscape

pollutants. By conducting improper inspections of waters, municipal services have caused the lake bed to be covered with a several-meter-thick layer of silt (Fig. 52.4).

Some of the measures implemented for the purpose of ensuring monument and nature protection are rendered ineffective by highly complex protection plans and mutually exclusive activities performed as part of different forms of protection (Fig. 52.5).

Fig. 52.3 Foam on the Służewiecki Stream (2 May 2020)

Fig. 52.4 Silt build-ups visible above Wilanowskie Lake water line (5 May 2020)

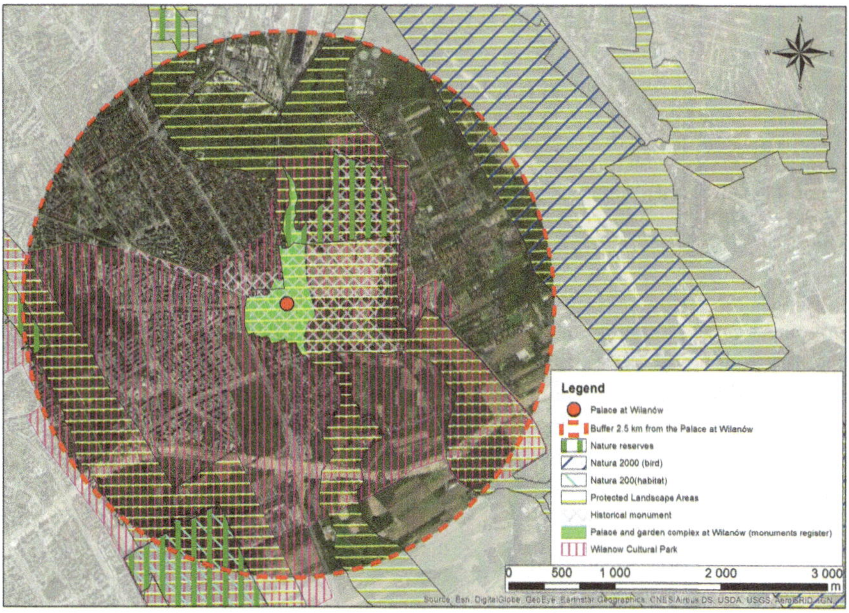

Fig. 52.5 Forms of nature and monument protection within 2.5 km of the Museum

52.3.2 Linking Statutory Objectives

The Museum could not remain passive in the face of growing and accumulating real threats to the landscape. It is also required to take action under statutory objectives, which largely coincide with those of landscape observatories. Thus, the elements

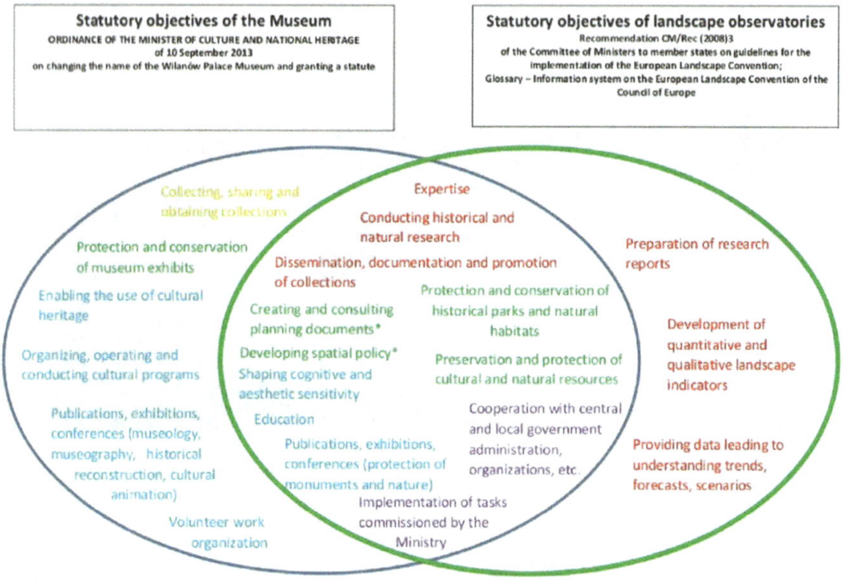

Fig. 52.6 Juxtaposition of statutory objectives of the Museum of King Jan III's Palace at Wilanów with those of landscape observatories (own study). *Legend* yellow—collection creation and management, red—research activity, methodology, data bank creation, green—protection of nature and culture, blue—social, educational and information activity. *—objectives implemented for the purpose of cultural and natural value protection which are not explicitly mentioned in the Museum's statute

linking these two types of institutions are: research, documentation, education, protection and management of the landscape (Fig. 52.6). The number of shared statutory objectives shows that the cooperation of such institutions is possible and even advisable, as common areas of research enable preserving natural and historical heritage resources. The landscape has a significant narrative potential which influences the perception of tourists. It also displays historical changes and contributes to the identity of a place. That is why it is so important to recognize, protect and connect it with the Museum.

52.3.3 Examples of Activities

For the Museum, July 9, 2016, was a watershed moment. The General Assembly of the International Council of Museums (ICOM) adopted Resolution No. 1 which underlined that museums have a responsibility toward landscape. At the time, however, the Museum had already begun working for the landscape around it. Many efforts undertaken by the Museum are consistent with Recommendation CM/Rec (2008)3 (Table 52.1).

Table 52.1 Juxtaposition of the scope of activities included in Recommendation CM/Rec (2008)3 and examples of activities undertaken by the Museum

No.	The scope of activities included in Recommendation CM/Rec (2008)3	Examples of activities undertaken by the Museum
1.	Describe the condition of landscapes at a given time	Drawing up analyses of landscape transformations together with creating databases (using historical data and ArcGIS 10.6.1 ESRI tools)
2.	Exchange information on policies and experience concerning protection, management and planning, public participation and implementation at different levels	Submitting comments to local spatial development plans and the Warsaw environmental protection plan, cooperating with the NGO social communication program "Once upon a Time at Wilanów" (interviews which record memories of Wilanów, archiving private photographs)
3.	Use and, if necessary, compile historical documents on landscapes which could be useful for knowing how the landscapes concerned have developed	Drafting analyses confirming the necessity of making entries to the register of monuments (form of protection), archiving maps, photos, texts, etc.
4.	Draw up quantitative and qualitative indicators to assess the effectiveness of landscape policies	Developing a method for evaluating the feasibility of the Wilanów Landscape Park protection plan
5.	Furnish data leading to an understanding of trends and to forecasts or forward-looking scenarios	Keeping a data bank on the temperature, humidity, air dustiness, salinity and water levels in Wilanowskie Lake and Służewiecki Stream

52.3.4 SWOT Analysis

Institutional and landscape opportunities are interrelated. Having carried out a multitude of professional activities, the Museum has come to be perceived in Poland as a landscape protection reference center making efforts to support other museums struggling with similar planning problems—which is particularly important for open-air museums. In the future, the Museum intends to prepare a set of good practices for other museums to implement (some have already been discussed in other articles). For the Museum, taking on the role of a landscape observatory is an opportunity to also strengthen its position among offices and universities (planned cooperation with the Warsaw University of Life Sciences). In turn, this would contribute to boosting the significance of the Wilanów landscape, an effect visible, for example, in Facebook posts. The biggest threat to this pioneering initiative is the absence of legal provisions governing the organization of landscape observatories in Poland. It should be noted that the Museum's failure to undertake this initiative would result in the intensification of landscape problems (Fig. 52.7).

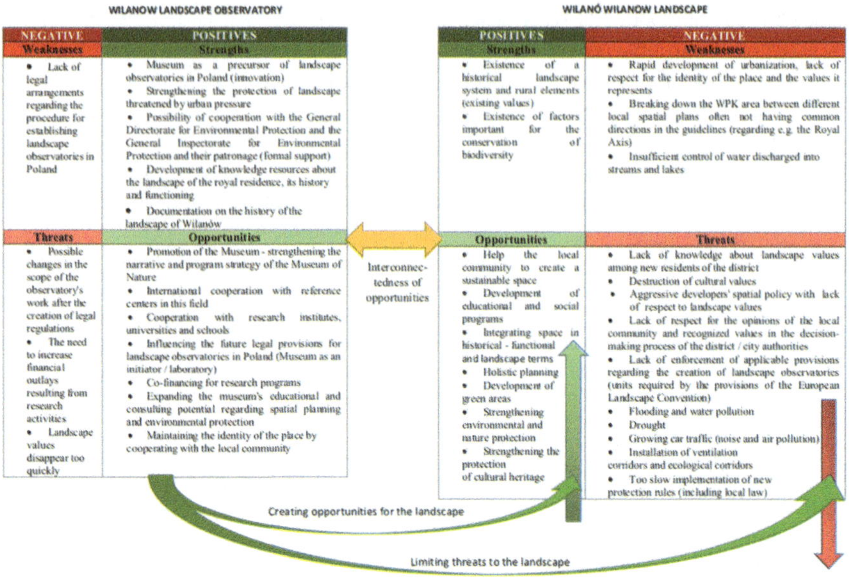

Fig. 52.7 Juxtaposition of SWOT analyses for Wilanów Landscape Observatory and Wilanów Landscape (own study)

52.4 Conclusions

The landscape heritage of Wilanów is under threat, and Poland's binding legal regulations on landscape protection and management are insufficient. Operating the Museum and a landscape observatory under the umbrella of one institution is possible and highly advisable. In the case of Wilanów, such a union would open up a significant opportunity for protecting landscape identity and creating synergy. The Wilanów Museum not only conducts research, creates databases, consults planning intentions and pursues educational activity but also strives to develop international legislation. Paradoxically, the biggest threat to the Wilanów initiative is posed by the failure of management institutions to fulfill their ELC obligations.

References

Kaczyńska M (2011) Wpływ zało±enia pałacowo-ogrodowego w Wilanowie i jego zało±eń filialnych na przemiany krajobrazu (Impact of the palace and garden complex in Wilanów and its affiliate premises on landscape changes). Wydawnictwo Sztuka Ogrodu Sztuka Krajobrazu, Warsaw (Language of publication: Polish)

Sikora D (2017) 'Krajobraz Wilanowa– jego przemiany i możliwości ochrony (Wilanów landscape—development and possibilities of protection)', Dissertations of Cultural Landscape Commission, No. 36, pp 9–20. (Language of publication: Polish)

Szpanowski P (2015) Warstwy krajobrazu klucza wilanowskiego (The layers of Wilanów property landscape). Museum of King Jan III's Palace at Wilanów, Warsaw (Language of publication: Polish)

Part V
Cultivating Innovative Perspectives

Pineforest, California, USA, 2010. ©Sebastião Salgado

Chapter 53
Landscape Continuity: Cultivating Different Allegories of the Inseparable

Saša Dobričič

Abstract As urgent as the call for action may appear, cultivating the continuity of European landscape imposes a different theory of action, as does any cultivation, one that is the opposite of the urgency to build, produce, interfere or perform at any cost. Cultivation is intended as a 'fine calibration' of the measure of doing, driven by care, attention or even devotion in order to maintain or promote the growth and sustainment of life. As its etymology indicates, cultivation is also about knowledge that is applied to 'refine inhabiting'. That is to say, if cultivated, inhabiting is performed at such a level it enables the spirit of care to nourish and even embellish life by rewarding us with 'fruits and flowers', with the 'useful and delightful'. And when inhabiting also nourishes beauty, in other words all the dimensions of existence beyond the simplistic function of dwelling, then it is also honoured as a celebration of life itself. After all, do we have to kneel in the garden to benefit from cut herbs or picked flowers? It is an everyday ritual of devotion to the realm of the earthly and domestic. Beyond any celestial or Edenic order, the reverence due comes from a deeply embedded and rooted earthly presence like the beautiful ancient tree in a picture by Sebastião Salgado, the epitome of the UNISCAPE initiative. Inhabiting as a 'deeply rooted presence' is thus far from being simply the occupation of soil or a step forward from the 'art of building' as architectural practice is typically defined; as a 'deeply rooted presence', inhabiting turns out to be an 'art of living' that is literally palpable with our own flesh because, in practice, it is experienced as art-of-life nourishment generated together with the multitude of living beings performing their living on a tiny earthly substrate.

Keywords Landscape · Landscape project · Aesthetic models · European landscape convention · Art of inhabiting

---

S. Dobričič (✉)
University of Nova Gorica, Nova Gorica, Slovenia
e-mail: sasa.dobricic@ung.si

T. Matteini et al. (eds.), *Cultivating Continuity of the European Landscape*, Environmental History 15, https://doi.org/10.1007/978-3-031-25713-1_53

53.1 Garden

As urgent as the call for action may appear, cultivating the continuity of European landscape imposes a different theory of action, as does any cultivation, one that is the opposite of the urgency to build, produce, interfere or perform at any cost. Cultivation is intended as a 'fine calibration' of the measure of doing, driven by care, attention or even devotion in order to maintain or promote the growth and sustainment of life. As its etymology indicates, cultivation is also about knowledge that is applied to 'refine inhabiting'.[1] That is to say, if cultivated, inhabiting is performed at such a level it enables the spirit of care to nourish and even embellish life by rewarding us with 'fruits and flowers', with the 'useful and delightful'.[2] And when inhabiting also nourishes beauty, in other words all the dimensions of existence beyond the simplistic function of dwelling, then it is also honoured as a celebration of life itself. After all, do we have to kneel in the garden to benefit from cut herbs or picked flowers? It is an everyday ritual of devotion to the realm of the earthly and domestic. Beyond any celestial or Edenic order, the reverence due comes from a deeply embedded and rooted earthly presence like the beautiful ancient tree in a picture by Sebastião Salgado, the epitome of the UNISCAPE initiative.

Inhabiting as a 'deeply rooted presence' is thus far from being simply the occupation of soil, or a step forward from the 'art of building' as architectural practice is typically defined; as a 'deeply rooted presence', inhabiting turns out to be an 'art of living' that is literally palpable with our own flesh because, in practice, it is experienced as art-of-life nourishment generated together with the multitude of living beings performing their living on a tiny earthly substrate.[3] The realm of this cultivation is the garden, *par excellence.* We could recognise gardening as a specific form of 'domestication of life', driven by the dialectic of care and control, that includes attention to detail, humility, maintenance as well as a particular design strategy that works together with a range of unpredictable factors.

In fact, every gardener knows that any tiny, even invisible, being can make the opus collapse; moreover, its production scenario may even count on it; meaning, a gardener is aware that collaborative design also includes undesirable collaborators. A gardener must keep all her senses alert in order to maintain a continuous relationship with different 'cultures' that are meant to grow, coexist and codesign on a common

---

[1] Cultivate (verb), from Medieval Latin *cultivatus*, past participle of *cultivare* 'to cultivate', from Late Latin *cultivus* 'tilled', from Latin *cultus* 'care, labour; cultivation', from past participle of *colere* 'to cultivate, to till; to inhabit; to frequent, practice, respect; tend, guard'. The noun cultus, past participle of the verb, indicates not only cultivating as nourishing growth, but also caring for something in general, and in a specific sense, both religious service towards the gods, as cult itself, as well as the cultivation of human beings, as the complex set of knowledge indicated as culture.

[2] For an understanding of the intimate relationship between cultivation, gardening and landscape see Lattini and Matteini (2017).

[3] In Illich's words the «... equation of dwelling and living goes back to times when the world was still habitable and humans were inhabitants. To dwell then meant to inhabit one's one traces, to let daily life write the webs and knots of one's biography into the landscape» (Illich 1984). https://www.atlasofplaces.com/essays/dwelling/.

plot of land. Delays, incompleteness, errancy or any other 'disturbance' act as the constitutional production forces of this art that demands dynamic knowledge which does not stop where problems begin and whose solution is inseparable from action.[4]

Maybe the cultivation of this different grammar of action, which unavoidably shifts into a common dimension of doing, using, producing and designing with other members of a living community, represents the 'continuity' to be cultivated in and 'for the sake' of landscape?

Still, landscape has an unsettled aesthetics that does not necessarily coincide with the idea of garden. Landscape is not necessarily a privileged life arena where an extended community of living beings 'invents life'[5] and trains in the art of inhabiting together; after all, landscapes do also happen when life proliferates magnificently without any direct human action, like in the 'wilderness'. On the other hand, human presence may totally overshadow the life performance of other living beings, like in highly anthropized urban areas (although we are currently experiencing that this might be only an apparent absence). Nevertheless, we can assert that landscape, being generated by seeing, has often been that fortunate arena for humanity to experience coexistence even as a contingent element simply observing the proliferation of life, both natural and social, and enjoying it, as actors out of the action. Thus, shifting cultivation out of the garden into landscape affects design attitude regarding both the practical wisdom of domestication through care and the different design grammar that includes the contingency of human action as an essential part of a codesign attitude. By including the 'unpredictable', doing 'almost nothing', or 'preferring not to act' at any cost, landscape also empowers a Project to finally contemplate (and not only execute) its own possibilities as a powerful thinking, and not merely implementation, tool. It is on this threshold of preference, as a nuance of the possible and impossible, that humanity finally enjoys its own impotence/contingency without the risk of slipping into inertia and contemplates its projects as a landscape option/preference, an imaginary that acts as a concrete earthy possibility.

53.2 Inseparable, Uncommon

However, the power of 'landscape's turn' should be also considered in relation to the obsolescence of other possible models of Earth, such as space, the main driver of the idea of the infinitely expandable, because measurable, modern state.[6] The Model of Space always delivers the face of the Earth on a map, on a white, infinite plane, but this has definitely expired today, together with the Project on it, resulting as an abstract vanishing certainty, distant from any earthly context of possibility.

---

[4] L. Lattini, T. Matteini, *Idem,* p. 12.

[5] Clément (2013).

[6] Fundamental to the comprehension of possible models of Earth intended as space, territory and landscape is the work of Farinelli (2003).

Although differently, territory too, as form of dwelling, another model of the face of Earth, the cornerstone of belonging to land 'as it is lived', has gradually rendered almost all conditions of subsistence and existence invisible.[7] Driven by pressuring norms, rules and codes, its generation has made us all feel globally and infinitely homeless '… literally without any land under our feet…' as Latour puts it; the world 'we live in' is not the world 'we live from'.[8]

Instead, surprisingly, landscape, the most subjective and aesthetic model of Earth due to its being generated subjectively 'in seeing', has secretly maintained and rendered the intimate relationship between us and that terrestrial materiality that we 'live from' visible, even when we do not necessarily 'live in' it. Landscape 'as perceived' has persistently cultivated our inseparability from (rather than our belonging to) all the existential as well as emotional components that life is made of, generating the coexistence of nature and culture as it functions in 'our eyes'. In fact, isn't landscape only generated, as testified to by its infinite representations throughout history, when cognitive and 'emotional engineering' instantaneously fuse nature's most human, social, scientific and cultural part—animated living beings—and its inanimate parts—the atmosphere, rocks, clouds—into a set of biological and emotional conditions of existence? It is in a landscape matrix that 'living beings melt with rocks and rocks melt with life'.[9]

This is the reason why 'if perceived, i.e. originated as landscape', even a piece of degraded and threatening land that clearly does not appear to be inhabitable is unavoidably experienced as part of us, of our life on Earth: a manifestation of our most intimate although frightening belonging. After all, landscape 'as perceived by' does not admit the expulsion of undesirable components, as Farinelli's fundamental contribution to the comprehension of the landscape model indicates: the subject and object of observation are indistinguishable, thus inseparable.[10]

In landscape, the power of context is always glorified as something that we share, that we have in common with other components of life, although not necessarily captured in any 'common household' commodifying notion. We perceive the facts of Earth as landscape just an instant before the realms of economy or ecology transform them into common household management or knowledge (as words economy and ecology reveal), even of all earthly living beings: hence, an instant before they appear like place, or are expelled as its opposite, non-place.[11] Landscape does represent a kind of dwelling model but it points towards inhabiting, which is always expressed and experienced as a contingency of household (although an existential one). Maybe it is because the landscape dwelling model is still an ecological possibility comprising

[7] In Raffestin's view, even the model of territory has today little or nothing to do with materiality: codes, models and simulated realities are the prevailing mediating principles of human being-in-the-world. From this point of view, it is not at all about 'dwelling' but about the (failing) attempt to 'dwell' in globally circulating monetised and consumable signs of reality (Raffestin 2012, pp. 21–14).

[8] Latour (2017).

[9] Todd (2021, pp. 159–163).

[10] F. Farinelli, *Idem.*

[11] Auge (1992).

all the possible proximities that different coexisting species can reciprocally enact while being cherished not because '*...of their functional, utilitarian value—that is, their usefulness—but studied and loved as scene, the stage of their possible fate...*'.[12]

Indeed, landscape is generated just before we realise how beings relate to their environment as habitat (ecology) and immediately after we define which components are out of place because they are not functional to the system of order that any good household imposes (economy). At its heart, landscape has contingency, that is, how things can stand simultaneously one next to the other, even 'unnecessarily' or out of common sense: how things can function beyond any necessary relationship, also stimulating other possible interdependences. Hence, it is a relational model of inhabiting, but so intense that it can also (or rather solely) persevere before any functional, operative, doctrinal or other accredited relationship has been established; it can function immediately after as well, irresistibly attracted to residuals and remains, as the 'third landscape' testifies. It is indeed a moment of conflict because interdependences are not taken for granted and not solely because my perception is different from yours (indeed humanity does not hold a common organ of perception), which, as the ELC indicates, entails a participatory exercise to establish what we have in common, but *because* this rivalry of voices easily risks slipping into the mechanisms which so often prevail in the generation of place. That is to say, although landscape can have the form of place, it is in landscape that 'intersectionality' goes beyond what all members have in common (which, at best, serves those that are less under-represented) and focuses on the uncommon, which still claims membership. It is in this sense that our 20th anniversary celebration call is to be understood:

> "... thinking of those forms of continuity that are cultivated entirely within and in-between diversity..." by rethinking the relationship between us and others, by shifting the focus from the single elements to the glue binding these elements together into an "ensemble" of many different voices that make up the mosaic of our "art of living".

In other words: Landscape is related to a process of assimilation (and not transformation) of differences, the uncommon, the residual. The focus is not so much on the process of transformation (of nature) under the action of humans, but on the process of assimilation (with nature) into a collective body that is our body of culture, an existential living mixture.

It turns out that something as non-functional, ephemeral and subjective as landscape can act as a reliable and mature model of envisaging those ways of inhabiting that enable us to melt with the life that we 'live from' although we do not 'live in'.

53.3 Continuity

The only international treaty that seriously considers something as ephemeral as a set of subjective perceptions and recognises it in law is the European Landscape Convention (ELC). Now, if we reflect upon the definition of landscape in the ELC

[12] Benjamin (1968, pp. 59–68).

'Landscape means an area, as perceived by people, whose character is the result of the action and interaction of natural and/or human factors' (ELC 2000, art. 1a), the objective base ground, the continuum, is the character of an area which is perceived as a network of discrete elements in a system of differences. Selection is driven by a distinctly subjective attitude deriving from experience, habit and the will of the observer. In other words, faced with the same character of an area, each of us will perceive different landscapes.[13] Hence, the manifestation of an area is indeed nonlinear, plastic and dynamic as a result of the strong interactions of the observer with a given context whose attribution of sense is driven by the relationships established between different elements. But how can these multiple sets of representations be of any use and represent a baseline, a continuum, for any political action within and for the sake of landscape? Or rather, following the ELC, could it be the other way round: that continuity in landscape achieves the character of a residual rather than of a base ground? That continuity in landscape, if 'as perceived' as a model and in practice, is not a standing background support upon which the traces of life are inscribed but is generated dynamically by the continuity of gestures and actions which leave traces of our common presence on Earth and is, therefore, an epiphenomenon,[14] a secondary symptom of our lifetimes lived in contingency next to other things and facts on Earth? It is indeed a secondary symptom in practice, following not preceding, because it always manifests after all the other processes of assimilation with a territory, its inhabiting, have left the ground. Landscape is what persistently remains, and it is exactly this residual and symptomatic character of landscape that allows the action (of inhabiting) to be repeated and continuity cultivated.

Because landscape, as indicated in the ELC, always returns and reiterates itself 'in our eyes' just before and after it lands 'under our feet', it is possible to think of those forms of coexistence where the aesthetic perfectly matches the political; it is in landscape that 'environment gains picture'[15] and we can establish our existential preferences (environmental, social, interspecies…) through the means that drive our most personal attention, our senses, our body, our special brains, without any mediation of a moral, legal or functional order. In fact, is it not true that through the eyes of landscape, things redefine their borders, limits, i.e. terms, that in landscape, and without any moral obligation, nature is much more than a means of survival, dwelling is more than shelter, place is more than homeland, and we are more than residents or citizens with the accommodating forms of belonging so often reclaimed as ownership? To conclude, it is because law is captured in landscape and not vice versa that the 'ELC is so much broader than a law', as Amy Strecker reminds us.

---

[13] For Turri (1979), this idea of landscape as a system of connections and networks is founded on the concept of *iconema*. An *iconema* is 'a sign within an organic set of signs' that indicates both the system and the individual signs (Turri 1979).

[14] For comprehension of the complex question of continuity, see the valuable exposition and readings explained and indicated in R. Fabbrichesi, G. Longo, «*Il 'mito del continuo' tra filosofia e scienza, Introduzione al convegno*» consultable on https://www.di.ens.fr/users/longo/files/PhilosophyAndCognition/discr-cont.pdf.

[15] George Monbiot (Guardian columnist and writer) expression relating to the uninspiring state of language of conservation.

53.4 Relative Measures for Accurate Sensing

It is in landscape that humanity can contemplate contingency as a powerful possibility. It is exactly because each twinkling variation of perception has the power of a decisive instant; and decisive here has nothing to do with the imperative of will but, again, is an outcome of seeing. Landscape is always irresistibly attracted to 'what is really here, but hardest to see'. It is in landscape that when something so constant in nature as the sky is perceived as complex diversity, it becomes an 'atmosphere': a complex set of nuances that make the difference (not yet definable in scientific terms but in the processes of alteration-complication). It was during his travels with Humboldt and his elaboration of the 'idea of landscape' in the eighteenth century that Saussure invented the cyanometer to comprehend the phenomenon of the sky's blueness.[16] The blueness of the sky has been revealed today in scientific terms but at the time of Saussure, it was a certain observational culture of thinking that was driving the comprehension of the facts of Earth, recognisable in tiny nuances of perception. Through the eyes of landscape, such an apparently certain 'really here' phenomenon as the colour of the sky redefines its borders and evades the lure of common acceptance, thus becoming multifaceted and complex. On the threshold of the scientific, still immersed in 'science's childhood' because linked to genuine (bodily) ways of perception, the colour of the sky is the outcome of altitude and time of day. The cyanometer is an example of painstaking and accurate measuring to provide only relative information, and Saussure was aware of that. So, what's the point? Not science yet, just a bit more than an 'empty poetic game' in nature? The cyanometer is an example of understanding continuity, in this case of the blueness of the sky, by detecting almost invisible diversity. Most importantly, Saussure made his own means of observation to understand the observation itself; he used the means of landscape to understand landscape, 'to see his own sensing' and his own chromatic space,[17] and moreover, to evaluate it.

These 'relative' measures for detecting a contingent problem reveal the secret vein of enjoyment of self as the fundamental component of the cultivating landscape matrix. Because at the heart of this enjoyment is 'how we do (evaluate) it' and it matters as much as 'what we are enjoying'; it is only when discovery and its means of evaluation embrace the object and the subject that responsibility can finally be

[16] Cyanometer was actually easy to use: the observer holds the cyanometer towards the desired direction and reads of the section with nuance of blue that most corresponds to the current sky. For the understanding of work of Saussure and invention of cyanometer we refer to Lindberg (2020). Original source in Horace Benedict de Saussure, *Description d'un cyanomètre ou d'un appareil destiné a mesurer l'intensité de la coleur bleue du ciel*« in Memoires de l'Academie Royale des Sciences, Libraire De L'A, 1790, Turin.

Another indicative opus is Charles Darwinn's Nomenclature of Colours, where each tiny difference of nuance indicates examples of vegetable, animal and mineral that occur in nature.

[17] «Seeing yourself sensing» is the installation by Olafur Eliasson conceived for the windows of MOMA's Garden Hall. Fifty sheets of striped transparent and mirrored glass act as a sort of phenomena maker to investigate perception as interaction between inside and outside. https://www.moma.org/calendar/exhibitions/161.

beyond any moral obligation. Isn't Humboldt's Naturgemälde, a groundbreaking representational effort on the threshold of information and entertainment, the best way to understand the network of intricate connections at work in landscape[18] at a single glance? Landscape indeed still represents one of the privileged earthly arenas where life can be experienced without the pressure of anthropocentric protagonism (also in the form of guilt), that is still hesitating on the threshold of an ecocentric resolution of self, the place where humanity can enjoy its own state of contingency and test habitability as interference, literally intrusion, with any, even the most diverse, ephemeral and 'insignificant' earthly presence, through the enjoyment of its own only relative measuring of the world.

It is in landscape that we still contemplate our belonging to Earth as an ever powerful 'unnecessary' potential (that turns out to be existential), rather than as the weak certainty that rights, duties, progress or easy localism persistently deliver to us. Let's remind ourselves, for the sake of place literacy, of the proliferation and accumulation of words melding with landscape as perceived by locals, scientist, poets, etc. The meticulous describing of each specific natural phenomena's or place's peculiarities or practices can indeed appear to be on the brink of fetishism. The language domain actually evades any common nationalised setting of Slovenian, Italian, English or Dutch vocabulary and shifts into a wonder realm of specific place lexicon to indicate all the possible distinctions that wind, water, forest, soil, woman, childhood, etc. can assume—a real empowerment of language (and thus also of attention) that might be unnecessary for common language to function, but that is certainly fundamental for a place 'to exist', keeping it safe from the risk of slipping into abstract notions of space, nation and identity. As if to say: 'this place speaks Slovenian'[19] and consequently, Slovenia is not a matter of borders as an abstract space category, but rather an exquisite landscape matter. And if this is true, the legitimacy over an area can be claimed only as the art (not the necessity) of the enjoyment and interaction between our bodily senses and words that all together approximately adhere to the 'character of an area as perceived by people'. There are no insiders nor outsiders, no passports in landscape, only clandestine observers.

Landscape critically engages with any easy or taken-for-granted form of inhabiting that immobilises or accommodates the observer. Being generated 'in sensing to give sense', it demands active and free agents that make their own means of evaluation, not only and solely to nourish subjectivity, but to subject themselves to the earthly presence out of pride and fear by calibrating their senses before they explode into 'empty poetic games'[20] or implode into 'assuring conviction'. To perceive landscape,

[18] https://learningfromlandscapes.com/2019/06/11/humboldt-the-invention-of-nature/.

[19] We refer here to an eye-opening article in the Guardian related to unusual lexicon for describing landscape and natural phenomena. The article points to the importance of cherishing place lexicon in an age when junior dictionaries radically reduce the number of words concerning nature to introduce words related mainly to the digital realm of social media. In R. Macfarlane, *The word-hoard: Robert Macfairlane on rewilding our language of landscape*, the Guardian, consultable on https://www.theguardian.com/books/2015/feb/27/robert-macfarlane-word-hoard-rewilding-landscape#comments.

[20] Farinelli (2014).

the observer is movable, on alert and under subjection. Although landscape is everywhere because it is 'just here' in our ordinary sensing of the world in everyday life, it is still not easy to perceive. Landscape is generated in the domain of the visible but has a distinctly clandestine character. No map, measure, perspective, selected point of view, nor privileged domain can capture landscape's character. Its detection has to evade all the meanders of accredited vision as it demands the skills of an accredited smuggler: skills that grasp tiny signs and nuances in the twinkling of the ordinary.... After all isn't it true that the enjoyment of discovery can be perpetuated only through the art of hiding, which turns out to be a distinctly cultural attitude of earthly nature in the landscape model. In landscape, as in any playground, it is not so much about measuring, evaluating or judging the facts of the Earth as it is about our ability to sense it. It is indeed a particular way of looking, where seeing does not lead to judgement but to recalibrating our distances and proximities to the other life forms present and to redefining strategies to inhabit it in common. Therefore, to detect landscape, it is not enough to simply sharpen our own gaze but also to perceive the gaze of others into our eyes.

53.5 Ultimate Gaze

To put it in the words of Didi Huberman,[21] it is possible that we are currently affected with some grave form of 'perceptive amnesia' towards landscape, a sort of observational cramp that prevents us from grasping the signs on the face of the Earth that are watching us here and announcing imminent change. And faced with the powerful gaze of the child in Sebastião Salgado's image we ask:

"What is the landscape in your eyes little child?" ... "What is the colour of the sky?" ... And with a trembling look of hope, we wait because we know that the real sentinel of the future is not in any digital, scientific, financial or legal futurology, but rather in front of us in this clandestine child that invites us to decide between possible change or imminent catastrophe.

References

Auge M (1992) Non lieux. Introduction à une anthropologie de la surmodernité, Seuil, Paris

Benjamin (1968) Unpacking my library: a talk about book collecting. In: Arendt H (ed), Zohn H (Tran) Illuminations: essays and reflections. Mariner Books, Houghton Mifflin Harcourt, Boston and New York

Council of Europe (2000) European Landscape Convention. Council of Europe Treaty Series – No. 176. Council of Europe, Strasbourg

Clément G (2013) Giardini, paesaggio e genio naturale, ed. Quodlibet

Didi-Huberman G (2021) Sentire il grisou, ed. Orthotes editrice, Napoli-salerno

[21] Didi-Huberman (2021).

Fabbrichesi R, Longo G. «Il 'mito del continuo' tra filosofia e scienza, Introduzione al convegno». https://www.di.ens.fr/users/longo/files/PhilosophyAndCognition/discr-cont.pdf
Farinelli F (2003) Geografia, un introduzione ai modelli del mondo, ed. Einaudi
Farinelli F, von Humboldt A (2014) Viaggio alle regioni equinoziali del Nuovo Continente, ed. Quodlibet
Illich I (1984) Dwelling. https://www.atlasofplaces.com/essays/dwelling/
Lattini L, Matteini T (2017) Manuale di coltivazione pratica e poetica, Per la cura dei luoghi storici ed archeologici nel Mediterraneo, il Poligrafo
Latour B (2017) Facing Gaia: eight lectures on the new climatic regime. Polity Press, Cambridge
Lindberg M (2020) How blue is the sky? Horace B. De Saussure and his cyanometer—its invention, use and legacy, Thesis, Umeå University
Macfarlane R, The word-hoard: Robert Macfairlane on rewolding our language of landscape. The Guardian. https://www.theguardian.com/books/2015/feb/27/robert-macfarlane-word-hoard-rewilding-landscape#comments
Raffestin C (2012) Space, territory, and territoriality. Environment and Planning D: Space and Society
Todd S (2021) Landing of earth: an educational project for the present. A response to Vanessa Andreotti. In: Ethics and education, vol 16
Turri E (1979) Semiologia del paesaggio italiano. Longanesi, Milano

Chapter 54
Cultivating Landscape Continuity. The Projects " Design" in the Architecture of the Landscape

Juan Manuel Palerm

Abstract This chapter addresses the landscape as a perception of a panorama of experiences and scenarios of our lives and their future in the territories we inhabit and cultivate. We use the term cultivate, not as an absence of a word, but as a preamble to a series of questions. Cultivate to reflect on what happens outside, to analyse the use and enjoyment of the spaces we produce and to interpret the new spatial reality in its different levels and stages. It is about entering into a close dimension, configured by cultural contexts that are marked by behaviours closely linked to the culture of the image and the project. To admit the condition of the Project and of the Architecture in the landscape, the landscape of Architecture, we must recognize that there are direct relationships between landscape and its territorial and environmental support, from diverse perspectives, capable of individualizing specific operations from different disciplines in the physical framework of the territory where it acts and generates a socio-cultural stratum that identifies it, with the interest of verifying its interdependencies and its ability to articulately activate refractions between distant dimensions, forms and materials. Through this text we question the mechanisms that make the landscape a booming cultural product and support the engine of its economy; the city with its parks, its gardens, its squares, its streets or its parasites; the beach, the forest, the mountains and the rivers; scenarios that are part of a cultural imaginary shaped by history; and places of enjoyment and recreation of a society that needs to constantly renew the activities that occur in its representative daily spaces. Scenarios where our actions make sense.

Keywords Landscape · Architecture · Project-Design · Sustainability · Beauty · Inclusion

---

J. M. Palerm (✉)
EA_ULPGC, Las Palmas, Spain
e-mail: jpalerm@paltab.com; jm.palerm@ulpgc.es; president@uniscape.eu

UNISCAPE, Florence, Italy

T. Matteini et al. (eds.), *Cultivating Continuity of the European Landscape*, Environmental History 15, https://doi.org/10.1007/978-3-031-25713-1_54

The topic that clearly and sharply reflects the objective of this book is the critical concept of "continuity" as the "future" of landscape, which offers us a new, effervescent "attentive look" at the landscape perspectives we will need to address in the near future.

On the 10th anniversary of the UNISCAPE network, celebrated in January 2018 in Florence, we discussed the transformations that landscape has been undergoing since the approval of the European Landscape Convention (ELC) in Florence (2000). Among the ideas put forward was the conviction that there was a need to revise certain topics, concepts and declarations in the Convention, a difficult but essential task for the European Landscape, and to discuss this at the celebration of the 20 years since the signing of the European Landscape Convention.

We ran into quite a few issues, but with effort and hard work, we got through them to draft a plan and the strategies to carry it out which were presented in the framework of the UNISCAPE congress in Florence on the 16 and 17 October 2020, which this book documents, gambling on its continuation until the 20 October 2021, also organized by UNISCAPE, to debate the conclusions.

But COVID-19 also arrived, entailing the alterity of the rational and critical continuity of being human, constructed each day as an evolutionary process, even commonly referred to as the new norm to deal with the acceptance of coexisting with an "other". I sincerely hope the impact of bearing up under the critical moments was not too great and we can continue on our way, approving, if possible, the new route we are discovering. In this new scenario, we must make an effort to overcome the new sensitivity towards landscape that has arisen from the conditions that COVID-19 has demanded, and continues to demand, from us and that is characterized by the opposition of "*Imagination and Reflection*".

Considering "Reflection and Imagination" as similar or even overlapping concepts entails going beyond the typical cliché of contemplating nature and critically externalizing it by looking attentively in order to understand the phenomena, agents and individuals interacting with landscape, whatever they might be. The recurrence of idyllic images of nature in this period of confinement is massively deteriorating the *raison d'être* of landscape.

This duality between "Reflection and Imagination" also implies being attentive and cautious with direct, obvious, radical views of the environment and territory, capable in recent years of ignoring "imagination" and supplanting "reflection" with preconceived actions insufficiently curtailed by man's analytical, scientific and proactive capacity, accommodating the obvious flow of subjectivity into a supposedly collective pragmatic praxis.

From this perspective, landscape is flourishing, and there seems to be a "continuity" of investigating and education based on imagination and reflection, precisely because in recent years this has contributed to the formation of a new generation of professionals attentive to understanding landscape.

All this has generated and activated new tools to better understand our territories, the individuals living there and their landscapes through the many investigative works of members that have been presented and discussed in numerous forums. All things considered, a unique European network (UNISCAPE) has offered a framework for the exchange of knowledge and experience in a continuous process that is able to identify and offer strategies for contemporary landscape change originating from diversity.

This should continue to be our main objective for the future as we take on new challenges through a structure able to respond to current demands and difficulties.

The indispensable continuous revision of landscape now permits us to recognize our differences and divergences, both internal and external, and to address political and cultural discussions from a critical position that is both thoughtful and imaginative towards the CDCPP of the Council of Europe and one's own CE as well as the landscape directives of the states in the EC and the CoE.

With these premises in mind, permit me to make reference to certain aspects and contents that have inspired this article.

54.1 Considerations Regarding the Landscape Situation in Europe and Its Territories

Anything to be said about the current state of "landscape projects" can certainly be found in all the sector analyses of the various departments and disciplines of local, regional, national and European administrations. They practically all qualify landscape interventions as being of great strategic potential for generating vital resources for the country and clearly express considerations of their responsibility towards the community regarding the beneficial effects this sector can have on the world of production and work, its social and economic utility, and its capacity to awaken a sense of "*civitas*" that is the basis of our concept of democracy.

Since landscape projects cannot be isolated from the universe of those living there as they give meaning and significance to their habitat, perceptions and desires, a project is a detached agent of environmental change, and culture is its basic adaptation mechanism. This is the contemporary dimension of landscape and its importance in boosting the value of both natural and cultural heritage.

Relaunching landscape quality should be an absolute priority in the interests of a country, not only because of its great cultural importance but also because of its little explored social and economic potential. And even more so in this dramatic moment when the pandemic seems to be bordering on an equilibrium, although for how long no one knows. Faced with a state of crisis that is even more tragic due to neglect, malice and apathy, we should involve landscape more forcefully, detailing its impact on the constitution of each territory and nation in the spirit of the European Charter as it is able to generate new synergy between the environment, territory and landscape in its strictest sense.

The question of landscape competence poses specific problems deriving from the concept itself. The concept of landscape does not appear in the competent areas of constitutional precepts, and if we admit that everything must be formally included, as jurists maintain, we must integrate some of the specified terms and competencies into constitutional texts regarding territory.

Parallel to this social-juridic and competence diagnosis, contradictions are masked and unmasked every day between the physical territory conventionally described as "real" or "in situ" and the views and perceptions of this reality in interpretations and thoughts described as "in visu". Every day fractures come to light between territory and landscape in architecture and urbanism, geography, or biology and planning, which makes understanding difficult and hinders the overlap in their meaning that we would assume. The rift between land ordinances and construction on or over planned land reveals once again the deficiency and obsolescence of operative, normative, political and cultural instruments.

54.1.1 Observing from the Indispensable "Attentive Look" (Observatories that Look Attentively)

When Brunelleschi directed his vision, or better, his inquiring eye, fixedly and unmovingly on familiar places in Florence, a "*Theocentric*" grid, he was motivated by illuminated and scientific consideration of the perspective system and could not have imagined that these new mathematically acquired constants would be supplanted in future by a new geocentrism represented by the culture of information technology.

Classic perspective as *"adult Vision"* (Merleau-Ponty) and at the same time as an invention dominated by instant synthesis is an inadequate device and instrument to represent contemporary spatiality where the accumulation and overlap of mandatory indications received in fragments stimulate perception related to sensations manifesting themselves in time and space, distinct from the pure perception of objects.

The fixed unmoving eye is substituted by the mobility of catching-capturing an image (camera) with its own dissolution, image composition and sequence impressed upon it; in using the new media tools, we have gotten used to a different way of interpreting the naturalness of abstract ideas and perception itself. Web pages, virtual worlds, virtual reality, multimedia, video games, interactive installations, animation, digital videos, films, etc. have cancelled the distance that allowed a domination of space through the open window frame of classic perspective and enlarged the door of visual culture to the point of making the spectator an active participant. From this vector, spectators must expand their sensorial sphere to be able to orient themselves in an architectural space conceived and constructed more with unanswered questions than certainties.

As Tomás Maldonado said: "*....Connected to the efforts to make the representation of reality more true (and so make the virtual more real) is the opposite position, making reality more virtual by questioning the materiality of objectives. In other words, Virtual forms through the dematerialization of materials*". As if to say, cut and paste, a sort of "collage" construction that can vilify the option of an "in situ" project's rigour and quality as a construction process of landscape and territories.

This dematerialization has undermined the traditional perception of objects, spaces and landscape by redirecting the attention of the observer towards the actual experience of perceiving phenomenon and thus forcing an "a priori" suspension of valuation or analysis of it since these cannot suitably nor adequately comprehend the contemporary labyrinth of complex signs;: in short, the difficulty of understanding people and the landscapes they live in.

"**Looking attentively**" was used by W. Benjamin when he affirmed that architecture is normally perceived distractedly through activity and not by attentive contemplation. We do not see the surrounding architecture and landscape that we inhabit and to which we have become accustomed. Observing it with attention "is typical of tourists in general and especially before famous buildings".

However, this "attentively looking" means "**looking critically**" at the mechanisms of reality perception and the collective and individual forms and traditions of living in our homes, cities or on land.

Man symbolically and experientially lives in a certain territory in a committed way if he has produced a representation of himself, as explained by F. Holderlin. Man's relationship with his territory meets his objectives not only in the material production of resources, in the form of his settlements, in the promotion of tourism and full employment, in his history, memory or footprint, but also in the production of his own representation, and not only for advertising, marketing or strategic commercial promotion or to secure votes for parliamentary representatives. It is not possible to experience the reality of a territory without a mental image of that reality, however fertile, harsh or bitter it might be.

If a territory is not offered the chance to become a landscape, it will not allow those who live there to have an identity as a social group, since, in the end, it is the representation of a territory that gives identity to a community of people, independently of its actual local, provincial, regional or national administration limits. Landscape, therefore, permits a population, the citizens, to establish a more attentive relationship with their environment and context so as to order and structure the elements and instruments regulating them in order to populate it with symbols, "....to construct a place of well-being and prosperity", to specify their heritage.

In our everyday life, a society that addresses and works on its capacity for self-representation has a consolidated image of itself, or better, it has developed landscape awareness and is a society able to attempt to determine its evolution with increasing naturalness. In a society where "images", "media" and "performances" normally mix with reality, landscape offers communities the opportunity to establish, consolidate and develop their own system of urban and territorial organization; a hierarchical system of relationships, with new instruments where **ARCHITECTURE PROJECTS** faced with orthodox urbanism and planning are able to produce critical reasoning and re-establish an "**attentive look at landscape**" by offering new challenges and future prospectives.

54.1.2 Observing Based on Law; the Instruments Governing Territories, the Environment and Landscape. Conflicts (Institutional Observatories)

Three actions relating to the landscape area could form the basis for structuring the interdepartmental, interdisciplinary and transversal shared work of an observatory.

The first is related to the construction of the «**OBSERVATORY CONCEPT**» itself. This action bears on the need to make landscape debate happen nationally and internationally on current issues. Therefore, we propose monitoring scientific and experimental areas to foster the formulation of investigations, actions and programs relating to landscape as well as interventions on specific territories, the construction of a **map** of important matters and critical points related to specific demands and issues which covers and expands the range of investigative topics that emerge from the debate so as to activate a process of synthesis regarding landscape;

The second action is born of the necessity to promote the convergence of investigative activities and the reality of the territory of reference. Summarizing, we could call this section «**OBSERVATORY OF TERRITORIES**»: linked to institutions with specific territorial competence. The objectives are: to investigate paradigmatic territories to put acquired knowledge to the test and experiment with new, even participatory, actions and practices; to identify areas of application in the protective and operative sense of the European Landscape Convention; and to evaluate the Observatory's capacity to foster the activation of cultural, social and economic revitalization processes as well as the valorization of landscape aspects.

The third action is to position the observatory in the debate on the construction of European Landscape observatories, according to the requisites of the European Landscape Convention and the indications of national legislation. In short, we could call this action «**OBSERVATORY INSTITUTION**». Starting from the most advanced experience of some European regions, we can outline and highlight issues regarding what proves to be especially important and thus put into play the actual role that observatories can have in landscape-territory transformation, their relationship with the planning and definition and carrying out of landscape quality objectives and their capacity to involve local actors in this process.

The construction of an institutional observatory should also be clearly seen as a cultural operator attending to juridical and administrative aspects, with the need to function as a link between the processes of understanding and ensuing territorial transformation. The experience of the Canary Island and Catalonian Observatories in Spain, or those of Veneto and Trentino in Italy, or that of other territories with their natural peculiarities can and must accept a European agreement on such a vital issue as landscape, probably one of the fundamental debates of the XXI century.

So, if this is the desire, a wish for the organization of landscape, reality has very different directions. Summarizing the instruments accepted juridically in the various European countries from the point of view of use and qualification of land through the classification diffusely accepted in Europe based on abiotic, biotic and cultural components, territorial ordinance and planning instruments will be structured on legislation and ordinance policies of territories and the environment in the following areas:

- CULTURAL. Territory, Land, Urbanization, Agriculture, Human Habitat.
- ENVIRONMENTAL. The environment. Climate change. Sustainability.
- NATURAL. Abiotic, Biotic.

Well, if we compare these parameters with the resolutions specified in the European Landscape Convention that are defined starting from:

- Management, planning.
- Protection, Conservation, Restoration, NEW Landscape (Project).
- Planning Ordinances.

We can understand the conflict of concepts, instruments and processes that come to the fore in proposing landscape projects as the new and necessary centre for managing and establishing sequential and parallel processes for acting on and in landscape.

The adjoining overview-diagram offers an alternative, in a European framework, that contemplates the possibility of a new concept compatible with current reality, introducing corrective factors into the appropriate process of diagnosis, of project design and of execution in an integrated operative intervention.

Only the operative and critical capacity of projects can reveal what factors need be considered in assessments, as well as the alternatives in integrated interventions. Without the value and quality of a project, the speculative declarations of land regulation and qualification instruments and environmental resolutions remain fixed and confined within parameters incapable of perceiving the actual issues and the opportunity and value of the intervention to be carried out even though it is for the conservation, protection, renovation or creation of a new landscape.

This diagram aims to offer a new intervention framework that will permit an articulated strategy of the actors and instruments participating in landscape and thus in territorial and environmental considerations.

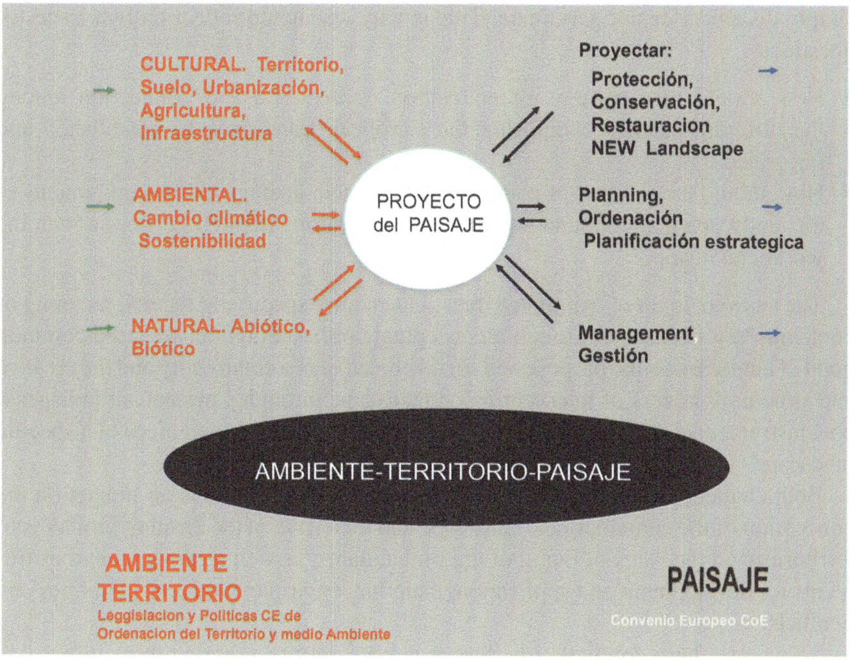

Diagrama realizado por JM Palerm

54.2 Cultivating Continuity

The combination and interaction of these two words, continuity and cultivating, that characterize the title of the UNISCAPE Congress in both 2020 and 2021, are indispensable for our strategy of enabling a landscape plan in Europe, a strategic plan directed at planting new challenges with innovative perspectives and instruments (new challenger / innovative perspective).

To explain the *raison d'être* of these words precisely and briefly, once they have been outlined and in line with the diagnosis of landscape given in the preceding sections, I will refer to the words of Victor Hugo in *Les Misérables* that perfectly describe what I mean by the word "cultivate":

> There are no bad plants or bad men; there are only bad cultivators.

Humans are only cultivators, facilitators in the process of working with landscape. This condition is also revealed in all the branches and knowledge of humankind by the need to understand it, study it and learn it from knowledge and to apply it properly.

This consideration means that actions on landscape and its planning must be careful not to confuse or impose laws or regulations on landscape interventions without a true understanding of the sense of "cultivating" which should permeate these laws or regulations originating from whatever department or discipline is aiming to act on landscape and its citizens.

Of course, this meaning of cultivating requires perfectly determined time and continuity; otherwise, the risk is that the fruit of the act of cultivating will wither or simply disappear or not germinate. This is why we should reflect on two issues or questions:

- How should new patterns of cultivation (cultivators) and connection among the different patches composing the European Landscape be established and integrated?
- How should these forms of continuity be adopted, keeping all the components of the landscape model bound together and in relation although in friction with the world and history?

The answers to these two linked questions require specifying the real meaning of contemporary landscape through a conceptual confrontation between the common good of landscape as the purpose and social objective of a community and the creative and structural process of intervention on landscape through a project, an indispensable instrument critical to specifying the context, intervention and place of a specific landscape.

Both challenges are affected by a series of conflicts that have an impact on the conception of the contemporary landscape and therefore on the ensuing problems of territories and the ways of living of their inhabitants, as well as the laws and instruments for development and application, including environmental considerations and conflicts.

To resolve these conflicts, I believe it is necessary to examine three topics that should act as mediators to respond to the conflicts and, in turn, create a body of thought accepting of the variety of positions and approaches without risking the loss of landscape "cultivators" and the continuity of cultivating. These topics are:

54.2.1 Time and Anachronism

The contemporary viewing of works of art (according to Lozano and Arrase), and of landscape (Palerm) is one of the major causes of anachronism. Today's view of a certain landscape of the past whose patina of many spectators' views and their various forms of addressing it in many different eras leads us to a ceremony of confusion.

We should remember the words of Hamlet (Act I Scene 5) "Time is Out of Joint" as Lozano proposes in his article on anachronism ... and which continues: *"Like in a museum room where objects and works from different epochs coexist, that, not in vain, served as metaphors for this semiotic space – and which, out of it, have no meaning".*

Landscapes whose characterizing elements have had a relationship throughout time coexist in a space outside of which they have no meaning. Here arose the landscape vocation, conceived in contemporary times from an anachronism by converting its temporal continuity/discontinuity into a reason for fortitude and comprehension.

Before this, and at the social and political apogee of ill-judged timing in landscape, while tradition is being invented, in this same horizon of hope flourish oracles, prophets, prognostics and predictions aligned with the present to decline the perfect future. The simple future is perfect.

To resolve these false and perverse conflicts, fewer chroniclers and more anachroniclers are needed, also in landscape.

54.2.2 Tradition and Investigation. Heritage and Landscape

Today, the city and its inseparable historical complex is a "collection of things"[1] of varying interest. This is a BRUTAL and modern definition (as a legacy of modernity) since it is incapable of controlling the "pile of things and forms" occupying the positions at ground level, nor their relationship to each other, nor the valence of one to the others in the interaction with citizens and in the acknowledgement of life forms.

---

[1] We refer to the differentiation between the architectural object and the urban plan/blueprint, the graphic of the territory where the objects are laid out. The term "collection of things" therefore implies the difficulty of logically coordinating the two.

This characteristic situation is based on the clear distinction between what is **material** and what is **immaterial.**[2] It is curious to note that before the aforementioned "modernity", the city was essentially immaterial.

We are bound and obliged to introduce the immaterial in the materiality of urban reality (in its urban form) and thus represent it. In historical complexes, this duality of the immaterial and material in the complex and its urbanity reaches extreme levels of singularity and representativeness. The current cartographic spatial model, an artefact of modernity, *does not permit this*; *it is not able to recognize and incorporate both worlds despite* the technological and computerized advances in the assessment and elaboration of data and the means of their representation.[3]

We need to go back and reflect with a different eye; that is, we are forced to understand it globally (globalization), to go back to seeing our territory as it has always been despite its profound modifications and wounds. And this is why it appears disparate, distinct and different.

Peter Sloterdijk,[4] novelist and philosopher, is not sufficiently precise when he says that the novel of globalization is Jules Verne's book "Around the World in Eighty Days". The problem of its central character, Phileas Fogg, is to make his journey quickly, and this is a spatial problem: in practice, space means reducing the world to the time it takes to go around it. More aptly, I prefer and admire the book " Around the Day in Eighty Worlds" by J.Cortazar.[5]

The New Crisis that started with the global (globalization) **requires giving up on Kant, on the time–space category, on the reduction or specialization of knowledge, and on the descriptive sense of geographic representation, of maps and blueprints**. Although these elements and instruments no longer help us, this does not mean popularizing catastrophism, nor taking refuge in legislative or regulatory protection, nor abandoning the city and heritage, as happens in many cases.

It implies **a crisis of the *spatial model*** as a conception of territory and the architecture that shapes it, its representation as a cartographic instrument; that which Brunelleschi himself described as "the transformation of the finite through the infinite" (linear perspective) is the profound contradiction at the basis of accepting modernity in our contemporary world and the corpus of laws, regulations and resolutions which descend from it and which are its instruments. The specific case of historical complexes[6] and the special blueprints and catalogues related to them is

---

[2] Jean Nogue in a book on the cartography of emotion and the social construction of landscape. Biblioteca Nueva 2007.

[3] Franco Farinelli.

[4] See the books by P. Sloterdijk: SPHERES I: Bubbles, Microspherology Siruela 2003 and SPHERES II: Globes, Macrospherology 2004 Siruela.

[5] *Around the Day in Eighty Worlds* (1967): In this work by Cortázar, instinct, chance, the pleasure of the senses, humour and play end up identifying themselves with the writing that is, in turn, a formulation of existence in the world. The rupture of chronological and spatial order deprives the reader of a conventional point of view, proposing various ways of participating so that the very act of reading is required to complete the narrative universe.

[6] In this sense, ICOMOS had already established the concept of heritage in 1999 and considered that (...) "*it encompasses landscapes, historic places, sites and built environments, as well*

evidence of the crisis of the model and the alienating absence of projects as an experimental alternative..

To understand the profound sense of this absence and model, it is indispensable to establish a new logic of the value of an architectural project doubly linked to the new paradigm of the **concept of heritage and landscape faced with the modern idea of space.**

In this sense, both the ICOMOS in 1999 and the Council of Europe in the European Landscape Convention 2000 established this correlation.

Establishing the duality **heritage-landscape** requires pointing out some contradictions that become apparent from the perception of **material and immaterial** (J Nogue) in contemporary reality that was already mentioned earlier and that will now be articulated in the following sections:

a. The idea of space in modernity is based on a **separation between subject and object**. In the historical artistic vanguards, a kinetic and flat vision or an expressive angle did not belong to the subject but rather to its contemplation; the subject stayed still and fixed so it could be perceived and thus represented. Delimited, partial, encapsulated.

   Nowadays, this separation is impossible; the perception models of the world, the environment and territories have fused. The attitude is one of contemplation of the scenic and fleeting, with no other possibility. Nor of course of a possible measure of the space. Space is that which stays/remains out of the logic of cartography (emotional cartography), of the reduction of the world to a plane (paper).
b. **The loss of limits** (the concept of delimiting historical complexes). The idea of enclosure, habitat, platform, covering (roof) and skin that G. Semper[7] proposed at the end of the XVIII century contrasts and opposes Vitruvius' terms (*utilitas, firmitas, venustas*) and the conceptual derivatives of modernity. This entails accepting the erroneous concept and/or term of functional architecture instead of

---

as biodiversity, collections, past and continuing cultural practices, and knowledge and living experiences".

In general, landscape means any area of the Earth's surface produced by the interaction of various factors present on it and that is visually reflected in this space.

According to the European Landscape Convention (2000) (formerly CEP): "*landscape means an area, as perceived by people, whose character is the result of the action and interaction of natural and/or human factors*".

Thus, it constitutes a favourable resource for economic activity and its protection and promotion can contribute to the creation of employment; it conserves elements of the collective memory of communities and contains natural and cultural heritage.

This Convention, promoted by the Council of Europe, has the fundamental objective of promoting the protection, management and planning of European landscapes.

[7] In 1851, G. Semper published ***The Four Elements** of Architecture* which was an attempt to explain the origin of architecture through anthropology. The book divides buildings into four elements: the hearth, ground, roof and enclosure. The origin of each element can be found in the artisanal tradition of primitive peoples.

that of A. Behne's[8] Modern Functional Construction (1919), producing **a simplification of the form of organization of space in modernity through the use of enclosures named by function and bounded by elements or parts or activities,** and calling zoning the instrumental process of shaping territory, parallel to the formal process of the avant-garde's abstract concept.

Why then can we not think or represent in terms of limits and boundaries which have characterized urban planning instruments since modernity?

Actually, in the end, because that is what modernity, through the cartographic vector, has convinced us: that reality is continuous, homogeneous and isotropic. BUT THAT IS NOT TRUE. IT IS NOT LIKE THAT.

So we must again come to terms with a world where certainty does not exist (completely). We are obliged to understand that today space means, once again and each time more energetically, the entirety, as F. Farinelli explains. The syntax of this new modernity is network; of course space–time in a network dilates, contrasts and confronts the theoretical concepts of Heidelberg's or Einstein's vision of the equation of gravity.[9]

c. **The link between "space and landscape in its determination of heritage in historical complexes" defines a relationship where objects, architecture, infrastructure and the territory with its fields and gardens, mountains or sea, should not be opposed to each other hierarchically by parameters fragmented into disciplines because the space presents as a harmonic aesthetic-sentimental entirety extraneous to any partial and/or rational analysis of the elements shaping it, be they of a scientific character or from an architectural tradition.** The formal actuality of our territory inherited from the spatial model of modernity is completely the opposite, where the concept of plans and planning (development) contrasts with protection and conservation and each object is autonomous and everything in turn presents a differentiation in concepts, disciplines and instruments. It is not that we are abandoning the city, we want to re-define it and to do this, it is useful to call it landscape because of its heritage value.

Thus, the horizon of space, the idea of space, is always "hazy". In the urban environment, no objectives capable of recognizing this dimension have been set nor are present although this is what the delimitation of territories in urban planning and its determination and instruments in current legislation are meant to do. There is indeed delimitation with clear, distinct boundaries but there is no formal reference to technical urban planning parameters. Today the city with its territory is one, a single unit in the space–time dimension and should be considered as landscape, all landscape.

[8] Adolf Behne. Modern Functional Construction. Behne, Adolf 1919. *Die Wiederkehr der Kunst*. Kurt Wolff, Leipzig. Reprint: Kraus, Nendeln/Liechtenstein, 1973; Gebr. Mann, Berlin, 1998.

[9] Javier Maderuelo. The Idea of Space in Contemporary Art and Architecture. 1960–1989. Ed Akal 2008.

Therefore, ruling out and re-abandoning the spatial model of modernity, there is a complete mutation. **The world begs to be considered in its entirety, in its uncertainty, globally as M. Castell explains (from an economic point of view) or Farinelli (from a geographic one), as a single, instantaneous unit, needing neither time nor space nor specification by function.**[10]

These angulations and reflections inherited from the past century[11] **have forced us to re-read the city and historical centres in a rather more sophisticated, articulated and cultured way by recognizing and abandoning the great myths of modernity, of modernist architecture and the postulates of historical complexes and heritage patrimony. It has helped us to see that the world is far more a collection of swaths, even contradictory ones, like the contradictory inert deposits that history has left in cities, increasingly more obvious since the 1960s, as a collage of juxtaposed pieces. Needing to eliminate what is based on this METRIC, on distance, on the separation between object and subject, the immaterial and the material, ... we must make, unmake and remake all this with a new frame of reference. This frame of reference must necessarily comprise acknowledgement of the experimental value of a unified Architecture and Landscape Project Process which includes people and social and economic opportunities in its development.**

Landscape and heritage projects as an attitude, poiesis, must proceed and respond to the intervention context not by generic determination of the general to the particular, nor project plan-planning, nor from morphology to typology, but rather by acting to promote thematic sequences defined by precise time–space directives. This is more the logic of activity and flow than of relationships between objects, with a strong vocation to welcome and integrate different kinds of knowledge and different opinions by being as open as possible to involving any available request in a responsible manner.

Of course, this strategy normally encounters insurmountable bureaucratic obstacles. Somewhere between philosophy and economy, the objectives of a landscape and heritage project are principles of orientation and new centrality while beauty, work and listening to the vocation of places are its central values. When a community is involved and is responsible for a place, it will play out its destiny there and a landscape project will draw it together.

---

[10] A comparative reading of writings by Jeremy Rifkin, Manuel Castell and François Ascher is recommended. This is what Franco Farinelli tells us in his new book *Geography. An Introduction to Models of the World:* «...books made up of maps and descriptions where the entire globe, starting from the Mediterranean, was divided into islands, that is, into something that before containing something was by definition contained in something else, the sea.»

[11] We refer to the ontological concept of Emmanuel Lévinas, a contemporary of Merleau-Ponty: «*...the objective perception of things semantically involves a shift in feeling, including that of the hands and legs and the whole body: everything that can be called life in the body as the body itself, as flesh embodying thought*».

We need a synergic process that transforms the concept of heritage from a historical-aesthetic vision of a monumental character to one of experimental experiences on a territory in order to open up a new ethical, social and pedagogical dimension.

54.2.3 Sustainability and the Environment Starting from Landscape. The Five Factors Affecting Landscape

The factors affecting territories in landscapes are ever-increasing and have become front page news in communication media. They need and require us to provide more and better scientific evidence regarding climate change, territorial ground structure and cadastres, migration, deregulation of competence and competition in production and industry, democratic governance, globalization, etc. But more than anything, we need to systematically work out the implications of these five decisive factors of change for landscape: energy, economy, ecology, ethics and aesthetics. Each of these and its implications will be decisive in future decisions involving and affecting landscape.

54.3 The Reason for an Architecture Project in Landscape

For over 70 years, we have been hoping for a resolute policy to rechannel landscape projects without betraying their essential value, and today, it is the culture of crisis that is calling us to rewrite this moral inheritance as an opportunity and a practical model of development. Extracting energy from crisis situations is what history and science have taught us.

Today, financial disaster threatens the system with collapse. Regarding landscape, the crisis itself should clearly be a pretext, a motor for betterment. For those weighing the costs and benefits of landscape projects, at least, the problem of their economic cost is bogus since not carrying it out is the biggest cost to society.

Marcel Duchamps said that modernity is what translates crisis into value, which is the equivalent of saying it is crisis that will translate the values of modernity into our contemporaneity. Crisis in landscape is like depression syndrome, and as such should be fought giving no quarter due to the harmful effects produced when places lose their human element, even before a physical departure. This happens when the map of our fundamental landscape values becomes murky, and it is like a chain: a lack of beauty, a lack of work and a lack of listening to vocations are all oppressive and depressing factors.

These considerations show us that the crisis of landscape is not so much one of the effects of the economic crisis, or at least only partially, but rather a critically decisive cause of it since the practices of territorial transformation debilitate confidence in our strength and seek compromises. The actual, enormously positive opportunities that spontaneously occur in response to the actions of a landscape project, in other words, the induced capacity that takes place between philosophical and economic reasoning on the vocation of a place, are a new field for evaluating options, for the most part still unknown, especially those of public interventions. But it is to the secondary effects that landscape projects must pay the most attention, to those processes in which our actions of protection, management and improvement gain or lose confirmation and clarity of meaning. We need to not only heal wounds but also build a new narrative with meaning encoded by citizens and not merely intervene with individual, punctual responses.

Of course, people must be centre stage: in contemporary thought, aware attentive landscape project designing evolves with great changes in what we consider fair and beauteous, and this is a fundamental step from an archaic to a modern democracy.

We need to question our habits and take note of the new needs of a world that has profoundly changed. We need quicker reasoning during the present wave of transformations whose dynamic is now unmatchable with the usual space–time categories and currently used methods and tools. We are straight-jacketed in obsolete disciplinary and professional frameworks and have difficulty dialoguing and communicating with citizens, and our vision of people's participation in territorial management and transformation is archaic in its most general sense. We are talking about citizens' concerns about security and economic recovery; public life is a battle over privileges and economic resources but to no avail without an impartial vision that is also spiritual, representative and symbolic.

If we recognize that beauty is a prerogative of excellence in Europe and on the planet, the problem then is to understand how and why it continues to exist given that it seems to disappear with subjective fortuity. The problem is knowing how to see, recognize and promote the symptoms of this reaction. A relaunch is necessary because landscape is an essential cultural, social and economic asset.

New approaches, tools and methods need to be defined to address these challenges. We need to establish actions, programs and procedures that can relaunch and, above all, instigate a practice of regenerating and re-examining landscape value.

Both the value of its assets and the great innate and induced economic and social potential of landscape are affected by our usual orthodox analyses and by the negative profile and impact of academic and institutional management, a major and wide-ranging issue. Consequently, this constant tension leads directly to a critical reconsideration of actions on and in landscape starting from the following concatenated sections conceived as an operational comprehensive plan project.

1. Re-negotiation of the processes of generating form in landscape.
2. Through the public and experimental dimension of projects. Conflicting disciplines and instruments.
3. A critical revision of conventional city-territory spatial models (legislative-urbanistic instruments). Archetype value and validity.

4. Space is no longer a map. Representation requires new mapmaking on the dimension of landscape without forced or preconceived boundaries.
5. Aesthetic rationale and norms in landscape projects as a cultural process.
6. Time and temporality in landscape projects. Designing with scale and motion. Specific management and realization.

These six points are the "*reason d'être*" of landscape research in the near future in its double dimensions of "in situ" and "in visu", that is, on specific territorial actions and realizations, and on our vision and social dreamscapes of the future.

Chapter 55
Cultivating the Landscape Dimension

Towards a New Landscape Language?

T. Matteini

> *Landscape is scene of life, cultivated construction, carrier of meaning. It is language.*
> *—A. Whiston Spirn, 1998.*

Abstract Specifically considering the field of landscape architecture and the art of "making places," the action of cultivating can be interpreted as the continuous taking care of places over time. In fact, the word cultivation also implicitly incorporates a chronological dimension in both its forms: evolutionary process (linear time) and seasonal cyclicity (circular time). Due to these aspects of continuity and evolution, the specific practices concerning cultivation (working the land, sowing and planting, pruning, watering, harvesting, in addition to the differences in crop care required by the various species) are necessarily guided by an overall strategic vision that is projective and experimental, and in this sense, strictly related to a planning/design/project approach.

Keywords Landscape architecture · Cultivating · Crossing borders · Landscape language

55.1 The Practice/Poetic of Cultivating

An intense and stimulating debate among the authors led to the decision to use the term (and the metaphor) "cultivating" for the title of this book. Therefore, exploring the different meanings of *cultivate* and highlighting its broad semantic extension as interpreted and proposed in recent works[1] is vital.

[1] Particularly Guideri and Matteini 2021 and Matteini 2021.

T. Matteini (✉)
Department of Architecture, University of Florence, Florence, Italy
e-mail: tessa.matteini@unifi.it

T. Matteini et al. (eds.), *Cultivating Continuity of the European Landscape*, Environmental History 15, https://doi.org/10.1007/978-3-031-25713-1_55

First of all, we should remember that the term *cultivation* derives from the Latin verb *còlere* which in turn can be traced to an Indo-European root, *kwel,* whose primary meaning is "turning the soil," "tilling the earth" (Angelini 2012, 39–46). Specifically, the Latin word expresses not only a wide range of cultivation and farming practices but also "to take care of," "to treat with respect," to "venerate" and by extension, "to dwell," "to inhabit" and "to live in." Hence, we can see how, above all, *cultivating* signifies becoming sedentary and appropriating a place to till and make fertile, thereby moving from the nomadism of hunter-gatherers to the sedentary life that generated the first settlements (Albertini 2009, 14, 45).

The term *culture* also derives from the same root, and from the future participle of *còlere* (Angelini 2012, 39–46). In affinity with this second germination of the Latin verb is the metaphorical *cultivation* of knowledge as well as the *cultivation* of future visions and projects. Thus, the practical and the poetic dimensions of cultivating are inseparably connected semantically (Latini and Matteini 2017).

Specifically considering the field of landscape architecture and the art of "making places," the action of cultivating can be interpreted as the continuous taking care of places over time (Boschiero et al. 2017). In fact, the word *cultivation* also implicitly incorporates a chronological dimension in both its forms: evolutionary process (linear time) and seasonal cyclicity (circular time). Due to these aspects of continuity and evolution, the specific practices concerning cultivation (working the land, sowing and planting, pruning, watering, harvesting, in addition to the differences in crop care required by the various species) are necessarily guided by an overall strategic vision that is projective and experimental, and in this sense, strictly related to a *planning/design/project* approach.

Thus, a proactive and strategic vision generated by that "forward-looking" attitude recommended by the European Landscape Convention (Florence, 2020) appears essential to all urban and rural landscape processes of *protection, management* and *planning* in which projects and the care of (and for) living components, be they plant, animal, or human, require an ability to gaze into the future in order to cultivate its various unpredictable possibilities.

55.2 Crossing Borders

The European Landscape Convention is usually described as a *groundbreaking* document but it is actually the outcome, or rather the maturation, or to continue with our metaphor, the gemmation, of many years of scientific and cultural cross-pollination across national and discipline borders (and not only European) of numerous experts in landscape-oriented studies.

The ELC is certainly the outcome of a long process[2] and was envisaged as an interpretative encapsulation of numerous charters and international agreements relating to *landscape issues*[3] but appears today as an essential area of mediation, able to combine viewpoints, approaches and languages that are complementary yet still strongly dialectical.

A quick look at the literature of the 1990s shows us that in those years the landscape dimension proposed by the European Landscape Convention was already present *in nuce* in the explorations of landscape theorists and project designers, and in the densely interconnected affinities, divergences and overlap created in academic disciplines and professional practice between the various modes of *reading/interpreting* landscape, on the one hand, and *designing/planning* for it, on the other. This is well documented, for instance, in the book by Bernard Lassus, *The Landscape Approach*, a fundamental point of contact between American and European culture, published in 1998 by the University of Pennsylvania Press and translated by Stephen Bann, Paul Buck and Catherine Petit (Lassus 1998).

The concept of a *landscape dimension* proposed by the Convention, however, truly brought about a "revolution" by framing numerous methodological issues that in previous years had put two sometimes irreconcilable interpretations of landscape in opposition: the cultural/aesthetic one and the scientific/ecological one, both with a long genealogy but dedicated to explorations with different aims (Zoppi, 2007, 29).

This dualism was well expressed by the term "reduction of landscape" (Roger 1997, 108–112),[4] in the sense of reducing the complexity of the historical and integrated concept of landscape to an exclusively scientific reading, depriving it of its strong and recognized cultural acceptation.[5]

---

[2] In September of 1994, a working group was set up to build a European Convention on landscape matters that would continue, integrate and summarize the various treaties and documents drafted on the topic over time. About the process, see Maniglio Calcagno in Ghersi (2007, 17–18).

[3] Besides the numerous conventions produced in previous years by the Council of Europe and by the Committee of Ministers and the UNESCO Convention (1972), there was also the *Mediterranean Landscape Charter* (Seville 1992–Montpellier 1993) by the Regions of Andalucia, Languedoc Roussillon and Tuscany, and the document *Parks for Life. Actions for Protected Areas in Europe* promoted by IUCN with the support of the WWF and a series of agencies and ministries of Sweden, Norway, France, Germany, Holland and the British *Countryside Commission* in 1994; also relevant the preparatory meeting- inter-Ministry Conference on Landscape promoted by the Council of Europe (Florence 1998). For Italy, the October 1999, *Naples Charter* (FEDAP) is also notable. For an exhaustive list, see Maniglio Calcagno, ibidem.

[4] And so reducting the landscape to the mere concept of environment: «Contre les écologues, je dirai que le paysage n'est jamais réductible à un écosystème. Contre les géographes, qu'il n'est pas davantage, à un géosystème [...]. Le paysage n'est pas un concept scientifique.»

[5] As well formulated by Le Dantec (2011, 66)« Voilà pourquoi tout paysage est un objet culturel 'au carré': à la puissance un parce que le pays dont il procède, aussi *sauvage* qu'il puisse parfois paraître, est un hybride de 'nature' et de travail humain; et à la puissance deux parce-que ce pays lui-même ne devient paysage qu'à travers l'émotion, individuelle et collective, qu'il provoque. C'est pourquoi la notion de paysage ne saurait être réduite à la matérialité du pays non plus qu'à ses représentations: elle est un mixte de ces deux dimensions.»

In 1982, Corajoud already affirmed that: « La difficulté actuelle d'éclaircir le concept de paysage n'est pas sans rapport avec une libération progressive et vraisemblablement illusoire des contingences territoriales. [...] Les sciences, et celles du paysage notamment, ont elles-mêmes largement contribué à cette désaffection; elles ont tranché dans la réalité pour constituer des isolats; elles ont, de ce fait, rompu les distributions, désarticulé les montages; elles ont tari la source de tous les indices du paysage, celle qui jaillit entre les phénomènes, dans l'intervalle où s'établissent les flux et les correspondances. Le paysage n'est pas réductible aux apparences et, sans doute, règne-t-il entre les choses comme principe de foisonnement et comme puissance nouante.» (Corajoud in Dagognet, 1982, 38–42). Bernard Lassus and Alain Roger dedicated numerous pages to the need to separate these two terms so they could be reconciled into a holistic vision of an alliance between culture and science (Lassus 1991; Roger 1997) in order to overcome the dichotomy and allow the construction of instruments to address the inescapable and integrated complexity of landscape. It is here that the Convention successfully carries out a courageous, nearly impossible operation as Franco Zagari notes: "It should be said that seldom does a legal text so neatly capture the state of affairs of a cultural debate." (Zagari 2006, p. 66). Likewise, Valerio Romani underlines the productivity and fertile "ambiguity" of the ELC, which allows the two traditionally separate acceptations of landscape (the subjective-perceptive and the physical-objective) to be received and combined "into an integrated form, ably overcoming every diatribe and also their sterile dichotomy" (Romani 2008, p. 174).

To integrate this already quite complex view of landscape which is sectioned (and sometimes "imprisoned") into various cages of interpretation proposed by the rigor of scientific thought, or investigated through the cultural lens of an artistic and perception matrix, we must not forget the project dimension which succeeds in reconciling and ordering the scattered fragments of the different analytical readings into an integrated, proactive vision, that "forward-looking" attitude that the Convention, and then the 2008 guidelines, remind us are necessary.

Michel Corajoud proposes an example of this type of approach in his essay "*Le paysage c'est l'endroit où le ciel e la terre se touchent*" contained in the volume "*Mort du paysage. Philosophie et éstétique du paysage*" edited by Dagognet and published in 1982 (Corajoud in Dagognet, 1982, 38–42)[6]. While acknowledging landscape's eminently relational nature and its inexhaustible, generative dimension which can and must be analyzed using the instruments of science, Corajoud maintains that for it to be understood in its entirety, a holistic, poetic and projectual reading is necessary: « Le paysage est le lieu du relationnel où toutes les localités ne sont compréhensibles que par référence à un ensemble qui s'intègre, à son tour en un ensemble plus vaste. Et ce qui fait qu'il n'y ait pas confusion ou éparpillement des données sensibles, c'est sans doute le fait que les choses qui le composent ne s'ignorent pas et qu'elles sont liées par un même pacte.»

[6] The text has been republished by Michel Corajoud n 2010 in the book *Le paysage c'est l'endroit où le ciel e la terre se touchent* printed by Actes Sud, pp. 9–22.

After the ratification of the Convention, the essentially project dimension of landscape and the fact that it is actively being written and re-written continuously and tirelessly, spontaneously or intentionally, by populations, groups, communities and individuals, gave rise to the winning expression "*Landscape as a project*" proposed by Franco Zagari. This characterized an important era of reflection also within UNISCAPE, viz. the 2010 book edited by Bas Pedroli and Tessa Goodman where members of the network reflect on a position paper by Franco Zagari from different disciplinary positions (Goodman and Pedroli 2010), or the *Manifesto of the Canary Islands for the European Landscape Projec*t (2011), coordinated by Juan Manuel Palerm whose eighteen points were clarified and discussed by numerous researchers from various disciplines and geographic areas[7]. More recently, Saša Dobričič has updated the slogan into "Landscape as a restless project" to restore the fertile complexity and the fundamental elusiveness of contemporary project attitudes and visions, specifically in reference to the Mediterranean area and countries in the south of the world (Dobricic, *infra*).

From the definition of landscape as it was developed in the '90s, and from the unavoidable complexity of the *landscape dimension* as proposed by the European Convention and filtered by interpretation, derives the multi/inter/trans disciplinary[8] attitude that is the only possible way to proceed in research *on* and *for* landscape, thus dismantling the certainties acquired in many theoretical or applied scientific and professional areas.

UNISCAPE was established in the villa di Careggi in 2008 precisely because the founders (a group of university lecturers and researchers from various parts of Europe) wanted to reunite disparate views and attitudes in the research and teaching of *landscape-oriented studies* across national and disciplinary boundaries, to dialectically and productively continue the operation of *brassage* and blending triggered by the content of the European Convention. In the words of one of its founder members (also first President from 2008 to 2016), Carlo Magnani: "Landscape can represent a provocation to the systems of knowledge and power. It is true that what we know is the fruit of scientific advancement, but how it is organized both academically and socially is the progressive result of the decomposition and fragmentation of knowledge and the division of labour[…]" (Magnani in Vanore 2013, p. 112).

Since its foundation, UNISCAPE has continued to pursue this "provocation," as testified by a series of online WDM seminaries "Landscape is where disciplines meet",[9] organized by Conor Newman[10] and held as part of the network's 2021 activities. But what knowledge areas are needed to address landscape matters? In the

[7] The Manifesto is available on www.uniscape.eu.

[8] Multidisciplinary means people from different disciplines working together, each drawing on their disciplinary knowledge; Interdisciplinary means integrating knowledge and methods from different disciplines, using a synthesis of approaches; Transdisciplinary means creating a unity of intellectual frameworks beyond the disciplinary perspectives.

[9] "Landscape is where disciplines meet" presented at the 2020 UNISCAPE Conference, *Cultivating Continuity of the European Landscape*, organized in Florence for the 20th anniversary of the CEP, 17 October 2020.

[10] With Tessa Matteini and Veerle van Eetvelde.

broadest sense and potentially, certainly all of them, if observed through the lens of the landscape dimension (Newman, *infra*). From a more specifically project-oriented vision, on the other hand, we can particularly focus on those disciplines that support the potential actions on landscape described in the ELC: *protection, management, planning*.

An important attempt to organize the constellation of landscape disciplines and professions is provided by Pierre Donadieu in his *Sciences du paysage. Entre théories et pratiques* published in 2012, a fundamental synthesis based on his many years of teaching at the *École Nationale Superieure du Paysage* in Versailles. Some of the objectives proposed by Donadieu: « Montrer d'abord qui la notion de paysage que n'est pas universelle, peut le devenir. Parce que elle est porteuse des idées d'engagement, de solidarité avec autrui, et de résistance [...] Suggérer enfin qu'il existe un domaine de sciences du paysage et du jardin, en tant que catégorie épistémique comparable à celles de la médecine, de l'agronomie ou de l'architecture» (Donadieu 2012, XV). Beyond these goals, fully achieved, he finally succeeds in integrating the scientific interpretation of landscapes with the cultural and aesthetic ones, so recomposing that dichotomy previously recalled and supporting the vision proposed by the ELC. Donadieu explores the various scientific and operational areas of landscape, constructing a coherently articulated mosaic of the various European countries that ably captures the situation at the end of the first decade of the new century, and ends up defining various categories of landscape sciences, once again adopting the theme "landscape as a project" and proposing a new "ethic of landscape and landscape architecture." And so, which paths and instruments can form new *landscape specialists* who consider the landscape ethic as a central theme? As the Declaration of Las Palmas (2019)[11] reminds us, we need to start from three indispensable common points: 1. *transdisciplinarity*; 2. *integration of scientific, normative and project approaches*; 3. *Landscape-based collective identity*.

55.3 Landscape as a Language?[12]

To innovatively address exploration of the landscape dimension in Higher Education, new networks and different interpretative categories need to be constructed in order to develop that "germinal trans-knowledge, where there are no longer disciplines but rather new ways of thinking and of making knowledge converge"(Brunon, Mosser in Pietrogrande 2006, p.22–23). In this sense, where can we start to recompose the various disciplinary and geographical areas, adopting the laudable synthesis proposed by the European Landscape Convention as our programmatic basis?

The project dimension, once again, comes to our aid. Point 9 of the previously cited *Landscape Project Manifesto* states: "The core aspect of a landscape project

[11] Proposed by UNISCAPE in November 2018 and signed by the Rectors of the Universities in the network. Available on www.uniscape.eu.

[12] Anne Whiston Spirn 1998,

is the transversality of knowledge in opposition to inflexible orthodox, disciplinary tools and rules. This process must be undertaken by associating diverse branches of knowledge [...] but also creating relationships beyond the sciences, using different scales of work and in pursuit of goals that need not be coincidental. In any case, renouncing any determinist procedure from the general to the particular."[13]

So how can we construct a different set of integrated educational instruments that responds to the needs proposed by the Manifesto without falling into determinist procedures? One potentially workable pathway can be found in the innovative *language of landscape* proposed by the European Convention. It should be remembered that since the 1990s, many theorists, project designers and researchers have investigated the affinities between landscape, text and language from various perspectives (Corner 1991, 129; Dixon Hunt 1992, 3–16; Whiston Spirn 1998; Potteiger and Purinton 1998)[14], particularly Ann Whiston Spirn in a series of affirmations in her book *The Language of Landscape* that were to have lasting effects, especially on the perception of the culture of landscape projects[15].

From a strictly linguistic point of view, in 2000, the Convention triggered a major revolution in the vocabulary related to landscape issues. Undoubtedly, the ELC has collected many of the existing landscape-oriented research lines and founded a completely new language. Initially drafted both in working English and working French, this new language is now pervading all the scientific and operational landscape communities that we would call *sociétés paysagistes* (Donadieu 2006[16]). In this sense, the glossaries prepared by the Council of Europe and the sharing of a set of already consolidated terms[17] appear fundamental, but in order to deeply and coherently apply the contents and values of the European Convention it seems important to rediscover the different landscape languages of countries in and beyond the European Union. In subsequent translations into the various languages, even the glossary drawn up by the Council of Europe has demonstrated a need to include and transmit

---

[13] *Manifesto of the Canary Islands for the European Landscape Project*, 2011, point 9. English version.

[14] These Authors are collected in and commented by Swaffield (2002, 123–143) in the IV part of his book (*Theory in Landscape Architecture*) dedicated to *Society, Language and Representation of Landscape.*

[15] The language of landscape can be spoken, written, read and imagined; [...]Landscape has all the features of language. It contains the equivalent of words and part of speech patterns of shape, structure, material and function [...] Landscape is pragmatic, poetic, rhetorical, polemical. Landscape is scene of life, cultivated construction, carrier of meaning. It is language.

[16] See the definition in Donadieu in P. Aubry, P. Donadieu, A.Laffage, J. P. Le Dantec, Y. Luginbühl, A. Roger, 2006, pp.94–95.

[17] The Council of Europe document presented in Strasbourg on 26/27 March 2013 (CE-CDCPP (2013) 4E) containing a draft version of the *glossary* supporting the reading of the Convention, subsequently published in 2018: *Glossary of the Information System of the Council of Europe Landscape Convention*, Spatial planning and Landscape, n.106, Council of Europe, Strasbourg, 2018.

all the shades of meaning that might be necessary to address the landscape dimension with the appropriate cultural, technical and scientific instruments[18].

While it is true that landscape is a formidable "teaching resource" as the 2008 Guidelines[19] remind us, the terms and expressions describing landscape, its components (living or mineral, material or immaterial, natural or cultural, physical or perceived…) and the strategies and instruments governing it (strategic, practical or poetic…) have various linguistic forms (different and peculiar to each European territory) and are just as indispensable for the creation of informed *landscape specialists*. These differences tell us that the common view is that a Mediterranean landscape cannot be described with the same lexical instruments as a Northern European one. And, that the terms adopted to describe the various *landscape features* are actually a parallel language that must be understood and compared.

In this sense, it seems quite important to recall the crucial educational experience of *Mouvance I* (1999) and *Mouvance II* (2006), a two-volume landscape glossary drafted by the teaching staff of DEA (Diplôme d'Etude Approfondies),[20] the PhD course in "Gardens, Landscapes and Territories" started in 1991 at the University of La Villette in Paris. The title, *Mouvance* (Mobility), "underlines what has brought together by the authors of this small publication, reflecting the fact that, for them, landscape is an ever-moving relationship (*relation paysagère*)" (Berque 1999[21]).

The authors of this glossary (in two different editions)[22] come from a broad range of disciplines and educational centers/universities. What is interesting and important is that authors with different formations, representing different academic organizations, have left the *comfort zone* of their own disciplines to accept a *métissage*, within the specific educational area of a doctorate. This experience testifies to the fact that no discipline, by itself, is able to exhaustively address the complexity of landscape matters and the dynamism of the *rélation paysagère.*

Another book worth mentioning that refers to an interesting research experience is: *Landscape in language: transdisciplinary perspectives* (Mark et al. 2011) where scholars from different backgrounds (anthropologists, geographers, information scientists, linguists, and philosophers) focus on how landscape is represented in various parts of the world in language and thought, thus reflecting on the relationships of people to places and land.

---

[18] There is a recent translation of the 2008 Guidelines edited by IASLA (the Italian Academic Scientific Association of Landscape Architecture).

[19] *Guidelines for the implementation of the European Landscape Convention.* CM(REC 2008)3. D. Education.

[20] On the retirement in 1998 of Bernard Lassus, founder of the Doctorate, the teaching staff prepared this book in homage to him.

[21] From the Introduction by Augustin Berque. An English translation can be found in "Lotus Navigator" 05, February 2002, pp.78–79.

[22] In the first phase: Augustin Berque, Michel Conan, Pierre Donadieu, Bernard Lassus, Alain Roger, respectively, geographer and orientalist; sociologist; agronomist and ecologist; artist and landscape architect; philosopher and writer. For the second book, published in 2006, Pascal Aubry, Arnauld Laffage, Jean-Pierre Le Dantec and Yves Luginbül were added.

We can therefore (not) conclude that to strengthen and structure the European network of education and research on landscape, it is necessary to agree on the premises for a new landscape language that goes beyond the functional and operative institutional use of English and French terms by integrating the various linguistic roots and vocations of the European Union and the contribution of the various disciplines.

Ann Whinston Spirn's 1998 exhortation seems to still validly orient our explorations: To read and write landscape is to learn and teach: to know the world, to express ideas and to influence others. Landscape, as language, makes thought tangible and imagination possible." (Whinston Spirn 1998).

References

Albertini L (2009) Agriculture Méditerranéennes. Agriculture et paysage des origines à nos jours, Actes Sud, Arles

Angelini M (2012) Dalla cultura al culto. Percorsi di antropologia filosofica intorno al simbolo, al sacro, al sapere vernacolare e alle comunanze, Nova Scripta Genova

Aubry P, Donadieu P, Laffage A, Le Dantec JP, Luginbühl Y, Roger A (2006) Sous la direction de A. Berque, Mouvance II, soixante-dix mots pour le paysage, Editions de la Villette, Paris

Boschiero P, Latini L, Zanon S (2017) Curare la terra/Caring for the land Luoghi, pratiche, esperienze/Places, practices, experiences, Fondazione Benetton Studi Ricerche-Antiga Edizioni, Treviso

Corajoud M (2010) Le paysage c'est l'endroit où le ciel e la terre se touchent, Actes Sud, Arles

Corner J (1991) A discourse on theory II: three tyrannies of contemporary theory and the alternative of hermeneutics. Landsc J 10(2)

Dagognet F (ed) (1982) Mort du Paysage? Philosophie et Esthétique du Paysage. Actes du Colloque de Lyon, Champ Vallon, Seyssel

Dixon Hunt J (1992) Introduction: reading and writing the site. In: Dixon Hunt J (ed) Gardens and the Picturesque: studies in the history of landscape architecture. MIT Press, Cambridge MA

Donadieu P (2012) Sciences du paysage. Entre théories et pratiques, Lavoisier, Paris

Ghersi A (ed) (2007) Politiche europee per il paesaggio. Proposte operative, Gangemi, Roma

Guideri S, Matteini T (2021) Cultivating archaeological landscape. Notes on a Mediterranean applied case study. JOELHO, Journal of Architectural Culture, 11/12. Archaeology, Landscape, Architecture: Crossings of Reciprocal Learnings pp 101–120

Lambertini A, Matteini T (2020) Exploring everyday landscapes of research, in Ri-Vista. Ricerche per La Progettazione Del Paesaggio 18:5–15

Lassus B (1991) Les continuités du paysage. Urbanisme 250:64–69

Lassus B (1998) The landscape approach. University of Pennsylvania Press, Philadelphia

Latini L, Matteini T (2017) Manuale di Coltivazione pratica e poetica per la cura dei luoghi storici e archeologici nel Mediterraneo, Poligrafo, Padova

Le Dantec JP (2011) Poétique des jardins, Actes Sud, Arles

Mark D, Turk AG, Burenhult N, Stea D (2011) Landscape in language: transdisciplinary perspectives. John Benjamins Publishing Company, Amsterdam

Matteini S (2021) Coltivazione. In: Criconia A, Cortesi I, Giovannelli A (eds) 40 parole per la cura della città. Roma/Macerata, Lessico dei paesaggi della salute, Quodlibet, pp 89–92

Pedroli B, Goodman T (eds) (2010) Landscape as a project. A survey of views among UNISCAPE members' reactions to a position paper of Franco Zagari. Libria, Melfi

Pietrogrande A (ed) Per un giardino della Terra, Leo S. Olschki, Florence

Potteiger M, Purinton J (1998) Landscape narratives: design practices for telling stories. Wiley, New York

Roger A (1997) Court Traité du paysage. Gallimard, Paris

Romani V (2008) Il paesaggio. Percorsi di studio, Milan, Franco Angeli

Swaffield S (ed) (2002) Theory in landscape architecture. A reader. University of Pennsylvania Press, Philadelphia

Vanore M (ed) (2013) Il paesaggio nel progetto. Il paesaggio come progetto. Aracne, Roma, p 112

Whiston Spirn A (1998) The language of landscape. Yale University Press, New Haven

Zagari F (2006) Questo è paesaggio. 48 definizioni, Mancosu, Roma

Zoppi M (2007) La via italiana all'architettura del paesaggio: dal Convegno di Bagni di Lucca alla Convenzione Europea dell'anno 2000. In: Ferrara G, Rizzo GG, Zoppi M (eds) Paesaggio. Didattica, ricerche, progetti. Florence University Press, Florence

Chapter 56
Living in an Anthropocene: Reconciling Culture and Nature for a Sustainable Future

Conor Newman

Abstract A founding principle of the Landscape Convention, the anthropocentric turn is potentially at odds with both contemporary material culture and agency theory, as well as public sentiment regarding human culpability for biodiversity loss, climate change and pollution. Reflecting on the phenomenological origin of the definition of landscape in the Convention, the paper considers how the new paradigmatic standing of culture and cultural heritage in Sustainable Development Goals indicates a context where both perspectives can align harmoniously and productively.

Keywords Phenomenology · Agency · Sustainability · Landscape

56.1 Behind the Landscape Convention

'Landscape means an area, *as* perceived by people' (CoE, European Landscape Convention 2000; emphasis my own). With this simple maxim the Landscape Convention asserts, with forcible conviction, that though we may be creatures of the earth, our *presence* is *cultural.* Inescapably so. We act in and on the earth, and even beyond it, *culturally.* No agnostic, disengaged observation, the 'area' in question is perceived, nay *apprehended*, *as* something, *as* a geometric, characterised *form.*[i] This form is a conception of the human imagination, and we call it 'landscape' (e.g. Freitas 2003; Newman 2009; Inanloo Dailoo and Pannekoek 2008, 26–28). In naming it, the Convention authorises and authenticates this uniquely human perception, but it also demands that we assume responsibility for all that ensues from it, all that is made and unmade.

In a paper for Uniscape's Where the Disciplines Meet online lecture series 2022, Glenn Deliège observed that the Swedish translation of the Landscape Convention uses 'conceived' rather than 'perceived'.

C. Newman (✉)
National University of Ireland, Galway, Ireland
e-mail: conor.newman@universityofgalway.ie

T. Matteini et al. (eds.), *Cultivating Continuity of the European Landscape*, Environmental History 15, https://doi.org/10.1007/978-3-031-25713-1_56

An important quality of this conception we call 'landscape' is how it transfixes and compels us to act; to, as it were, *bind* the boundless. Thus, we mark, measure and map to position and locate ourselves in a space–time continuum of our own invention (e.g. Smith et al. 1998; Malpas 2014; Hershok and Ames 2019). To be human is to be a place-maker as well as a time traveller, to live simultaneously in the past, present and future, moving freely and seamlessly between the time zones of then, now and if, to be here and yet capable of imagining there.

Melding time and space together *culturally*—and culture is the *only* medium where time and space can meet—establishes the essential geometry of 'landscape': everything happens at some place and at some time. However, it takes more than just space and time to make a place. Places, to recycle a term used by Tim Ingold (2010), are 'entanglements' of phenomena that make up the messy, colourful fabrics of our lives, in the weave and loose threads of which are snagged and suspended our values, our stories, our work, our dreams and our very selves. This then is what attachment to places looks like—cultural beings and cultural matter (a differentiation that is quite suspect) caught up in the evolving knot-work of the places where we live and work, plan, play, dream and rage.

As entanglements, and not mere coordinates, when we say that everything happens some*where* and some*time*, we acknowledge that a great mixture of phenomena is implicated. 'Context' seems too small a word when what really transpires is a gatherum, a happenstance, fleeting perhaps but historically resonant because, like archaeological strata, happenings, gathera (to coin a term) are laid down in layers, each the antecessor, in some way, of what follows.

One of the most brilliant essays on the subject is Edward S. Casey's (1996) 'How to Get from Space to Place in a Fairly Short Stretch of Time: Phenomenological Prolegomna' in *Senses of Place* edited by Steven Feld and Keith Basso. Basso too is one of the luminaries of landscape studies, and he begins his essay with a quotation from Archytas of Tarentum's commentary on Aristotle's *Categories*:

> Place is the first of all beings, since everything that exists is in a place and cannot exist without a place.

For a fourth-century BC Greek, credited with the most compelling argument for the infinity of space, this is a revealing point of departure, or arrival, depending.

56.2 Holistic Approaches: Landscape and Heritage

Of course, all of this is under-the-bonnet, so to speak, of the Landscape Convention; it is the thinking that is behind its genesis, and it is foundation stone of the thinking that gave rise to the Faro Framework Convention on the Value of Cultural Heritage for Society (Council of Europe 2005). They are both from an epistemological stable that underpins the growing body of literature linking landscape and heritage together—Fairclough uses the term 'twinned' (Fairclough 2019, 1150; see also Fairclough and Grau Møller 2008; Scazzosi 2018). If both conventions apply the phenomenological turn, and heritage is defined in its widest sense as the co-memoration, the remembering-together, of the record of actions, tangible and intangible, arising from the human perspective, it becomes genuinely difficult to separate them out. As we have seen, everything occurs some where and at some time, including every element of the historical/heritage record, which is inscribed on the landscape just as surely as every tangible monument is. But there is more to it than that. It turns out that heritage is tailor-made for communal activity on landscape matters, where landscape philosophy and the 'landscape approach' yield tangible socio-economic results in a fairly short space of time. The strong connections that exist between heritage and place, and the depth to which people and communities are invested in heritage and place, close the gap between people and the slightly more abstract concept of landscape, resulting in the sort of landscape literacy which, as we shall see, is needed to support and drive the sustainability paradigm.

Whereas in principle heritage invokes the record of all that has issued from the human perspective, on a day-to-day level, *heritages* are more atomised, more local, communal and personal: this or that aspect of the heritage is valorised, and as values evolve and change, aspects of the heritage move in and out of public consciousness. Similarly, though landscape too is a totality beyond measure, on a day-to-day basis, we speak of this or that landscape and, more often, of this or that place. And, both heritage *and* landscape are in a state of eternal becoming, they are, by nature, processual.

This is where things get tricky. While the core, universal principle underpinning the two conventions claims the authority of existential truth, actioning the conventions means scaling down, radically, from principle to particle, species to specimen. However, paradigms that rely on so many moving parts don't shrink easily, at least not without the risk of disintegration-by-fragmentation. The first thing to go, usually, is holistic, joined-up-thinking, only to be replaced by localism and sectoralism, both of which are anathema to what is required to take on subject-matter as complex and protean as an entanglement. It is a genuine challenge for practitioners and decision-makers to manage this transition, and there are no easy answers. Everyone is feeling their way into it, which is why it is so important that we network and share experiences.

Multidisciplinarity is the leitmotif of a metamorphosis affecting not just research and education but large swathes of the emerging employment landscape where multidisciplinary teams are mobilised to tackle not component parts but whole systems of complex interconnections and relationships, precisely the sort of thing that landscape is. Although clearly not all are ecosystems per se, the ecological metaphor is, nevertheless, routinely applied because it is a genuinely good analogy and, more importantly, being part of public discourse on global warming and biodiversity loss, *it has currency*. These are the challenges of our time, and polls demonstrate that they are drivers too. Propelled by good conscience, genuine concern, wall-to-wall media coverage and the amplifier/multiplier effect of digital platforms, students are gravitating towards academic courses that speak to sustainability and solutions, teamwork and, above all, holistic approaches that *combine* the so-called hard and soft sciences in a humanistic modality. It should be obvious that all of these are qualities of what might be termed the 'landscape approach' (see Cameron 2004). The point is that the Convention is just one aspect from this way of thinking.

56.3 Beyond Anthropocentrism: Lessons from Material Culture Studies

As systems think grows, however, so too does unease with the Eurocentric template of the subjectified object which has for so long been the foundation stone of Western thinking and which lurks behind that part of the definition of landscape quoted at the start of this paper. In short, the domination of human agency and the human gaze has become an increasingly uncomfortable standpoint, redolent of the Age of Empires, the Industrial Revolution, White Man History and the rhetoric of conquest and subjugation of nature and native, often accompanied by religious exoneration (e.g. Young 1991, 8–12; Inanloo Dailoo and Pannekoek 2008). The second element of the definition of landscape '[an area] whose character is the result of the action and interaction of natural and/or human factors' seems to appease, or at least, *anticipate* some of these concerns. However, to reject the ontological standing of the first part of the definition would be to throw the baby out with the bathwater: metaphysics *can* be separated out from ethics. Colonialism pits two or more perspectives against each other, but irrespective of which one wins out (and colonialism has been and continues to be responsible for immeasurable global harm) both are projections of the human imagination. The Landscape Convention may authenticate and authorise our cultural presence but that is not the same thing as decriminalising our excesses just because they are 'in our nature'. Rather, as far as I can tell, the Convention quietly, i.e. sub-textually, espouses the axiom that self-knowledge, knowing *how* we are in the world, is the first step towards reigning in our more rapacious tendencies, and starting to live in a way that is less injurious to the planet and all its passengers. Decolonialisation may not have been to the forefront of the mind's of the drafters of the Convention in the 1990s but undoubtedly it would feature it were being drafted today.

Material culture studies and in particular materiality theory, which looks at the nature of our relationships with material culture or 'made things' (i.e. artefacts, from the proverbial needle to necropolis) and how these relationships find expression, are important elements in the archaeological toolbox and are relevant here too, not least because so many of the world's problems, from the environment to the economy, revolve around our obsession with material goods. A great deal of attention has been paid to social agency and how artefacts 'perform' in social contexts for or on behalf of their owners, makers, guardians and audiences. Accordingly, and perpetuating the phenomenological turn, agency seemed to have become firmly established as originating with the human beholder, the animator of artefacts, until scholars began to reflect on how artefacts appear, in some lights, to enjoy a sort of reciprocal agency in the way they arouse our interest, prick us into action and mediate social intercourse.

As in the case of landscape theory, commentators have tried to find a vocabulary to distinguish the thing-perceived from the thing-in-itself (Note here echoes of Aristotelian form and matter, and the definitive differentiation of Sartrean still-becoming, etre-pour-soi, and historical finitude, etre-en-soi.), the latter being what Bill Brown would describe as 'outside the field of phenomenological attention' (2001, 4; see also Brown 2004). Brown expresses it thus: 'we look through objects (to see what they disclose about history, society, nature, or culture—above all, what they disclose about us), but we only catch a glimpse of things'. Accordingly, chance interruptions in the phenomenological continuum occur when, for instance, something breaks create a sort of phenomenological wormhole through which we get to look *at* rather than *through* a thing. Quoting Merleau-Ponty, Brown describes these as moments that teach you that you're 'caught up in things' and that the 'body is a thing among things'. He has been criticised for reinforcing an anthropocentric model of man-the-meaning-giver, and of bias in favour of material presence over absence, substance over void (Fowles 2010). Valid concerns to be sure but perhaps not fully deserved.

As these various considerations converge, we can ask: is there a way to preserve the ontological integrity of the cultural *as* (in the definition of landscape in the Convention), from which issues forth the decree to honour the human perspective *because* it is the very font of our wellbeing and self-realisation—we literally owe it to ourselves—while at the same time embrace a more holistic, inclusive response to the global challenges that also acknowledges the independent existential autonomy *and agency* of all living things?

What for us may turn out to be an influential departure comes in the form of Tim Ingold's motif of the 'creative-entanglement' which aligns with Merleau-Ponty, viz. objects are turned into things in the matrix of becoming rather than the other way around. (Note: Ingold's 'thing' is to 'object' what 'landscape' is to 'environment', if, for present purposes, we are permitted to define environment as that which *is* rather than that which is *perceived*.) His concern is to repatriate agency to all living things (see also Harman 2018). Challenging the sole authority of the human gaze—a canonical cul-de-sac he traces back to the Aristotelian template of form + matter—his is a thesis that will appeal to anyone who finds the traditional, anthropocentric view unpalatable for the reasons outlined above. Shot with bullets from a different war, the mythopoeic worldview, which always aspires to holism, fell a long time ago to scientific rationalism but is enjoying something of a comeback amidst concerns about global warming, biodiversity loss and a renewed interest in non-institutionalised spirituality. His proposal may also find purchase in the ostensibly equitable distribution of agency between 'natural *and/or* human factors' (emphasis my own) advocated in the Landscape Convention. Whether or not this is precisely what the drafters of the Convention had in mind is an open question (I suspect it is not). The important point for us is that, countering the prevailing view, Ingold is also championing the rights of all other living things, pressing home the fact that we are things among things, co-mingling, inextricably and inseparably, with all that is alive in a dance (my word not his) that is subject to greater randomness than the term 'system' would typically allow.

A poetic image to be sure: not only are climate change and biodiversity loss the great concerns of our time, they are also the great battle grounds. In a post-truth world, where vested interests wrestle with science (e.g. Nasiritousi 2017), we are subjected, once again, to the spectre of first world culture squaring up to nature. This is the realpolitik, and it makes debates about agency and phenomenology seem rarefied, élitist and irrelevant. But they are not, because culturally induced problems require culturally led solutions, which is why culture and cultural heritage are now firmly established as the fourth pillar of sustainable development. Like the concept of landscape, the relevance of culture and heritage to sustainable development arose in academic discourse (see Nurse 2006) before making its way into the political arena in policy announcements like that from the World Summit of United Cities and Local Governments in Mexico (UCLG 2010), and the Namur Declaration (CoE 2015) where culture *and* cultural heritage join to make up the fourth pillar. Nurse (2006, 34–5) reminds us that an important part of the journey towards this moment involved exposing and critiquing the hegemonic hold neo-liberal, Western ideologies had on the concept of sustainable development at the expense of alternative value systems, particularly those of the so-called developing countries. We can see the same sort of thinking in the Landscape Convention.

56.4 Postscript

Whether nature or nurture, we have perhaps always lived in an Anthropocene. From the moment our earliest ancestors turned an object into a thing, banging crudely at a pebble to re-shape it into a tool for chopping and slicing, the world changed and a new, human world was born, for it was not only the pebble lying on the ground that was reimagined, the whole environment was, and the person in whose hands the tool was held was irrevocably transformed. Irrespective of what has led from that moment of blinding brilliance to islands of discarded plastic in the ocean, and regardless of where agency resides, this is our story and our truth. We rely on this imagined world just as surely as we depend on and co-create the natural world.

References

Basso K (1996) Wisdom sits in places: notes on a western Apache landscape. In: Feld S, Basso KH (eds) Senses of place. School of American Research Press, Santa Fe, pp 53–90

Brown B (2001) Thing theory. Crit Inq 28(1):1–22

Brown B (2004) Things. university of Chicago Press

Cameron JI (2004) Some implications of Malpas' place and experience for place ethics and education. Environ Archit Phenomenol 15(1):5–9

Casey ES (1996) How to get from space to place in a fairly short stretch of time: phenomenological prolegomna. In: Feld S, Basso KH (eds) Senses of place. School of American Research Press, Santa Fe, pp 13–52
CheriScape: https://cheriscape.eu/
Council of Europe, European Landscape Convention (2000) Council of Europe Treaty Series—No. 176. Council of Europe, Strasbourg
Council of Europe, Framework Convention on the Value of Cultural Heritage for Society (2005) Council of Europe Treaty Series—No. 199. Council of Europe, Strasbourg
Council of Europe, Namur Declaration (2015) The Ministers of the States Parties to the European Cultural Convention Meeting in Namur on 23–24 April 2015. Namur. Microsoft Word - Namur Declaration_EN final.doc (coe.int). Accessed 20 May 2021
Fairclough G (2019) Landscape and heritage: ideas from Europe for culturally based solutions in rural environments. J Environ Planning Manage 62(7):1149–1195
Fairclough G, Grau Møller P (2008) Landscape as heritage. The management and protection of landscape in Europe: a summary by the Action COST A27 "Landmarks". Geographica Bernensia G79. University of Berne
Fowles S (2010) People without things. In: Bille M, Hastrup F, Soerensen TF (eds) An anthropology of absence: materializations of transcendence and loss. Springer-Verlag, New York, pp 23–41
Freitas SR (2003) Landscape: where geography and ecology converge. Holos Environ 3(2)
Harman G (2018) Object-orientated ontology. a new theory of everything. Penguin Random House
Hershock PD, Ames RT (2019) Philosophies of place: an intercultural conversation. University of Hawai'i Press
Inanloo Dailoo S, Pannekoek F (2008) Nature and culture: a new world heritage context. Int J Cult Prop 15:25–47
Ingold T (2010) Bringing things to life. Creative entanglements in a world of materials. Realities National Centre for Research Methods Working Paper 15, pp 2–14
Malpas J (2014) On human being as placed being. J Environ Archit Phenomenol 25(3):11–12
Nasiritousi NF (2017) Fuel emitters and climate change: unpacking the governance activities of large oil and gas companies. Environ Politics 26(4):621–647
Newman C (2009) Landscapes 'Я' Us. In: The Irish landscape: looking around, looking ahead. Irish National Landscape Conference Published Papers. Heritage Council, pp 8–14
Nurse K (2006) Culture as the fourth pillar of sustainable development. Report prepared for the Commonwealth Secretariat, London. http://placemakers.wdfiles.com/local--files/theoretical-analysis-examined/Cultureas4thPillarSD.pdf. Accessed 20 May 2021
Smith JM, Light A, Roberts D (1998) Introduction: philosophies and geographies of place. In: Light A, Smith JM (eds) Philosophies of place, philosophy and geography lii. Rowman and Littlefield, Landham, pp 1–7
Scazzosi L (2018) Rural landscape as heritage: reasons for and implications of principles concerning rural landscapes as heritage ICOMOS-IFLA 2017. Built Heritage 3:39–52
United Cities and Local Governments, Culture: Fourth Pillar of Sustainable Development (2010) 3rd UCLG World Congress—The World summit of local and regional leaders. Mexico. http://www.agenda21culture.net/sites/default/files/files/documents/en/zz_culture4pillarsd_eng.pdf. Accessed 20 May 2021
Young D (1991) Origins of the sacred. Ecstasies of love and war. St Martin's Press, New York

Chapter 57
The European Landscape as a Task: Understanding the Past, Engaging in the Present, Living the Future?

Bas Pedroli

Abstract Landscape stories based on memories and dreams are essential means to realise how we identify with our landscapes and with what makes up the value of its past. However, the ever more dominating global market orientation of the agricultural sector (and of the many other policy sectors impacting landscape development) makes the European landscape undergo serious and at times radical changes. Landscape change is normal, but it can work out very detrimental if it has not been consciously envisaged. Engaging in today's landscape and in the functions it can have for society is a precondition for a living landscape. Such engagement will not come from policy. Engaging means getting involved, based on shared interests, using local as well as scientific knowledge and paying respect to the land managers of today and to the unique character of the landscape. If we manage, we can responsibly live the future of the European landscape.

57.1 Understanding the Past: About Memories and Dreams

Landscape is something most people do not think about, it is just there. Sometimes you remember its features, smells and atmosphere when you recall a nice holiday or a remarkable picnic shared with dear people in the countryside. Of course, we take loads of pictures of landscapes, but that is not landscape, it is just pictures. The landscape is in our memory, and is characterised by odours, sounds and colours and by warm or cold feelings rather than by pixels.

Still, when asking people for their favourite landscape they will immediately start thinking of beautiful landscapes: romantic mountain valleys, wide views or enchanting forests. And they will start sketching them in words and feelings. Landscapes are as stories (Box 57.1, from Pedroli (2020, pp. 42–43)).

B. Pedroli (✉)
Land Use Planning Group, Wageningen University, Wageningen, The Netherlands
e-mail: bas.pedroli@wur.nl

JustLandscape, Amsterdam, The Netherlands

T. Matteini et al. (eds.), *Cultivating Continuity of the European Landscape*, Environmental History 15, https://doi.org/10.1007/978-3-031-25713-1_57

Consciously dreaming away in a landscape—be it on-site or in memory—you can experience a great deal of its essential features. One could call this way of observing the landscape with all of the senses, active wondering. It provides a dynamic picture, perhaps more real than the factual landscape (cfr. Jacobs (2006)).

Box 57.1: Home

The silvery spots on the grey lake seem to glow. The landscape breathes silence. Clouds hang motionless around the intense green hills on the other side of the water. On a small peninsula a simple chapel. I walk there. Royal fern and fragrant gale along the banks. From the loosely stacked thick slates of the ancient wall around the churchyard, oak ferns and lilac foxgloves grow, patches of moss hang down. Many graves bear Irish crosses. There is a shrine against the wall. The inscription tells Saint Finbarr is buried here. He lived here as a hermit in the sixth century. He then moved downstream along the River Lee to the coast, where he founded Cork.

Later, I walk a coast path near Cork. The ocean swell creates raging surf and foaming swirls of seawater between the jagged rocks. In the hazy distance lie low islands under a grey sky. Every now and then a ray of sun lights up some rocks close by, covered with rust brown and light grey lichens where the sea water no longer reaches. The vegetation begins a little higher, on the edge of the open agricultural landscape. I suddenly realise that it is already June. Everywhere the pink flower heads of thrift swing back and forth. There are many yellow and white flowers and even the heather is already blooming here and there.

Terry says he has made maps of the environment with people from his village, just as far as they perceive a connection with it. Such 'landscape circles' help to actively experience the landscape again, and also to discover the 'uninhabited' places. Nature easily takes over from people here. But even in the midst of these primal natural forces, inhabitants have always created their place. It was not for nothing that they brought Finbarr back to the spot where he had seen God's greatness reflected in the lake, and in the mountains his monastery. Landscape is where we can feel at home.

Landscapes evolve. When you return to a landscape, or start making study of it through maps and literature, it will change its appearance. Often this is a little disappointing, because in the meantime it had evolved in your memory as much more beautiful and complete than it could be out there, more like the great landscape paintings of the seventeenth and eighteenth centuries. Moreover, there are always reasons for decline in their perceived beauty, such as new roads and constructions, prolonged rainy weather, unexpected fences or high-rise buildings on the horizon. On the other hand, sometimes one can experience completely new exciting aspects of a landscape even after many visits. After deep digging in archives, the remnants of a forlorn water mill you always overlooked tells a new story, an unexploited trail brings you to hidden meadows full of flowers you never imagined, or the discovery of a beaver dam leaves you astonished at the sophistication of land management by beavers.

The landscapes of Europe that give rise to such stories show a surprising diversity. The features of today's European landscape can perhaps be symbolised by its ensemble of walls, trees, cows and fields (Pinto-Correia and Vos 2004). Grazing livestock on permanent grassland represents a distinctive character of the historical European rural landscape (Luginbühl 2012, pp. 43–46). Clusters of farms were centrally situated in the landscape, surrounded by fields, meadows and woodland further away. As claimed in Pinto Correia et al. (2018, p. 11), Europe can well be characterised as the continent of cheese! Animal husbandry—generally using common land for grazing—not only guaranteed the provision of some milk and meat, but also enhanced the fertility of the fields through manuring. The mixed family farm was well-suited to manage this array of functions, which dominated European agriculture for many centuries. The remnants of the mixed family farm system are gradually vanishing today, giving way to rationalised farming systems, but they still contribute to the memory of the European landscape. And nonetheless, very good cheeses—and many other famous regional farming products—continue to be produced by dedicated farmers.

57.2 Engaging in the Present: Whose Landscape is This?

Since the European Landscape Convention was launched in 2000, it has been ratified by 40 of the 47 member states of the Council of Europe. In the present volume, the philosophy of the European Landscape Convention is discussed extensively. Reaching beyond the mere protection, management and planning of landscape from a national policy point of view—which as such formed a substantial policy innovation in 2000—landscape is now increasingly also approached as a vehicle itself for urban and spatial planning, rather than as an object to be protected, managed or planned (Nassauer 2012; van Rooij et al. 2021). After all, the European Landscape Convention emphasises that every citizen should have the opportunity to participate in decision processes relevant to landscape. This is, however, almost impossible given the indirect effects of almost all sectoral national and regional policies on landscape. Landscape-based visions can remedy that incongruence, focussing on the landscape as a comprehensive principle, to which all spatial processes are inherently related (van Rooij et al. 2021). One could claim that landscape should not be considered the inevitably remaining problem inherent to the interaction of different sectoral interests, but rather the solution in complex urban and spatial planning projects. Well-facilitated participative processes can mobilise the inhabitants and other stakeholders to take responsibility for co-creation instead of embarking on polarisation of interests. Land can be owned, not landscape (Pedroli 2019, p. 19)! Although land rights can pose a fundamental problem here, this implies that landscape is a common good, where everyone with clear interests has the right to participate in developing visions for the future landscape and pathways to reach this future.

Engaging in the present means, on the other hand, that we assume responsibility to get acquainted with the landscape: without people, no landscape (Priore 2009). This can only be realised if we identify ourselves with the landscapes we live in, we work in, we recreate in or we visit on holiday. A vivid interest for landscape history, for the layers of geology, soil, vegetation and fauna, and for the present land use and the actors in the landscape should at least complement the more consumptive mood that is generally promoted by advertisements for cars and mountain bikes.

57.3 Living the Future?

This essay began with memories and dreams, with landscape stories. It is clear that these are essential elements to realise how we identify with our landscapes and with what makes up the value of their past. Yet, these elements apparently are not enough to guarantee a living landscape for the future. The ever more dominating global market orientation of the agricultural sector (and of the many other policy sectors impacting landscape development) makes the European landscape undergo serious and at times radical changes, because the current land use is different from the use that led to the landscape character as we can still perceive it in many places. Landscape change is normal, but it can turn out to be very detrimental if it has not been consciously envisaged. Engaging in today's landscape, and in the functions that it can have for society, is a precondition for a living landscape. Such engagement will not come from policy. This challenge is increasingly a task for urban people that now form the majority anyway in Europe, whether living in cities or in the countryside. Engaging means getting involved, based on shared interests, using local as well as scientific knowledge and paying respect to the land managers of today and for the unique character of the landscape. If we manage, we can responsibly live the future of the European landscape.

References

Jacobs M (2006) The production of mindscapes: a comprehensive theory of landscape experience (PhD). Wageningen University, Wageningen

Luginbühl Y (2012) La mise en scène du monde: La construction du paysage Européen. CNRS Éditions, Paris

Nassauer JI (2012) Landscape as medium and method for synthesis in urban ecological design. Landsc Urban Plan 106(3):221–229

Pedroli B (2019) Natural heritage management, or is it cultural heritage after all? Towards new commons and sharing interests in the landscape. Ex Novo J Archaeol 4:13–22

Pedroli B (2020) Luisterend naar het landschap. Nachtwind, Hilversum

Pinto-Correia T, Vos W (2004) Multifunctionality in Mediterranean landscapes–past and future. In: Jongman RGH (ed) The new dimensions of the European landscape. Springer, Dordrecht, pp 135–164

Pinto Correia T, Primdahl J, Pedroli B (2018) European landscapes in transition. Implications for policy and practice. Cambridge University Press, Cambridge

Priore R (2009) No people, no landscape. La Convenzione europea del paesaggio: luci e ombre nel processo di attuazione in Italia. Franco Angeli, Milano

van Rooij S, Timmermans W, Roosenschoon O, Keesstra S, Sterk M, Pedroli B (2021) Landscape-based visions as powerful boundary objects in spatial planning: lessons from three Dutch projects. Land 10(1):16. https://doi.org/10.3390/land10010016

Chapter 58
Where the Disciplines Meet: Uniscape as a Laboratory

Conor Newman

Abstract Like landscape itself, the landscape paradigm is not in the sole ownership of any academic discipline or community-of-interest. Instead, irrespective of what name it goes by, the principles underpinning it inform a lot of current discourses across a very broad range of disciplines and creative and performative activities. More than that, the roll-out of the landscape paradigm has given rise to new sub-disciplines. Above all, the landscape paradigm challenges traditional disciplinary divisions. Uniscape offers a laboratory-like space where colleagues from different backgrounds encounter each other's approaches and perspectives and share experiences from different teaching environments and traditions.

Keywords Landscape education and training · Transdisciplinarity · Landscape heritage

The title of this short contribution comes from one of many engaging conversations among the members of the Uniscape network on the challenges facing those of us at the coalface of landscape research and, above all, education. A quick glance at the range of topics in this volume alone reveals the kaleidoscopic nature of landscape, spread across areas as diverse and yet interrelated as governance, design, usage, futures and pasts, identity and of course sustainability. The metaphor of shards of coloured glass looked at through a tiny lens, twisted this way and that to make a bewildering variety of patterns is, in some ways, regrettably apt—landscape is not always treated with the seriousness it demands, too often its management is splintered and sectoralised, and it is reduced to aesthetics and the notion of the picturesque—and undoubtedly contributes to the opinion that it is too complex for the non-professional to grasp, and too abstract and protean to fit into the architecture of government departments.

Teaching landscape is equally challenging when the faculty and disciplinary architecture of universities and their pedagogical traditions, which vary hugely across

C. Newman (✉)
University of Galway, Galway, Ireland
e-mail: conor.newman@universityofgalway.ie

T. Matteini et al. (eds.), *Cultivating Continuity of the European Landscape*, Environmental History 15, https://doi.org/10.1007/978-3-031-25713-1_58

Europe, also place barriers in the path of multi-, inter- and cross-disciplinarity. And, while there are disciplines that fall naturally into the landscape envelope, many others do not commonly identify their expertise and research interests with landscape. Disciplines like Landscape Architecture, Geography, Environmental Science, Archaeology and Heritage could be described as seasoned landscapers with bodies of literature to prove it, but really important contributions have also come from literary criticism, philosophy, art and art-history and sociology to name but a few. Moreover, there are exciting emerging specialisms too, such as landscape and well-being, landscape law and democracy and, of course, landscape, sustainability and the circular economy. Diversification such as this speaks to the canonical standing of landscape theory: it is not so much a discipline as a way of thinking, and can be applied in almost any context. In fact, it is already widely dispersed: the position on culture and cultural heritage espoused, for instance, in the Namur Declaration speaks to that.

Discussed and dissected in an impressive corpus of academic literature dating back to at least the 1970s, and defined in the Florence Convention, landscape represents the convergence of culture and nature, valorised and championed for all its unique qualities in a Convention that validates us as placed beings. With deep roots in phenomenology, landscape is the ultimate phenomenon, a space–time continuum of the human imagination where *everything* meets, including academic disciplines. It is, therefore, the ultimate interdisciplinary platform, a motif that Uniscape has employed in its online lecture series 2021 and 2022 *Landscape: Where the Disciplines Meet*, which has borne witness to a remarkable diversity of papers, ranging from psychoanalysis to the subterranean tunnels of post-industrial mining landscapes, the heritage of cheese-making (pasture dairying really, and its impact on shaping the Eurasian landscape), ecomuseums, community and the green circular economy and low-carbon, renewable energy.

One of many initiatives organised or supported by Uniscape, the series gives us a glimpse of a possible future for landscape education across the network, an online training and knowledge-exchange for students, researchers and colleagues from the four corners of Europe and beyond. Of course, this has been possible for some time but the transition *en masse* to remote teaching and learning forced upon us by the COVID-19 pandemic has accelerated the evolution of this modality resulting in a generation of teachers and students who are habituated to the online classroom and the liberated by the freedom it creates around timetabling and venues, not to mention the dramatic reduction in travel miles which is environment- and family-friendly. Online lectures may be a compromise on the real thing; especially place-based learning which is such an important tool in landscape education; but they have earned their place, and with a little effort it should be possible for Uniscape members to develop and offer modules taught by experts from across the network. In fact, such is already in train with the 2021 launch of the Enlight (**E**uropean university **N**etwork to promote equitable quality of **L**ife, **S**ustainability and **G**lobal engagement through **H**igher education **T**ransformation) Programme which is being piloted by nine universities (University of the Basque Country; University of Bordeaux; Comenius University Bratislava; National University of Ireland Galway; Ghent University; University of

Göttingen; Univerisity of Gronigen; University of Tartu), including three who are already in the Uniscape network.

58.1 Landscape and Heritage

The growing body of literature on landscape and heritage signals a deep and important connection and junction between past, present and future, that has shone a spotlight on an important theatre of activity that has been instrumental in the designation of culture and cultural heritage as the fourth pillar of the Sustainable Development Agenda. This nexus has been explored, for instance, in a JPI-funded project called CHeriScape: Cultural Heritage in Landscape (2014–2016). With the close involvement of members and associates of the Uniscape network, CHeriScape ran a five-conference series on the theme of landscape-as-heritage. Though one might debate the statement on the website asserting that landscape is a category of heritage, and maintain that it ought to be the other way around, here is not the place for that. Instead, we should note simply that the principles at work here operate in the Framework Convention on The Value of Cultural Heritage for Society (Council of Europe, Faro 2005) which has bound the two conventions together.

Linkages between specialist and society built into the Landscape Convention are reinforced in the Faro Convention, the common political trajectory here being to hard-wire into policy and practice tangible benefits the citizen-participant can *experience* in terms of quality of life, well-being and participation in the democratic process. As testified to in other essays in this volume, landscape observatories have a proven track-record as intermediaries between policy, specialist and public, and are a vital public resource. Experience has demonstrated, for instance, that landscape observatories and their equivalent (e.g. landscape laboratories) are capacity-builders, mobilising people and communities and, more importantly, generating how-to training in mobilisation and communal projects. This is the sort of social capital that is needed, not only to deliver the democratic and sustainable vision common to the Florence and Faro conventions but a slew of other national and international instruments as well.

The manufacturer's authorised representative in the EU is Springer Nature Customer Service Centre GmbH, Europaplatz 3, 69115 Heidelberg, Germany. If you have any concerns regarding our products, please contact ProductSafety@springernature.com

Printed and bound by CPI Group (UK) Ltd, Croydon, CR0 4YY
30/09/2026
02225561-0018